T

Redgr[illegible] [illegible]afety

Redgrave's Health and Safety

Redgrave's Health and Safety

Eighth edition, second (cumulative) supplement 2014

Jonathan Clarke
Barrister, Old Square Chambers

EU consultant

Michael Ford LLB, MA
Barrister, Old Square Chambers
Visiting Fellow, London School of Economics

Scottish case editor

Astrid Smart MA, LLB
Advocate, Compass Chambers

LexisNexis®

Members of the LexisNexis Group worldwide

United Kingdom	LexisNexis, a Division of Reed Elsevier (UK) Ltd, Lexis House, 30 Farringdon Street, London EC4A 4HH, and 9–10 St Andrew Square, Edinburgh EH2 2AF
Australia	Reed International Books Australia PTy Ltd trading as LexisNexis, Chatswood, New South Wales
Austria	LexisNexis Verlag ARD Orac GmbH & Co KG, Vienna
Benelux	LexisNexis Benelux, Amsterdam
Canada	LexisNexis Canada, Markham, Ontario
China	LexisNexis China, Beijing and Shanghai
France	LexisNexis SA, Paris
Germany	LexisNexis GmbH, Dusseldorf
Hong Kong	LexisNexis Hong Kong, Hong Kong
India	LexisNexis India, New Delhi
Italy	Giuffrè Editore, Milan
Japan	LexisNexis Japan, Tokyo
Malaysia	Malayan Law Journal Sdn Bhd, Kuala Lumpur
New Zealand	LexisNexis New Zealand Ltd, Wellington
Singapore	LexisNexis Singapore, Singapore
South Africa	LexisNexis, Durban
USA	LexisNexis, Dayton, Ohio

Published by LexisNexis
This is a Butterworths title

A CIP Catalogue record for this book is available from the British Library.

Printed and bound by CPI Group (UK) Ltd, Croydon CR0 4YY

Visit LexisNexis at www.lexisnexis.co.uk

PUBLISHER'S NOTE

This supplement contains all necessary amendments and additions to the statutory contents of the Eighth edition of *Redgrave's Health and Safety*. Statutory materials are set out as amended by statute or statutory instrument, and should be taken as replacements for the existing paragraphs in the main work. However, the annotations to these paragraphs show only the new amendments: this is intended to guide readers to the latest additions.

New statutory materials have been added as demanded by statute. These have been numbered as appropriate and should be read in conjunction with the main work.

This supplement is cumulative in that it incorporates the contents of the previous supplement published in December 2013 which it now replaces.

CONTENTS

PART 3
LEGISLATION OF GENERAL APPLICATION

Employers' Liability (Compulsory Insurance) Act 1969

1969 chapter 57

as amended

[3.1]

1 Insurance against liability for employees

(1) Except as otherwise provided by this Act, every employer carrying on any business in Great Britain shall insure, and maintain insurance, under one or more approved policies with an authorised insurer or insurers against liability for bodily injury or disease sustained by his employees, and arising out of and in the course of their employment in Great Britain in that business, but except in so far as regulations otherwise provide not including injury or disease suffered or contracted outside Great Britain.

(2) Regulations may provide that the amount for which an employer is required by this Act to insure and maintain insurance shall, either generally or in such cases or classes of case as may be prescribed by the regulations, be limited in such manner as may be so prescribed.

(3) For the purposes of this Act—

- (a) 'approved policy' means a policy of insurance not subject to any conditions or exceptions prohibited for those purposes by regulations;
- [(b) 'authorised insurer' means—
  - (i) a person who has permission under [Part 4A] of the Financial Services and Markets Act 2000 to effect and carry out contracts of insurance of a kind required by this Act and regulations made under this Act, or
  - (ii) an EEA firm of the kind mentioned in paragraph 5(d) of Schedule 3 to the Financial Services and Markets Act 2000, which has permission under paragraph 15 of that Schedule to effect and carry out contracts of insurance of a kind required by this Act and regulations made under this Act;]
- (c) 'business' includes a trade or profession, and includes any activity carried on by a body of persons, whether corporate or unincorporate;
- (d) except as otherwise provided by regulations, an employer not having a place of business in Great Britain shall be deemed not to carry on business there.

[(3A) Subsection (3)(b) must be read with—

- (a) section 22 of the Financial Services and Markets Act 2000;
- (b) any relevant order under that section; and
- (c) Schedule 2 to that Act.]

[3.2]

Statutory annotations

Amendment

Sub-s (3): in para (b)(i) words 'Part 4A' in square brackets substituted by the Financial Services Act 2012, reg 114(1), Sch 18, Pt 2, para 34.

[3.5]

3 Employers exempted from insurance

(1) This Act shall not require any insurance to be effected by—

- (a) any such authority as is mentioned in subsection (2) below; or

(b) any body corporate established by or under any enactment for the carrying on of any industry or part of an industry, or of any undertaking, under national ownership or control; or

(c) in relation to any such cases as may be specified in the regulations, any employer exempted by regulations.

(2) The authorities referred to in subsection (1)(a) above are

[(a) a health service body, as defined in section 60(7) of the National Health Service and Community Care Act 1990, [the National Health Service Commissioning Board, a clinical commissioning group established under section 14D of the National Health Service Act 2006,] . . . a National Health Service trust established under [section 25 of the National Health Service Act 2006, section 18 of the National Health Service (Wales) Act 2006] or the National Health Service (Scotland) Act [1978,] [an NHS foundation trust . . .] [and a Local Health Board established under [section 11 of the National Health Service (Wales) Act 2006]]; and

(b)] the Common Council of the City of London, . . . , the council of a London borough, the council of a county, . . . or county district in England [the council of a county or county borough in Wales], [the Broads Authority] [a National Park authority] [a council constituted under section 2 of the Local Government etc (Scotland) Act 1994 in] Scotland, any joint board or joint committee in England and Wales or joint committee in Scotland which is so constituted as to include among its members representatives of any such council [. . .] [. . . , a joint authority established by Part IV of the Local Government Act 1985], [an economic prosperity board established under section 88 of the Local Democracy, Economic Development and Construction Act 2009, a combined authority established under section 103 of that Act,] [an authority established for an area in England by an order under section 207 of the Local Government and Public Involvement in Health Act 2007 (joint waste authorities),] [the London Fire and Emergency Planning Authority] [, any local policing body, any chief constable established under section 2 of the Police Reform and Social Responsibility Act 2011, and the Commissioner of Police of the Metropolis]][, the Scottish Police Authority and the Scottish Fire and Rescue Service][; and

(c) the Commission for Equality and Human Rights].

[3.6]

Statutory annotations

Amendment

Sub-s (2): in para (a) words from 'the National Health Service Commissioning Board,' to 'National Health Service Act 2006,' in square brackets inserted by the Health and Social Care Act 2012, s 55(2), Sch 5, para 14(a); in para (a) second words omitted repealed by the Health and Social Care Act 2012, s 55(2), Sch 5, para 14(b).

Sub-s (2), para (b): words ', the Scottish Police Authority and the Scottish Fire and Rescue Service' in square brackets inserted by SI 2013/602, art 26, Sch 2, Pt 1, para 6.

Health and Safety at Work etc Act 1974

1974 chapter 37

PART I

HEALTH, SAFETY AND WELFARE IN CONNECTION WITH WORK, AND CONTROL OF DANGEROUS SUBSTANCES AND CERTAIN EMISSIONS INTO THE ATMOSPHERE

The Health and Safety Commission and the Health and Safety Executive

[3.33]

[11 Functions of the Executive]

[1) It shall be the general duty of the Executive to do such things and make such arrangements as it considers appropriate for the general purposes of this Part.

(2) In connection with the general purposes of this Part, the Executive shall—

(a) assist and encourage persons concerned with matters relevant to those purposes to further those purposes;

(b) make such arrangements as it considers appropriate for the carrying out of research, the publication of the results of research and the provision of training and information, and encourage research and the provision of training and information by others;

(c) make such arrangements as it considers appropriate to secure that the following persons are provided with an information and advisory service on matters relevant to those purposes and are kept informed of and are adequately advised on such matters—

(i) government departments,

(ii) local authorities,

(iii) employers,

(iv) employees,

(v) organisations representing employers or employees, and

(vi) other persons concerned with matters relevant to the general purposes of this Part.

(3) The Executive shall submit from time to time to the Secretary of State such proposals as the Executive considers appropriate for the making of regulations under any of the relevant statutory provisions.

(4) In subsections (1) to (3)—

(a) references to the general purposes of this Part do not include references to [any of the transferred purposes]; and

(b) the reference to the making of regulations under the relevant statutory provisions does not include a reference so far as the regulations are made[—

(i) for any of the transferred purposes, or

(ii) under section 43 and concern fees relating to nuclear site regulation].

[(4A) In subsection (4)—

(a) 'the transferred purposes' means—

(i) the railway safety purposes;

(ii) the nuclear safety purposes;

(iii) the nuclear security purposes;

(iv) the nuclear safeguards purposes;

(v) the radioactive material transport purposes;

(b) 'fees relating to nuclear site regulation' means fees payable for or in connection with the performance of a function by or on behalf of—
(i) the Office for Nuclear Regulation, or
(ii) any inspector appointed by the Office for Nuclear Regulation.

(4B) The Executive may submit to the Secretary of State any proposal submitted to it by the Office for Nuclear Regulation under section 81 of the Energy Act 2013 (proposals about orders and regulations).]

(5) It shall be the duty of the Executive—
(a) to submit to the Secretary of State from time to time particulars of what it proposes to do for the purpose of performing of its functions;
(b) to ensure that its activities are in accordance with proposals approved by the Secretary of State; and
(c) to give effect to any directions given to it by the Secretary of State.

(6) The Executive shall provide a Minister of the Crown on request—
(a) with information about its activities in connection with any matter with which the Minister is concerned; and
(b) with advice on any matter with which he is concerned, where relevant expert advice is obtainable from any of the officers or servants of the Executive, but which is not relevant to the general purposes of this Part.]

[3.34]

Statutory annotations

Amendment

Sub-s (4): in para (a) words 'any of the transferred purposes' in square brackets substituted by the Energy Act 2013, s 116, Sch 12, Pt 1, paras 1, 2(1), (2)(a); para (b)(i), (ii) substituted by the Energy Act 2013, s 116, Sch 12, Pt 1, paras 1, 2(1), (2)(b).

Sub-ss (4A), (4B): inserted by the Energy Act 2013, s 116, Sch 12, Pt 1, paras 1, 2(1), (3).

[3.37]

[13 Powers of the Executive]

[(1) Subject to subsection (2), the Executive shall have power to do anything which is calculated to facilitate, or is conducive or incidental to, the performance of its functions, including a function conferred on it under this subsection.

(2) The power in subsection (1) shall not include the power to borrow money.

(3) The Executive may make agreements with a government department or other person for that department or person to perform any of its functions, with or without payment.

(4) Subject to subsections (5) and (6), the Executive may make agreements with a Minister of the Crown, with a government department or with a public authority to perform functions exercisable by that Minister, department or authority, with or without payment.

(5) The functions referred to in subsection (4)—
(a) in the case of a Minister of the Crown, include functions not conferred by an enactment;
(b) shall be functions which the Secretary of State considers can be appropriately performed by the Executive; and
(c) do not include any power to make regulations or other instruments of a legislative character.

(6) The Executive may provide services or facilities, with or without payment, otherwise than for the general purposes of this Part, to a government department or public authority in connection with the exercise of that department's or authority's functions.

[(6A) The reference in subsection (6) to the general purposes of this Part does not include a reference to any of the following—

(a) the nuclear safety purposes;

(b) the nuclear security purposes;

(c) the nuclear safeguards purposes;

(d) the radioactive material transport purposes.]

(7) The Executive may appoint persons or committees of persons to provide it with advice in connection with any of its functions and, without prejudice to subsection (8), it may remunerate these persons.

(8) The Executive may, in connection with the performance of its functions, pay to any person—

(a) travelling and subsistence allowances, and

(b) compensation for loss of remunerative time.

(9) Any amounts paid under subsections (7) and (8) shall be such as may be determined by the Secretary of State, with the approval of the Minister for the Civil Service.

(10) The Executive may—

(a) carry out, arrange for, or make payments for the carrying out of, research into any matter connected with its functions, and

(b) disseminate or arrange for or make payments for the dissemination of information derived from this research.

(11) The Executive may include, in any arrangements made for the provision of services or facilities under subsection (6), provision for the making of payments to the Executive, or any person acting on its behalf, by other parties to the arrangements and by persons using those services or facilities.]

[3.38]

Statutory annotations

Amendment

Sub-s (6A): inserted by the Energy Act 2013, s 116, Sch 12, Pt 1, paras 1, 3.

[3.39]

14 Power of [the Executive] to direct investigations and inquiries

(1) This section applies to the following matters, that is to say any accident, occurrence, situation or other matter whatsoever which [the Executive] thinks it necessary or expedient to investigate for any of the general purposes of this Part or with a view to the making of regulations for those purposes; and for the purposes of this subsection

[(a) hose general purposes shall be treated as not including the railway safety purposes [or the ONR's purposes]; but

(b) it is otherwise]

immaterial whether the Executive is or is not responsible for securing the enforcement of such (if any) of the relevant statutory provisions as relate to the matter in question.

[(2) The Executive may at any time—

(a) investigate and make a special report on any matter to which this section applies; or

(b) authorise another person to investigate and make a special report into any such matter.

(2A) The Executive may at any time, with the consent of the Secretary of State, direct an inquiry to be held into any matter to which this section applies.]

(3) Any inquiry held by virtue of [subsection (2A)] above shall be held in accordance with regulations made for the purposes of this subsection by the

Secretary of State, and shall be held in public except where or to the extent that the regulations provide otherwise.

(4) Regulations made for the purposes of subsection (3) above may in particular include provision—

(a) conferring on the person holding any such inquiry, and any person assisting him in the inquiry, powers of entry and inspection;

(b) conferring on any such person powers of summoning witnesses to give evidence or produce documents and power to take evidence on oath and administer oaths or require the making of declarations;

(c) requiring any such inquiry to be held otherwise than in public where or to the extent that a Minister of the Crown so directs.

[(4A) Provision that may be made by virtue of subsection (4)(a) includes, in particular, provision conferring functions on the Office for Nuclear Regulation in relation to powers of entry and inspection in relation to any premises for which it is an enforcing authority.]

[(5) In the case of a special report made by virtue of subsection (2), or a report made by the person holding an inquiry by virtue of subsection (2A), the Executive may cause the report, or so much of it as the Executive thinks fit, to be made public at such time and in such manner as it thinks fit.]

(6) [The Executive]—

(a) in the case of an investigation and special report made by virtue of [subsection (2)] above (otherwise than by an officer or servant of the Executive), may pay to the person making it such remuneration and expenses as the Secretary of State may, with the approval of the Minister for the Civil Service, determine;

(b) in the case of an inquiry held by virtue of [subsection (2A)] above, may pay to the person holding it and to any assessor appointed to assist him such remuneration and expenses, and to persons attending the inquiry as witnesses such expenses, as the Secretary of State may, with the like approval, determine; and

(c) may, to such extent as the Secretary of State may determine, defray the other costs, if any, of any such investigation and special report or inquiry.

(7) Where an inquiry is directed to be held by virtue of [subsection (2A)] above into any matter to which this section applies arising in Scotland, being a matter which causes the death of any person, no inquiry with regard to that death shall, unless the Lord Advocate otherwise directs, be held in pursuance of the [Fatal Accidents and Sudden Deaths Inquiry (Scotland) Act 1976].

[3.40]

Statutory annotations

Amendment

Sub-s (1): in para (a) words 'or the ONR's purposes' in square brackets inserted by the Energy Act 2013, s 116, Sch 12, Pt 1, paras 1, 4(1), (2).

Sub-s (4A): inserted by the Energy Act 2013, s 116, Sch 12, Pt 1, paras 1, 4(1), (3).

Health and safety regulations and approved codes of practice

[3.41]

15 Health and safety regulations

[(1) Subject to the provisions of section 50, the Secretary of State . . . shall have power to make regulations under this section for any of the general purposes of this Part (and regulations so made are in this Part referred to as 'health and safety regulations').]

[(1A) In subsection (1), the reference to the general purposes of this Part does not include a reference to any of the following—

(a) the nuclear safety purposes;

(b) the nuclear security purposes;

(c) the nuclear safeguards purposes;

(d) the radioactive material transport purposes.

(1B) Subsection (1A) does not preclude health and safety regulations from including provision merely because the provision could be made for any of the purposes mentioned in paragraphs (a) to (d) of that subsection.]

(2) Without prejudice to the generality of [subsection (1)], health and safety regulations may for any of the general purposes of this Part make provision for any of the purposes mentioned in Schedule 3.

(3) Health and safety regulations—

(a) may repeal or modify any of the existing statutory provisions;

(b) may exclude or modify in relation to any specified class of case any of the provisions of sections 2 to 9 or any of the existing statutory provisions;

(c) may make a specified authority or class of authorities responsible, to such extent as may[, subject to subsection (3A),] be specified, for the enforcement of any of the relevant statutory provisions.

[(3A) Nothing in this section is to be taken to permit health and safety regulations to make provision about responsibility for the enforcement of any of the relevant statutory provisions as they apply in relation to any GB nuclear site.

(3B) Subsection (3A) does not prevent health and safety regulations providing for the Office of Rail Regulation to be responsible for the enforcement, in relation to GB nuclear sites, of any of the relevant statutory provisions that are made for the railway safety purposes.

(3C) In subsections (3A) and (3B), 'GB nuclear site' has the same meaning as in section 68 of the Energy Act 2013 (nuclear safety purposes).]

(4) Health and safety regulations—

(a) may impose requirements by reference to the approval of [the Executive] or any other specified body or person;

(b) may provide for references in the regulations to any specified document to operate as references to that document as revised or re-issued from time to time.

(5) Health and safety regulations—

(a) may provide (either unconditionally or subject to conditions, and with or without limit of time) for exemptions from any requirement or prohibition imposed by or under any of the relevant statutory provisions;

(b) may enable exemptions from any requirement or prohibition imposed by or under any of the relevant statutory provisions to be granted (either unconditionally or subject to conditions, and with or without limit of time) by any specified person or by any person authorised in that behalf by a specified authority.

(6) Health and safety regulations—

(a) may specify the persons or classes of persons who, in the event of a contravention of a requirement or prohibition imposed by or under the regulations, are to be guilty of an offence, whether in addition to or to the exclusion of other persons or classes of persons;

(b) may provide for any specified defence to be available in proceedings for any offence under the relevant statutory provisions either generally or in specified circumstances;

(c) may exclude proceedings on indictment in relation to offences consisting of a contravention of a requirement or prohibition imposed by or under any of the existing statutory provisions, sections 2 to 9 or health and safety regulations;

(d) may restrict the punishments [(other than the maximum fine on conviction on indictment)] which can be imposed in respect of any such offence as is mentioned in paragraph (c) above;

[(e) . . .

(7) Without prejudice to section 35, health and safety regulations may make provision for enabling offences under any of the relevant statutory provisions to be treated as having been committed at any specified place for the purpose of bringing any such offence within the field of responsibility of any enforcing authority or conferring jurisdiction on any court to entertain proceedings for any such offence.

(8) Health and safety regulations may take the form of regulations applying to particular circumstances only or to a particular case only (for example, regulations applying to particular premises only).

(9) If an Order in Council is made under section 84 (3) providing that this section shall apply to or in relation to persons, premises or work outside Great Britain then, notwithstanding the Order, health and safety regulations shall not apply to or in relation to aircraft in flight, vessels, hovercraft or offshore installations outside Great Britain or persons at work outside Great Britain in connection with submarine cables or submarine pipelines except in so far as the regulations expressly so provide.

(10) In this section 'specified' means specified in health and safety regulations.

[3.42]

Statutory annotations

Amendment

Sub-ss (1A), (1B): inserted by the Energy Act 2013, s 116, Sch 12, Pt 1, paras 1, 5(1), (2).

Sub-s (2): words 'subsection (1)' in square brackets substituted by the Energy Act 2013, s 116, Sch 12, Pt 1, paras 1, 5(1), (3).

Sub-s (3): in para (c) words ', subject to subsection (3A),' in square brackets inserted by the Energy Act 2013, s 116, Sch 12, Pt 1, paras 1, 5(1), (4).

Sub-ss (3A)–(3C): inserted by the Energy Act 2013, s 116, Sch 12, Pt 1, paras 1, 5(1), (5).

[3.43]

16 Approval of codes of practice by the Commission

(1) For the purpose of providing practical guidance with respect to the requirements of any provision of [any of the enactments or instruments mentioned in subsection (1A) below], [the Executive] may, subject to the following subsection . . .—

(a) approve and issue such codes of practice (whether prepared by it or not) as in its opinion are suitable for that purpose;

(b) approve such codes of practice issued or proposed to be issued otherwise than by [the Executive] as in its opinion are suitable for that purpose.

[(1A) Those enactments and instruments are—

(a) sections 2 to 7 above;

(b) health and safety regulations, except so far as they make provision exclusively in relation to transport systems falling within paragraph 1(3) of Schedule 3 to the Railways Act 2005; and

(c) the existing statutory provisions that are not such provisions by virtue of section 117(4) of the Railways Act 1993.]

(2) [The Executive] shall not approve a code of practice under subsection (1) above without the consent of the Secretary of State, and shall, before seeking his consent, consult—

(a) any government department or other body that appears to [the Executive] to be appropriate . . . ; and

(b) such government departments and other bodies, if any, as in relation to any matter dealt with in the code, [the Executive] is required to consult under this section by virtue of directions given to it by the Secretary of State.

(3) Where a code of practice is approved by [the Executive] under subsection (1) above, [the Executive] shall issue a notice in writing—

(a) identifying the code in question and stating the date on which its approval by [the Executive] is to take effect; and

(b) specifying for which of the provisions mentioned in subsection (1) above the code is approved.

(4) [The Executive] may—

(a) from time to time revise the whole or any part of any code of practice prepared by it in pursuance of this section;

(b) approve any revision or proposed revision of the whole or any part of any code of practice for the time being approved under this section;

and the provisions of subsections (2) and (3) above shall, with the necessary modifications, apply in relation to the approval of any revision under this subsection as they apply in relation to the approval of a code of practice under subsection (1) above.

(5) [The Executive] may at any time with the consent of the Secretary of State withdraw its approval from any code of practice approved under this section, but before seeking his consent shall consult the same government departments and other bodies as it would be required to consult under subsection (2) above if it were proposing to approve the code.

(6) Where under the preceding subsection [the Executive] withdraws its approval from a code of practice approved under this section, [the Executive] shall issue a notice in writing identifying the code in question and stating the date on which its approval of it is to cease to have effect.

(7) References in this part to an approved code of practice are references to that code as it has effect for the time being by virtue of any revision of the whole or any part of it approved under this section.

(8) The power of [the Executive] under subsection (1)(b) above to approve a code of practice issued or proposed to be issued otherwise than by [the Executive] shall include power to approve a part of such a code of practice; and accordingly in this Part 'code of practice' may be read as including a part of such a code of practice.

[3.44]

Statutory annotations

Amendment

Sub-s (2): in para (a) words omitted repealed by the Health and Social Care Act 2012, s 56(4), Sch 7, paras 4, 5.

Enforcement

[3.47]

18 Authorities responsible for enforcement of the relevant statutory provisions

(1) It shall be the duty of the Executive to make adequate arrangements for the enforcement of the relevant statutory provisions except to the extent that some other

authority or class of authorities is by any of those provisions or by regulations under subsection (2) below made responsible for their enforcement.

[(1A) The Office for Nuclear Regulation is responsible for the enforcement of the relevant statutory provisions as they apply in relation to GB nuclear sites (within the meaning given in section 68 of the Energy Act 2013 (nuclear safety purposes)).

(1B) Subsection (1A) is subject to any provision of health and safety regulations making the Office of Rail Regulation responsible for the enforcement of any of the relevant statutory provisions to any extent in relation to such sites.]

(2) The Secretary of State may by regulations—

[(za) make the Office for Nuclear Regulation responsible for the enforcement of the relevant statutory provisions to such extent as may be prescribed (and may in particular provide for any site or matter in relation to which the Office for Nuclear Regulation is made so responsible to be determined by the Secretary of State or the Office for Nuclear Regulation under the regulations);]

(a) make local authorities responsible for the enforcement of the relevant statutory provisions to such extent as may be prescribed;

(b) make provision for enabling responsibility for enforcing any of the relevant statutory provisions to be, to such extent as may be determined under the regulations—

[(zi) transferred from the Executive or local authorities to the Office for Nuclear Regulation, or from the Office for Nuclear Regulation to the Executive or local authorities;]

(i) transferred from the Executive to local authorities or from local authorities to the Executive; or

(ii) assigned to the Executive[, to the Office for Nuclear Regulation] or to local authorities for the purpose of removing any uncertainty as to what are by virtue of [subsection (1A) or] this subsection their respective responsibilities for the enforcement of those provisions;

[(iii) assigned to the Office of Rail Regulation or the Office for Nuclear Regulation for the purpose of removing any uncertainty as to what are by virtue of any of the relevant statutory provisions their respective responsibilities for the enforcement of any of those provisions;]

and any regulations made in pursuance of paragraph (b) above shall include provision for securing that any transfer or assignment effected under the regulations is brought to the notice of persons affected by it.

(3) Any provision made by regulations under the preceding subsection shall have effect subject to any provision made by health and safety regulations . . . in pursuance of section 15(3)(c).

[(3A) Regulations under subsection (2)(a) may not make local authorities enforcing authorities in relation to any site in relation to which the Office for Nuclear Regulation is an enforcing authority.

(3B) Where the Office for Nuclear Regulation is, by or under subsection (1A) or (2), made responsible for the enforcement of any of the relevant statutory provisions to any extent, it must make adequate arrangements for the enforcement of those provisions to that extent.]

(4) It shall be the duty of every local authority—

(a) to make adequate arrangements for the enforcement within their area of the relevant statutory provisions to the extent that they are by any of those provisions or by regulations under subsection (2) above made responsible for their enforcement; and

(b) to perform the duty imposed on them by the preceding paragraph and any other functions conferred on them by any of the relevant statutory provisions in accordance with such guidance as [the Executive] may give them.

[(4A) Before the Executive gives guidance under subsection (4)(b) it shall consult the local authorities.

(4B) It shall be the duty of the Executive and the local authorities—

(a) to work together to establish best practice and consistency in the enforcement of the relevant statutory provisions;

(b) to enter into arrangements with each other for securing cooperation and the exchange of information in connection with the carrying out of their functions with regard to the relevant statutory provisions; and

(c) from time to time to review those arrangements and to revise them when they consider it appropriate to do so.]

(5) Where any authority other than . . . the Executive[, the Office for Nuclear Regulation] or a local authority is by any of the relevant statutory provisions . . . made responsible for the enforcement of any of those provisions to any extent, it shall be the duty of that authority—

(a) to make adequate arrangements for the enforcement of those provisions to that extent; and

(b) [except where that authority is the Office of Rail Regulation,] to perform the duty imposed on the authority by the preceding paragraph and any other functions conferred on the authority by any of the relevant statutory provisions in accordance with such guidance as [the Executive] may give to the authority.

(6) Nothing in the provisions of this Act or of any regulations made thereunder charging any person in Scotland with the enforcement of any of the relevant statutory provisions shall be construed as authorising that person to institute proceedings for any offence.

(7) In this Part—

(a) 'enforcing authority' means the Executive or any other authority which is by any of the relevant statutory provisions or by regulations under subsection (2) above made responsible for the enforcement of any of those provisions to any extent; and

(b) any reference to an enforcing authority's field of responsibility is a reference to the field over which that authority's responsibility for the enforcement of those provisions extends for the time being;

but where by virtue of [subsection (3) of section 13] [of this Act or section 95 of the Energy Act 2013 (power for Office for Nuclear Regulation to arrange for exercise of functions by others)] the performance of any function of . . . the Executive [or the Office for Nuclear Regulation] is delegated to a government department or person, references to . . . the Executive [or the Office for Nuclear Regulation (as the case may be)] (or to an enforcing authority where that authority is the Executive [or the Office for Nuclear Regulation]) in any provision of this Part which relates to that function shall, so far as may be necessary to give effect to any agreement [or arrangements under the provision in question], be construed as references to that department or person; and accordingly any reference to the field of responsibility of an enforcing authority shall be construed as a reference to the field over which that department or person for the time being performs such a function.

[3.48]

Statutory annotations

Amendment

Sub-ss (1A), (1B): inserted by the Energy Act 2013, s 116, Sch 12, Pt 1, paras 1, 6(1), (2).

Sub-s (2): para (za) inserted by the Energy Act 2013, s 116, Sch 12, Pt 1, paras 1, 6(1), (3)(a); para (b)(zi) inserted by the Energy Act 2013, s 116, Sch 12, Pt 1, paras 1, 6(1), (3)(b); in para (b)(ii) words ', to the Office for Nuclear Regulation' in square brackets inserted by the Energy Act 2013, s 116, Sch 12, Pt 1, paras 1, 6(1), (3)(c)(i); in para (b)(ii) words 'subsection (1A) or' in square brackets inserted by the Energy Act 2013, s 116, Sch 12, Pt 1, paras 1, 6(1), (3)(c)(ii); para (b)(iii) inserted by the Energy Act 2013, s 116, Sch 12, Pt 1, paras 1, 6(1), (3)(d).

Sub-ss (3A), (3B): inserted by the Energy Act 2013, s 116, Sch 12, Pt 1, paras 1, 6(1), (4).

Sub-s (5): words ', the Office for Nuclear Regulation' in square brackets inserted by the Energy Act 2013, s 116, Sch 12, Pt 1, paras 1, 6(1), (5).

Sub-s (7): words from 'of this Act' to 'functions by others)' in square brackets inserted by the Energy Act 2013, s 116, Sch 12, Pt 1, paras 1, 6(1), (6)(a); words 'or the Office for Nuclear Regulation' in square brackets in both places they occur inserted by the Energy Act 2013, s 116, Sch 12, Pt 1, paras 1, 6(1), (6)(b); words 'or the Office for Nuclear Regulation (as the case may be)' in square brackets inserted by the Energy Act 2013, s 116, Sch 12, Pt 1, paras 1, 6(1), (6)(c); words 'or arrangements under the provision in question' in square brackets substituted by the Energy Act 2013, s 116, Sch 12, Pt 1, paras 1, 6(1), (6)(d).

Obtaining and disclosure of information

[3.67]

27 Obtaining of information by . . . the Executive, enforcing authorities etc

(1) For the purpose of obtaining—

(a) any information which [the Executive] needs for the discharge of its functions; or

(b) any information which an enforcing authority [other than the Office for Nuclear Regulation] needs for the discharge of the authority's functions,

[the Executive] may, with the consent of the Secretary of State, serve on any person a notice requiring that person to furnish to [the Executive] or, as the case may be, to the enforcing authority in question such information about such matters as may be specified in the notice, and to do so in such form and manner and within such time as may be so specified.

In this subsection 'consent' includes a general consent extending to cases of any stated description.

(2) Nothing in section 9 of the Statistics of Trade Act 1947 (which restricts the disclosure of information obtained under that Act) shall prevent or penalise—

(a) the disclosure by a Minister of the Crown to . . . the Executive of information obtained under that Act about any undertaking within the meaning of that Act, being information consisting of the names and addresses of the persons carrying on the undertaking, the nature of the undertaking's activities, the numbers of persons of different descriptions who work in the undertaking, the addresses or places where activities of the undertaking are or were carried on, the nature of the activities carried on there, or the numbers of persons of different descriptions who work or worked in the undertaking there; . . .

(b) . . .

[(3) In the preceding subsection any reference to a Minister of the Crown or the Executive includes respectively a reference to an officer of that person or of that body and also, in the case of a reference to the Executive, includes a reference to—

(a) a person performing any functions of the Executive on its behalf by virtue of section 13(3);

(b) an officer of a body which is so performing any such functions; and

(c) an adviser appointed under section 13(7).]

(4) A person to whom information is disclosed in pursuance of subsection (2) above shall not use the information for a purpose other than a purpose . . . of the Executive.

[3.68]

Statutory annotations

Amendment

Sub-s (1): in para (b) words 'other than the Office for Nuclear Regulation' in square brackets inserted by the Energy Act 2013, s 116, Sch 12, Pt 1, paras 1, 7.

[3.69]

[27A Information communicated by Commissioners of Revenue and Customs]

[(1) If they think it appropriate to do so for the purpose of facilitating the exercise or performance by any person to whom subsection (2) below applies of any of that person's powers or duties under any of the relevant statutory provisions, [the Commissioners for Her Majesty's Revenue and Customs] may authorise the disclosure to that person of any information obtained [or held] for the purposes of the exercise [by Her Majesty's Revenue and Customs] of their functions in relation to imports.

(2) This subsection applies to an enforcing authority and to an inspector[, other than the Office for Nuclear Regulation or an inspector appointed by the Office for Nuclear Regulation].

(3) A disclosure of information made to any person under subsection (1) above shall be made in such manner as may be directed by [the Commissioners for Her Majesty's Revenue and Customs] and may be made through such persons acting on behalf of that person as may be so directed.

(4) Information may be disclosed to a person under subsection (1) above whether or not the disclosure of the information has been requested by or on behalf of that person.]

[3.70]

Statutory annotations

Amendment

Sub-s (2): words ', other than the Office for Nuclear Regulation or an inspector appointed by the Office for Nuclear Regulation' in square brackets inserted by the Energy Act 2013, s 116, Sch 12, Pt 1, paras 1, 8.

[3.71]

28 Restrictions on disclosure of information

(1) In this and the two following subsections—

(a) 'relevant information' means information obtained by a person under section 27(1) or furnished to any person[, other than the Office for Nuclear Regulation (or an inspector appointed by it),] [under section 27A above[, by virtue of section 43A(6) below] or] in pursuance of a requirement imposed by any of the relevant statutory provisions; and

(b) 'the recipient', in relation to any relevant information, means the person by whom that information was so obtained or to whom that information was so furnished, as the case may be.

(2) Subject to the following subsection, no relevant information shall be disclosed without the consent of the person by whom it was furnished.

(3) The preceding subsection shall not apply to—

(a) disclosure of information to . . . , the Executive[, the Office for Nuclear Regulation], [the Environment Agency, [the Natural Resources Body for Wales,] the Scottish Environment Protection Agency,] a government department or any enforcing authority;

(b) without prejudice to paragraph (a) above, disclosure by the recipient of information to any person for the purpose of any function conferred on the recipient by or under any of the relevant statutory provisions;

(c) without prejudice to paragraph (a) above, disclosure by the recipient of information to—

(i) an officer of a local authority who is authorised by that authority to receive it,

[(ii) an officer . . . of a water undertaker, sewerage undertaker, water authority or water development board who is authorised by that . . . undertaker, authority or board to receive it,]

(iii) . . .

(iv) a constable authorised by a chief officer of police to receive it;

(d) disclosure by the recipient of information in a form calculated to prevent it from being identified as relating to a particular person or case;

(e) disclosure of information for the purposes of any legal proceedings or any investigation or inquiry held by virtue of [section 14(2) or (2A)], or for the purposes of a report of any such proceedings or inquiry or of a special report made by virtue of [section 14(2) or (2A)];

[(f) any other disclosure of information by the recipient, if—

(i) the recipient is, or is acting on behalf of a person who is, a public authority for the purposes of the Freedom of Information Act 2000 [or a Scottish public authority for the purposes of the Freedom of Information (Scotland) Act 2002], and

(ii) the information is not held by the authority on behalf of another person].

[(4) In the preceding subsection, any reference to the Executive[, the Office for Nuclear Regulation], the Environment Agency, [the Natural Resources Body for Wales,] the Scottish Environment Protection Agency, a government department or an enforcing authority includes respectively a reference to an officer of that body or authority (including, in the case of an enforcing authority, any inspector appointed by it), and also, in the case of a reference to the Executive [or the Office for Nuclear Regulation], includes a reference to—

(a) a person performing any functions of the Executive [or Office for Nuclear Regulation] on its behalf by virtue of section 13(3) [of this Act or, as the case may be, section 95 of the Energy Act 2013];

(b) an officer of a body which is so performing any such functions; and

(c) an adviser appointed under section 13(7) [or, in the case of the Office for Nuclear Regulation, a person providing advice to that body].]

(5) A person to whom information is disclosed in pursuance of [any of paragraphs (a) to (e) of] subsection (3) above shall not use the information for a purpose other than—

(a) in a case falling within paragraph (a) of that subsection, a purpose . . . of the Executive[, of the Office for Nuclear Regulation] or [of the Environment Agency [or of the Natural Resources Body for Wales] or of the Scottish Environment Protection Agency or] of the government department in question, or the purposes of the enforcing authority in question in connection with the relevant statutory provisions, as the case may be;

(b) in the case of information given to an officer of a [body which is a local authority, . . . , a water undertaker, a sewerage undertaker, a water authority, a river purification board or a water development board, the purposes of the body] in connection with the relevant statutory provisions or any enactment whatsoever relating to public health, public safety or the protection of the environment;

(c) in the case of information given to a constable, the purposes of the police in connection with the relevant statutory provisions or any enactment whatsoever relating to public health, public safety or the safety of the State.

[(6) References in subsections (3) and (5) above to a local authority include . . . a joint authority established by Part IV of the Local Government Act 1985[, [an economic prosperity board established under section 88 of the Local Democracy, Economic Development and Construction Act 2009, a combined authority established under section 103 of that Act,] an authority established for an area in England by an order under section 207 of the Local Government and Public Involvement in Health Act 2007 (joint waste authorities)] [and the London Fire and Emergency Planning Authority].]

(7) A person shall not disclose any information obtained by him as a result of the exercise of any power conferred by section 14(4)(a) or 20 (including, in particular, any information with respect to any trade secret obtained by him in any premises entered by him by virtue of any such power) except—

(a) for the purposes of his functions; or

(b) for the purposes of any legal proceedings or any investigation or inquiry held by virtue of [section 14(2) or (2A)] or for the purposes of a report of any such proceedings or inquiry or of a special report made by virtue of [section 14(2) or (2A)]; or

(c) with the relevant consent.

In this subsection 'the relevant consent' means, in the case of information furnished in pursuance of a requirement imposed under section 20, the consent of the person who furnished it, and, in any other case, the consent of a person having responsibilities in relation to the premises where the information was obtained.

(8) Notwithstanding anything in the preceding subsection an inspector shall, in circumstances in which it is necessary to do so for the purpose of assisting in keeping persons (or the representatives of persons) employed at any premises adequately informed about matters affecting their health, safety and welfare, give to such persons or their representatives the following descriptions of information, that is to say—

(a) factual information obtained by him as mentioned in that subsection which relates to those premises or anything which was or is therein or was or is being done therein; and

(b) information with respect to any action which he has taken or proposes to take in or in connection with those premises in the performance of his functions;

and, where an inspector does as aforesaid, he shall give the like information to the employer of the first-mentioned persons.

[(9) Notwithstanding anything in subsection (7) above, a person who has obtained such information as is referred to in that subsection may furnish to a person who appears to him to be likely to be a party to any civil proceedings arising out of any accident, occurrence, situation or other matter, a written statement of relevant facts observed by him in the course of exercising any of the powers referred to in that subsection.]

[(9A) Subsection (7) above does not apply if—

(a) the person who has obtained any such information as is referred to in that subsection is, or is acting on behalf of a person who is, a public authority for the purposes of the Freedom of Information Act 2000 [or a Scottish public authority for the purposes of the Freedom of Information (Scotland) Act 2002], and

(b) the information is not held by the authority on behalf of another person.]

[(9B) Nothing in subsection (7) or (9) applies to a person appointed as an inspector by the Office for Nuclear Regulation in relation to functions which the person has by virtue of that appointment.]

[(10) The Broads Authority and every National Park authority shall be deemed to be local authorities for the purposes of this section.]

[3.72]

Statutory annotations

Amendment

Sub-s (1): words ', other than the Office for Nuclear Regulation (or an inspector appointed by it),' in square brackets inserted by the Energy Act 2013, s 116, Sch 12, Pt 1, paras 1, 9(1), (2).

Sub-s (3): in para (a) words ', the Office for Nuclear Regulation' in square brackets inserted by the Energy Act 2013, s 116, ch 12, Pt 1, paras 1, 9(1), (3); in para (a) words 'the Natural Resources Body for Wales,' in square brackets inserted by SI 2013/755, art 4(1), Sch 2, Pt 1, paras 111, 112(1), (2).

Sub-s (4): words ', the Office for Nuclear Regulation' in square brackets inserted by the Energy Act 2013, s 116, Sch 12, Pt 1, paras 1, 9(1), (4)(a)(i); words 'the Natural Resources Body for Wales,' in square brackets inserted by SI 2013/755, art 4(1), Sch 2, Pt 1, paras 111, 112(1), (3); words 'or the Office for Nuclear Regulation' in square brackets inserted by the Energy Act 2013, s 116, Sch 12, Pt 1, paras 1, 9(1), (4)(a)(ii); in para (a) words 'or Office for Nuclear Regulation' and 'of this Act or, as the case may be, section 95 of the Energy Act 2013' in square brackets inserted by the Energy Act 2013, s 116, Sch 12, Pt 1, paras 1, 9(1), (4)(b); in para (c) words 'or, in the case of the Office for Nuclear Regulation, a person providing advice to that body' in square brackets inserted by the Energy Act 2013, art 4(2), Sch 12, Pt 1, paras 1, 9(1), (4)(c).

Sub-s (5): in para (a) words ', of the Office for Nuclear Regulation' in square brackets inserted by the Energy Act 2013, art 4(2), Sch 12, Pt 1, paras 1, 9(1), (5); in para (a) words 'or of the Natural Resources Body for Wales' in square brackets inserted by SI 2013/755, art 4(1), Sch 2, Pt 1, paras 111, 112(1), (4).

Sub-s (9B): inserted by the Energy Act 2013, art 4(2), Sch 12, Pt 1, paras 1, 9(1), (6).

Provisions as to offences

[3.76]

34 Extension of time for bringing summary proceedings

(1) Where—

(a) a special report on any matter to which section 14 of this Act applies is made by virtue of subsection [(2)] of that section; or

(b) a report is made by the person holding an inquiry into any such matter by virtue of subsection [(2A)] of that section; or

(c) [an investigation under Part 1 of the Coroners and Justice Act 2009 is conducted into] the death of any person whose death may have been caused by an accident which happened while he was at work or by a disease which he contracted or probably contracted at work or by any accident, act or omission which occurred in connection with the work of any person whatsoever; or

(d) . . .

and it appears [from the report or investigation or, in a case falling within paragraph (d) above, from the proceedings at the] inquiry, that any of the relevant statutory provisions was contravened at a time which is material in relation to the subject-matter of the [report, investigation or inquiry], summary proceedings against any person liable to be proceeded against in respect of the contravention may be commenced at any time within three months of the making of the report or, in a case falling within paragraph (c) or (d) above, within three months of the [conclusion of the investigation] or inquiry.

(2) Where an offence under any of the relevant statutory provisions is committed by reason of a failure to do something at or within a time fixed by or under any of those provisions, the offence shall be deemed to continue until that thing is done.

(3) Summary proceedings for an offence to which this subsection applies may be commenced at any time within six months from the date on which there comes to the knowledge of a responsible enforcing authority evidence sufficient in the opinion of that authority to justify a prosecution for that offence; and for the purposes of this subsection—

- (a) a certificate of an enforcing authority stating that such evidence came to its knowledge on a specified date shall be conclusive evidence of that fact; and
- (b) a document purporting to be such a certificate and to be signed by or on behalf of the enforcing authority in question shall be presumed to be such a certificate unless the contrary is proved.

(4) The preceding subsection applies to any offence under any of the relevant statutory provisions which a person commits by virtue of any provision or requirement to which he is subject as the designer, manufacturer, importer or supplier of any thing; and in that subsection 'responsible enforcing authority' means an enforcing authority within whose field of responsibility the offence in question lies, whether by virtue of section 35 or otherwise.

(5) In the application of subsection (3) above to Scotland—

- (a) for the words from 'there comes' to 'that offence' there shall be substituted the words 'evidence, sufficient in the opinion of the enforcing authority to justify a report to the Lord Advocate with a view to consideration of the question of prosecution, comes to the knowledge of the authority';
- (b) at the end of paragraph (b) there shall be added the words

  'and

  - (c) [section 331(3) of the Criminal Procedure (Scotland) Act 1975] (date of commencement of proceedings) shall have effect as it has effect for the purposes of that section.'

[(6) In the application of subsection (4) above to Scotland, after the words 'applies to' there shall be inserted the words 'any offence under section 33(1)(c) above where the health and safety regulations concerned were made for the general purpose mentioned in section 18(1) of the Gas Act 1986 and'.]

[3.77]

Statutory annotations

Amendment

Sub-s (1): in para (c) words 'an investigation under Part 1 of the Coroners and Justice Act 2009 is conducted into' in square brackets substituted by the Coroners and Justice Act 2009, s 177(1), Sch 21, Pt 1, para 25(1), (2); words from 'from the report' to 'proceedings at the' in square brackets substituted by the Coroners and Justice Act 2009, s 177(1), Sch 21, Pt 1, para 25(1), (3); words 'from the report or investigation or, in a case falling within paragraph (d) above, from the proceedings at the' in square brackets substituted by the Coroners and Justice Act 2009, s 177(1), Sch 21, Pt 1, para 25(1), (4); words 'conclusion of the investigation' in square brackets substituted by the Coroners and Justice Act 2009, s 177(1), Sch 21, Pt 1, para 25(1), (5).

Miscellaneous and supplementary

[3.98]

44 Appeals in connection with licensing provisions in the relevant statutory provisions

(1) Any person who is aggrieved by a decision of an authority having power to issue licences . . . under any of the relevant statutory provisions—

(a) refusing to issue him a licence, to renew a licence held by him, or to transfer to him a licence held by another;

(b) issuing him a licence on or subject to any term, condition or restriction whereby he is aggrieved;

(c) varying or refusing to vary any term, condition or restriction on or subject to which a licence is held by him; or

(d) revoking a licence held by him,

may appeal to the Secretary of State.

(2) The Secretary of State may, in such cases as he considers it appropriate to do so, having regard to the nature of the questions which appear to him to arise, direct that an appeal under this section shall be determined on his behalf by a person appointed by him for that purpose.

(3) Before the determination of an appeal the Secretary of State shall ask the appellant and the authority against whose decision the appeal is brought whether they wish to appear and be heard on the appeal and—

(a) the appeal may be determined without a hearing of the parties if both of them express a wish not to appear and be heard as aforesaid;

(b) the Secretary of State shall, if either of the parties expresses a wish to appear and be heard, afford to both of them an opportunity of so doing.

(4) The Tribunals and Inquiries Act [1992] shall apply to a hearing held by a person appointed in pursuance of subsection (2) above to determine an appeal as it applies to a statutory inquiry held by the Secretary of State, but as if in [section 10(1)] of that Act (statement of reasons for decisions) the reference to any decision taken by the Secretary of State included a reference to a decision taken on his behalf by that person.

[(4A) . . .]

(5) A person who determines an appeal under this section on behalf of the Secretary of State and the Secretary of State, if he determines such an appeal, may give such directions as he considers appropriate to give effect to his determination.

(6) The Secretary of State may pay to any person appointed to hear or determine an appeal under this section on his behalf such remuneration and allowances as the Secretary of State may with the approval of the Minister for the Civil Service determine.

(7) In this section—

(a) 'licence' means a licence under any of the relevant statutory provisions . . . ;

(b) . . .

(8) . . .

[3.99]

Statutory annotations

Amendment

Sub-s (1): words omitted repealed by the Energy Act 2013, s 116, Sch 12, Pt 1, paras 1, 10(1), (2).

Sub-s (4A): repealed by SI 2013/2042, art 2(2), Schedule, para 2.

Sub-s (7): in para (a) words omitted repealed by the Energy Act 2013, s 116, Sch 12, Pt 1, paras 1, 10(1), (3)(a); para (b) repealed by the Energy Act 2013, art 4(2), Sch 12, Pt 1, paras 1, 10(1), (3)(b).

Sub-s (8): repealed by the Energy Act 2013, s 116, Sch 12, Pt 1, paras 1, 10(1), (4).

[3.104]

47 Civil liability

(1) Nothing in this Part shall be construed—

(a) as conferring a right of action in any civil proceedings in respect of any failure to comply with any duty imposed by sections 2 to 7 or any contravention of section 8; or

(b) . . .

(c) as affecting the operation of section 12 of the Nuclear Installations Act 1965 (right to compensation by virtue of certain provisions of that Act).

[(2) Breach of a duty imposed by a statutory instrument containing (whether alone or with other provision) health and safety regulations shall not be actionable except to the extent that regulations under this section so provide.

(2A) Breach of a duty imposed by an existing statutory provision shall not be actionable except to the extent that regulations under this section so provide (including by modifying any of the existing statutory provisions).

(2B) Regulations under this section may include provision for—

(a) a defence to be available in any action for breach of the duty mentioned in subsection (2) or (2A);

(b) any term of an agreement which purports to exclude or restrict any liability for such a breach to be void.]

(3) No provision made by virtue of section 15(6)(b) shall afford a defence in any civil proceedings . . . .

(4) Subsections (1)(a)[, (2) and (2A)] above are without prejudice to any right of action which exists apart from the provisions of this Act, and subsection [(2B)(a)] above is without prejudice to any defence which may be available apart from the provisions of the regulations there mentioned.

(5) . . .

(6) . . .

[(7) The power to make regulations under this section shall be exercisable by the Secretary of State.]

[3.105]

Statutory annotations

Amendment

Sub-s (1): para (b) (omitted) repealed by the Enterprise and Regulatory Reform Act 2013, s 69(1), (2), as from 1 October 2013 (see SI 2013/2227).

Sub-ss (2), (2A), (2B): substituted, for sub-s (2) as originally enacted, by the Enterprise and Regulatory Reform Act 2013, s 69(1), (3), as from 1 October 2013 (see SI 2013/2227).

Sub-s (3): words omitted repealed by the Enterprise and Regulatory Reform Act 2013, s 69(1), (4), as from 1 October 2013 (see SI 2013/2227).

Sub-s (4): words in square brackets substituted by the Enterprise and Regulatory Reform Act 2013, s 69(1), (5), as from 1 October 2013 (see SI 2013/2227).

Sub-ss (5), (6): repealed by the Enterprise and Regulatory Reform Act 2013, s 69(1), (6), as from 1 October 2013 (see SI 2013/2227).

Sub-s (7): inserted by the Enterprise and Regulatory Reform Act 2013, s 69(1), (7), as from 1 October 2013 (see SI 2013/2227).

[3.110]

50 Regulations under the relevant statutory provisions

[(1) Where any power to make regulations under any of the relevant statutory provisions is exercisable by the Secretary of State, that power may be exercised either—

(a) so as to give effect (with or without modifications) to proposals submitted by the Executive under section 11(3); or

(b) subject to subsection (1AA), independently of such proposals.

(1AA) The Secretary of State shall not exercise the power referred to in subsection (1) independently of proposals from the Executive unless [the Secretary of State has consulted—

(a) the Executive,

(b) the Office for Nuclear Regulation, and

(c) such other bodies as appear to the Secretary of State to be appropriate].]

[(1A) Subsection (1) does not apply to the exercise of a power to make regulations so far as it is exercised—

(a) for giving effect (with or without modifications) to proposals submitted by the Office of Rail Regulation under paragraph 2(5) of Schedule 3 to the Railways Act 2005; or

(b) otherwise for or in connection with the railway safety purposes.]

(2) Where the [authority who is to exercise any such power as is mentioned in subsection (1) above proposes to exercise that power] so as to give effect to any such proposals as are there mentioned with modifications, he shall, before making the regulations, consult[—

(a) the Executive, and

(b) the Office for Nuclear Regulation].

(3) Where [the Executive] proposes to submit [under section [11(3)]] any such proposals as are mentioned in subsection (1) above except proposals for the making of regulations under section 43(2), it shall, before so submitting them, consult—

[(za) the Office for Nuclear Regulation;]

(a) any government department or other body that appears to [the Executive] to be appropriate (and, in particular, in the case of proposals for the making of regulations under section 18(2), any body representing local authorities that so appears . . .);

(b) such government departments and other bodies, if any, as, in relation to any matter dealt with in the proposals, [the Executive] is required to consult under this subsection by virtue of directions given to it by the Secretary of State.

[(4) If the Executive has consulted the Office for Nuclear Regulation under subsection (3) in relation to a proposal under section 11(3) for regulations under any of the relevant statutory provisions, it must, when it submits the proposal (with or without modification) to the Secretary of State, also submit—

(a) any representations made by the Office for Nuclear Regulation in response to the consultation, and

(b) any response to those representations given by the Executive to the Office for Nuclear Regulation.

(5) The preceding provisions of this section do not apply to the exercise of the power in section 43 to make ONR fees regulations, but the Secretary of State must consult the Office for Nuclear Regulation before—

(a) making ONR fees regulations independently of any proposals submitted by the Office for Nuclear Regulation under section 81(1) of the Energy Act 2013, or

Part 3

(b) making ONR fees regulations which give effect to such proposals but with modifications.

(6) In subsection (5) 'ONR fees regulations' means regulations under section 43 so far as they make provision in relation to fees payable for or in connection with the performance of a function by or on behalf of—

(a) the Office for Nuclear Regulation, or

(b) any inspector appointed by the Office for Nuclear Regulation.]

(4), (5) . . .

[3.111]

Statutory annotations

Amendment

Sub-s (1AA): words from 'the Secretary of' to 'to be appropriate' in square brackets substituted by the Energy Act 2013, s 116, Sch 12, Pt 1, paras 1, 11(1), (2).

Sub-ss (2): paras (a), (b) substituted by the Energy Act 2013, s 116, Sch 12, Pt 1, paras 1, 11(1), (3).

Sub-s (3): para (za) inserted by the Energy Act 2013, s 116, Sch 12, Pt 1, paras 1, 11(1), (4); in para (a) words omitted repealed by the Health and Social Care Act 2012, s 56(4), Sch 7, paras 4, 6.

Sub-ss (4)–(6): inserted by the Energy Act 2013, s 116, Sch 12, Pt 1, paras 1, 11(1), (5).

[3.114]

[51A Application of Part to police

(1) For the purposes of this Part, a person who, otherwise than under a contract of employment, holds the office of constable or an appointment as police cadet shall be treated as an employee of the relevant officer.

(2) In this section 'the relevant officer'—

(a) in relation to a member of a police force or a special constable or police cadet appointed for a police area, [means—

(i) the chief officer of police of that force, or

(i) in the case of a member of the force or a special constable who is, by virtue of a collaboration agreement under section 22A of the Police Act 1996, under the direction and control of a chief officer (within the meaning given by section 23I of that Act), that chief officer],

[(b) in relation to a member of a police force seconded to the [National Crime Agency to serve as a National Crime Agency officer], means that Agency, and]

(c) in relation to any other person holding the office of constable or an appointment as police cadet, [means—

(i) the person who has the direction and control of the body of constables or cadets in question, or

(i) in the case of a constable who is, by virtue of a collaboration agreement under section 22A of the Police Act 1996, under the direction and control of a chief officer (within the meaning given by section 23I of that Act), that chief officer].

[(2A) For the purposes of this Part the relevant officer, as defined by subsection (2)(a) or (c) above, shall[, if not a corporation sole,] be treated as a corporation sole.

(2B) Where, in a case in which the relevant officer, as so defined, is guilty of an offence by virtue of this section, it is proved—

(a) that the officer-holder personally consented to the commission of the offence,

(b) that he personally connived in its commission, or

(c) that the commission of the offence was attributable to personal neglect on his part,

the office-holder (as well as the corporation sole) shall be guilty of the offence and shall be liable to be proceeded against and punished accordingly.

(2C) In subsection (2B) above 'the office-holder', in relation to the relevant officer, means an individual who, at the time of the consent, connivance or neglect—

(a) held the office or other position mentioned in subsection (2) above as the office or position of that officer; or

(b) was for the time being responsible for exercising and performing the powers and duties of that office or position.

(2D) The provisions mentioned in subsection (2E) below (which impose the same liability for unlawful conduct of constables on persons having their direction or control as would arise if the constables were employees of those persons) do not apply to any liability by virtue of this Part.

(2E) Those provisions are—

[(a) section 24 of the Police and Fire Reform (Scotland) Act 2012 (asp 8);]

(b) section 88(1) of the Police Act 1996;

(c) . . .

(d) . . .

(e) paragraph 14(1) of Schedule 3 to the Criminal Justice and Police Act 2001;

[(f) paragraph 2 of Schedule 4 to the Crime and Courts Act 2013;]

[(g) paragraph 20 of Schedule 1 to the Police and Justice Act 2006].

(2F) In the application of this section to Scotland—

(a) subsection (2A) shall have effect as if for the words 'corporation sole' there were substituted 'distinct juristic person (that is to say, as a juristic person distinct from the individual who for the time being is the office-holder)';

(b) subsection (2B) shall have effect as if for the words 'corporation sole' there were substituted 'juristic person'; and

(c) subsection (2C) shall have effect as if for the words 'subsection (2B)' there were substituted 'subsections (2A) and (2B)'.]

(3) For the purposes of regulations under section 2(4) above—

(a) the Police Federation for England and Wales shall be treated as a recognised trade union recognised by each chief officer of police in England and Wales,

(b) the Police Federation for Scotland shall be treated as a recognised trade union recognised by [the chief constable of the Police Service of Scotland], and

(c) any body recognised by the Secretary of State for the purposes of section 64 of the Police Act 1996 shall be treated as a recognised trade union recognised by each chief officer of police in England, Wales and Scotland.

(4) Regulations under section 2(4) above may provide, in relation to persons falling within subsection (2)(b) or (c) above, that a body specified in the regulations is to be treated as a recognised trade union recognised by such person as may be specified.]

[3.115]

Statutory annotations

Amendment

Sub-s (2): in para (a) words from 'means—' to the end in square brackets substituted by the Police Reform and Social Responsibility Act 2011, s 99, Sch 16, Pt 3, para 119(1), (2)(a); for transitional provisions see s 98, Sch 15, Pts 3, 4 thereto; para (b) substituted by the Serious Organised Crime and Police Act 2005, s 59, Sch 4, para 20; in para (b) words 'National Crime

Agency to serve as a National Crime Agency officer' in square brackets substituted by the Crime and Courts Act 2013, s 15(3), Sch 8, Pt 2, para 21(1), (2); in para (c) words from 'means—' to the end in square brackets substituted by the Police Reform and Social Responsibility Act 2011, s 99, Sch 16, Pt 3, para 119(1), (2)(b); for transitional provisions see s 98, Sch 15, Pts 3, 4 thereto.

Sub-s (2A): words ', if not a corporation sole,' in square brackets inserted by the Police Reform and Social Responsibility Act 2011, s 99, Sch 16, Pt 3, para 119(1), (3); for transitional provisions see s 98, Sch 15, Pts 3, 4 thereto.

Sub-s (2E): para (a) substituted in relation to Scotland by the Police and Fire Reform (Scotland) Act 2012, s 128(1), Sch 7, Pt 1, para 2(a) and in relation to England and Wales by SI 2013/602, art 25, Sch 1, para 2(a); paras (c), (d) repealed by the Police and Justice Act 2006, s 52, Sch 15, Pt 1(A); para (f) substituted by the Crime and Courts Act 2013, s 15(3), Sch 8, Pt 2, para 21(1), (3); para (g) inserted by the Police and Justice Act 2006, s 1(3), Sch 1, Pt 7, para 54.

Sub-s (3): in para (b) words 'the chief constable of the Police Service of Scotland' in square brackets substituted by the Police and Fire Reform (Scotland) Act 2012, s 128(1), Sch 7, Pt 1, para 2(b).

[3.118]

53 General interpretation of Part I

(1) In this Part, unless the context otherwise requires—

. . .

'article for use at work' means—

(a) any plant designed for use or operation (whether exclusively or not) by persons at work, and

(b) any article designed for use as a component in any such plant;

['article of fairground equipment' means any fairground equipment or any article designed for use as a component in any such equipment;]

'code of practice' (without prejudice to section 16(8)) includes a standard, a specification and any other documentary form of practical guidance;

. . .

'conditional sale agreement' means an agreement for the sale of goods under which the purchase price or part of it is payable by instalments, and the property in the goods is to remain in the seller (notwithstanding that the buyer is to be in possession of the goods) until such conditions as to the payment of instalments or otherwise as may be specified in the agreement are fulfilled;

'contract of employment' means a contract of employment or apprenticeship (whether express or implied and, if express, whether oral or in writing);

'credit-sale agreement' means an agreement for the sale of goods, under which the purchase price or part of it is payable by instalments, but which is not a conditional sale agreement;

['customs officer' means an officer within the meaning of the Customs and Excise Management Act 1979;]

'domestic premises' means premises occupied as a private dwelling (including any garden, yard, garage, outhouse or other appurtenance of such premises which is not used in common by the occupants of more than one such dwelling), and 'non-domestic premises' shall be construed accordingly;

'employee' means an individual who works under a contract of employment [or is treated by section 51A as being an employee], and related expressions shall be construed accordingly;

'enforcing authority' has the meaning assigned by section 18(7);

'the Executive' has the meaning assigned by section [10(1)];

'the existing statutory provisions' means the following provisions while and to the extent that they remain in force, namely the provisions of the Acts mentioned in Schedule 1 which are specified in the third column of that Schedule and of the regulations, orders or other instruments of a legislative character made or having effect under any provision so specified;

. . .

['fairground equipment' means any fairground ride, any similar plant which is designed to be in motion for entertainment purposes with members of the public on or inside it or any plant which is designed to be used by members of the public for entertainment purposes either as a slide or for bouncing upon, and in this definition the reference to plant which is designed to be in motion with members of the public on or inside it includes a reference to swings, dodgems and other plant which is designed to be in motion wholly or partly under the control of, or to be put in motion by, a member of the public;]

'the general purposes of this Part' has the meaning assigned by section 1;

'health and safety regulations' has the meaning assigned by section 15(1);

'hire-purchase agreement' means an agreement other than a conditional sale agreement, under which—

(a) goods are bailed or (in Scotland) hired in return for periodical payments by the person to whom they are bailed or hired; and

(b) the property in the goods will pass to that person if the terms of the agreement are complied with and one or more of the following occurs:

(i) the exercise of an option to purchase by that person;

(ii) the doing of any other specified act by any party to the agreement;

(iii) the happening of any other event;

and 'hire-purchase' shall be construed accordingly;

'improvement notice' means a notice under section 21;

'inspector' means an inspector appointed under section 19;

. . .

'local authority' means—

(a) in relation to England . . . , a county council, . . . , a district council, a London borough council, the Common Council of the City of London, the Sub-Treasurer of the Inner Temple or the Under-Treasurer of the Middle Temple,

[(aa) in relation to Wales, a county council or a county borough council,]

(b) . . .

['micro-organism' includes any microscopic biological entity which is capable of replication;]

['nuclear safeguards purposes' has the same meaning as in Part 3 of the Energy Act 2013 (nuclear regulation etc) (see section 72 of that Act);]

['nuclear safety purposes' has the same meaning as in that Part of that Act (see section 68 of that Act);]

['nuclear security purposes' has the same meaning as in that Part of that Act (see section 70 of that Act);]

'offshore installation' means any installation which is intended for underwater exploitation of mineral resources or exploration with a view to such exploitation;

['the ONR's purposes' has the same meaning as in Part 3 of the Energy Act 2013 (see section 67 of that Act);']

'personal injury' includes any disease and any impairment of a person's physical or mental condition;

'plant' includes any machinery, equipment or appliance;

'premises' includes any place and, in particular, includes—

(a) any vehicle, vessel, aircraft or hovercraft,

(b) any installation on land (including the foreshore and other land intermittently covered by water), any offshore installation, and any other installation (whether floating, or resting on the seabed or the subsoil thereof, or resting on other land covered with water or the subsoil thereof), and

(c) any tent or movable structure;

'prescribed' means prescribed by regulations made by the Secretary of State;

'prohibition notice' means a notice under section 22;

['the radioactive material transport purposes' means the transport purposes within the meaning of Part 3 of the Energy Act 2013 (see section 73 of that Act);]

['railway safety purposes' has the same meaning as in Schedule 3 to the Railways Act 2005;]

. . .

'the relevant statutory provisions' means—

(a) the provisions of this Part and of any health and safety regulations . . . ; and

(b) the existing statutory provisions;

'self-employed person' means an individual who works for gain or reward otherwise than under a contract of employment, whether or not he himself employs others;

'substance' means any natural or artificial substance [(including micro-organisms)], whether in solid or liquid form or in the form of a gas or vapour;

. . .

'supply', where the reference is to supplying articles or substances, means supplying them by way of sale, lease, hire or hire-purchase, whether as principal or agent for another.

(2)–(6) . . .

[3.119]

Statutory annotations

Amendment

Sub-s (1): definition 'nuclear safeguards purposes' inserted by the Energy Act 2013, s 116, Sch 12, Pt 1, paras 1, 12(a); definition 'nuclear safety purposes' inserted by the Energy Act 2013, s 116, Sch 12, Pt 1, paras 1, 12(a); definition 'nuclear security purposes' inserted by the Energy Act 2013, s 116, Sch 12, Pt 1, paras 1, 12(a); definition 'the ONR's purposes' inserted by the Energy Act 2013, s 116, Sch 12, Pt 1, paras 1, 12(b); definition 'the radioactive material transport purposes' inserted by the Energy Act 2013, s 116, Sch 12, Pt 1, paras 1, 12(c).

PART II

THE EMPLOYMENT MEDICAL ADVISORY SERVICE

[3.132]

60 Supplementary

(1) It shall be the duty of the Secretary of State to secure that [the National Health Service Commissioning Board or each clinical commissioning group (in relation to England) and each Local Health Board (in relation to Wales)] arranges for] a fully registered medical practitioner [who holds a licence to practise] to furnish, on the application of an employment medical adviser, such particulars of the school medical record of a person who has not attained the age of eighteen and such other information relating to his medical history as the adviser may reasonably require for the efficient performance of his functions; but no particulars or information about any person which may be furnished to an adviser in pursuance of this subsection shall (without the consent of that person) be disclosed by the adviser otherwise than for the efficient performance of his functions.

(2), (3) . . .

(4) References to the chief employment medical adviser or a deputy chief employment medical adviser in any provision of an enactment or instrument made under an enactment shall be read as references to a person appointed for the purposes of that provision by the authority responsible for maintaining the employment medical advisory service.

(5) The following provisions of the Employment Medical Advisory Service Act 1972 (which are superseded by the preceding provisions of this Part or rendered unnecessary by provisions contained in Part I), namely sections 1 and 6 and Schedule 1, shall cease to have effect; but—

(a) in so far as anything done under or by virtue of the said section 1 or Schedule 1 could have been done under or by virtue of a corresponding provision of Part I or this Part, it shall not be invalidated by the repeal of that section and Schedule by this Act but shall have effect as if done under or by virtue of that corresponding provision; and

(b) any order made under the said section 6 which is in force immediately before the repeal of that section by this Act shall remain in force notwithstanding that repeal, but may be revoked or varied by regulations under section 43(2) or 57, as if it were an instrument containing regulations made under section 43(2) or 57 as the case may require.

(6) Where any Act (whether passed before, or in the same Session as, this Act) or any document refers, either expressly or by implication, to or to any enactment contained in any of the provisions of the said Act of 1972 which are mentioned in the preceding subsection, the reference shall, except where the context otherwise requires, be construed as, or as including, a reference to the corresponding provision of this Act.

(7) Nothing in subsection (5) or (6) above shall be taken as prejudicing the operation of section 38 of the Interpretation Act 1889 (which relates to the effect of repeals).

[3.133]

Statutory annotations

Amendment

Sub-s (1): words from 'the National Health' to 'relation to Wales)' in square brackets substituted by the Health and Social Care Act 2012, s 55(2), Sch 5, para 19(1), (2).

PART IV

MISCELLANEOUS AND GENERAL

[3.140]

82 General provisions as to interpretation and regulations

(1) In this Act—

(a) 'Act' includes a provisional order confirmed by an Act;

(b) 'contravention' includes failure to comply, and 'contravene' has a corresponding meaning;

(c) 'modifications' includes additions, omissions and amendments, and related expressions shall be construed accordingly;

(d) any reference to a Part, section or Schedule not otherwise identified is a reference to that Part or section of, or Schedule to, this Act.

(2) Except in so far as the context otherwise requires, any reference in this Act to an enactment is a reference to it as amended, and includes a reference to it as applied, by or under any other enactment, including this Act.

(3) Any power conferred by Part I or II or this Part to make regulations—

(a) includes power to make different provision by the regulations for different circumstances or cases and to include in the regulations such incidental, supplemental and transitional provisions as the authority making the regulations considers appropriate in connection with the regulations; and

(b) shall be exercisable by statutory instrument, which [(unless subsection [(3A) or] (4) applies)] shall be subject to annulment in pursuance of a resolution of either House of Parliament.

[(3A) In the case of a statutory instrument which also contains regulations under section 74 of the Energy Act 2013 (nuclear regulations), subsection (3) is subject to section 113 of that Act (subordinate legislation).]

[(4) The first regulations under section 43A(1) shall not be made unless a draft has been laid before and approved by resolution of each House of Parliament.]

[3.141]

Amendment

Sub-s (3): in para (b) words '(3A) or' in square brackets inserted by the Energy Act 2013, s 116, Sch 12, Pt 1, paras 1, 13(1), (2).

Sub-s (3A): inserted by the Energy Act 2013, s 116, Sch 12, Pt 1, paras 1, 13(1), (3).

SCHEDULE 1

EXISTING ENACTMENTS WHICH ARE RELEVANT STATUTORY PROVISIONS

Sections 1, 53

[3.147]

Chapter	Short title	Provisions which are relevant statutory provisions
1875 c 17	The Explosives Act 1875	The whole Act except sections *30 to 32, 80 and* 116 to 121.
1882 c 22	The Boiler Explosions Act 1882	The whole Act.
1890 c 35	The Boiler Explosions Act 1890	The whole Act.
1906 c 14	The Alkali, &c Works Regulation Act 1906	The whole Act.
1909 c 43	The Revenue Act 1909	Section 11.
. . .	. . .	. . .
1920 c 65	The Employment of Women, Young Persons and Children Act 1920	The whole Act.
. . .	. . .	. . .
. . .	. . .	. . .
1926 c 43	The Public Health (Smoke Abatement) Act 1926	The whole Act.
1928 c 32	The Petroleum (Consolidation) Act 1928	The whole Act.
1936 c 22	The Hours of Employment (Conventions) Act 1936	The whole Act except section 5.
1936 c 27	The Petroleum (Transfer of Licences) Act 1936	The whole Act.

1937 c 45	The Hydrogen Cyanide (Fumigation) Act 1937	The whole Act.
1945 c 19	The Ministry of Fuel and Power Act 1945	Section 1(1) so far as it relates to maintaining and improving the safety, health and welfare of persons employed in or about mines and quarries in Great Britain.
1946 c 59	The Coal Industry Nationalisation Act 1946	Section 42(1) and (2).
1948 c 37	The Radioactive Substances Act 1948	Section 5(1)(a).
. . .	. . .	. . .
. . .	. . .	. . .
1952 c 60	The Agriculture (Poisonous Substances) Act 1952	The whole Act.
. . .	. . .	. . .
[. . .	. . .	. . .]
1954 c 70	The Mines and Quarries Act 1954	The whole Act except section 151.
1956 c 49	The Agriculture (Safety, Health and Welfare Provisions) Act 1956	The whole Act.
1961 c 34	The Factories Act 1961	The whole Act except section 135.
1961 c 64	The Public Health Act 1961	Section 73.
1962 c 58	The Pipe-lines Act 1962	Sections 20 to 26, 33, 34 and 42, Schedule 5.
1963 c 41	The Offices, Shops and Railway Premises Act 1963	The whole Act.
. . .	. . .	. . .
1969 c 10	The Mines and Quarries (Tips) Act 1969	Sections 1 to 10.
1971 c 20	The Mines Management Act 1971	The whole Act.
1972 c 28	The Employment Medical Advisory Service Act 1972	The whole Act except sections 1 and 6 and Schedule 1.

[3.148]

Statutory annotations

Amendment

Entry relating to 'The Celluloid and Cinematograph Film Act 1922' (omitted) repealed by SI 2013/448, reg 2, Sch; entry relating to 'The Nuclear Installations Act 1965' (omitted) repealed by the Energy Act 2013, s 116, Sch 12, Pt 1, paras 1, 14.

[SCHEDULE 2

ADDITIONAL PROVISIONS RELATING TO THE CONSTITUTION ETC OF THE HEALTH AND SAFETY EXECUTIVE]

[Section 10]

[The Health and Safety Executive

[3.149]

1

The Executive shall consist of—

(a) the Chair of the Executive, and

(b) at least seven and no more than [twelve] other members (referred to in this Schedule as 'members').

2

(1) The Secretary of State shall appoint the Chair of the Executive.

(2) [Subject to sub-paragraph (3A),] the Secretary of State shall appoint the other members of the Executive according to sub-paragraph (3).

(3) The Secretary of State—

(a) shall appoint three members after consulting such organisations representing employers as he considers appropriate;

(b) shall appoint three members after consulting such organisations representing employees as he considers appropriate;

(c) shall appoint one member after consulting such organisations representing local authorities as he considers appropriate; and

(d) may appoint up to four other members after consulting, as he considers appropriate—

(i) the Scottish Ministers,

(ii) the Welsh Ministers, or

(iii) such organisations as he considers appropriate, including professional bodies, whose activities are concerned with matters relating to the general purposes of this Part.

[(3A) The Office for Nuclear Regulation may appoint a member from among the non-executive members of the Office for Nuclear Regulation ('an ONR member').

(3B) The Office for Nuclear Regulation must notify the Executive and the Secretary of State whenever it appoints an ONR member.]

(4) Service as the Chair or as another member of the Executive is not service in the civil service of the State.

(5) The Secretary of State, with the approval of the Chair, may appoint one of the other members appointed under sub-paragraph (2) to be the deputy chair of the Executive.

Terms of Appointment of the Executive

3

Subject to paragraphs 4[, 4A] and 5, a person shall hold and vacate office as the Chair or as another member according to the terms of the instrument appointing him to that office.

4

The Chair or any other member of the Executive[, other than an ONR member,] may at any time resign his office by giving notice in writing to the Secretary of State.

[4A

(1) An ONR member may at any time resign from office by giving notice in writing to the Office for Nuclear Regulation.

(2) An ONR member ceases to be a member of the Executive upon ceasing to be a non-executive member of the Office for Nuclear Regulation.

(3) The Office for Nuclear Regulation may remove an ONR member from office by giving notice in writing.

(4) The Office for Nuclear Regulation must notify the Executive and the Secretary of State whenever an ONR member—

(a) resigns from office,

(b) ceases to be a non-executive member of the Office for Nuclear Regulation, or

(c) is removed from office.]

5

The Secretary of State may remove a Chair or other member[, other than an ONR member,] who—

(a) has been absent from meetings of the Executive for a period longer than six months without the permission of the Executive;

(b) has become bankrupt or [has had a debt relief order (under Part 7A of the Insolvency Act 1986) made in respect of him or] has made an arrangement with his creditors;

(c) in Scotland, has had his estate sequestrated or has made a trust deed for creditors or a composition contract;

(d) has become incapacitated by physical or mental illness; or

(e) is otherwise, in the opinion of the Secretary of State, unable or unfit to carry out his functions.

Remuneration of Members

6

(1) The Executive shall pay[—

(a) to each member, other than an ONR member, such remuneration, and

(b) to each member such travelling and other allowances, as may be determined by the Secretary of State].

(2) The Executive shall pay to, or in respect of, any member [other than an ONR member], such sums by way of pension, superannuation allowances and gratuities as the Secretary of State may determine.

(3) Where a person ceases to be a member [other than an ONR member] otherwise than on the expiry of his term of office, and the Secretary of State determines that there are special circumstances which make it right that he should receive compensation, the Executive shall pay to him such amount by way of compensation as the Secretary of State may determine.

[(4) Where—

(a) a member appointed under paragraph 4(4)(a) of Schedule 7 to the Energy Act 2013 to be a member of the Office for Nuclear Regulation (the 'HSE member of the ONR')—

(i) ceases to be the HSE member of the ONR otherwise than on the expiry of his or her term of office as HSE member of the ONR, but

(ii) does not cease to be a member of the Executive, and

(b) it appears to the Executive that there are special circumstances that make it right for that person to receive compensation, the Executive may pay the member such amount by way of compensation as the Secretary of State may determine.]

Proceedings of the Executive

7

(1) The Executive may regulate its own procedure.

(2) The validity of any proceedings of the Executive shall not be affected by any vacancy among the members or by any defect in the appointment of a member.

(3) The Executive shall consult with the Secretary of State before making or revising its rules and procedures for dealing with conflicts of interest.

(4) The Executive shall from time to time publish a summary of its rules and procedures.

Staff

8

(1) The Executive shall, with the consent of the Secretary of State, appoint a person to act as Chief Executive on such terms and conditions as the Secretary of State may determine.

(2) The Executive shall appoint such other staff to the service of the Executive as it may determine, with the consent of the Secretary of State as to numbers of persons appointed and as to the terms and conditions of their service.

(3) The Executive shall pay to the Minister for the Civil Service at such times as that Minister may direct, such sums as the Minister may determine in respect of any increase attributable to this paragraph in the sums payable out of monies provided by Parliament under the Superannuation Act 1972.

(4) A person appointed to the staff of the Executive may not at the same time be a member of the Executive.

(5) Service as a member of staff of the Executive is service in the civil service of the State.

Performance of functions

9

(1) Subject to sub-paragraphs (2) to (4), anything authorised or required to be done by the Executive (including exercising the powers under this paragraph) may be done by—

- (a) such members of the Executive or members of staff of the Executive as the Executive considers fit to authorise for that purpose, whether generally or specifically; or
- (b) any committee of the Executive which has been so authorised.

(2) Sub-paragraph (1)(b) does not apply to a committee whose members include a person who is neither a member of the Executive nor a member of staff of the Executive.

(3) The Executive—

- (a) shall authorise such of its members of staff as it considers fit to authorise for that purpose, to perform on its behalf those of its functions which consist of the enforcement of the relevant statutory provisions in any particular case; but
- (b) shall not authorise any member or committee of the Executive to make decisions concerning the enforcement of the relevant statutory provisions in any particular case.

(4) The Executive shall not authorise any person to legislate by subordinate instrument.

(5) The Executive shall publish any authorisations which it makes under this paragraph.

Accounts and reports

10

(1) It shall be the duty of the Chief Executive—

(a) to keep proper accounts and proper records in relation to the accounts;

(b) to prepare in respect of each accounting year a statement of accounts in such form as the Secretary of State may direct with the approval of the Treasury; and

(c) to send copies of the statement to the Secretary of State and the Comptroller and Auditor General before the end of November next following the accounting year to which the statement relates.

(2) The Comptroller and the Auditor General shall examine, certify and report on each statement referred to in sub-paragraph (1)(c) and shall lay copies of each statement and his report before each House of Parliament.

(3) As soon as possible after the end of the accounting year, the Executive shall make to the Secretary of State a report on the performance of the Executive's functions during the year.

(4) The Secretary of State shall lay the report referred to in sub-paragraph (3) before each House of Parliament.

(5) In this paragraph, 'accounting year' means the period of 12 months ending with 31st March in any year; but the first accounting year of the Executive shall, if the Secretary of State so directs, be of such other period not exceeding 2 years as may be specified in the direction.

Supplemental

11

The Secretary of State shall not make any determination or give his consent under paragraph 6 or 8 of this Schedule except with the approval of the Minister for the Civil Service.

12

(1) The fixing of the common seal of the Executive shall be authenticated by the signature of the Chair or some other person authorised by the Executive to act for that purpose.

(2) A document purporting to be duly executed under the seal of the Executive shall be received in evidence and shall be deemed to be so executed unless the contrary is proved.

(3) This paragraph does not apply to Scotland.]

[3.150]

Statutory annotations

Amendment

Para 1: in sub-para (b) word 'twelve' in square brackets substituted by the Energy Act 2013, s 116, Sch 12, Pt 1, paras 1, 15(1), (5).

Para 2: in sub-para (2) words 'Subject to sub-paragraph (3A),' in square brackets inserted by the Energy Act 2013, s 116, Sch 12, Pt 1, paras 1, 15(1), (3); sub-paras (3A), (3B) inserted by the Energy Act 2013, s 116, Sch 12, Pt 1, paras 1, 15(1), (4).

Para 3: reference to ', 4A' in square brackets inserted by the Energy Act 2013, s 116, Sch 12, Pt 1, paras 1, 15(1), (5).

Para 4: words ', other than an ONR member,' in square brackets inserted by the Energy Act 2013, s 116, Sch 12, Pt 1, paras 1, 15(1), (6);

Para 4A: inserted by the Energy Act 2013, s 116, Sch 12, Pt 1, paras 1, 15(1), (7).

Para 5: words ', other than an ONR member,' in square brackets inserted by the Energy Act 2013, s 116, Sch 12, Pt 1, paras 1, 15(1), (8); in sub-para (b) words from 'has had a' to 'of him or' in square brackets inserted by SI 2012/2404, art 3(2), Sch 2, para 6.

Para 6: words from '— (a) to each' to 'Secretary of State' in square brackets substituted by the Energy Act 2013, s 116, Sch 12, Pt 1, paras 1, 15(1), (9), (10); in sub-para (2) words 'other than an ONR member' in square brackets inserted by the Energy Act 2013, s 116, Sch 12, Pt 1, paras 1, 15(1), (9), (11); in sub-para (3) words 'other than an ONR member' in square brackets inserted by the Energy Act 2013, s 116, Sch 12, Pt 1, paras 1, 15(1), (9), (12); sub-para (4) inserted by the Energy Act 2013, ss 116, Sch 12, Pt 1, paras 1, 15(1), (9), (13).

Deregulation and Contracting Out Act 1994

1994 chapter 40

as amended

PART I

DEREGULATION

CHAPTER II

MISCELLANEOUS DEREGULATORY PROVISIONS

[3.158]

37 Power to repeal certain health and safety provisions

(1) The appropriate authority may by regulations repeal or, as the case may be, revoke—

(a) any provision which is an existing statutory provision for the purposes of Part I of the Health and Safety at Work etc Act 1974 ('the 1974 Act'),

(b) any provision of regulations under section 15 of the 1974 Act (health and safety regulations) which has effect in place of a provision which was an existing statutory provision for the purposes of that Part,

[(ba) any of the relevant nuclear provisions,

(bb) any provision of regulations under section 74 of the Energy Act 2013 which has effect in place of any of the relevant nuclear provisions,]

(c) any provision which is an existing statutory provision for the purposes of the Health and Safety at Work (Northern Ireland) Order 1978 ('the 1978 Order'), or

(d) any provision of regulations under Article 17 of the 1978 Order (health and safety regulations) which has effect in place of a provision which has an existing statutory provision for the purposes of that Order.

(2) Before making regulations under subsection (1) above, the appropriate authority shall consult—

[(a) in the case of regulations under paragraph (a) of that subsection repealing or revoking a provision specified in section 117(4) of the Railways Act 1993, the Office of Rail Regulation,

(aa) in the case of regulations under paragraph (a) of that subsection not falling within paragraph (a) of this subsection, [the Health and Safety Executive],

(ab) in the case of regulations under paragraph (b) of that subsection revoking a provision of regulations which make provision exclusively in relation to transport systems falling within paragraph 1(3) of Schedule 3 to the Railways Act 2005, the Office of Rail Regulation,

(ac) in the case of regulations under paragraph (b) of that subsection not falling within paragraph (ab) of this subsection, [the Health and Safety Executive],]

[(ad) in the case of regulations under paragraph (ba) or (bb) of that subsection, the Office for Nuclear Regulation,]

(b) in the case of regulations under paragraph (c) or (d) of that subsection, the Health and Safety Agency for Northern Ireland,

and, in [any] case, such other persons as the appropriate authority considers appropriate.

(3) Instead of consulting such other persons as the appropriate authority considers it appropriate to consult under subsection (2) above, the authority may require [the required consultee] to consult such persons as it considers appropriate for the purpose of deciding how it should respond to consultation under that subsection.

(4) Instead of consulting a person whom the appropriate authority considers it appropriate to consult under subsection (2) above, the authority may require [the required consultee] to consult the person for the purpose of deciding how it should respond to consultation under that subsection.

[(4A) In subsections (3) and (4), 'the required consultee' means the body which the appropriate authority is required to consult by virtue of subsection (2)(a) to (b).]

(5) The appropriate authority may require consultation under subsection (3) or (4) above to be carried out in accordance with the authority's directions.

(6) Regulations under subsection (1) above may contain such transitional provisions and savings as the appropriate authority considers appropriate.

(7) Regulations under paragraph (a) [(b), (ba) or (bb)] of subsection (1) above shall be made by statutory instrument, and no instrument shall be made under that paragraph unless a draft of it has been laid before, and approved by a resolution of, each House of Parliament.

(8) Regulations under subsection (1)(c) or (d) above—

(a) shall be statutory rules for the purposes of the Statutory Rules (Northern Ireland) Order 1979, and

(b) shall be subject to affirmative resolution, as defined in section 41(4) of the Interpretation Act (Northern Ireland) 1954, as if they were statutory instruments within the meaning of that Act.

(9) In this section, 'appropriate authority'—

(a) in relation to regulations under subsection (1)(a)[(b), (ba) or (bb)] above, means the Secretary of State, and

(b) in relation to regulations under subsection (1)(c) or (d) above, means the Department concerned (within the meaning of the 1978 Order, but as if any reference to that Order included a reference to this section).

[(10) In subsection (1), 'the relevant nuclear provisions' means—

(a) sections 1, 3 to 6, 22 and 24A of the Nuclear Installations Act 1965, and

(b) any regulations made under any of those sections,

so far as they have effect in England and Wales or Scotland.]

[3.159]

Statutory annotations

Amendment

Sub-s (1): paras (ba), (bb) inserted by the Energy Act 2013, s 116(1), Sch 12, Pt 5, para 71(1), (2).

Sub-s (2): para (ad) inserted by the Energy Act 2013, s 116(1), Sch 12, Pt 5, para 71(1), (3).

Sub-s (7): words '(b), (ba) or (bb)' in square brackets substituted by the Energy Act 2013, s 116(1), Sch 12, Pt 5, para 71(1), (4).

Sub-s (9): words '(b), (ba) or (bb)' in square brackets substituted by the Energy Act 2013, s 116(1), Sch 12, Pt 5, para 71(1), (5).

Sub-s (10): inserted by the Energy Act 2013, s 116(1), Sch 12, Pt 5, para 71(1), (6).

Compensation Act 2006

2006 chapter 29

PART 1

STANDARD OF CARE

[3.175]

3 Mesothelioma: damages

(1) This section applies where—

(a) a person ('the responsible person') has negligently or in breach of statutory duty caused or permitted another person ('the victim') to be exposed to asbestos,

(b) the victim has contracted mesothelioma as a result of exposure to asbestos,

(c) because of the nature of mesothelioma and the state of medical science, it is not possible to determine with certainty whether it was the exposure mentioned in paragraph (a) or another exposure which caused the victim to become ill, and

(d) the responsible person is liable in tort, by virtue of the exposure mentioned in paragraph (a), in connection with damage caused to the victim by the disease (whether by reason of having materially increased a risk or for any other reason).

(2) The responsible person shall be liable—

(a) in respect of the whole of the damage caused to the victim by the disease (irrespective of whether the victim was also exposed to asbestos—

(i) other than by the responsible person, whether or not in circumstances in which another person has liability in tort, or

(ii) by the responsible person in circumstances in which he has no liability in tort), and

(b) jointly and severally with any other responsible person.

(3) Subsection (2) does not prevent—

(a) one responsible person from claiming a contribution from another, or

(b) a finding of contributory negligence.

(4) In determining the extent of contributions of different responsible persons in accordance with subsection (3)(a), a court shall have regard to the relative lengths of the periods of exposure for which each was responsible; but this subsection shall not apply—

(a) if or to the extent that responsible persons agree to apportion responsibility amongst themselves on some other basis, or

(b) if or to the extent that the court thinks that another basis for determining contributions is more appropriate in the circumstances of a particular case.

(5) In subsection (1) the reference to causing or permitting a person to be exposed to asbestos includes a reference to failing to protect a person from exposure to asbestos.

(6) In the application of this section to Scotland—

(a) a reference to tort shall be taken as a reference to delict, and

(b) a reference to a court shall be taken to include a reference to a jury.

(7) The Treasury may make regulations about the provision of compensation to a responsible person where—

(a) he claims, or would claim, a contribution from another responsible person in accordance with subsection (3)(a), but
(b) he is unable or likely to be unable to obtain the contribution, because an insurer of the other responsible person is unable or likely to be unable to satisfy the claim for a contribution.

(8) The regulations may, in particular—
(a) . . .
(b) replicate or apply (with or without modification) a transitional compensation provision;
(c) provide for a specified person to assess and pay compensation;
(d) provide for expenses incurred (including the payment of compensation) to be met out of levies collected in accordance with section 213(3)(b) of the Financial Services and Markets Act 2000 (c 8) (the Financial Services Compensation Scheme);
(e) modify the effect of a transitional compensation provision;
(f) enable the [Financial Conduct Authority or the Prudential Regulation Authority] to amend the Financial Services Compensation Scheme;
(g) modify the Financial Services and Markets Act 2000 in its application to an amendment pursuant to paragraph (f);
(h) make, or require the making of, provision for the making of a claim by a responsible person for compensation whether or not he has already satisfied claims in tort against him;
(i) make, or require the making of, provision which has effect in relation to claims for contributions made on or after the date on which this Act is passed.

(9) . . .

(10) In subsections (7) and (8)—
(a) a reference to a responsible person includes a reference to an insurer of a responsible person, and
(b) 'transitional compensation provision' means a provision of an enactment which is made under the Financial Services and Markets Act 2000 and—
(i) preserves the effect of the Policyholders Protection Act 1975 (c 75), or
(ii) applies the Financial Services Compensation Scheme in relation to matters arising before its establishment.

(11) Regulations under subsection (7)—
(a) may include consequential or incidental provision,
(b) may make provision which has effect generally or only in relation to specified cases or circumstances,
(c) may make different provision for different cases or circumstances,
(d) shall be made by statutory instrument, and
(e) may not be made unless a draft has been laid before and approved by resolution of each House of Parliament.

[3.176]

Statutory annotations

Amendment

Sub-s (8): para (a) repealed by the Financial Services Act 2012, s 114(1), Sch 18, Pt 2, paras 109(1), (2)(a); in para (f) words 'Financial Conduct Authority or the Prudential Regulation Authority' in square brackets substituted by the Financial Services Act 2012, s 114(1), Sch 18, Pt 2, paras 109(1), (2)(b).

Sub-s (9): repealed by the Financial Services Act 2012, s 114(1), Sch 18, Pt 2, paras 109(1), (3).

Corporate Manslaughter and Corporate Homicide Act 2007

2007 chapter 19

Relevant duty of care

[3.243]

6 Emergencies

(1) Any duty of care owed by an organisation within subsection (2) in respect of the way in which it responds to emergency circumstances is not a 'relevant duty of care' unless it falls within section 2(1)(a) or (b).

(2) The organisations within this subsection are—

(a) a fire and rescue authority in England and Wales;

[(b) the Scottish Fire and Rescue Service;]

(c) the Northern Ireland Fire and Rescue Service Board;

(d) any other organisation providing a service of responding to emergency circumstances either—

(i) in pursuance of arrangements made with an organisation within paragraph (a), (b) or (c), or

(ii) (if not in pursuance of such arrangements) otherwise than on a commercial basis;

(e) a relevant NHS body;

(f) an organisation providing ambulance services in pursuance of arrangements—

(i) made by, or at the request of, a relevant NHS body, or

(ii) made with the Secretary of State or with the Welsh Ministers;

(g) an organisation providing services for the transport of organs, blood, equipment or personnel in pursuance of arrangements of the kind mentioned in paragraph (f);

(h) an organisation providing a rescue service;

(i) the armed forces.

(3) For the purposes of subsection (1), the way in which an organisation responds to emergency circumstances does not include the way in which—

(a) medical treatment is carried out, or

(b) decisions within subsection (4) are made.

(4) The decisions within this subsection are decisions as to the carrying out of medical treatment, other than decisions as to the order in which persons are to be given such treatment.

(5) Any duty of care owed in respect of the carrying out, or attempted carrying out, of a rescue operation at sea in emergency circumstances is not a 'relevant duty of care' unless it falls within section 2(1)(a) or (b).

(6) Any duty of care owed in respect of action taken—

(a) in order to comply with a direction under Schedule 3A to the Merchant Shipping Act 1995 (c 21) (safety directions), or

(b) by virtue of paragraph 4 of that Schedule (action in lieu of direction),

is not a 'relevant duty of care' unless it falls within section 2(1)(a) or (b).

(7) In this section—

'emergency circumstances' means circumstances that are present or imminent and—

(a) are causing, or are likely to cause, serious harm or a worsening of such harm, or

(b) are likely to cause the death of a person;

'medical treatment' includes any treatment or procedure of a medical or similar nature;

'relevant NHS body' means—

[(za) the National Health Service Commissioning Board;]

(a) [a clinical commissioning group,] . . . . . . *NHS trust*, Special Health Authority or NHS foundation trust in England;

(b) a Local Health Board, NHS trust or Special Health Authority in Wales;

(c) a Health Board or Special Health Board in Scotland, or the Common Services Agency for the Scottish Health Service;

(d) a Health and Social Services trust or Health and Social Services Board in Northern Ireland;

'serious harm' means—

(a) serious injury to or the serious illness (including mental illness) of a person;

(b) serious harm to the environment (including the life and health of plants and animals);

(c) serious harm to any building or other property.

(8) A reference in this section to emergency circumstances includes a reference to circumstances that are believed to be emergency circumstances.

[3.244]

Statutory annotations

Amendment

Sub-s (2): para (b) substituted by the Police and Fire Reform (Scotland) Act 2012, s 128(1), Sch 7, Pt 2, para 71.

Sub-s (7): in definition 'relevant NHS body' para (za) inserted by the Health and Social Care Act 2012, s 55(2), Sch 5, para 147(a); in definition 'relevant NHS body' in para (a) words 'a clinical commissioning group,' in square brackets inserted by the Health and Social Care Act 2012, s 55(2), Sch 5, para 147(b)(i); in definition 'relevant NHS body' in para (a) first words omitted repealed by the Health and Social Care Act 2012, s 55(2), Sch 5, para 147(b)(i); in definition 'relevant NHS body' in para (a) second words omitted repealed by the Health and Social Care Act 2012, s 55(2), Sch 5, para 147(b)(iii); in definition 'relevant NHS body' in para (a) words 'NHS trust,' in italics repealed by the Health and Social Care Act 2012, s 179(6), Sch 14, Pt 2, para 102 from a date to be appointed: see the Health and Social Care Act 2012, s 306(4).

[3.245]

7 Child-protection and probation functions

(1) A duty of care to which this section applies is not a 'relevant duty of care' unless it falls within section 2(1)(a), (b) or (d).

(2) This section applies to any duty of care that a local authority or other public authority owes in respect of the exercise by it of functions conferred by or under—

(a) Parts 4 and 5 of the Children Act 1989 (c 41),

(b) Part 2 of the Children (Scotland) Act 1995 (c 36), or

[(ba) the Children's Hearings (Scotland) Act 2011,] or

(c) Parts 5 and 6 of the Children (Northern Ireland) Order 1995 (SI 1995/755 (NI 2)).

(3) This section also applies to any duty of care that a local probation board[, a provider of probation services] or other public authority owes in respect of the exercise by it of functions conferred by or under—

(a) Chapter 1 of Part 1 of the Criminal Justice and Court Services Act 2000 (c 43),

[(aa) section 13 of the Offender Management Act 2007 (c 21),]
(b) section 27 of the Social Work (Scotland) Act 1968 (c 49), or
(c) Article 4 of the Probation Board (Northern Ireland) Order 1982 (SI 1982/713 (NI 10)).

[(4) This section also applies to any duty of care that a provider of probation services owes in respect of the carrying out by it of activities in pursuance of arrangements under section 3 of the Offender Management Act 2007.]

[3.246]

Statutory annotations

Amendment

Sub-s (2): para (ba) inserted by SI 2013/1465, art 17(1), Sch 1, Pt 1, para 11.

Application to particular categories of organisation

[3.257]

13 Application to police forces

(1) In this Act 'police force' means—
(a) a police force within the meaning of—
(i) the Police Act 1996 (c 16), . . .
(ii) . . .
[(aa) the Police Service of Scotland;]
(b) the Police Service of Northern Ireland;
(c) the Police Service of Northern Ireland Reserve;
(d) the British Transport Police Force;
(e) the Civil Nuclear Constabulary;
(f) the Ministry of Defence Police.

(2) For the purposes of this Act a police force is to be treated as owing whatever duties of care it would owe if it were a body corporate.

(3) For the purposes of section 2—
(a) a member of a police force is to be treated as employed by that force;
(b) a special constable appointed for a police area in England and Wales is to be treated as employed by the police force maintained by the [local policing body] for that area;
(c) a special constable appointed for a police force mentioned in paragraph (d) or (f) of subsection (1) is to be treated as employed by that force;
(d) a police cadet undergoing training with a view to becoming a member of a police force mentioned in paragraph (a)[, (aa)] or (d) of subsection (1) is to be treated as employed by that force;
(e) a police trainee appointed under section 39 of the Police (Northern Ireland) Act 2000 (c 32) or a police cadet appointed under section 42 of that Act is to be treated as employed by the Police Service of Northern Ireland;
(f) a police reserve trainee appointed under section 40 of that Act is to be treated as employed by the Police Service of Northern Ireland Reserve;
(g) a member of a police force [seconded to the National Crime Agency to serve as a National Crime Agency officer is to be treated] as employed by that Agency.

(4) A reference in subsection (3) to a member of a police force is to be read, in the case of [the Police Service of Scotland, as a reference to a constable of that Service].

(5) For the purposes of section 2 any premises occupied for the purposes of a police force are to be treated as occupied by that force.

(6) For the purposes of sections 2 to 7 anything that would be regarded as done by a police force if the force were a body corporate is to be so regarded.

(7) Where—

(a) by virtue of subsection (3) a person is treated for the purposes of section 2 as employed by a police force, and

(b) by virtue of any other statutory provision (whenever made) he is, or is treated as, employed by another organisation,

the person is to be treated for those purposes as employed by both the force and the other organisation.

[3.258]

Statutory annotations

Amendment

Sub-s (1): in para (a)(i) word omitted repealed by SI 2013/602, art 26, Sch 2, Pt 1, para 57(1), (2)(a); para (a)(ii) repealed by SI 2013/602, art 26, Sch 2, Pt 1, para 57(1), (2)(b); para (aa) inserted by SI 2013/602, art 26, Sch 2, Pt 1, para 57(1), (2)(c).

Sub-s (3): in para (d) reference to ', (aa)' in square brackets inserted by SI 2013/602, art 26, Sch 2, Pt 1, para 57(1), (2)(d); in para (g) words from 'seconded to the' to 'to be treated' in square brackets substituted by the Crime and Courts Act 2013, s 15(3), Sch 8, Pt 2, paras 173, 174.

Sub-s (4): words 'the Police Service of Scotland, as a reference to a constable of that Service' in square brackets substituted by SI 2013/602, art 26, Sch 2, Pt 1, para 57(1), (2)(e).

General and supplementary

[3.278]

25 Interpretation

In this Act—

'armed forces' has the meaning given by section 12(1);

'corporation' does not include a corporation sole but includes any body corporate wherever incorporated;

'employee' means an individual who works under a contract of employment or apprenticeship (whether express or implied and, if express, whether oral or in writing), and related expressions are to be construed accordingly; see also sections 11(3)(a), 12(2) and 13(3) (which apply for the purposes of section 2);

'employers' association' has the meaning given by section 122 of the Trade Union and Labour Relations (Consolidation) Act 1992 (c 52) or Article 4 of the Industrial Relations (Northern Ireland) Order 1992 (SI 1992/807 (NI 5));

'enforcement authority' means an authority responsible for the enforcement of any health and safety legislation;

'health and safety legislation' means any statutory provision dealing with health and safety matters, including in particular provision contained in the Health and Safety at Work etc Act 1974 (c 37) or the Health and Safety at Work (Northern Ireland) Order 1978 (SI 1978/1039 (NI 9)) [and provision dealing with health and safety matters contained in Part 3 of the Energy Act 2013 (nuclear regulation)];

'member', in relation to the armed forces, is to be read in accordance with section 12(3);

'partnership' means—

(a) a partnership within the Partnership Act 1890 (c 39), or

(b) a limited partnership registered under the Limited Partnerships Act 1907 (c 24),

or a firm or entity of a similar character formed under the law of a country or territory outside the United Kingdom;

'police force' has the meaning given by section 13(1);

'premises' includes land, buildings and moveable structures;

'public authority' has the same meaning as in section 6 of the Human Rights Act 1998 (c 42) (disregarding subsections (3)(a) and (4) of that section);

'publicity order' means an order under section 10(1);

'remedial order' means an order under section 9(1);

'statutory provision', except in section 15, means provision contained in, or in an instrument made under, any Act, any Act of the Scottish Parliament or any Northern Ireland legislation;

'trade union' has the meaning given by section 1 of the Trade Union and Labour Relations (Consolidation) Act 1992 (c 52) or Article 3 of the Industrial Relations (Northern Ireland) Order 1992 (SI 1992/807 (NI 5)).

[3.279]

Statutory annotations

Amendment

In definition 'health and safety legislation' words from 'and provision dealing' to 'Energy Act 2013 (nuclear regulation)' in square brackets inserted by the Energy Act 2013, s 116(1), Sch 12, Pt 5, para 93.

SCHEDULE 1

LIST OF GOVERNMENT DEPARTMENTS ETC

Section 1

[3.285]

. . .

Attorney General's Office

Cabinet Office

Central Office of Information

Crown Office and Procurator Fiscal Service

Crown Prosecution Service

[. . .

[Department for Business, Innovation and Skills]

. . .]

Department for Communities and Local Government

. . .

Department for Culture, Media and Sport

. . .

[Department for Education]

[Department of Energy and Climate Change]

Department for Environment, Food and Rural Affairs

[. . . .]

Department for International Development

Department for Transport

Department for Work and Pensions

Department of Health

. . .

Export Credits Guarantee Department
Foreign and Commonwealth Office
Forestry Commission
General Register Office for Scotland
Government Actuary's Department
Her Majesty's Land Registry
Her Majesty's Revenue and Customs
Her Majesty's Treasury
Home Office
Ministry of Defence
[Ministry of Justice (including the Scotland Office and the Wales Office)]
National Archives
National Archives of Scotland

. . .

[National Crime Agency]
National Savings and Investments
National School of Government
Northern Ireland Audit Office
Northern Ireland Court Service
Northern Ireland Office
Office for National Statistics

. . .

Office of Her Majesty's Chief Inspector of Education and Training in Wales
Ordnance Survey

. . .

Public Prosecution Service for Northern Ireland
Registers of Scotland Executive Agency
Revenue and Customs Prosecutions Office
Royal Mint
Scottish Executive
Serious Fraud Office
Treasury Solicitor's Department
UK Trade and Investment
Welsh Assembly Government

[3.286]

Statutory annotations

Amendment

Entry 'National Audit Office' (omitted) repealed by the Budget Responsibility and National Audit Act 2011, s 26(2), Sch 5, Pt 2, para 32.

Entry 'National Crime Agency' inserted by the Crime and Courts Act 2013, s 15(3), Sch 8, Pt 2, paras 173, 175.

Regulatory Enforcement and Sanctions Act 2008

2008 chapter 13

SCHEDULE 5

DESIGNATED REGULATORS

Section 37(1)

[3.344]

British Hallmarking Council
Charity Commission for England and Wales
Coal Authority
Competition Commission
Countryside Council for Wales
Environment Agency
[Financial Conduct Authority]
Food Standards Agency
. . .
Forestry Commissioners
Gambling Commission
Gangmasters Licensing Authority
Health and Safety Executive
Hearing Aid Council
Historic Buildings and Monuments Commission for England ('English Heritage')
Housing Corporation
Human Fertilisation and Embryology Authority
Human Tissue Authority
Information Commissioner
. . .
Natural England
Office of Communications
Office of Fair Trading
Office of Rail Regulation
Pensions Regulator
[Prudential Regulation Authority]
Security Industry Authority
Statistics Board

[3.344A]

Statutory annotations

Amendment

Entry 'Financial Conduct Authority' substituted, for entry 'Financial Services Authority' as originally enacted, by the Financial Services Act 2012, s 114(1), Sch 18, Pt 2, para 126(a).

Entry 'Prudential Regulation Authority' inserted by the Financial Services Act 2012, s 114(1), Sch 18, Pt 2, para 126(b).

SCHEDULE 6

ENACTMENTS SPECIFIED FOR THE PURPOSES OF ORDERS UNDER PART 3
Section 37(2)

[3.345]

Accommodation Agencies Act 1953 (c 23), section 1
Administration of Justice Act 1970 (c 31), section 40
Agriculture Act 1947 (c 48), section 100
Agriculture Act 1967 (c 22), section 14(2)
Agriculture Act 1970 (c 40)
Agriculture and Horticulture Act 1964 (c 28), sections 14 and 15
Ancient Monuments and Archaeological Areas Act 1979 (c 46)
Animal Boarding Establishments Act 1963 (c 43), section 1(8)
Animal Health Act 1981 (c 22)
Animal Health and Welfare Act 1984 (c 40), section 10(6)
Animals (Scientific Procedures) Act 1986 (c 14)
Animal Welfare Act 2006 (c 45)
Anti-social Behaviour Act 2003 (c 38), sections 54 and 75
Aviation and Maritime Security Act 1990 (c 31), sections 15 and 48 and Part 3
Aviation Security Act 1982 (c 36), Part 2
Bees Act 1980 (c 12), section 1(7)
Breeding of Dogs Act 1973 (c 60)
Breeding of Dogs Act 1991 (c 64)
Building Act 1984 (c 55)
Cancer Act 1939 (c 13), section 4
Children and Young Persons Act 1933 (c 12), section 7(1)
Children and Young Persons Act 1963 (c 37), section 40
Children and Young Persons (Protection from Tobacco) Act 1991 (c 23)
Christmas Day (Trading) Act 2004 (c 26), sections 2 and 3
Chronically Sick and Disabled Persons Act 1970 (c 44), section 21
Clean Air Act 1993 (c 11)
Clean Neighbourhoods and Environment Act 2005 (c 16), sections 3 to 7 and Part 7
Coast Protection Act 1949 (c 74), sections 18(7) and 36
Commons Act 2006 (c 26), section 34(1)
Conservation of Seals Act 1970 (c 30)
Consumer Credit Act 1974 (c 39)
Consumer Protection Act 1987 (c 43), section 12
Control of Pollution Act 1974 (c 40), Part 3
Control of Pollution (Amendment) Act 1989 (c 14), sections 1, 5 and 7
Criminal Justice and Public Order Act 1994 (c 33), sections 77(3), 78(4), 166 and 167
Dangerous Dogs Act 1989 (c 30)
Dangerous Dogs Act 1991 (c 65)
Dangerous Vessels Act 1985 (c 22), section 5
Dangerous Wild Animals Act 1976 (c 38)
Deer Act 1991 (c 54)
Destructive Imported Animals Act 1932 (c 12), section 6
. . .
Diseases of Fish Act 1937 (c 33)
Diseases of Fish Act 1983 (c 30)

Education Reform Act 1988 (c 40), section 214
Employment Agencies Act 1973 (c 35), sections 5(2), 6(2), and 10(2)
Energy Act 1976 (c 76), section 18
Enterprise Act 2002 (c 40), section 227E
Environment Act 1995 (c 25), section 110
Environmental Protection Act 1990 (c 43), Parts 2, 2A and 3, and sections 118 and 150(5)
Estate Agents Act 1979 (c 38)
Explosives Act 1875 (c 17), sections 30 to 32, 43, 73 and 80
Fair Trading Act 1973 (c 41), sections 23, 30 and 120
Farriers (Registration) Act 1975 (c 35)
Fire Safety and Safety of Places of Sport Act 1987 (c 27)
Fireworks Act 2003 (c 22), section 11
Fisheries Act 1981 (c 29)
Fishery Limits Act 1976 (c 86)
Food and Environment Protection Act 1985 (c 48), Parts 1, 2 and 3
Food Safety Act 1990 (c 16), Parts 2 and 3
Food Standards Act 1999 (c 28)
Gambling Act 2005 (c 19)
Game Act 1831 (c 32), sections 3, 3A, and 24
Goods Vehicles (Licensing of Operators) Act 1995 (c 23)
Ground Game Act 1880 (c 47), section 6
Hallmarking Act 1973 (c 43), sections 1, 3, 5 to 7 and 11
Hares Preservation Act 1892 (c 8), section 2
Health and Safety at Work etc Act 1974 (c 37), section 33
Highways Act 1980 (c 66), Part 9
Horticultural Produce Act 1986 (c 20)
Import of Live Fish (England and Wales) Act 1980 (c 27), section 3
International Carriage of Perishable Foodstuffs Act 1976 (c 58), sections 6 to 11
Intoxicating Substances (Supply) Act 1985 (c 26), section 1
Knives Act 1997 (c 21)
Licensing Act 2003 (c 17)
Local Government (Miscellaneous Provisions) Act 1976 (c 57), Part 2
Local Government (Miscellaneous Provisions) Act 1982 (c 30), Part 8 and Schedules 3 and 4
London Local Authorities Act 1995 (c x), sections 24 and 25
London Local Authorities Act 2004 (c i), sections 22, 26 and 29
Medicines Act 1968 (c 67)
Merchant Shipping Act 1995 (c 21), Parts 1 to 5 and 9 to 12
Mines and Quarries (Tips) Act 1969 (c 10), Part 2
Motor Cycle Noise Act 1987 (c 34)
New Roads and Street Works Act 1991 (c 22), section 51
Noise Act 1996 (c 37)
Olympic Symbol etc (Protection) Act 1995 (c 32), section 8
Pedlars Act 1871 (c 96), sections 4, 12 and 14
Pests Act 1954 (c 68), sections 8, 9 and 12
Pet Animals Act 1951 (c 35)
Petroleum (Consolidation) Act 1928 (c 32), sections 1, 5 and 18
Pilotage Act 1987 (c 21)
Planning (Listed Buildings and Conservation Areas) Act 1990 (c 9)
Plant Varieties Act 1997 (c 66), sections 19, 31 and 32

Plant Varieties and Seeds Act 1964 (c 14), sections 16 and 25
Poisons Act 1972 (c 66)
Prevention of Damage by Pests Act 1949 (c 55), sections 3(4), 17 and 22(4)
Prices Act 1974 (c 24)

. . .

Protection from Eviction Act 1977 (c 43), section 1
Protection of Badgers Act 1992 (c 51), sections 1 to 5 and 10
Protection of Wrecks Act 1973 (c 33)
Public Health Act 1936 (c 49), Part 2 and sections 269, 288 and 290
Public Health (Control of Disease) Act 1984 (c 22), sections 15 to 55
Public Passenger Vehicles Act 1981 (c 14), Part 2 and section 67

. . .

Railways Act 1993 (c 43), sections 119, 120 and 121
Railways and Transport Safety Act 2003 (c 20), sections 78 to 80
Refuse Disposal (Amenity) Act 1978 (c 3), sections 2(1) and 2B(2)
Rent Act 1977 (c 42), sections 77 and 81
Road Traffic Act 1988 (c 52)
Road Traffic (Driver Licensing and Information Systems) Act 1989 (c 22), sections 9 and 11
Road Traffic (Foreign Vehicles) Act 1972 (c 27), section 3
Road Traffic (New Drivers) Act 1995 (c 13), section 3
Road Traffic Offenders Act 1988 (c 53), sections 25, 62, 67, 90D and 91
Road Traffic Regulation Act 1984 (c 27), sections 5, 16, 47, 115 and 117
Safety of Sports Grounds Act 1975 (c 52), section 12(1) and (6)
Salmon Act 1986 (c 62), section 32(1)
Salmon and Freshwater Fisheries Act 1975 (c 51)
Scrap Metal Dealers Act 1964 (c 69), sections 1 and 5
Sea Fish (Conservation) Act 1967 (c 84)
Sea Fisheries Act 1968 (c 77)

. . .

Sea Fisheries (Shellfish) Act 1967 (c 83)
Solicitors Act 1974 (c 47), section 20
Sunday Trading Act 1994 (c 20)

. . .

Trade Descriptions Act 1968 (c 29)
Transport Act 1968 (c 73), Part 6
Transport Act 1985 (c 67), sections 23, 30 and 101
Transport Act 2000 (c 38), section 148 and Part 3
Unsolicited Goods and Services Act 1971 (c 30), sections 2 and 3
Vehicles (Crime) Act 2001 (c 3)
Vehicle Excise and Registration Act 1994 (c 22), section 28A, Part 3 and section 59
Veterinary Surgeons Act 1966 (c 36)
Video Recordings Act 1984 (c 39), sections 9 to 14
Video Recordings Act 1993 (c 24)
Water Industry Act 1991 (c 56), sections 70(1), 86, 109, 111 and 118
Water Resources Act 1991 (c 57)
Weeds Act 1959 (c 54), section 2
Weights and Measures Act 1985 (c 72), Parts 2 to 7
Wildlife and Countryside Act 1981 (c 69), Parts 1 and 2
Zoo Licensing Act 1981 (c 37), section 19

[3.345A]–[3.500]

Statutory annotations

Amendment

Entry relating 'Property Misdescriptions Act 1991 (c 29)' (omitted) repealed by SI 2013/1575, art 3, Schedule, Pt 1, paras 12, 14.

Entry relating to 'Radioactive Material (Road Transport) Act 1991 (c 27)' (omitted) repealed by the Energy Act 2013, s 116(1), Sch 12, Pt 5, para 94.

Control of Major Accident Hazards Regulations 1999

SI 1999/743

as amended by SIs 1999/2597, 2002/2469, 2005/676, 2005/1088, 2008/736, 2008/960, 2008/1087, 2008/2323, 2013/235, 2013/766 and the Greater London Authority Act 1999

PART 1

INTRODUCTION

[3.754]

2 Interpretation

(1) In these Regulations, unless the context otherwise requires—

'the 1974 Act' means the Health and Safety at Work etc Act 1974;

'the 1984 Regulations' means the Control of Industrial Major Accident Hazards Regulations 1984;

'the Agency' in relation to an establishment in—

(a) England and Wales, means the Environment Agency,

(b) Scotland, means the Scottish Environment Protection Agency;

'CIMAH report' means a report sent to the Executive pursuant to regulations 7 or 8 of the 1984 regulations;

'competent authority' means the Executive and the Agency acting jointly;

'control' in relation to a person means control in the course of a trade, business or other undertaking carried on by him;

'dangerous substance' means (subject to regulation 3(3)(b)) a substance, mixture or preparation—

(a) listed in column 1 of Part 2 of Schedule 1, or,

(b) within a category specified in column 1 of Part 3 of Schedule 1,

and present as a raw material, product, by-product, residue or intermediate;

'the Directive' means Council Directive 96/82/EC on the control of major accident hazards involving dangerous substances [as amended by Directive 2003/105/EC of the European Parliament and of the Council of 16 December 2003];

'emergency services' means—

(a) those police, fire and ambulance services who are liable to be required to respond to an emergency at the establishment,

(b) where appropriate, Her Majesty's Coastguard;

'establishment' means the whole area under the control of the same person where dangerous substances are present in one or more installations, and for this purpose two or more areas under the control of the same person and separated only by a road, railway or inland waterway shall be treated as one whole area;

'the Executive' means the Health and Safety Executive;

'existing establishment' shall be construed in accordance with paragraph (7);

'hazardous substances consent' means a hazardous substances consent granted under the Planning (Hazardous Substances) Act 1990 or the Planning (Hazardous Substances) (Scotland) Act 1997;

['health authority' means—

- (a) in relation to England, [a clinical commissioning group established under section 14D of the National Health Service Act 2006];
- (b) in relation to Wales, a Health Authority established under section 8 of that Act; and
- (c) in relation to Scotland, a Health Board established under section 2 of the National Health Service (Scotland) Act 1978;]

'industrial activity' has the same meaning as in the 1984 Regulations;

'installation' means a unit in which dangerous substances present are, or are intended to be, produced, used, handled or stored, and it includes—

- (a) equipment, structures, pipework, machinery and tools,
- (b) railway sidings, docks and unloading quays serving the unit, and
- (c) jetties, warehouses or similar structures, whether floating or not,

which are necessary for the operation of the unit;

'local authority' means—

- (a) for the purposes of regulation 14, in relation to—
  - (i) the City of London, the Common Council for the City of London;
  - (ii) an area in the rest of London, the London Borough Council for that area;
  - (iii) the isles of Scilly, the Council of the Isles of Scilly;
  - (iv) an area in the rest of England, the district council for that area or where there is no district council for that area, the county council for that area;
  - (v) an area in Scotland, the council for the local government area, and
  - (vi) an area in Wales, the county council or the county borough council for that area;
- (b) for the purposes of other regulations, in relation to—
  - (i) London, the [London Fire and Emergency Planning Authority];
  - (ii) an area where there is a fire and civil defence authority, that authority;
  - (iii) the Isles of Scilly, the Council of the Isles of Scilly;
  - (iv) an area in the rest of England, the county council for that area, or where there is no county council for that area, the district council for that area;
  - (v) an area in Scotland, the council for the local government area;
  - (vi) an area in Wales, the county council or the county borough council for that area;

'major accident' means an occurrence (including in particular, a major emission, fire or explosion) resulting from uncontrolled developments in the course of the operation of any establishment and leading to serious danger to human health or the environment, immediate or delayed, inside or outside the establishment, and involving one or more dangerous substances;

'major accident prevention policy document' shall be construed in accordance with regulation 5(1);

['notify' means notify—

- (a) in writing, including in an email; or

(b) by such other means as the recipient may allow,

and 'notification' shall be construed accordingly;]

'off-site emergency plan' shall be construed in accordance with regulation 10(1);

'on-site emergency plan' shall be construed in accordance with regulation 9(1);

'operator' shall be construed in accordance with paragraph (2);

'pipeline' means a pipeline to which the Pipelines Safety Regulations 1996 applies;

'road' means—

(a) in relation to England and Wales, a road within the meaning of section 192(1) of the Road Traffic Act 1988;

(b) in relation to Scotland, a road within the meaning of the Roads (Scotland) Act 1984;

'safety report' means a report sent to the competent authority pursuant to regulation 7 or a part of a report sent to the competent authority pursuant to regulation 7(10) except that where any such report or part has been revised pursuant to regulation 8, it means the report or part as so revised.

(2) Any reference in these Regulations to an operator is a reference to a person who is in control of the operation of an establishment or installation (or in relation to an establishment or installation which is to be constructed or operated, the person who proposes to control its operation or, if that person is not known, the person who in the course of a trade, business or other undertaking carried on by him has commissioned its design and construction); and any duty imposed by these Regulations on him shall extend only in relation to that establishment or installation.

(3) Any reference in these Regulations to the presence of dangerous substances includes a reference to the anticipated presence of such substances and the presence of those which it is reasonable to believe may be generated during the loss of control of an industrial chemical process.

(4) Any reference in these Regulations to the storage of dangerous substances includes a reference to the presence of dangerous substances for the purposes of warehousing, depositing in safe custody or keeping in stock.

(5) Any reference in these Regulations to the start of construction or operation of an establishment is, subject to paragraph (6), a reference to the start of construction or operation, as the case may be, of the installation in the establishment, or where there is or is to be more than one installation in the establishment, the one whose construction or operation, as the case may be, is first started, and where an installation in an establishment has been constructed, or its construction has been started, before the coming into force of these Regulations, regulations 6(1) and 7(1) shall not apply in respect of that establishment.

(6) Where after the coming into force of these Regulations an establishment becomes subject to any of these Regulations by reason of any increase in the quantity of dangerous substances present there, any reference in that regulation to the start of operation of the establishment, is a reference to the time when the establishment first becomes so subject.

(7) Any reference in a regulation to an existing establishment is a reference to an establishment whose operation commenced before the coming into force of these Regulations and which, on the coming into force of these Regulations, is an establishment to which that regulation applies.

(8) The columns in Parts 2 and 3 of Schedule 1 shall be applied in accordance with the provisions of Part I of that Schedule and notes set out in each of those Parts.

(9) Any reference in these Regulations to—

(a) a numbered regulation or Schedule is a reference to the regulation in or Schedule to these Regulations so numbered, and
(b) a numbered paragraph is a reference to the paragraph so numbered in the Regulation or Schedule in which the reference appears.

[3.755]

Statutory annotations

Amendment

Para (1): in definition 'health authority' in para (a) words from 'a clinical commissioning group' to 'National Health Service Act 2006' in square brackets substituted by SI 2013/235, art 11, Sch 2, Pt 1, para 36(1), (2).

[3.756]

3 Application

(1) These Regulations shall apply to an establishment where a dangerous substance listed in column 1 of Parts 2 or 3 of Schedule 1 is present in a quantity equal to or exceeding the quantity listed in the entry for that substance in column 2 of those Parts, except that regulations 7 to 14 shall apply only to an establishment where such a dangerous substance is present in a quantity equal to or exceeding the quantity listed in the entry for that substance in column 3 of those Parts.

(2) The reference in paragraph (1) to the presence of dangerous substances shall not include the presence of dangerous substances in the following activities—
(a) the transport of those substances and their intermediate temporary storage by road, rail, inland waterways, sea or air, including their loading and unloading and transport to and from another means of transport at docks, wharves and marshalling yards; or
(b) the transport of those substances in a pipeline or pumping station.

(3) These Regulations shall not apply to—
(a) an establishment which is under the control of—
(i) the Secretary of State for the purposes of the Ministry of Defence,
(ii) a headquarters or organisation designated for the purposes of the International Headquarters and Defence Organisations Act 1964 or the service authorities of a visiting force within the meaning of any of the provisions of Part 1 of the Visiting Forces Act 1952;
(b) substances which create a hazard from ionising radiation if present on a site for which a nuclear site licence has been granted or is required for the purposes of section 1 of the Nuclear Installations Act 1965;
[(c) the exploration, extraction and processing of minerals in mines, quarries or by means of boreholes, except—
(i) chemical and thermal processing operations; and
(ii) storage relating to those operations,

which involve dangerous substances; and
[(d) waste land-fill sites, except—
(i) tailing ponds or dams and other operational tailings disposal facilities containing dangerous substances, in particular when any such facilities are used in connection with the chemical and thermal processing of minerals;
(ii) sites used for the storage of metallic mercury pursuant to Article 3(1)(b) of Regulation (EC) No 1102/2008 of the European Parliament and of the Council on the banning of exports of metallic mercury and certain mercury compounds and mixtures and the safe storage of metallic mercury].

(4) . . .

(5) These Regulations shall not apply in Northern Ireland.

[3.757]

Statutory annotations

Amendment

Para (3): sub-para (d) substituted by SI 2013/766, reg 3.

PART 4

EMERGENCY PLANS

[3.767]

9 On-site emergency plan

(1) Every operator of an establishment shall prepare an emergency plan (in these Regulations referred to as an 'on-site emergency plan') which shall be adequate for securing the objectives specified in Part 1 of Schedule 5 and shall contain the information specified in Part 2 of that Schedule.

(2) The on-site emergency plan shall be prepared—

(a) in the case of an existing establishment where the industrial activity carried on there was, immediately before the coming into force of these Regulations, subject to the requirements of regulation 10 of the 1984 Regulations, by 3 February 2001;

(b) in the case of any other existing establishment, by 3 February 2002;

[(c) in the case of an establishment which has not started to operate, before it starts to operate;

(d) in any other case, without delay but at all events within 1 year after the establishment becomes subject to this regulation].

(3) The operator shall consult—

(a) persons [working] in the establishment;

(b) the Agency;

[(ba) if the establishment is situated in England, the National Health Service Commissioning Board and Public Health England, an executive agency of the Department of Health;]

(c) the emergency services; and

(d) the health authority for the area where the establishment is situated;

on the preparation of the on-site emergency plan.

(4) The operator shall consult the local authority in whose area the establishment is situated on the preparation of an on-site emergency plan, except this shall not apply where the local authority has been exempted from the requirement to prepare an off-site emergency plan in respect of the establishment pursuant to regulation 10(7).

[3.768]

Statutory annotations

Amendment

Para (3): sub-para (ba) inserted by SI 2013/235, art 11, Sch 2, Pt 1, para 36(1), (3).

[3.769]

10 Off-site emergency plan

(1) The local authority, in whose area there is an establishment, shall prepare an emergency plan (in these Regulations referred to as an 'off-site emergency plan') in respect of that establishment, and such a plan shall be adequate for securing the objectives specified in Part 1 of Schedule 5 and shall contain the information specified in Part 3 of that Schedule.

(2) The off-site emergency plan shall be prepared no later than 6 months (or such longer period, not exceeding 9 months, as the competent authority may agree in writing) after—

(a) the receipt by the local authority of a notice from the competent authority informing the local authority of the need to prepare an off-site emergency plan in respect of the establishment;

(b) the time an on-site emergency plan is required to be prepared for the establishment pursuant to regulation 9; or

(c) the receipt by the local authority of the information referred to in paragraphs (3) and (5);

whichever is later.

(3) An operator shall supply to the local authority in whose area the establishment is situated the information necessary for the purpose of enabling the authority to prepare the off-site emergency plan.

(4) The information referred to in paragraph (3) shall be supplied no later than the time an on-site emergency plan is required to be prepared for the establishment pursuant to regulation 9.

(5) The operator shall supply to the local authority any additional information it may reasonably request in writing to enable the off-site emergency plan to be prepared, and the information shall be so provided within such period as the local authority specifies in the request.

(6) The local authority shall consult the operator, the competent authority, [the Agency,] the emergency services, each health authority for the area in the vicinity of the establishment and such members of the public as it considers appropriate on the preparation of the off-site emergency plan.

[(6A) If the establishment is situated in England, the local authority shall also consult the National Health Service Commissioning Board and Public Health England, an executive agency of the Department of Health, on the preparation of the emergency off-site plan.]

(7) The competent authority may in view of the information contained in a safety report exempt a local authority from the requirement to prepare an off-site emergency plan in respect of an establishment, and any such exemption shall be in writing and state the reasons for granting it.

(8) Where an exemption has been given under paragraph (7), the local authority shall, for the purposes of these Regulations and while the exemption is in force, have no function in relation to the preparation, review, testing and putting into effect of an off-site emergency plan for the establishment concerned.

[3.770]

Statutory annotations

Amendment

Para (6A): inserted by SI 2013/235, art 11, Sch 2, Pt 1, para 36(1), (4).

PART 5
PROVISION OF INFORMATION BY OPERATOR

[3.777]

15 Provision of information to competent authority

(1) Every operator of an establishment shall, when requested to do so by the competent authority provide sufficient information to the authority to demonstrate that he has taken all measures necessary to comply with these Regulations, and the information shall be so provided within such period as the competent authority specifies in the request.

(2) Without prejudice to the generality of paragraph (1), the operator shall when requested to do so by the competent authority, provide the authority with any information necessary to enable the authority—

- (a) fully to assess the possibility of a major accident and to determine the scope of possible increased probability or aggravation of a major accident;
- (b) to take substances into account which, due to their physical form, particular conditions or location, may require additional consideration; or
- (c) to perform its functions of obtaining or collecting information under regulation 19(4);

and the information shall be so provided within such period as the competent authority specifies in the request.

(3) Where a major accident has occurred at an establishment the operator shall forthwith inform the competent authority of that accident.

(4) Where the operator has notified a major accident to the Executive in accordance with the requirements of [the Reporting of Injuries, Diseases and Dangerous Occurrences Regulations 2013], he shall be deemed to have complied with the requirement to inform the competent authority of that accident under paragraph (3).

(5) Anything required to be sent by an operator of an establishment to the competent authority pursuant to these Regulations shall be sent to the authority at an office of the Executive.

[3.777A]

Statutory annotations

Amendment

Para (4): words in square brackets substituted by SI 2013/1471, Sch 4.

[SCHEDULE 1

DANGEROUS SUBSTANCES TO WHICH THE REGULATIONS APPLY

Regulations 2(1) and (8) and 3(1)]

PART 2

NAMED SUBSTANCES

[3.791]

Where a substance or group of substances listed in this Part also falls within a category of Part 3, the qualifying quantities set out in this Part must be used.

Column 1	***Column 2***	***Column 3***
Dangerous substances	***Quantity in tonnes***	
Ammonium nitrate (as described in Note 1 of this Part; see also Note 8(1) and (2))	5,000	10,000
Ammonium nitrate (as described in Note 2 of this Part; see also Note 8)	1,250	5,000
Ammonium nitrate (as described in Note 3 of this Part; see also Note 8(2) and (3))	350	2,500
Ammonium nitrate (as described in Note 4 of this Part; see also Note 8)	10	50
Potassium nitrate (as described in Note 5 of this Part)	5,000	10,000

Column 1 *Dangerous substances*	*Column 2* *Quantity in tonnes*	*Column 3*
Potassium nitrate (as described in Note 6 of this Part)	1,250	5,000
Arsenic pentoxide, arsenic (V) acid and/or salts	1	2
Arsenic trioxide, arsenious (III) acid and/or salts	0.1	0.1
Bromine	20	100
Chlorine	10	25
Nickel compounds in inhalable powder form (nickel monoxide, nickel dioxide, nickel sulphide, trinickel disulphide, dinickel trioxide)	1	1
Ethyleneimine	10	20
Fluorine	10	20
Formaldehyde (concentration ≥ 90%)	5	50
Hydrogen	5	50
Hydrogen chloride (liquefied gas)	25	250
Lead alkyls	5	50
Liquefied extremely flammable gases (including LPG) and natural gas (whether liquefied or not)	50	200
Acetylene	5	50
Ethylene oxide	5	50
Propylene oxide	5	50
Methanol	500	5,000
4, 4-Methylenebis (2-chloraniline) and/or salts, in powder form	0.01	0.01
Methylisocyanate	0.15	0.15
Oxygen	200	2,000
Toluene diisocyanate	10	100
Carbonyl dichloride (phosgene)	0.3	0.75
Arsenic trihydride (arsine)	0.2	1
Phosphorus trihydride (phosphine)	0.2	1
Sulphur dichloride	1	1
Sulphur trioxide	15	75
Polychlorodibenzofurans and polychlorodibenzodioxins (including TCDD), calculated in TCDD equivalent	0.001	0.001
The following CARCINOGENS at concentrations above 5% by weight:	0.5	2
4-Aminobiphenyl and/or its salts, Benzotrichloride, Benzidine and/or salts, Bis (chloromethyl) ether, Chloromethyl methyl ether, 1,2-Dibromoethane, Diethyl sulphate, Dimethyl sulphate, Dimethylcarbamoyl chloride, 1,2-Dibromo-3-chloropropane, 1, 2-Dimethylhydrazine, Dimethylnitrosamine, Hexamethylphosphoric triamide, Hydrazine, 2-Naphthylamine and/or salts, 4-Nitrodiphenyl and 1,3-Propanesultone		
Petroleum products:	2,500	25,000
(a) gasolines and naphthas		

Column 1 Dangerous substances	Column 2 Quantity in tonnes	Column 3
(b) kerosenes (including jet fuels)		
(c) gas oils (including diesel fuels, home heating oils and gas oil blending streams)		
[(d) heavy fuel oils]		

NOTES

1

Ammonium nitrate (5,000/10,000): fertilisers capable of self-sustaining decomposition.

This applies to ammonium nitrate-based compound/composite fertilisers (compound or composite fertilisers containing ammonium nitrate with phosphate and/or potash) in which the nitrogen content as a result of ammonium nitrate is—

(a) between 15.75% and 24.5% by weight and either with not more than 0.4% total combustible or organic materials or which satisfy the detonation resistance test described in Schedule 2 to the Ammonium Nitrate Materials (High Nitrogen Content) Safety Regulations 2003 'the detonation resistance test'; or

(b) 15.75% or less by weight and unrestricted combustible materials,

and which are capable of self-sustaining decomposition according to the UN Trough Test specified in United Nations Recommendations on the Transport of Dangerous Goods: Manual of Tests and Criteria (3rd revised Edition), Part III, subsection 38.2.

2

Ammonium nitrate (1,250/5,000): fertiliser grade.

This applies to straight ammonium nitrate-based fertilisers and to ammonium nitrate-based compound/composite fertilisers in which the nitrogen content as a result of ammonium nitrate is—

(a) more than 24.5% by weight, except for mixtures of ammonium nitrate with dolomite, limestone and/or calcium carbonate with a purity of at least 90%;

(b) more than 15.75% by weight for mixtures of ammonium nitrate and ammonium sulphate; or

(c) more than 28% by weight for mixtures of ammonium nitrate with dolomite, limestone and/or calcium carbonate with a purity of at least 90%,

and which satisfy the detonation resistance test.

3

Ammonium nitrate (350/2,500): technical grade.

This applies to—

(a) ammonium nitrate and preparations of ammonium nitrate in which the nitrogen content as a result of the ammonium nitrate is—

  (i) between 24.5% and 28% by weight, and which contain not more than 0.4% combustible substances; or

  (ii) more than 28% by weight, and which contain not more than 0.2% combustible substances; and

(b) aqueous ammonium nitrate solutions in which the concentration of ammonium nitrate is more than 80% by weight.

4

Ammonium nitrate (10/50): 'off-specs' material and fertilisers not satisfying the detonation resistance test.

This applies to—

(a) material rejected during the manufacturing process and to ammonium nitrate and preparations of ammonium nitrate, straight ammonium nitrate-based fertilisers and ammonium nitrate-based compound/composite fertilisers referred to in Notes 2 and 3, that are being or have been returned from the final user to a manufacturer, temporary storage or reprocessing plant for reworking, recycling or treatment for safe use, because they no longer comply with the specifications of Notes 2 and 3; or

(b) fertilisers which do not fall within Notes 1(a) and 2 because they do not satisfy the detonation resistance test, other than fertilisers which—

(i) at the time of delivery to a final user satisfied the detonation resistance test; but

(ii) later became degraded or contaminated; and

(iii) are temporarily present at the establishment of the final user prior to their return for reworking, recycling or treatment for safe use or to their being applied as fertiliser.

5

Potassium nitrate (5,000/10,000): composite potassium nitrate-based fertilisers composed of potassium nitrate in prilled/granular form.

6

Potassium nitrate (1,250/5,000): composite potassium nitrate-based fertilisers composed of potassium nitrate in crystalline form.

7

Polychlorodibenzofurans and polychlorodibenzodioxins.

The quantities of polychlorodibenzofurans and polychlorodibenzodioxins are calculated using the following factors:

International Toxic Equivalent Factors (ITEF) for the congeners of concern (NATO/CCMS)*

2, 3, 7, 8-TCDD	1	2, 3, 7, 8-TCDF	0.1
1, 2, 3, 7, 8-PeCDD	0.5	2, 3, 4, 7, 8-PeCDF	0.5
		1, 2, 3, 7, 8-PeCDF	0.05
1, 2, 3, 4, 7, 8-HxCDD	0.1		
1, 2, 3, 6, 7, 8-HxCDD	0.1	1, 2, 3, 4, 7, 8-HxCDF	0.1
1, 2, 3, 7, 8, 9-HxCDD	0.1	1, 2, 3, 7, 8, 9-HxCDF	0.1
		1, 2, 3, 6, 7, 8-HxCDF	0.1
1, 2, 3, 4, 6, 7, 8-HpCDD	0.01	2, 3, 4, 6, 7, 8-HxCDF	0.1
		1, 2, 3, 4, 6, 7, 8-HpCDF	0.01
OCDD	0.001	1, 2, 3, 4, 7, 8, 9-HpCDF	0.01
		OCDF	0.001

*(T = tetra, Pe = penta, Hx = hexa, Hp = hepta, O = octa)

8

(1) 15.75% nitrogen content by weight as a result of ammonium nitrate corresponds to 45% ammonium nitrate.

(2) 24.5% nitrogen content by weight as a result of ammonium nitrate corresponds to 70% ammonium nitrate.

(3) 28% nitrogen content by weight as a result of ammonium nitrate corresponds to 80% ammonium nitrate.

[3.791A]

Statutory annotations

Amendment

In entry relating to 'Petroleum Products' para (d) inserted by SI 2014/162, reg 2.

Management of Health and Safety at Work Regulations 1999

SI 1999/3242

as amended by SI 2003/2457, 2005/1541, 2006/438, 2007/320, 2010/93, 2013/1667 and SSI 2006/457

[3.875]

[22 Restriction of civil liability for breach of statutory duty

[(1) Breach of a duty imposed by regulation 16, 16A, 17 or 17A shall, so far as it causes damage, be actionable by the new or expectant mother.

(2) Any term of an agreement which purports to exclude or restrict any liability for such a breach is void.]

[3.876]

Statutory annotations

Amendment

Substituted by SI 2013/1667, reg 3.

Health and Safety (Fees) Regulations 2012

SI 2012/1652

as amended by SI 2013/448

Made 25th June 2012

[3.1175]

21

[*Revoked by SI 2013/448, reg 2, Sch.*]

SCHEDULE 16

[3.1206]

[*Revoked by SI 2013/448, reg 2, Sch.*]

Health and Safety at Work etc Act 1974 (Application Outside Great Britain) Order 2013

SI 2013/240

Made 12th February 2013

[3.1210]

General note

This Order extends the 'prescribed provisions' of the HSWA 1974 to various offshore areas and activities.

[12.627]

Her Majesty, in exercise of the powers conferred by section 84(3) and (4) of the Health and Safety at Work etc Act 1974, is pleased, by and with the advice of Her Privy Council, to order, and it is hereby ordered as follows:

[3.1211]

1 Citation, commencement and revocation

(1) This Order may be cited as the Health and Safety at Work etc Act 1974 (Application outside Great Britain) Order 2013 and shall come into force on 6th April 2013.

(2) The Health and Safety at Work etc Act 1974 (Application outside Great Britain) Order 2001 and the Health and Safety at Work etc Act 1974 (Application outside Great Britain) (Variation) Order 2011 are hereby revoked.

[3.1212]

2 Interpretation

(1) In this Order—

'the 1974 Act' means the Health and Safety at Work etc Act 1974;

'activity' includes, unless the context otherwise requires, a diving project and standing a vessel by;

'designated area' means any area designated by order under section 1(7) of the Continental Shelf Act 1964 and 'within a designated area' includes over and under it;

'diving project' has the same meaning as it has in the Diving at Work Regulations 1997 save that it includes an activity in which a person takes part as a diver wearing an atmospheric pressure suit and without breathing in air or other gas at a pressure greater than atmospheric pressure;

'gas importation and storage zone' has the meaning given by section 1(5) of the Energy Act 2008 and 'within a gas importation and storage zone' includes over and under it;

'offshore installation' has the meaning given in article 4(2) and (3) of this Order;

'the prescribed provisions of the 1974 Act' means sections 1 to 59 and 80 to 82 of the 1974 Act;

'supplementary unit' means a fixed or floating structure, other than a vessel, for providing energy, information or substances to an offshore installation;

'territorial sea' means the territorial sea adjacent to Great Britain and 'within the territorial sea' includes on, over and under it;

'vessel' includes a hovercraft and any floating structure which is capable of being navigated.

(2) For the purposes of this Order, any structures and devices on top of a well are treated as forming part of the well.

[3.1213]

3 Application of the 1974 Act outside Great Britain

(1) The prescribed provisions of the 1974 Act, to the extent specified in the following articles of this Order, apply to and in relation to the premises and activities outside Great Britain which are so specified as those provisions apply within Great Britain.

(2) The reference in paragraph (1) of this article to premises and activities includes a reference to any person, article or substance on those premises or engaged in or, as the case may be, used or for use in connection with any such activity, but does not include a reference to an aircraft which is airborne.

[3.1214]

Great Britain

See para [2.51].

[3.1215]

4 Offshore installations

(1) The prescribed provisions of the 1974 Act apply within the territorial sea, a designated area or a gas importation and storage zone to and in relation to—

- (a) any offshore installation and any activity on it;
- (b) any activity in connection with, or any activity which is immediately preparatory to an activity in connection with, an offshore installation, whether carried on from the installation itself, in or from a vessel or in any manner, other than an activity falling within paragraph (4);
- (c) a diving project involving—
  - (i) the survey and preparation of the sea bed for an offshore installation;
  - (ii) the survey and restoration of the sea bed consequent on the removal of an offshore installation.

(2) In this Order 'offshore installation' means, subject to paragraph (3) of this article, a structure which is, or is to be, or has been, used while standing or stationed in water, or on the foreshore or other land intermittently covered with water—

- (a) for the exploitation, or exploration with a view to exploitation, of mineral resources by means of a well;
- (b) for undertaking activities falling within article 8(2);
- (c) for the conveyance of things by means of a pipe; or
- (d) for undertaking activities that involve mechanically entering the pressure containment boundary of a well; or
- (e) primarily for the provision of accommodation for persons who work on or from a structure falling within any of sub-paragraphs (a) to (d),

together with any supplementary unit which is ordinarily connected to it and all the connections.

(3) Any reference in paragraph (2) to a structure or supplementary unit does not include—

(a) a structure which is connected with dry land by a permanent structure providing access at all times and for all purposes;
(b) a well;
(c) a mobile structure which has been taken out of use and is not yet being moved with a view to its being used for any of the purposes specified in paragraph (2) of this article;
(d) any part of a pipeline; and
(e) a structure falling within article 10(c).

(4) Subject to paragraph (5), the following activities fall within this paragraph—
(a) transporting, towing or navigating an installation;
(b) any of the following activities carried on in or from a vessel—
(i) giving assistance in the event of an emergency;
(ii) training in relation to the giving of assistance in the event of an emergency;
(iii) testing equipment for use in giving assistance in the event of an emergency;
(iv) putting or maintaining a vessel on stand-by ready for an activity referred to in any of sub-subparagraphs (i) to (iii).

(5) Paragraph (4)(b) does not apply in respect of a vessel in or from which an activity is carried on or in connection with, or any activity that is immediately preparatory to an activity in connection with, an offshore installation other than an activity falling within paragraph 4(b).

[3.1216]

5 Wells

(1) Subject to paragraph (2) of this article, the prescribed provisions of the 1974 Act shall apply within the territorial sea or a designated area to and in relation to—
(a) a well and any activity in connection with it; and
(b) an activity which is immediately preparatory to any activity in sub-paragraph (a) above.

(2) Paragraph (1) of this article includes keeping a vessel on station for the purpose of working on a well but otherwise does not include navigation or an activity connected with navigation.

[3.1217]

6 Pipelines

(1) The prescribed provisions of the 1974 Act shall apply within the territorial sea or a designated area to and in relation to—
(a) any pipeline;
(b) any pipeline works;
(c) the following activities in connection with pipeline works—
(i) the loading, unloading, fuelling or provisioning of a vessel engaged in pipeline works;
(ii) the loading, unloading, fuelling, repair and maintenance of an aircraft on a vessel engaged in pipeline works; or
(iii) the moving, supporting, laying or retrieving of anchors attached to a pipe-laying vessel including the supervision of those activities and giving of instruction in connection with them.

(2) In this article—

'pipeline' means a pipe or system of pipes for the conveyance of any thing, together with—

(a) any apparatus for inducing or facilitating the flow of any thing through, or through part of, the pipe or system;
(b) any apparatus for treating or cooling any thing which is to flow through, or through part of, the pipe or system;
(c) valves, valve chambers and similar works which are annexed to, or incorporated in the course of, the pipe or system;
(d) apparatus for supplying energy for the operation of any such apparatus or works as are mentioned in the preceding paragraphs;
(e) apparatus for the transmission of information for the operation of the pipe or system;
(f) apparatus for the cathodic protection of the pipe or system; and
(g) a structure used or to be used solely for the support of a part of the pipe or system;

but not including a pipeline of which no initial or terminal point is situated in the United Kingdom, within the territorial sea adjacent to the United Kingdom, or within a designated area;

'pipeline works' means—

(a) assembling or placing a pipeline or length of pipeline including the provision of internal or external protection for it;
(b) inspecting, testing, maintaining, adjusting, repairing, altering or renewing a pipeline or length of pipeline;
(c) changing the position of or dismantling or removing a pipeline or length of pipeline;
(d) opening the bed of the sea for the purposes of the works mentioned in sub-paragraphs (a) to (c) of this definition, and tunnelling or boring for those purposes;
(e) any activities incidental to the activities described in sub-paragraphs (a) to (d) of this definition;
(f) a diving project in connection with any of the works mentioned in sub-paragraphs (a) to (e) of this definition or for the purpose of determining whether a place is suitable as part of the site of a proposed pipeline and the carrying out of surveying operations for settling the route of a proposed pipeline.

[3.1218]

7 Mines

(1) The prescribed provisions of the 1974 Act apply to and in relation to a mine within the territorial sea or extending beyond it, and any activity in connection with it, while it is being worked.

(2) In this article 'mine' has the same meaning as in the Mines and Quarries Act 1954 ('the 1954 Act').

(3) For the purposes of this article a mine shall be treated as being worked when it is to be so treated for the purposes of the 1954 Act.

[3.1219]

8 Gas importation and storage

(1) Subject to paragraph (3), the prescribed provisions of the 1974 Act apply within the territorial sea or a gas importation and storage zone to and in relation to any activities connected with or immediately preparatory to the activities set out in paragraph (2).

(2) The activities are—

(a) the unloading of gas to an installation or pipeline;
(b) the storage of gas, whether temporary or permanent, in or under the shore or bed of any water;
(c) the conversion of any natural feature for the purpose of storing gas, whether temporarily or permanently;
(d) the recovery of gas stored;
(e) exploration with a view to, or in connection with, the carrying on of activities within sub-paragraphs (a) to (d).

(3) Paragraph (1) does not apply to an activity falling within paragraph (2) if the prescribed provisions of the 1974 Act apply to or in relation to that activity by virtue of article 4(1).

(4) In this article—

'gas' means any substance which is gaseous at a temperature of 15°C and a pressure of 101.325 kPa (1013.25 mb); and

'installation' means any floating structure or device maintained on a station by whatever means.

(5) For the purposes of paragraphs (2) and (4), references to gas include any substance which consists wholly or mainly of gas.

[3.1220]

9 Production of energy from water or wind

(1) The prescribed provisions of the 1974 Act apply within the territorial sea or a renewable energy zone to and in relation to any energy structure or activities connected with or preparatory to—

(a) the exploitation of those areas for the production of energy from water or wind;
(b) the exploration of such areas with a view to, or in connection with, the production of energy from water or wind, or
(c) the operation of a cable for transmitting electricity from an energy structure.

(2) In this article—

'energy structure' means a fixed or floating structure or machine, other than a vessel, which is, or is to be, or has been, used for producing energy from water or wind; and

'renewable energy zone' has the meaning given by section 84(4) of the Energy Act 2004 and 'within a renewable energy zone' includes over and under it.

[3.1221]

10 Underground coal gasification

The prescribed provisions of the 1974 Act apply within the territorial sea or a designated area and in relation to—

(a) underground coal gasification and any activity in connection with it;
(b) any activity which is immediately preparatory to any activity in paragraph (a); and
(c) any fixed or floating structure which is, or is to be, or has been, used in connection with the carrying on of activities within paragraphs (a) and (b).

[3.1222]

11 Other activities within the territorial sea

(1) Subject to paragraph (2), the prescribed provisions of the 1974 Act apply within the territorial sea to and in relation to—

(a) the construction, reconstruction, alteration, repair, maintenance, cleaning, use, operation, demolition and dismantling of any building, or other structure, not being in any case a vessel, or any preparation for any such activity;

(b) the transfer of people or goods between a vessel or aircraft and a structure (including a building) mentioned in sub-paragraph (a) of this paragraph;

(c) the loading, unloading, fuelling or provisioning of a vessel;

(d) a diving project;

(e) the laying, installation, inspection, maintenance, operation, recovery or repair of a cable;

(f) the construction, reconstruction, finishing, refitting, repair, maintenance, cleaning or breaking up of a vessel except when carried out by the master or any officer or member of the crew of that vessel;

(g) the maintaining on a station of a vessel which would be an offshore installation were it not a structure to which article 4(3)(c) of this Order applies;

(h) the transfer of people or goods between a vessel or aircraft and a structure mentioned in sub-paragraph (g) of this paragraph.

(2) This article does not apply—

(a) to a case where article 4, 5, 6, 7, 8, 9 or 10 of this Order applies; or

(b) to vessels which are registered outside the United Kingdom and are on passage through the territorial sea.

[3.1223]

12 Legal proceedings

(1) Proceedings for any offence under section 33 of the 1974 Act, being an offence to which that section applies by virtue of this Order, may be taken, and the offence may for all incidental purposes be treated as having been committed, in any place in Great Britain.

(2) Section 3 of the Territorial Waters Jurisdiction Act 1878 (which requires certain consents for the institution of proceedings) does not apply to proceedings for any offence to which paragraph (1) of this article relates.

[3.1224]

Great Britain

See para [2.51].

[3.1225]

13 Miscellaneous provisions

The prescribed provisions of the 1974 Act shall apply in accordance with this Order to individuals whether or not they are British subjects, and to bodies corporate whether or not they are incorporated under the law of any part of the United Kingdom.

[3.1226]–[3.1230]

14

Nothing in this Order except article 12(2) limits or prejudices the operation which any Act or legislative instrument may, apart from this Order, have in the territorial sea or elsewhere.

Employment Tribunals (Constitution and Rules of Procedure) Regulations 2013

SI 2013/1237

Made 31st May 2013

[3.1231]

General note

These Regulations replace, with amendments, the Employment Tribunals (Constitution and Rules of Procedure) Regulations 2004. Regulations 1, 3 and 11 came into force on 1 July 2013 and the remainder on 29 July 2013. These Regulations apply generally to employment tribunal procedure on unfair dismissal etc. The rules of procedure are set out in Schedule 1. Rule 105 contains the provisions specific to dealing with appeals against improvement and prohibition notices under the Health and Safety at Work etc Act 1974. Schedules 2 and 3 (which relate to national security and to equal value claims) are not reproduced in this work.

The Secretary of State, in exercise of the powers conferred by section 24(2) of the Health and Safety at Work etc Act 1974, sections 1(1), 4(6) and (6A), 7(1), (3), (3ZA), (3A), (3AA), (3AB), (3B), (3C) and (5), 7A(1) and (2), 7B(1) and (2), 9(1) and (2), 10(2), (5), (6) and (7), 10A(1), 11(1), 12(2), 13, 13A, 19, and 41(4) of the Employment Tribunals Act 1996, paragraph 37 of Schedule 6 to the Scotland Act 1998, and paragraph 32 of Schedule 9 to the Government of Wales Act 2006, makes the following Regulations.

The Secretary of State has consulted with the Administrative Justice and Tribunals Council, and that Council has consulted with the Scottish Committee and the Welsh Committee, in accordance with paragraph 24 of Schedule 7 to the Tribunals, Courts and Enforcement Act 2007.

[3.1232]

1 Citation and commencement

(1) These Regulations may be cited as the Employment Tribunals (Constitution and Rules of Procedure) Regulations 2013 and the Rules of Procedure contained in Schedules 1, 2 and 3 may be referred to, respectively, as—

(a) the Employment Tribunals Rules of Procedure 2013;

(b) the Employment Tribunals (National Security) Rules of Procedure 2013, and

(c) the Employment Tribunals (Equal Value) Rules of Procedure 2013.

(2) This regulation and regulations 3 and 11 come into force on 1st July 2013 and the remainder of these Regulations (including the Schedules) come into force on 29th July 2013.

[3.1233]

2 Revocation

Subject to the savings in regulation 15 the Employment Tribunals (Constitution and Rules of Procedure) Regulations 2004 are revoked.

[3.1234]

3 Interpretation

Except in the Schedules which are subject to the definitions contained in the Schedules, in these Regulations—

"2004 Regulations" means the Employment Tribunals (Constitution and Rules of Procedure) Regulations 2004;

"appointing office holder" means, in England and Wales, the Lord Chancellor, and in Scotland, the Lord President;

"Employment Tribunals Act" means the Employment Tribunals Act 1996;

"Lord President" means the Lord President of the Court of Session;

"national security proceedings" means proceedings in relation to which a direction is given, or an order is made, under rule 94 of Schedule 1;

"President" means either of the two presidents appointed from time to time in accordance with regulation 5(1);

"Regional Employment Judge" means a person appointed or nominated in accordance with regulation 6(1) or (2);

"Senior President of Tribunals" means the person appointed in accordance with section 2 of the Tribunals, Courts and Enforcement Act 2007;

"Tribunal" means an employment tribunal established in accordance with regulation 4 and, in relation to any proceedings, means the Tribunal responsible for the proceedings in question, whether performing administrative or judicial functions;

"Vice President" means a person appointed or nominated in accordance with regulation 6(3) or (4).

[3.1235]

4 Establishment of employment tribunals

There are to be tribunals known as employment tribunals.

[3.1236]

5 President of Employment Tribunals

(1) There shall be a President of Employment Tribunals, responsible for Tribunals in England and Wales, and a President of Employment Tribunals, responsible for Tribunals in Scotland, appointed by the appointing office holder.

(2) A President shall be—

(a) a person who satisfies the judicial-appointment eligibility condition within the meaning of section 50 of the Tribunals, Courts and Enforcement Act 2007 on a 5-year basis;

(b) an advocate or solicitor admitted in Scotland of at least five years standing; or

(c) a member of the Bar of Northern Ireland or solicitor of the Supreme Court of Northern Ireland of at least five years standing.

(3) A President may at any time resign from office by giving the appointing officer holder notice in writing to that effect.

(4) The appointing officer holder may remove a President from office on the ground of inability or misbehaviour, or if the President is adjudged to be bankrupt or makes a composition or arrangement with his creditors.

(5) Where a President is unable to carry out the functions set out in these Regulations, those functions may be discharged by a person nominated by the appointing office holder (save that any nomination in relation to England and Wales shall be made by the Lord Chief Justice following consultation with the Senior President of Tribunals, rather than by the Lord Chancellor).

(6) The Lord Chief Justice may nominate a judicial office holder (as defined in section 109(4) of the Constitutional Reform Act 2005) to exercise his functions under this regulation.

[3.1237]

6 Regional Employment Judges and the Vice President

(1) The Lord Chancellor may appoint Regional Employment Judges.

(2) The President (England and Wales) or the Regional Employment Judge for an area may nominate an Employment Judge to discharge the functions of the Regional Employment Judge for that area.

(3) The Lord President may appoint a Vice President.

(4) The President (Scotland) or the Vice President may nominate an Employment Judge to discharge the functions of the Vice President.

(5) Appointments and nominations under this regulation shall be from the [salaried] Employment Judges on the panel referred to in regulation 8(2)(a).

[3.1237A]

Statutory annotations

Amendment

Para (5): word 'salaried' in square brackets substituted by SI 2014/271, reg 2.

[3.1238]

7 Responsibilities of the Presidents, Regional Employment Judges and Vice President

(1) The President shall, in relation to the area for which the President is responsible, use the resources available to—

(a) secure, so far as practicable, the speedy and efficient disposal of proceedings;

(b) determine the allocation of proceedings between Tribunals; and

(c) determine where and when Tribunals shall sit.

(2) The President (England and Wales) may direct Regional Employment Judges, and the President (Scotland) may direct the Vice President, to take action in relation to the fulfilment of the responsibilities in paragraph (1) and the Regional Employment Judges and Vice President shall follow such directions.

[3.1239]

8 Panels of members for tribunals

(1) There shall be three panels of members for the Employment Tribunals (England and Wales) and three panels of members for the Employment Tribunals (Scotland).

(2) The panels of members shall be—

(a) a panel of [Employment Judges] who satisfy the criteria set out in regulation 5(2) and are appointed by the appointing office holder . . . ;

(b) a panel of persons appointed by the Lord Chancellor after consultation with organisations or associations representative of employees; and

(c) a panel of persons appointed by the Lord Chancellor after consultation with organisations or associations representative of employers.

(3) Members of the panels shall hold and vacate office in accordance with the terms of their appointment, but may resign from office by written notice to the person who appointed them under paragraph (2), and any member who ceases to hold office shall be eligible for reappointment.

(4) The President may establish further specialist panels of members referred to in paragraph (2) and may select persons from those panels to deal with proceedings in which particular specialist knowledge would be beneficial.

[3.1239A]

Statutory annotations

Amendment

Para (2): in sub-para (a) words 'Employment Judges' in square brackets substituted by SI 2014/271, reg 3(a); in sub-para (a) words omitted revoked by SI 2014/271, reg 3(b).

[3.1240]

9 Composition of tribunals

(1) Where proceedings are to be determined by a Tribunal comprising an Employment Judge and two other members, the President, Vice President or a Regional Employment Judge shall select—

(a) an Employment Judge; and

(b) one member from each of the panels referred to in regulation 8(2)(b) and (c),

and for all other proceedings shall select an Employment Judge.

(2) The President, Vice President or a Regional Employment Judge may select him or herself as the Employment Judge required under paragraph (1).

(3) The President, Vice President or a Regional Employment Judge may select from the appropriate panel a substitute for a member previously selected to hear any proceedings.

(4) This regulation does not apply in relation to national security proceedings (see regulation 10(2)).

[3.1241]

10 National security proceedings—panel of members and composition of tribunals

(1) The President shall select—

(a) a panel of persons from the panel referred to in regulation 8(2)(a);

(b) a panel of persons from the panel referred to in regulation 8(2)(b); and

(c) a panel of persons from the panel referred to in regulation 8(2)(c),

who may act in national security proceedings.

(2) Where proceedings become national security proceedings, the President, Vice President or a Regional Employment Judge shall—

(a) select an Employment Judge from the panel referred to in paragraph (1)(a) and may select him or herself; and

(b) where the proceedings are to be determined by a Tribunal comprising an Employment Judge and two other members, select in addition one member from each of the panels referred to in sub-paragraphs (b) and (c) of paragraph (1).

[3.1242]

11 Practice directions

(1) The President may make, vary or revoke practice directions about the procedure of the Tribunals in the area for which the President is responsible, including—

(a) practice directions about the exercise by Tribunals of powers under these Regulations (including the Schedules); and

(b) practice directions about the provision by Employment Judges of mediation, in relation to disputed matters in a case that is the subject of proceedings, and may permit an Employment Judge to act as mediator in a case even though they have been selected to decide matters in that case.

(2) Practice directions may make different provision for different cases, different areas, or different types of proceedings.

(3) Any practice direction made, varied or revoked shall be published by the President in an appropriate manner to bring it to the attention of the persons to whom it is addressed.

[3.1243]

12 Power to prescribe

(1) The Secretary of State may prescribe—

(a) one or more versions of a form which shall be used by claimants to start proceedings in a Tribunal;

(b) one or more versions of a form which shall be used by respondents to respond to a claim before a Tribunal; and

(c) that the provision of certain information on the prescribed forms is mandatory.

(2) It is not necessary to use a form prescribed under paragraph (1) if the proceedings are—

(a) referred to a Tribunal by a court;

(b) proceedings in which a Tribunal will be exercising its appellate jurisdiction; or

(c) proceedings brought by an employer under section 11 of the Employment Rights Act 1996.

(3) The Secretary of State shall publish the prescribed forms in an appropriate manner to bring them to the attention of prospective claimants, respondents and their advisers.

[3.1244]

13 Application of Schedules 1 to 3

(1) Subject to paragraph (2), Schedule 1 applies to all proceedings before a Tribunal except where separate rules of procedure made under the provisions of any enactment are applicable.

(2) Schedules 2 and 3 apply to modify the rules in Schedule 1 in relation, respectively, to proceedings which are—

(a) national security proceedings; or

(b) proceedings which involve an equal value claim (as defined in rule 1 of Schedule 3).

[3.1245]

14 Register and proof of judgments

(1) The Lord Chancellor shall maintain a register containing a copy of all judgments and written reasons issued by a Tribunal which are required to be entered in the register under Schedules 1 to 3.

(2) . . .

(3) A document purporting to be certified by a member of staff of a Tribunal to be a true copy of an entry of a judgment in the register shall, unless the contrary is proved, be sufficient evidence of the document and its contents.

[3.1245A]

Statutory annotations

Amendment

Para (2): revoked by SI 2014/611, reg 2.

[3.1246]

15 Transitional provisions

(1) Subject to paragraphs (2) and (3), these Regulations and the Rules of Procedure contained in Schedules 1 to 3 apply in relation to all proceedings to which they relate.

(2) Where a respondent receives from a Tribunal a copy of the claim form before 29th July 2013, rules 23 to 25 of Schedule 1 do not apply to the proceedings and rule 7 of Schedule 1 to the 2004 Regulations continues to apply.

(3) Where in accordance with Schedules 3 to 5 of the 2004 Regulations, a notice of appeal was presented to a Tribunal before 29th July 2013, Schedule 1 does not apply to the proceedings and Schedule 3, 4 or 5, as appropriate, of the 2004 Regulations continues to apply.

SCHEDULE 1

THE EMPLOYMENT TRIBUNALS RULES OF PROCEDURE

Regulation 13(1)

Introductory and general

[3.1247]

1 Interpretation

(1) In these Rules—

'ACAS' means the Advisory, Conciliation and Arbitration Service referred to in section 247 of the Trade Union and Labour Relations (Consolidation) Act 1992;

'claim' means any proceedings before an Employment Tribunal making a complaint;

'claimant' means the person bringing the claim;

'Commission for Equality and Human Rights' means the body established under section 1 of the Equality Act 2006;

'complaint' means anything that is referred to as a claim, complaint, reference, application or appeal in any enactment which confers jurisdiction on the Tribunal;

['early conciliation certificate' means a certificate issued by ACAS in accordance with the Employment Tribunals (Early Conciliation: Exemptions and Rules of Procedure) Regulations 2013;]

['early conciliation exemption' means an exemption contained in regulation 3(1) of the Employment Tribunals (Early Conciliation: Exemptions and Rules of Procedure) Regulations 2014;]

['early conciliation number' means the unique reference number which appears on an early conciliation certificate;]

'Employment Appeal Tribunal' means the Employment Appeal Tribunal established under section 87 of the Employment Protection Act 1975 and continued in existence under section 135 of the Employment Protection (Consolidation) Act 1978 and section 20(1) of the Employment Tribunals Act;

'electronic communication' has the meaning given to it by section 15(1) of the Electronic Communications Act 2000;

'employee's contract claim' means a claim brought by an employee in accordance with articles 3 and 7 of the Employment Tribunals Extension of Jurisdiction (England and Wales) Order 1994 or articles 3 and 7 of the Employment Tribunals Extension of Jurisdiction (Scotland) Order 1994;

'employer's contract claim' means a claim brought by an employer in accordance with articles 4 and 8 of the Employment Tribunals Extension of Jurisdiction (England and Wales) Order 1994 or articles 4 and 8 of the Employment Tribunals Extension of Jurisdiction (Scotland) Order 1994;

'Employment Tribunal' or 'Tribunal' means an employment tribunal established in accordance with regulation 4, and in relation to any proceedings means the Tribunal responsible for the proceedings in question, whether performing administrative or judicial functions;

'Employment Tribunals Act' means the Employment Tribunals Act 1996;

'Equality Act' means the Equality Act 2010;

'full tribunal' means a Tribunal constituted in accordance with section 4(1) of the Employment Tribunals Act;

'Health and Safety Act' means the Health and Safety at Work etc Act 1974;

'improvement notice' means a notice under section 21 of the Health and Safety Act;

'levy appeal' means an appeal against an assessment to a levy imposed under section 11 of the Industrial Training Act 1982;

'Minister' means Minister of the Crown;

'prescribed form' means any appropriate form prescribed by the Secretary of State in accordance with regulation 12;

'present' means deliver (by any means permitted under rule 85) to a tribunal office;

'President' means either of the two presidents appointed from time to time in accordance with regulation 5(1);

'prohibition notice' means a notice under section 22 of the Health and Safety Act;

'Regional Employment Judge' means a person appointed or nominated in accordance with regulation 6(1) or (2);

'Register' means the register of judgments and written reasons kept in accordance with regulation 14;

['relevant proceedings' means those proceedings listed in section 18(1) of the Employment Tribunals Act 1996;]

'remission application' means any application which may be made under any enactment for remission or part remission of a Tribunal fee;

'respondent' means the person or persons against whom the claim is made;

'Tribunal fee' means any fee which is payable by a party under any enactment in respect of a claim, employer's contract claim, application or judicial mediation in an Employment Tribunal;

'tribunal office' means any office which has been established for any area in either England and Wales or Scotland and which carries out administrative functions in support of the Tribunal, and in relation to particular proceedings it is the office notified to the parties as dealing with the proceedings;

'unlawful act notice' means a notice under section 21 of the Equality Act 2006;

'Vice President' means a person appointed or nominated in accordance with regulation 6(3) or (4);

'writing' includes writing delivered by means of electronic communication.

(2) Any reference in the Rules to a Tribunal applies to both a full tribunal and to an Employment Judge acting alone (in accordance with section 4(2) or (6) of the Employment Tribunals Act).

(3) An order or other decision of the Tribunal is either—

(a) a "case management order", being an order or decision of any kind in relation to the conduct of proceedings, not including the determination of any issue which would be the subject of a judgment; or

(b) a "judgment", being a decision, made at any stage of the proceedings (but not including a decision under rule 13 or 19), which finally determines—
  (i) a claim, or part of a claim, as regards liability, remedy or costs (including preparation time and wasted costs); . . .
  (ii) any issue which is capable of finally disposing of any claim, or part of a claim, even if it does not necessarily do so (for example, an issue whether a claim should be struck out or a jurisdictional issue);
  [(iii) the imposition of a financial penalty under section 12A of the Employment Tribunals Act].

2 Overriding objective

The overriding objective of these Rules is to enable Employment Tribunals to deal with cases fairly and justly. Dealing with a case fairly and justly includes, so far as practicable—

(a) ensuring that the parties are on an equal footing;
(b) dealing with cases in ways which are proportionate to the complexity and importance of the issues;
(c) avoiding unnecessary formality and seeking flexibility in the proceedings;
(d) avoiding delay, so far as compatible with proper consideration of the issues; and
(e) saving expense.

A Tribunal shall seek to give effect to the overriding objective in interpreting, or exercising any power given to it by, these Rules. The parties and their representatives shall assist the Tribunal to further the overriding objective and in particular shall co-operate generally with each other and with the Tribunal.

3 Alternative dispute resolution

A Tribunal shall wherever practicable and appropriate encourage the use by the parties of the services of ACAS, judicial or other mediation, or other means of resolving their disputes by agreement.

4 Time

(1) Unless otherwise specified by the Tribunal, an act required by these Rules, a practice direction or an order of a Tribunal to be done on or by a particular day may be done at any time before midnight on that day. If there is an issue as to whether the act has been done by that time, the party claiming to have done it shall prove compliance.

(2) If the time specified by these Rules, a practice direction or an order for doing any act ends on a day other than a working day, the act is done in time if it is done on the next working day. "Working day" means any day except a Saturday or Sunday, Christmas Day, Good Friday or a bank holiday under section 1 of the Banking and Financial Dealings Act 1971.

(3) Where any act is required to be, or may be, done within a certain number of days of or from an event, the date of that event shall not be included in the calculation. (For example, a response shall be presented within 28 days of the date on which the respondent was sent a copy of the claim: if the claim was sent on 1st October the last day for presentation of the response is 29th October.)

(4) Where any act is required to be, or may be, done not less than a certain number of days before or after an event, the date of that event shall not be included in the calculation. (For example, if a party wishes to present representations in writing for consideration by a Tribunal at a hearing, they shall be presented not less than 7 days before the hearing: if the hearing is fixed for 8th October, the representations shall be presented no later than 1st October.)

(5) Where the Tribunal imposes a time limit for doing any act, the last date for compliance shall, wherever practicable, be expressed as a calendar date.

(6) Where time is specified by reference to the date when a document is sent to a person by the Tribunal, the date when the document was sent shall, unless the contrary is proved, be regarded as the date endorsed on the document as the date of sending or, if there is no such endorsement, the date shown on the letter accompanying the document.

5 Extending or shortening time

The Tribunal may, on its own initiative or on the application of a party, extend or shorten any time limit specified in these Rules or in any decision, whether or not (in the case of an extension) it has expired.

6 Irregularities and non-compliance

A failure to comply with any provision of these Rules (except rule 8(1), 16(1), 23 or 25) or any order of the Tribunal (except for an order under rules 38 or 39) does not of itself render void the proceedings or any step taken in the proceedings. In the case of such non-compliance, the Tribunal may take such action as it considers just, which may include all or any of the following—

- (a) waiving or varying the requirement;
- (b) striking out the claim or the response, in whole or in part, in accordance with rule 37;
- (c) barring or restricting a party's participation in the proceedings;
- (d) awarding costs in accordance with rules 74 to 84.

7 Presidential Guidance

The Presidents may publish guidance for England and Wales and for Scotland, respectively, as to matters of practice and as to how the powers conferred by these Rules may be exercised. Any such guidance shall be published by the Presidents in an appropriate manner to bring it to the attention of claimants, respondents and their advisers. Tribunals must have regard to any such guidance, but they shall not be bound by it.

Starting a claim

8 Presenting the claim

(1) A claim shall be started by presenting a completed claim form (using a prescribed form) in accordance with any practice direction made under regulation 11 which supplements this rule.

(2) A claim may be presented in England and Wales if—

- (a) the respondent, or one of the respondents, resides or carries on business in England and Wales;
- (b) one or more of the acts or omissions complained of took place in England and Wales;
- (c) the claim relates to a contract under which the work is or has been performed partly in England and Wales; or
- (d) the Tribunal has jurisdiction to determine the claim by virtue of a connection with Great Britain and the connection in question is at least partly a connection with England and Wales.

(3) A claim may be presented in Scotland if—

- (a) the respondent, or one of the respondents, resides or carries on business in Scotland;
- (b) one or more of the acts or omissions complained of took place in Scotland;
- (c) the claim relates to a contract under which the work is or has been performed partly in Scotland; or

(d) the Tribunal has jurisdiction to determine the claim by virtue of a connection with Great Britain and the connection in question is at least partly a connection with Scotland.

9 Multiple claimants

Two or more claimants may make their claims on the same claim form if their claims are based on the same set of facts. Where two or more claimants wrongly include claims on the same claim form, this shall be treated as an irregularity falling under rule 6.

10 Rejection: form not used or failure to supply minimum information

(1) The Tribunal shall reject a claim if—

(a) it is not made on a prescribed form; . . .

(b) it does not contain all of the following information—

(i) each claimant's name;

(ii) each claimant's address;

(iii) each respondent's name;

(iv) each respondent's address[; or

(c) it does not contain one of the following—

(i) an early conciliation number;

(ii) confirmation that the claim does not institute any relevant proceedings; or

(iii) confirmation that one of the early conciliation exemptions applies].

(2) The form shall be returned to the claimant with a notice of rejection explaining why it has been rejected. The notice shall contain information about how to apply for a reconsideration of the rejection.

11 Rejection: absence of Tribunal fee or remission application

(1) The Tribunal shall reject a claim if it is not accompanied by a Tribunal fee or a remission application.

(2) Where a claim is accompanied by a Tribunal fee but the amount paid is lower than the amount payable for the presentation of that claim, the Tribunal shall send the claimant a notice specifying a date for payment of the additional amount due and the claim, or part of it in respect of which the relevant Tribunal fee has not been paid, shall be rejected by the Tribunal if the amount due is not paid by the date specified.

(3) If a remission application is refused in part or in full, the Tribunal shall send the claimant a notice specifying a date for payment of the Tribunal fee and the claim shall be rejected by the Tribunal if the Tribunal fee is not paid by the date specified.

(4) If a claim, or part of it, is rejected, the form shall be returned to the claimant with a notice of rejection explaining why it has been rejected.

12 Rejection: substantive defects

(1) The staff of the tribunal office shall refer a claim form to an Employment Judge if they consider that the claim, or part of it, may be—

(a) one which the Tribunal has no jurisdiction to consider; . . .

(b) in a form which cannot sensibly be responded to or is otherwise an abuse of the process;

[(c) one which institutes relevant proceedings and is made on a claim form that does not contain either an early conciliation number or confirmation that one of the early conciliation exemptions applies;

(d) one which institutes relevant proceedings, is made on a claim form which contains confirmation that one of the early conciliation exemptions applies, and an early conciliation exemption does not apply;

(e) one which institutes relevant proceedings and the name of the claimant on the claim form is not the same as the name of the prospective claimant on the early conciliation certificate to which the early conciliation number relates; or

(f) one which institutes relevant proceedings and the name of the respondent on the claim form is not the same as the name of the prospective respondent on the early conciliation certificate to which the early conciliation number relates].

(2) The claim, or part of it, shall be rejected if the Judge considers that the claim, or part of it, is of a kind described in sub-paragraphs (a)[, (b), (c) or (d)] of paragraph (1).

(2A) The claim, or part of it, shall be rejected if the Judge considers that the claim, or part of it, is of a kind described in sub-paragraph (e) or (f) of paragraph (1) unless the Judge considers that the claimant made a minor error in relation to a name or address and it would not be in the interests of justice to reject the claim.]

(3) If the claim is rejected, the form shall be returned to the claimant together with a notice of rejection giving the Judge's reasons for rejecting the claim, or part of it. The notice shall contain information about how to apply for a reconsideration of the rejection.

13 Reconsideration of rejection

(1) A claimant whose claim has been rejected (in whole or in part) under rule 10 or 12 may apply for a reconsideration on the basis that either—

(a) the decision to reject was wrong; or

(b) the notified defect can be rectified.

(2) The application shall be in writing and presented to the Tribunal within 14 days of the date that the notice of rejection was sent. It shall explain why the decision is said to have been wrong or rectify the defect and if the claimant wishes to request a hearing this shall be requested in the application.

(3) If the claimant does not request a hearing, or an Employment Judge decides, on considering the application, that the claim shall be accepted in full, the Judge shall determine the application without a hearing. Otherwise the application shall be considered at a hearing attended only by the claimant.

(4) If the Judge decides that the original rejection was correct but that the defect has been rectified, the claim shall be treated as presented on the date that the defect was rectified.

14 Protected disclosure claims: notification to a regulator

If a claim alleges that the claimant has made a protected disclosure, the Tribunal may, with the consent of the claimant, send a copy of any accepted claim to a regulator listed in Schedule 1 to the Public Interest Disclosure (Prescribed Persons) Order 1999. "Protected disclosure" has the meaning given to it by section 43A of the Employment Rights Act 1996.

The response to the claim

15 Sending claim form to respondents

Unless the claim is rejected, the Tribunal shall send a copy of the claim form, together with a prescribed response form, to each respondent with a notice which includes information on—

(a) whether any part of the claim has been rejected; and

(b) how to submit a response to the claim, the time limit for doing so and what will happen if a response is not received by the Tribunal within that time limit.

16 Response

(1) The response shall be on a prescribed form and presented to the tribunal office within 28 days of the date that the copy of the claim form was sent by the Tribunal.

(2) A response form may include the response of more than one respondent if they are responding to a single claim and either they all resist the claim on the same grounds or they do not resist the claim.

(3) A response form may include the response to more than one claim if the claims are based on the same set of facts and either the respondent resists all of the claims on the same grounds or the respondent does not resist the claims.

17 Rejection: form not used or failure to supply minimum information

(1) The Tribunal shall reject a response if—

(a) it is not made on a prescribed form; or

(b) it does not contain all of the following information—

(i) the respondent's full name;

(ii) the respondent's address;

(iii) whether the respondent wishes to resist any part of the claim.

(2) The form shall be returned to the respondent with a notice of rejection explaining why it has been rejected. The notice shall explain what steps may be taken by the respondent, including the need (if appropriate) to apply for an extension of time, and how to apply for a reconsideration of the rejection.

18 Rejection: form presented late

(1) A response shall be rejected by the Tribunal if it is received outside the time limit in rule 16 (or any extension of that limit granted within the original limit) unless an application for extension has already been made under rule 20 or the response includes or is accompanied by such an application (in which case the response shall not be rejected pending the outcome of the application).

(2) The response shall be returned to the respondent together with a notice of rejection explaining that the response has been presented late. The notice shall explain how the respondent can apply for an extension of time and how to apply for a reconsideration.

19 Reconsideration of rejection

(1) A respondent whose response has been rejected under rule 17 or 18 may apply for a reconsideration on the basis that the decision to reject was wrong or, in the case of a rejection under rule 17, on the basis that the notified defect can be rectified.

(2) The application shall be in writing and presented to the Tribunal within 14 days of the date that the notice of rejection was sent. It shall explain why the decision is said to have been wrong or rectify the defect and it shall state whether the respondent requests a hearing.

(3) If the respondent does not request a hearing, or the Employment Judge decides, on considering the application, that the response shall be accepted in full, the Judge shall determine the application without a hearing. Otherwise the application shall be considered at a hearing attended only by the respondent.

(4) If the Judge decides that the original rejection was correct but that the defect has been rectified, the response shall be treated as presented on the date that the defect was rectified (but the Judge may extend time under rule 5).

20 Applications for extension of time for presenting response

(1) An application for an extension of time for presenting a response shall be presented in writing and copied to the claimant. It shall set out the reason why the extension is sought and shall, except where the time limit has not yet expired, be accompanied by a draft of the response which the respondent wishes to present or an explanation of why that is not possible and if the respondent wishes to request a hearing this shall be requested in the application.

(2) The claimant may within 7 days of receipt of the application give reasons in writing explaining why the application is opposed.

(3) An Employment Judge may determine the application without a hearing.

(4) If the decision is to refuse an extension, any prior rejection of the response shall stand. If the decision is to allow an extension, any judgment issued under rule 21 shall be set aside.

21 Effect of non-presentation or rejection of response, or case not contested

(1) Where on the expiry of the time limit in rule 16 no response has been presented, or any response received has been rejected and no application for a reconsideration is outstanding, or where the respondent has stated that no part of the claim is contested, paragraphs (2) and (3) shall apply.

(2) An Employment Judge shall decide whether on the available material (which may include further information which the parties are required by a Judge to provide), a determination can properly be made of the claim, or part of it. To the extent that a determination can be made, the Judge shall issue a judgment accordingly. Otherwise, a hearing shall be fixed before a Judge alone.

(3) The respondent shall be entitled to notice of any hearings and decisions of the Tribunal but, unless and until an extension of time is granted, shall only be entitled to participate in any hearing to the extent permitted by the Judge.

22 Notification of acceptance

Where the Tribunal accepts the response it shall send a copy of it to all other parties.

Employer's contract claim

23 Making an employer's contract claim

Any employer's contract claim shall be made as part of the response, presented in accordance with rule 16, to a claim which includes an employee's contract claim. An employer's contract claim may be rejected on the same basis as a claimant's claim may be rejected under rule 12, in which case rule 13 shall apply.

24 Notification of employer's contract claim

When the Tribunal sends the response to the other parties in accordance with rule 22 it shall notify the claimant that the response includes an employer's contract claim and include information on how to submit a response to the claim, the time limit for doing so, and what will happen if a response is not received by the Tribunal within that time limit.

25 Responding to an employer's contract claim

A claimant's response to an employer's contract claim shall be presented to the tribunal office within 28 days of the date that the response was sent to the claimant. If no response is presented within that time limit, rules 20 and 21 shall apply.

Initial consideration of claim form and response

26 Initial consideration

(1) As soon as possible after the acceptance of the response, the Employment Judge shall consider all of the documents held by the Tribunal in relation to the claim, to confirm whether there are arguable complaints and defences within the jurisdiction of the Tribunal (and for that purpose the Judge may order a party to provide further information).

(2) Except in a case where notice is given under rule 27 or 28, the Judge conducting the initial consideration shall make a case management order (unless made already), which may deal with the listing of a preliminary or final hearing, and may propose judicial mediation or other forms of dispute resolution.

27 Dismissal of claim (or part)

(1) If the Employment Judge considers either that the Tribunal has no jurisdiction to consider the claim, or part of it, or that the claim, or part of it, has no reasonable prospect of success, the Tribunal shall send a notice to the parties—

(a) setting out the Judge's view and the reasons for it; and

(b) ordering that the claim, or the part in question, shall be dismissed on such date as is specified in the notice unless before that date the claimant has presented written representations to the Tribunal explaining why the claim (or part) should not be dismissed.

(2) If no such representations are received, the claim shall be dismissed from the date specified without further order (although the Tribunal shall write to the parties to confirm what has occurred).

(3) If representations are received within the specified time they shall be considered by an Employment Judge, who shall either permit the claim (or part) to proceed or fix a hearing for the purpose of deciding whether it should be permitted to do so. The respondent may, but need not, attend and participate in the hearing.

(4) If any part of the claim is permitted to proceed the Judge shall make a case management order.

28 Dismissal of response (or part)

(1) If the Employment Judge considers that the response to the claim, or part of it, has no reasonable prospect of success the Tribunal shall send a notice to the parties—

(a) setting out the Judge's view and the reasons for it;

(b) ordering that the response, or the part in question, shall be dismissed on such date as is specified in the notice unless before that date the respondent has presented written representations to the Tribunal explaining why the response (or part) should not be dismissed; and

(c) specifying the consequences of the dismissal of the response, in accordance with paragraph (5) below.

(2) If no such representations are received, the response shall be dismissed from the date specified without further order (although the Tribunal shall write to the parties to confirm what has occurred).

(3) If representations are received within the specified time they shall be considered by an Employment Judge, who shall either permit the response (or part) to stand or fix a hearing for the purpose of deciding whether it should be permitted to do so. The claimant may, but need not, attend and participate in the hearing.

(4) If any part of the response is permitted to stand the Judge shall make a case management order.

(5) Where a response is dismissed, the effect shall be as if no response had been presented, as set out in rule 21 above.

Case management orders and other powers

29 Case management orders

The Tribunal may at any stage of the proceedings, on its own initiative or on application, make a case management order. The particular powers identified in the following rules do not restrict that general power. A case management order may vary, suspend or set aside an earlier case management order where that is necessary in the interests of justice, and in particular where a party affected by the earlier order did not have a reasonable opportunity to make representations before it was made.

30 Applications for case management orders

(1) An application by a party for a particular case management order may be made either at a hearing or presented in writing to the Tribunal.

(2) Where a party applies in writing, they shall notify the other parties that any objections to the application should be sent to the Tribunal as soon as possible.

(3) The Tribunal may deal with such an application in writing or order that it be dealt with at a preliminary or final hearing.

31 Disclosure of documents and information

The Tribunal may order any person in Great Britain to disclose documents or information to a party (by providing copies or otherwise) or to allow a party to inspect such material as might be ordered by a county court or, in Scotland, by a sheriff.

32 Requirement to attend to give evidence

The Tribunal may order any person in Great Britain to attend a hearing to give evidence, produce documents, or produce information.

33 Evidence from other EU Member States

The Tribunal may use the procedures for obtaining evidence prescribed in Council Regulation (EC) No [1206/2001] of 28 May 2001 on cooperation between the courts of the Member States in the taking of evidence in civil or commercial matters.

34 Addition, substitution and removal of parties

The Tribunal may on its own initiative, or on the application of a party or any other person wishing to become a party, add any person as a party, by way of substitution or otherwise, if it appears that there are issues between that person and any of the existing parties falling within the jurisdiction of the Tribunal which it is in the interests of justice to have determined in the proceedings; and may remove any party apparently wrongly included.

35 Other persons

The Tribunal may permit any person to participate in proceedings, on such terms as may be specified, in respect of any matter in which that person has a legitimate interest.

36 Lead cases

(1) Where a Tribunal considers that two or more claims give rise to common or related issues of fact or law, the Tribunal or the President may make an order specifying one or more of those claims as a lead case and staying, or in Scotland sisting, the other claims ("the related cases").

(2) When the Tribunal makes a decision in respect of the common or related issues it shall send a copy of that decision to each party in each of the related cases and, subject to paragraph (3), that decision shall be binding on each of those parties.

(3) Within 28 days after the date on which the Tribunal sent a copy of the decision to a party under paragraph (2), that party may apply in writing for an order that the decision does not apply to, and is not binding on the parties to, a particular related case.

(4) If a lead case is withdrawn before the Tribunal makes a decision in respect of the common or related issues, it shall make an order as to—

- (a) whether another claim is to be specified as a lead case; and
- (b) whether any order affecting the related cases should be set aside or varied.

37 Striking out

(1) At any stage of the proceedings, either on its own initiative or on the application of a party, a Tribunal may strike out all or part of a claim or response on any of the following grounds—

- (a) that it is scandalous or vexatious or has no reasonable prospect of success;

(b) that the manner in which the proceedings have been conducted by or on behalf of the claimant or the respondent (as the case may be) has been scandalous, unreasonable or vexatious;

(c) for non-compliance with any of these Rules or with an order of the Tribunal;

(d) that it has not been actively pursued;

(e) that the Tribunal considers that it is no longer possible to have a fair hearing in respect of the claim or response (or the part to be struck out).

(2) A claim or response may not be struck out unless the party in question has been given a reasonable opportunity to make representations, either in writing or, if requested by the party, at a hearing.

(3) Where a response is struck out, the effect shall be as if no response had been presented, as set out in rule 21 above.

38 Unless orders

(1) An order may specify that if it is not complied with by the date specified the claim or response, or part of it, shall be dismissed without further order. If a claim or response, or part of it, is dismissed on this basis the Tribunal shall give written notice to the parties confirming what has occurred.

(2) A party whose claim or response has been dismissed, in whole or in part, as a result of such an order may apply to the Tribunal in writing, within 14 days of the date that the notice was sent, to have the order set aside on the basis that it is in the interests of justice to do so. Unless the application includes a request for a hearing, the Tribunal may determine it on the basis of written representations.

(3) Where a response is dismissed under this rule, the effect shall be as if no response had been presented, as set out in rule 21.

39 Deposit orders

(1) Where at a preliminary hearing (under rule 53) the Tribunal considers that any specific allegation or argument in a claim or response has little reasonable prospect of success, it may make an order requiring a party ("the paying party") to pay a deposit not exceeding £1,000 as a condition of continuing to advance that allegation or argument.

(2) The Tribunal shall make reasonable enquiries into the paying party's ability to pay the deposit and have regard to any such information when deciding the amount of the deposit.

(3) The Tribunal's reasons for making the deposit order shall be provided with the order and the paying party must be notified about the potential consequences of the order.

(4) If the paying party fails to pay the deposit by the date specified the specific allegation or argument to which the deposit order relates shall be struck out. Where a response is struck out, the consequences shall be as if no response had been presented, as set out in rule 21.

(5) If the Tribunal at any stage following the making of a deposit order decides the specific allegation or argument against the paying party for substantially the reasons given in the deposit order—

(a) the paying party shall be treated as having acted unreasonably in pursuing that specific allegation or argument for the purpose of rule 76, unless the contrary is shown; and

(b) the deposit shall be paid to the other party (or, if there is more than one, to such other party or parties as the Tribunal orders),

otherwise the deposit shall be refunded.

(6) If a deposit has been paid to a party under paragraph (5)(b) and a costs or preparation time order has been made against the paying party in favour of the party

who received the deposit, the amount of the deposit shall count towards the settlement of that order.

40 Non-payment of fees

(1) Subject to rule 11, where a party has not paid a relevant Tribunal fee or presented a remission application in respect of that fee the Tribunal will send the party a notice specifying a date for payment of the Tribunal fee or presentation of a remission application.

(2) If at the date specified in a notice sent under paragraph (1) the party has not paid the Tribunal fee and no remission application in respect of that fee has been presented—

- (a) where the Tribunal fee is payable in relation to a claim, the claim shall be dismissed without further order;
- (b) where the Tribunal fee is payable in relation to an employer's contract claim, the employer's contract claim shall be dismissed without further order;
- (c) where the Tribunal fee is payable in relation to an application, the application shall be dismissed without further order;
- (d) where the Tribunal fee is payable in relation to judicial mediation, the judicial mediation shall not take place.

(3) Where a remission application is refused in part or in full, the Tribunal shall send the claimant a notice specifying a date for payment of the Tribunal fee.

(4) If at the date specified in a notice sent under paragraph (3) the party has not paid the Tribunal fee, the consequences shall be those referred to in sub-paragraphs (a) to (d) of paragraph (2).

(5) In the event of a dismissal under paragraph (2) or (4) a party may apply for the claim or response, or part of it, which was dismissed to be reinstated and the Tribunal may order a reinstatement. A reinstatement shall be effective only if the Tribunal fee is paid, or a remission application is presented and accepted, by the date specified in the order.

Rules common to all kinds of hearing

41 General

The Tribunal may regulate its own procedure and shall conduct the hearing in the manner it considers fair, having regard to the principles contained in the overriding objective. The following rules do not restrict that general power. The Tribunal shall seek to avoid undue formality and may itself question the parties or any witnesses so far as appropriate in order to clarify the issues or elicit the evidence. The Tribunal is not bound by any rule of law relating to the admissibility of evidence in proceedings before the courts.

42 Written representations

The Tribunal shall consider any written representations from a party, including a party who does not propose to attend the hearing, if they are delivered to the Tribunal and to all other parties not less than 7 days before the hearing.

43 Witnesses

Where a witness is called to give oral evidence, any witness statement of that person ordered by the Tribunal shall stand as that witness's evidence in chief unless the Tribunal orders otherwise. Witnesses shall be required to give their oral evidence on oath or affirmation. The Tribunal may exclude from the hearing any person who is to appear as a witness in the proceedings until such time as that person gives evidence if it considers it in the interests of justice to do so.

44 Inspection of witness statements

Subject to rules 50 and 94, any witness statement which stands as evidence in chief shall be available for inspection during the course of the hearing by members of the public attending the hearing unless the Tribunal decides that all or any part of the statement is not to be admitted as evidence, in which case the statement or that part shall not be available for inspection.

45 Timetabling

A Tribunal may impose limits on the time that a party may take in presenting evidence, questioning witnesses or making submissions, and may prevent the party from proceeding beyond any time so allotted.

46 Hearings by electronic communication

A hearing may be conducted, in whole or in part, by use of electronic communication (including by telephone) provided that the Tribunal considers that it would be just and equitable to do so and provided that the parties and members of the public attending the hearing are able to hear what the Tribunal hears and see any witness as seen by the Tribunal.

47 Non-attendance

If a party fails to attend or to be represented at the hearing, the Tribunal may dismiss the claim or proceed with the hearing in the absence of that party. Before doing so, it shall consider any information which is available to it, after any enquiries that may be practicable, about the reasons for the party's absence.

48 Conversion from preliminary hearing to final hearing and vice versa

A Tribunal conducting a preliminary hearing may order that it be treated as a final hearing, or vice versa, if the Tribunal is properly constituted for the purpose and if it is satisfied that neither party shall be materially prejudiced by the change.

49 Majority decisions

Where a Tribunal is composed of three persons any decision may be made by a majority and if it is composed of two persons the Employment Judge has a second or casting vote.

50 Privacy and restrictions on disclosure

(1) A Tribunal may at any stage of the proceedings, on its own initiative or on application, make an order with a view to preventing or restricting the public disclosure of any aspect of those proceedings so far as it considers necessary in the interests of justice or in order to protect the Convention rights of any person or in the circumstances identified in section 10A of the Employment Tribunals Act.

(2) In considering whether to make an order under this rule, the Tribunal shall give full weight to the principle of open justice and to the Convention right to freedom of expression.

(3) Such orders may include—

- (a) an order that a hearing that would otherwise be in public be conducted, in whole or in part, in private;
- (b) an order that the identities of specified parties, witnesses or other persons referred to in the proceedings should not be disclosed to the public, by the use of anonymisation or otherwise, whether in the course of any hearing or in its listing or in any documents entered on the Register or otherwise forming part of the public record;
- (c) an order for measures preventing witnesses at a public hearing being identifiable by members of the public;
- (d) a restricted reporting order within the terms of section 11 or 12 of the Employment Tribunals Act.

(4) Any party, or other person with a legitimate interest, who has not had a reasonable opportunity to make representations before an order under this rule is

made may apply to the Tribunal in writing for the order to be revoked or discharged, either on the basis of written representations or, if requested, at a hearing.

(5) Where an order is made under paragraph (3)(d) above—

- (a) it shall specify the person whose identity is protected; and may specify particular matters of which publication is prohibited as likely to lead to that person's identification;
- (b) it shall specify the duration of the order;
- (c) the Tribunal shall ensure that a notice of the fact that such an order has been made in relation to those proceedings is displayed on the notice board of the Tribunal with any list of the proceedings taking place before the Tribunal, and on the door of the room in which the proceedings affected by the order are taking place; and
- (d) the Tribunal may order that it applies also to any other proceedings being heard as part of the same hearing.

(6) "Convention rights" has the meaning given to it in section 1 of the Human Rights Act 1998.

Withdrawal

51 End of claim

Where a claimant informs the Tribunal, either in writing or in the course of a hearing, that a claim, or part of it, is withdrawn, the claim, or part, comes to an end, subject to any application that the respondent may make for a costs, preparation time or wasted costs order.

52 Dismissal following withdrawal

Where a claim, or part of it, has been withdrawn under rule 51, the Tribunal shall issue a judgment dismissing it (which means that the claimant may not commence a further claim against the respondent raising the same, or substantially the same, complaint) unless—

- (a) the claimant has expressed at the time of withdrawal a wish to reserve the right to bring such a further claim and the Tribunal is satisfied that there would be legitimate reason for doing so; or
- (b) the Tribunal believes that to issue such a judgment would not be in the interests of justice.

Preliminary hearings

53 Scope of preliminary hearings

(1) A preliminary hearing is a hearing at which the Tribunal may do one or more of the following—

- (a) conduct a preliminary consideration of the claim with the parties and make a case management order (including an order relating to the conduct of the final hearing);
- (b) determine any preliminary issue;
- (c) consider whether a claim or response, or any part, should be struck out under rule 37;
- (d) make a deposit order under rule 39;
- (e) explore the possibility of settlement or alternative dispute resolution (including judicial mediation).

(2) There may be more than one preliminary hearing in any case.

(3) "Preliminary issue" means, as regards any complaint, any substantive issue which may determine liability (for example, an issue as to jurisdiction or as to whether an employee was dismissed).

54 Fixing of preliminary hearings

A preliminary hearing may be directed by the Tribunal on its own initiative following its initial consideration (under rule 26) or at any time thereafter or as the result of an application by a party. The Tribunal shall give the parties reasonable notice of the date of the hearing and in the case of a hearing involving any preliminary issues at least 14 days notice shall be given and the notice shall specify the preliminary issues that are to be, or may be, decided at the hearing.

55 Constitution of tribunal for preliminary hearings

Preliminary hearings shall be conducted by an Employment Judge alone, except that where notice has been given that any preliminary issues are to be, or may be, decided at the hearing a party may request in writing that the hearing be conducted by a full tribunal in which case an Employment Judge shall decide whether that would be desirable.

56 When preliminary hearings shall be in public

Preliminary hearings shall be conducted in private, except that where the hearing involves a determination under rule 53(1)(b) or (c), any part of the hearing relating to such a determination shall be in public (subject to rules 50 and 94) and the Tribunal may direct that the entirety of the hearing be in public.

Final hearing

57 Scope of final hearing

A final hearing is a hearing at which the Tribunal determines the claim or such parts as remain outstanding following the initial consideration (under rule 26) or any preliminary hearing. There may be different final hearings for different issues (for example, liability, remedy or costs).

58 Notice of final hearing

The Tribunal shall give the parties not less than 14 days' notice of the date of a final hearing.

59 When final hearing shall be in public

Any final hearing shall be in public, subject to rules 50 and 94.

Decisions and reasons

60 Decisions made without a hearing

Decisions made without a hearing shall be communicated in writing to the parties, identifying the Employment Judge who has made the decision.

61 Decisions made at or following a hearing

(1) Where there is a hearing the Tribunal may either announce its decision in relation to any issue at the hearing or reserve it to be sent to the parties as soon as practicable in writing.

(2) If the decision is announced at the hearing, a written record (in the form of a judgment if appropriate) shall be provided to the parties (and, where the proceedings were referred to the Tribunal by a court, to that court) as soon as practicable. (Decisions concerned only with the conduct of a hearing need not be identified in the record of that hearing unless a party requests that a specific decision is so recorded.)

(3) The written record shall be signed by the Employment Judge.

62 Reasons

(1) The Tribunal shall give reasons for its decision on any disputed issue, whether substantive or procedural (including any decision on an application for reconsideration or for orders for costs, preparation time or wasted costs).

(2) In the case of a decision given in writing the reasons shall also be given in writing. In the case of a decision announced at a hearing the reasons may be given orally at the hearing or reserved to be given in writing later (which may, but need not, be as part of the written record of the decision). Written reasons shall be signed by the Employment Judge.

(3) Where reasons have been given orally, the Employment Judge shall announce that written reasons will not be provided unless they are asked for by any party at the hearing itself or by a written request presented by any party within 14 days of the sending of the written record of the decision. The written record of the decision shall repeat that information. If no such request is received, the Tribunal shall provide written reasons only if requested to do so by the Employment Appeal Tribunal or a court.

(4) The reasons given for any decision shall be proportionate to the significance of the issue and for decisions other than judgments may be very short.

(5) In the case of a judgment the reasons shall: identify the issues which the Tribunal has determined, state the findings of fact made in relation to those issues, concisely identify the relevant law, and state how that law has been applied to those findings in order to decide the issues. Where the judgment includes a financial award the reasons shall identify, by means of a table or otherwise, how the amount to be paid has been calculated.

63 Absence of Employment Judge

If it is impossible or not practicable for the written record or reasons to be signed by the Employment Judge as a result of death, incapacity or absence, it shall be signed by the other member or members (in the case of a full tribunal) or by the President, Vice President or a Regional Employment Judge (in the case of a Judge sitting alone).

64 Consent orders and judgments

If the parties agree in writing or orally at a hearing upon the terms of any order or judgment a Tribunal may, if it thinks fit, make such order or judgment, in which case it shall be identified as having been made by consent.

65 When a judgment or order takes effect

A judgment or order takes effect from the day when it is given or made, or on such later date as specified by the Tribunal.

66 Time for compliance

A party shall comply with a judgment or order for the payment of an amount of money within 14 days of the date of the judgment or order, unless—

- (a) the judgment, order, or any of these Rules, specifies a different date for compliance; or
- (b) the Tribunal has stayed (or in Scotland sisted) the proceedings or judgment.

67 The Register

Subject to rules 50 and 94, a copy shall be entered in the Register of any judgment and of any written reasons for a judgment.

68 Copies of judgment for referring court

Where the proceedings were referred to the Tribunal by a court a copy of any judgment and of any written reasons shall be provided to that court.

69 Correction of clerical mistakes and accidental slips

An Employment Judge may at any time correct any clerical mistake or other accidental slip or omission in any order, judgment or other document produced by a Tribunal. If such a correction is made, any published version of the document shall also be corrected. If any document is corrected under this rule, a copy of the corrected version, signed by the Judge, shall be sent to all the parties.

Reconsideration of judgments

70 Principles

A Tribunal may, either on its own initiative (which may reflect a request from the Employment Appeal Tribunal) or on the application of a party, reconsider any judgment where it is necessary in the interests of justice to do so. On reconsideration, the decision ("the original decision") may be confirmed, varied or revoked. If it is revoked it may be taken again.

71 Application

Except where it is made in the course of a hearing, an application for reconsideration shall be presented in writing (and copied to all the other parties) within 14 days of the date on which the written record, or other written communication, of the original decision was sent to the parties or within 14 days of the date that the written reasons were sent (if later) and shall set out why reconsideration of the original decision is necessary.

72 Process

(1) An Employment Judge shall consider any application made under rule 71. If the Judge considers that there is no reasonable prospect of the original decision being varied or revoked (including, unless there are special reasons, where substantially the same application has already been made and refused), the application shall be refused and the Tribunal shall inform the parties of the refusal. Otherwise the Tribunal shall send a notice to the parties setting a time limit for any response to the application by the other parties and seeking the views of the parties on whether the application can be determined without a hearing. The notice may set out the Judge's provisional views on the application.

(2) If the application has not been refused under paragraph (1), the original decision shall be reconsidered at a hearing unless the Employment Judge considers, having regard to any response to the notice provided under paragraph (1), that a hearing is not necessary in the interests of justice. If the reconsideration proceeds without a hearing the parties shall be given a reasonable opportunity to make further written representations.

(3) Where practicable, the consideration under paragraph (1) shall be by the Employment Judge who made the original decision or, as the case may be, chaired the full tribunal which made it; and any reconsideration under paragraph (2) shall be made by the Judge or, as the case may be, the full tribunal which made the original decision. Where that is not practicable, the President, Vice President or a Regional Employment Judge shall appoint another Employment Judge to deal with the application or, in the case of a decision of a full tribunal, shall either direct that the reconsideration be by such members of the original Tribunal as remain available or reconstitute the Tribunal in whole or in part.

73 Reconsideration by the Tribunal on its own initiative

Where the Tribunal proposes to reconsider a decision on its own initiative, it shall inform the parties of the reasons why the decision is being reconsidered and the decision shall be reconsidered in accordance with rule 72(2) (as if an application had been made and not refused).

Costs, preparation time orders and wasted costs orders

74 Definitions

(1) "Costs" means fees, charges, disbursements or expenses incurred by or on behalf of the receiving party (including expenses that witnesses incur for the purpose of, or in connection with, attendance at a Tribunal hearing). In Scotland all references to costs (except when used in the expression "wasted costs") shall be read as references to expenses.

(2) "Legally represented" means having the assistance of a person (including where that person is the receiving party's employee) who—

(a) has a right of audience in relation to any class of proceedings in any part of the Senior Courts of England and Wales, or all proceedings in county courts or magistrates' courts;

(b) is an advocate or solicitor in Scotland; or

(c) is a member of the Bar of Northern Ireland or a solicitor of the Court of Judicature of Northern Ireland.

(3) "Represented by a lay representative" means having the assistance of a person who does not satisfy any of the criteria in paragraph (2) and who charges for representation in the proceedings.

75 Costs orders and preparation time orders

(1) A costs order is an order that a party ("the paying party") make a payment to—

(a) another party ("the receiving party") in respect of the costs that the receiving party has incurred while legally represented or while represented by a lay representative;

(b) the receiving party in respect of a Tribunal fee paid by the receiving party; or

(c) another party or a witness in respect of expenses incurred, or to be incurred, for the purpose of, or in connection with, an individual's attendance as a witness at the Tribunal.

(2) A preparation time order is an order that a party ("the paying party") make a payment to another party ("the receiving party") in respect of the receiving party's preparation time while not legally represented. "Preparation time" means time spent by the receiving party (including by any employees or advisers) in working on the case, except for time spent at any final hearing.

(3) A costs order under paragraph (1)(a) and a preparation time order may not both be made in favour of the same party in the same proceedings. A Tribunal may, if it wishes, decide in the course of the proceedings that a party is entitled to one order or the other but defer until a later stage in the proceedings deciding which kind of order to make.

76 When a costs order or a preparation time order may or shall be made

(1) A Tribunal may make a costs order or a preparation time order, and shall consider whether to do so, where it considers that—

(a) a party (or that party's representative) has acted vexatiously, abusively, disruptively or otherwise unreasonably in either the bringing of the proceedings (or part) or the way that the proceedings (or part) have been conducted; or

(b) any claim or response had no reasonable prospect of success.

(2) A Tribunal may also make such an order where a party has been in breach of any order or practice direction or where a hearing has been postponed or adjourned on the application of a party.

(3) Where in proceedings for unfair dismissal a final hearing is postponed or adjourned, the Tribunal shall order the respondent to pay the costs incurred as a result of the postponement or adjournment if—

(a) the claimant has expressed a wish to be reinstated or re-engaged which has been communicated to the respondent not less than 7 days before the hearing; and

(b) the postponement or adjournment of that hearing has been caused by the respondent's failure, without a special reason, to adduce reasonable evidence as to the availability of the job from which the claimant was dismissed or of comparable or suitable employment.

(4) A Tribunal may make a costs order of the kind described in rule 75(1)(b) where a party has paid a Tribunal fee in respect of a claim, employer's contract claim or application and that claim, counterclaim or application is decided in whole, or in part, in favour of that party.

(5) A Tribunal may make a costs order of the kind described in rule 75(1)(c) on the application of a party or the witness in question, or on its own initiative, where a witness has attended or has been ordered to attend to give oral evidence at a hearing.

77 Procedure

A party may apply for a costs order or a preparation time order at any stage up to 28 days after the date on which the judgment finally determining the proceedings in respect of that party was sent to the parties. No such order may be made unless the paying party has had a reasonable opportunity to make representations (in writing or at a hearing, as the Tribunal may order) in response to the application.

78 The amount of a costs order

(1) A costs order may—

(a) order the paying party to pay the receiving party a specified amount, not exceeding £20,000, in respect of the costs of the receiving party;

(b) order the paying party to pay the receiving party the whole or a specified part of the costs of the receiving party, with the amount to be paid being determined, in England and Wales, by way of detailed assessment carried out either by a county court in accordance with the Civil Procedure Rules 1998, or by an Employment Judge applying the same principles; or, in Scotland, by way of taxation carried out either by the auditor of court in accordance with the Act of Sederunt (Fees of Solicitors in the Sheriff Court)(Amendment and Further Provisions) 1993, or by an Employment Judge applying the same principles;

(c) order the paying party to pay the receiving party a specified amount as reimbursement of all or part of a Tribunal fee paid by the receiving party;

(d) order the paying party to pay another party or a witness, as appropriate, a specified amount in respect of necessary and reasonably incurred expenses (of the kind described in rule 75(1)(c)); or

(e) if the paying party and the receiving party agree as to the amount payable, be made in that amount.

(2) Where the costs order includes an amount in respect of fees charged by a lay representative, for the purposes of the calculation of the order, the hourly rate applicable for the fees of the lay representative shall be no higher than the rate under rule 79(2).

(3) For the avoidance of doubt, the amount of a costs order under sub-paragraphs (b) to (e) of paragraph (1) may exceed £20,000.

79 The amount of a preparation time order

(1) The Tribunal shall decide the number of hours in respect of which a preparation time order should be made, on the basis of—

(a) information provided by the receiving party on time spent falling within rule 75(2) above; and

(b) the Tribunal's own assessment of what it considers to be a reasonable and proportionate amount of time to spend on such preparatory work, with reference to such matters as the complexity of the proceedings, the number of witnesses and documentation required.

(2) The hourly rate is £33 and increases on 6 April each year by £1.

(3) The amount of a preparation time order shall be the product of the number of hours assessed under paragraph (1) and the rate under paragraph (2).

80 When a wasted costs order may be made

(1) A Tribunal may make a wasted costs order against a representative in favour of any party ("the receiving party") where that party has incurred costs—

- (a) as a result of any improper, unreasonable or negligent act or omission on the part of the representative; or
- (b) which, in the light of any such act or omission occurring after they were incurred, the Tribunal considers it unreasonable to expect the receiving party to pay.

Costs so incurred are described as "wasted costs".

(2) "Representative" means a party's legal or other representative or any employee of such representative, but it does not include a representative who is not acting in pursuit of profit with regard to the proceedings. A person acting on a contingency or conditional fee arrangement is considered to be acting in pursuit of profit.

(3) A wasted costs order may be made in favour of a party whether or not that party is legally represented and may also be made in favour of a representative's own client. A wasted costs order may not be made against a representative where that representative is representing a party in his or her capacity as an employee of that party.

81 Effect of a wasted costs order

A wasted costs order may order the representative to pay the whole or part of any wasted costs of the receiving party, or disallow any wasted costs otherwise payable to the representative, including an order that the representative repay to its client any costs which have already been paid. The amount to be paid, disallowed or repaid must in each case be specified in the order.

82 Procedure

A wasted costs order may be made by the Tribunal on its own initiative or on the application of any party. A party may apply for a wasted costs order at any stage up to 28 days after the date on which the judgment finally determining the proceedings as against that party was sent to the parties. No such order shall be made unless the representative has had a reasonable opportunity to make representations (in writing or at a hearing, as the Tribunal may order) in response to the application or proposal. The Tribunal shall inform the representative's client in writing of any proceedings under this rule and of any order made against the representative.

83 Allowances

Where the Tribunal makes a costs, preparation time, or wasted costs order, it may also make an order that the paying party (or, where a wasted costs order is made, the representative) pay to the Secretary of State, in whole or in part, any allowances (other than allowances paid to members of the Tribunal) paid by the Secretary of State under section 5(2) or (3) of the Employment Tribunals Act to any person for the purposes of, or in connection with, that person's attendance at the Tribunal.

84 Ability to pay

In deciding whether to make a costs, preparation time, or wasted costs order, and if so in what amount, the Tribunal may have regard to the paying party's (or, where a wasted costs order is made, the representative's) ability to pay.

Delivery of documents

85 Delivery to the Tribunal

(1) Subject to paragraph (2), documents may be delivered to the Tribunal—

- (a) by post;
- (b) by direct delivery to the appropriate tribunal office (including delivery by a courier or messenger service); or

(c) by electronic communication.

(2) A claim form may only be delivered in accordance with the practice direction made under regulation 11 which supplements rule 8.

(3) The Tribunal shall notify the parties following the presentation of the claim of the address of the tribunal office dealing with the case (including any fax or email or other electronic address) and all documents shall be delivered to either the postal or the electronic address so notified. The Tribunal may from time to time notify the parties of any change of address, or that a particular form of communication should or should not be used, and any documents shall be delivered in accordance with that notification.

86 Delivery to parties

(1) Documents may be delivered to a party (whether by the Tribunal or by another party)—

(a) by post;

(b) by direct delivery to that party's address (including delivery by a courier or messenger service);

(c) by electronic communication; or

(d) by being handed personally to that party, if an individual and if no representative has been named in the claim form or response; or to any individual representative named in the claim form or response; or, on the occasion of a hearing, to any person identified by the party as representing that party at that hearing.

(2) For the purposes of sub-paragraphs (a) to (c) of paragraph (1), the document shall be delivered to the address given in the claim form or response (which shall be the address of the party's representative, if one is named) or to a different address as notified in writing by the party in question.

(3) If a party has given both a postal address and one or more electronic addresses, any of them may be used unless the party has indicated in writing that a particular address should or should not be used.

87 Delivery to non-parties

Subject to the special cases which are the subject of rule 88, documents shall be sent to non-parties at any address for service which they may have notified and otherwise at any known address or place of business in the United Kingdom or, if the party is a corporate body, at its registered or principal office in the United Kingdom or, if permitted by the President, at an address outside the United Kingdom.

88 Special cases

Addresses for serving the Secretary of State, the Law Officers, and the Counsel General to the Welsh Assembly Government, in cases where they are not parties, shall be issued by practice direction.

89 Substituted service

Where no address for service in accordance with the above rules is known or it appears that service at any such address is unlikely to come to the attention of the addressee, the President, Vice President or a Regional Employment Judge may order that there shall be substituted service in such manner as appears appropriate.

90 Date of delivery

Where a document has been delivered in accordance with rule 85 or 86, it shall, unless the contrary is proved, be taken to have been received by the addressee—

(a) if sent by post, on the day on which it would be delivered in the ordinary course of post;

(b) if sent by means of electronic communication, on the day of transmission;

(c) if delivered directly or personally, on the day of delivery.

91 Irregular service

A Tribunal may treat any document as delivered to a person, notwithstanding any non-compliance with rules 86 to 88, if satisfied that the document in question, or its substance, has in fact come to the attention of that person.

92 Correspondence with the Tribunal: copying to other parties

Where a party sends a communication to the Tribunal (except an application under rule 32) it shall send a copy to all other parties, and state that it has done so (by use of "cc" or otherwise). The Tribunal may order a departure from this rule where it considers it in the interests of justice to do so.

Miscellaneous

93 ACAS

(1) Where proceedings concern an enactment which provides for conciliation, the Tribunal shall—

(a) send a copy of the claim form and the response to an ACAS conciliation officer; and

(b) inform the parties that the services of an ACAS conciliation officer are available to them.

(2) Subject to rules 50 and 94, a representative of ACAS may attend any preliminary hearing.

94 National security proceedings

(1) Where in relation to particular Crown employment proceedings a Minister considers that it would be expedient in the interests of national security, the Minister may direct a Tribunal to—

(a) conduct all or part of the proceedings in private;

(b) exclude a person from all or part of the proceedings;

(c) take steps to conceal the identity of a witness in the proceedings.

(2) Where the Tribunal considers it expedient in the interests of national security, it may order—

(a) in relation to particular proceedings (including Crown employment proceedings), anything which can be required to be done under paragraph (1);

(b) a person not to disclose any document (or the contents of any document), where provided for the purposes of the proceedings, to any other person (save for any specified person).

Any order made must be kept under review by the Tribunal.

(3) Where the Tribunal considers that it may be necessary to make an order under paragraph (2) in relation to particular proceedings (including Crown employment proceedings), the Tribunal may consider any material provided by a party (or where a Minister is not a party, by a Minister) without providing that material to any other person. Such material shall be used by the Tribunal solely for the purposes of deciding whether to make that order (unless that material is subsequently used as evidence in the proceedings by a party).

(4) Where a Minister considers that it would be appropriate for the Tribunal to make an order under paragraph (2), the Minister may make an application for such an order.

(5) Where a Minister has made an application under paragraph (4), the Tribunal may order—

(a) in relation to the part of the proceedings preceding the outcome of the application, anything which can be required to be done under paragraph (1);

(b) a person not to disclose any document (or the contents of any document) to any other person (save for any specified person), where provided for the purposes of the proceedings preceding the outcome of the application.

(6) Where a Minister has made an application under paragraph (4) for an order to exclude any person from all or part of the proceedings, the Tribunal shall not send a copy of the response to that person, pending the decision on the application.

(7) If before the expiry of the time limit in rule 16 a Minister makes a direction under paragraph (1) or makes an application under paragraph (4), the Minister may apply for an extension of the time limit in rule 16.

(8) A direction under paragraph (1) or an application under paragraph (4) may be made irrespective of whether or not the Minister is a party.

(9) Where the Tribunal decides not to make an order under paragraph (2), rule 6 of Schedule 2 shall apply to the reasons given by the Tribunal under rule 62 for that decision, save that the reasons will not be entered on the Register.

(10) The Tribunal must ensure that in exercising its functions, information is not disclosed contrary to the interests of national security.

95 Interim relief proceedings

When a Tribunal hears an application for interim relief (or for its variation or revocation) under section 161 or section 165 of the Trade Union and Labour Relations (Consolidation) Act 1992 or under section 128 or section 131 of the Employment Rights Act 1996, rules 53 to 56 apply to the hearing and the Tribunal shall not hear oral evidence unless it directs otherwise.

96 Proceedings involving the National Insurance Fund

The Secretary of State shall be entitled to appear and be heard at any hearing in relation to proceedings which may involve a payment out of the National Insurance Fund and shall be treated as a party for the purposes of these Rules.

97 Collective agreements

Where a claim includes a complaint under section 146(1) of the Equality Act relating to a term of a collective agreement, the following persons, whether or not identified in the claim, shall be regarded as the persons against whom a remedy is claimed and shall be treated as respondents for the purposes of these Rules—

(a) the claimant's employer (or prospective employer); and

(b) every organisation of employers and organisation of workers, and every association of or representative of such organisations, which, if the terms were to be varied voluntarily, would be likely, in the opinion of an Employment Judge, to negotiate the variation.

An organisation or association shall not be treated as a respondent if the Judge, having made such enquiries of the claimant and such other enquiries as the Judge thinks fit, is of the opinion that it is not reasonably practicable to identify the organisation or association.

98 Devolution issues

(1) Where a devolution issue arises, the Tribunal shall as soon as practicable send notice of that fact and a copy of the claim form and response to the Advocate General for Scotland and the Lord Advocate, where it is a Scottish devolution issue, or to the Attorney General and the Counsel General to the Welsh Assembly Government, where it is a Welsh devolution issue, unless they are a party to the proceedings.

(2) A person to whom notice is sent may be treated as a party to the proceedings, so far as the proceedings relate to the devolution issue, if that person sends notice to the Tribunal within 14 days of receiving a notice under paragraph (1).

(3) Any notices sent under paragraph (1) or (2) must at the same time be sent to the parties.

(4) "Devolution issue" has the meaning given to it in paragraph 1 of Schedule 6 to the Scotland Act 1998 (for the purposes of a Scottish devolution issue), and in paragraph 1 of Schedule 9 to the Government of Wales Act 2006 (for the purposes of a Welsh devolution issue).

99 Transfer of proceedings between Scotland and England & Wales

(1) The President (England and Wales) or a Regional Employment Judge may at any time, on their own initiative or on the application of a party, with the consent of the President (Scotland), transfer to a tribunal office in Scotland any proceedings started in England and Wales which could (in accordance with rule 8(3)) have been started in Scotland and which in that person's opinion would more conveniently be determined there.

(2) The President (Scotland) or the Vice President may at any time, on their own initiative or on the application of a party, with the consent of the President (England and Wales), transfer to a tribunal office in England and Wales any proceedings started in Scotland which could (in accordance with rule 8(2)) have been started in England and Wales and in that person's opinion would more conveniently be determined there.

100 References to the Court of Justice of the European Union

Where a Tribunal decides to refer a question to the Court of Justice of the European Union for a preliminary ruling under Article 267 of the Treaty on the Functioning of the European Union, a copy of that decision shall be sent to the registrar of that court.

101 Transfer of proceedings from a court

Where proceedings are referred to a Tribunal by a court, these Rules apply as if the proceedings had been presented by the claimant.

102 Vexatious litigants

The Tribunal may provide any information or documents requested by the Attorney General, the Solicitor General or the Lord Advocate for the purpose of preparing an application or considering whether to make an application under section 42 of the Senior Courts Act 1981, section 1 of the Vexatious Actions (Scotland) Act 1898 or section 33 of the Employment Tribunals Act.

103 Information to the Commission for Equality and Human Rights

The Tribunal shall send to the Commission for Equality and Human Rights copies of all judgments and written reasons relating to complaints under section 120, 127 or 146 of the Equality Act. That obligation shall not apply in any proceedings where a Minister of the Crown has given a direction, or a Tribunal has made an order, under rule 94; and either the Security Service, the Secret Intelligence Service or the Government Communications Headquarters is a party to the proceedings.

104 Application of this Schedule to levy appeals

For the purposes of a levy appeal, references in this Schedule to a claim or claimant shall be read as references to a levy appeal or to an appellant in a levy appeal respectively.

105 Application of this Schedule to appeals against improvement and prohibition notices

(1) A person ("the appellant") may appeal an improvement notice or a prohibition notice by presenting a claim to a tribunal office—

- (a) before the end of the period of 21 days beginning with the date of the service on the appellant of the notice which is the subject of the appeal; or
- (b) within such further period as the Tribunal considers reasonable where it is satisfied that it was not reasonably practicable for an appeal to be presented within that time.

(2) For the purposes of an appeal against an improvement notice or a prohibition notice, this Schedule shall be treated as modified in the following ways—

(a) references to a claim or claimant shall be read as references to an appeal or to an appellant in an appeal respectively;

(b) references to a respondent shall be read as references to the inspector appointed under section 19(1) of the Health and Safety Act who issued the notice which is the subject of the appeal.

106 Application of this Schedule to appeals against unlawful act notices

For the purposes of an appeal against an unlawful act notice, this Schedule shall be treated as modified in the following ways—

(a) references in this Schedule to a claim or claimant shall be read as references to a notice of appeal or to an appellant in an appeal against an unlawful act notice respectively;

(b) references to a respondent shall be read as references to the Commission for Equality and Human Rights.

[3.1248]–[3.1250]

Statutory annotations

Amendment

Rule 1: in para (1) definition 'early conciliation certificate' inserted by SI 2014/271, regs 4, 5; in para (1) definition 'early conciliation exemption' inserted by SI 2014/271, regs 4, 5; in para (1) definition 'early conciliation number' inserted by SI 2014/271, regs 4, 5; in para (1) definition 'relevant proceedings' inserted by SI 2014/271, regs 4, 5; in para (3)(b)(i) word omitted revoked by SI 2014/271, regs 4, 6(a); para (3)(b)(iii) inserted by SI 2014/271, regs 4, 6(b).

Rule 10: in para (1)(a) word omitted revoked by SI 2014/271, regs 4, 7(a); para (1)(c) and word '; or' immediately preceding it inserted by SI 2014/271, regs 4, 7(b), (c).

Rule 12: in para (1)(a) word omitted revoked by SI 2014/271, regs 4, 8(a); para (1)(c)–(f) inserted by SI 2014/271, regs 4, 8(b); in para (2) words ', (b), (c) or (d)' in square brackets substituted by SI 2014/271, regs 4, 8(c); para (2A) inserted by SI 2014/271, regs 4, 8(d).

Rule 33: reference to '1206/2001' in square brackets substituted by SI 2014/271, regs 4, 9.

Reporting of Injuries, Diseases and Dangerous Occurrences Regulations 2013

SI 2013/1471

[3.1251]

The Secretary of State makes these Regulations in exercise of the powers conferred by sections 15(1), (2), (3)(a), (4)(a), (5)(a) and (b), (6)(b), (9) and 52(2) of, and paragraphs 15(1), 16 and 20 of Schedule 3 to, the Health and Safety at Work etc Act 1974 ("the 1974 Act").

The Regulations give effect without modification to proposals submitted to the Secretary of State by the Health and Safety Executive ("the Executive") under section 11(3) of the 1974 Act.

Before submitting those proposals to the Secretary of State, the Executive consulted the bodies that appeared to it to be appropriate as required by section 50(3) of the 1974 Act.

[3.1252]

1 Citation and commencement

(1) These Regulations may be cited as the Reporting of Injuries, Diseases and Dangerous Occurrences Regulations 2013.

(2) Subject to paragraph (3), these Regulations come into force on 1st October 2013.

(3) Regulation 18(2) comes into force immediately after the coming into force of the other regulations.

[3.1253]

2 Interpretation

(1) In these Regulations—

"the 1954 Act" means the Mines and Quarries Act 1954;

"the 1969 Act" means the Mines and Quarries (Tips) Act 1969;

"the 1974 Act" means the Health and Safety at Work etc Act 1974;

"the 1999 Regulations" means the Quarries Regulations 1999;

"the 2002 Regulations" means the Control of Substances Hazardous to Health Regulations 2002;

"the 2006 Regulations" means the Railways and Other Guided Transport Systems (Safety) Regulations 2006;

"the 2013 Order" means the Health and Safety at Work etc Act 1974 (Application Outside Great Britain) Order 2013;

"accident" includes an act of non-consensual physical violence done to a person at work;

"approved manner" means published in a form considered appropriate and approved for the time being for the purposes of these Regulations—

(a) by the Executive; or

(b) in relation to activities covered by regulation 3 of the Health and Safety (Enforcing Authority for Railways and Other Guided Transport Systems) Regulations 2006, by the ORR;

"biological agent" has the meaning given by regulation 2(1) of the 2002 Regulations;

"carcinogen" has the meaning given by regulation 2(1) of the 2002 Regulations;

"consecutive days" includes any days which are not or would not have been working days;

"construction site" has the meaning given by regulation 2(1) of the Construction (Design and Management) Regulations 2007;

"dangerous occurrence" means an occurrence which arises out of or in connection with work and is of a class specified in—

(a) Part 1 of Schedule 2;

(b) Part 2 of Schedule 2 and takes place anywhere except an offshore workplace;

(c) Part 3 of Schedule 2 and takes place at a mine;

(d) Part 4 of Schedule 2 and takes place at a quarry;

(e) Part 5 of Schedule 2 and takes place where a relevant transport system is operated; or

(f) Part 6 of Schedule 2 and takes place at an offshore workplace;

"disease" includes a medical condition;

"diagnosis" means a registered medical practitioner's identification (in writing, where it pertains to an employee) of—

(a) new symptoms; or

(b) symptoms which have significantly worsened;

"diving contractor" and "diving project" have the meanings they are given by regulation 2(1) of the Diving at Work Regulations 1997;

"dock" means any place to which section 125(1) of the Factories Act 1961 applies;

"the Executive" means the Health and Safety Executive;

"explosives" has the meaning given by regulation 2(1) of the Manufacture and Storage of Explosives Regulations 2005;

"factory" has the meaning given by section 175 of the Factories Act 1961;

"flammable gas" and "flammable liquid" have the meanings associated with those hazard classes in Part 2 of Annex I of Regulation (EC) No 1272/2008 (the CLP Regulation);

"mine" has the meaning given by section 180 of the 1954 Act and for the purposes of these Regulations includes a closed tip within the meaning of section 2(2)(b) of the 1969 Act which is associated with such a mine;

"mutagen" has the meaning given by regulation 2(1) of the 2002 Regulations;

"nominated person" means, in relation to a mine or quarry, the person (if any) who is for the time being nominated—

(a) in a case where there is an association or body representative of a majority of the total number of persons employed at a mine or quarry, by that association or body; or

(b) in any other case, jointly by associations or bodies which are together representative of such a majority,

to receive notices under paragraph 4 of Part 1 of Schedule 1 on behalf of the persons employed at that mine or quarry;

"non-passenger train" means any train except a passenger train;

"offshore installation" has the meaning given by article 4(2) of the 2013 Order;

"offshore workplace" means any place where activities are carried on, or any premises, such that prescribed provisions of the 1974 Act are applied to those activities or premises by article 4, 5 or 6 of the 2013 Order (which for this purpose are deemed to apply to activities or premises within Great Britain which are in tidal waters or on the foreshore or other land intermittently covered by such waters as they apply to activities or premises within territorial waters or a designated area within the meaning of article 2(1) of that Order);

"operator" means—

(a) in relation to a pipeline, the person identified as such by regulation 2(1) of the Pipelines Safety Regulations 1996; and
(b) in relation to a quarry, the person in overall control of the working of the quarry;

"the ORR" means the Office of Rail Regulation;

"owner" in relation to a mine means the person who is for the time being entitled to work it;

"passenger train" means a train carrying passengers or made available for that purpose;

"pipeline" and "pipeline works" have the meanings given by article 6(2) of the 2013 Order;

"quarry" means a quarry to which the 1999 Regulations apply;

"railway" has the meaning given by regulation 2(1) of the 2006 Regulations;

"relevant transport system" means—

(a) a railway;
(b) a tramway as defined by regulation 2(1) of the 2006 Regulations;
(c) a trolley vehicle system as defined by section 67 of the Transport and Works Act 1992, except when it operates on a road; or
(d) any other system using guided transport as defined by regulation 2(1) of the 2006 Regulations,

except at a factory, dock, construction site, mine or quarry, and does not include a guided bus system as defined by regulation 2(1) of the 2006 Regulations;

"reportable incident" means an incident giving rise to a notification or reporting requirement under these Regulations;

"reporting procedure" means, in relation to—

(a) an injury, death or dangerous occurrence (except at a mine or quarry), the procedure described in paragraph 1 of Part 1 of Schedule 1;
(b) an occupational disease or a disease offshore, the procedure described in paragraph 2 of Part 1 of Schedule 1;
(c) exposure to a carcinogen, mutagen or biological agent, the procedure described in paragraph 3 of Part 1 of Schedule 1; or
(d) an injury, death or dangerous occurrence at a mine or quarry, the procedure described in paragraph 4 of Part 1 of Schedule 1;

"responsible person" means the person identified in accordance with regulation 3;

"road" includes bridges over which a road passes, and—

(a) in relation to England and Wales, means any highway and any other road to which the public has access;
(b) in relation to Scotland, means any road within the meaning of the Roads (Scotland) Act 1984 and any other way to which the public has access;

"road vehicle" means any vehicle on a road, other than a train;

"routine work" means work which a person might reasonably be expected to do, either under that person's contract of employment, or, if there is no such contract, in the normal course of that person's work;

"running line" means any line ordinarily used for the passage of trains which is not a siding;

"specified injury" means any injury or condition specified in regulation 4(1)(a) to (h);

"train" includes a locomotive, tramcar or other power unit, and any vehicle used on a relevant transport system;

"well" includes any structures and devices on top of a well;

"workmen's inspectors" means workmen's inspectors exercising the powers conferred on them by either section 123 of the 1954 Act or regulation 40 of the 1999 Regulations;

"work-related accident" means an accident arising out of or in connection with work.

(2) In these Regulations, any reference to a work-related accident or dangerous occurrence includes an accident or dangerous occurrence attributable to—

(a) the manner of conducting an undertaking;

(b) the plant or substances used for the purposes of an undertaking; or

(c) the condition of the premises used for the purposes of an undertaking or any part of them.

(3) For the purposes of these Regulations, a person at an offshore workplace is deemed to be at work at all times when that person is at that workplace in connection with that person's work.

[3.1254]

3 Responsible person

(1) In these Regulations, the "responsible person" is—

(a) in relation to an injury, death or dangerous occurrence reportable under regulation 4, 5, 6 or 7 or recordable under regulation 12(1)(b) involving—

(i) an employee, that employee's employer; or

(ii) a person not at work or a self-employed person, or in relation to any other dangerous occurrence, the person who by means of their carrying on any undertaking was in control of the premises where the reportable or recordable incident happened, at the time it happened; or

(b) in relation to a diagnosis reportable under regulation 8, 9 or 10 in respect of—

(i) an employee, that employee's employer; or

(ii) a self-employed person, that self-employed person.

(2) Despite paragraph (1), in these Regulations the "responsible person" is—

(a) in relation to a mine, the manager of that mine;

(b) in relation to a closed tip, the owner of the mine with which that tip is associated;

(c) in relation to a quarry, the operator of that quarry;

(d) in relation to a dangerous occurrence—

(i) at a pipeline, the operator of that pipeline; or

(ii) at a well, the person appointed to organise and supervise the drilling of, and operations using, that well by any person granted a licence under section 3 of the Petroleum Act 1998, or where no such person is appointed, that licensee; or

(e) except in relation to a diagnosis reportable under regulation 8, 9 or 10—

(i) at an offshore installation, the duty holder for the purposes of the Offshore Installations and Pipeline Works (Management and Administration) Regulations 1995 (provided that for the purposes of this provision regulation 3(2)(c) of those Regulations is deemed not to apply); or

(ii) in relation to a diving project, the diving contractor.

[3.1255]

4 Non-fatal injuries to workers

(1) Where any person at work, as a result of a work-related accident, suffers—

(a) any bone fracture diagnosed by a registered medical practitioner, other than to a finger, thumb or toe;

(b) amputation of an arm, hand, finger, thumb, leg, foot or toe;

(c) any injury diagnosed by a registered medical practitioner as being likely to cause permanent blinding or reduction in sight in one or both eyes;

(d) any crush injury to the head or torso causing damage to the brain or internal organs in the chest or abdomen;

(e) any burn injury (including scalding) which—

(i) covers more than 10% of the whole body's total surface area; or

(ii) causes significant damage to the eyes, respiratory system or other vital organs;

(f) any degree of scalping requiring hospital treatment;

(g) loss of consciousness caused by head injury or asphyxia; or

(h) any other injury arising from working in an enclosed space which—

(i) leads to hypothermia or heat-induced illness; or

(ii) requires resuscitation or admittance to hospital for more than 24 hours,

the responsible person must follow the reporting procedure.

(2) Where any person at work is incapacitated for routine work for more than seven consecutive days (excluding the day of the accident) because of an injury resulting from an accident arising out of or in connection with that work, the responsible person must send a report to the relevant enforcing authority in an approved manner as soon as practicable and in any event within 15 days of the accident.

(3) This regulation is subject to regulations 14 and 15.

[3.1256]

5 Non-fatal injuries to non-workers

Where any person not at work, as a result of a work-related accident, suffers—

(a) an injury, and that person is taken from the site of the accident to a hospital for treatment in respect of that injury; or

(b) a specified injury on hospital premises,

the responsible person must follow the reporting procedure, subject to regulations 14 and 15.

[3.1257]

6 Work-related fatalities

(1) Where any person dies as a result of a work-related accident, the responsible person must follow the reporting procedure.

(2) Where any person dies as a result of occupational exposure to a biological agent, the responsible person must follow the reporting procedure.

(3) Where an employee has suffered an injury reportable under regulation 4 which is a cause of his death within one year of the date of the accident, the employer must notify the relevant enforcing authority of the death in an approved manner without delay, whether or not the injury has been reported under regulation 4.

(4) This regulation is subject to regulations 14 and 15, and does not apply to a self-employed person who suffers a fatal accident or fatal exposure on premises controlled by that self-employed person.

[3.1258]

7 Dangerous occurrences

Where there is a dangerous occurrence, the responsible person must follow the reporting procedure, subject to regulations 14 and 15.

[3.1259]

8 Occupational diseases

Where, in relation to a person at work, the responsible person receives a diagnosis of—

(a) Carpal Tunnel Syndrome, where the person's work involves regular use of percussive or vibrating tools;

(b) cramp in the hand or forearm, where the person's work involves prolonged periods of repetitive movement of the fingers, hand or arm;

(c) occupational dermatitis, where the person's work involves significant or regular exposure to a known skin sensitizer or irritant;

(d) Hand Arm Vibration Syndrome, where the person's work involves regular use of percussive or vibrating tools, or the holding of materials which are subject to percussive processes, or processes causing vibration;

(e) occupational asthma, where the person's work involves significant or regular exposure to a known respiratory sensitizer; or

(f) tendonitis or tenosynovitis in the hand or forearm, where the person's work is physically demanding and involves frequent, repetitive movements,

the responsible person must follow the reporting procedure, subject to regulations 14 and 15.

[3.1260]

9 Exposure to carcinogens, mutagens and biological agents

Where, in relation to a person at work, the responsible person receives a diagnosis of—

(a) any cancer attributed to an occupational exposure to a known human carcinogen or mutagen (including ionising radiation); or

(b) any disease attributed to an occupational exposure to a biological agent,

the responsible person must follow the reporting procedure, subject to regulations 14 and 15.

[3.1261]

10 Diseases offshore

Where, in relation to a person at an offshore workplace, the responsible person receives a diagnosis of any of the diseases listed in Schedule 3, the responsible person must follow the reporting procedure, subject to regulations 14 and 15.

[3.1262]

11 Gas-related injuries and hazards

(1) Where a conveyor of flammable gas through a fixed pipe distribution system, or a filler, importer or supplier (except by retail) of a refillable container containing liquefied petroleum gas, receives notification of the death, loss of consciousness or taking to hospital of a person because of an injury arising in connection with that gas, that person must—

(a) notify the Executive of the incident without delay; and

(b) send a report of the incident to the Executive in an approved manner within 14 days of the incident.

(2) Where an approved person has sufficient information to decide that the design, construction, manner of installation, modification or servicing of a gas fitting is or could have been likely to cause the death, loss of consciousness or taking to hospital of a person because of—

(a) the accidental leakage of gas;

(b) the incomplete combustion of gas; or

(c) the inadequate removal of the products of combustion of gas,

the approved person must send a report of that information to the Executive in an approved manner within 14 days of acquiring that information.

(3) Nothing is reportable—

(a) under this regulation, if it is notifiable or reportable elsewhere in these Regulations;

(b) under paragraph (2), in relation to any gas fitting undergoing testing or examination at a place set aside for that purpose; or

(c) under paragraph (2), if the approved person has previously reported that information.

(4) In this regulation—

"approved person" means an employer or self-employed person who is a member of a class of persons approved by the Executive for the purposes of regulation 3(3) of the Gas Safety (Installation and Use) Regulations 1998;

"gas fitting" means a gas fitting defined in those Regulations or any flue or ventilation used in connection with that fitting; and

"liquefied petroleum gas" means commercial butane (that is, a hydrocarbon mixture consisting predominantly of butane, butylene or any mixture of them) or commercial propane (that is, a hydrocarbon mixture consisting predominantly of propane, propylene or any mixture of them) or any mixture of commercial butane and commercial propane.

[3.1263]

12 Recording and record-keeping

(1) The responsible person must keep a record of any—

(a) reportable incident under regulation 4, 5, 6 or 7, which contains the particulars specified in paragraphs 5 to 11 of Part 2 of Schedule 1;

(b) diagnosis reportable under regulation 8, 9 or 10, which contains the particulars specified in paragraphs 12 to 17 of Part 2 of Schedule 1;

(c) injury to a person at work resulting from an accident arising out of or in connection with that work, incapacitating that person for routine work for more than three consecutive days (excluding the day of the accident), which contains the particulars specified in paragraphs 18 to 21 of Part 2 of Schedule 1; and

(d) other particulars approved by the Executive or the ORR for demonstrating compliance with the approved manner of reporting under Part 1 of Schedule 1.

(2) An entry in the record referred to in paragraph (1) must be kept for at least three years from the date on which it was made, and the record must be—

(a) kept at the place where the work to which it relates is carried on, or at the usual place of business of the responsible person; and

(b) in the case of a mine or quarry, available for inspection by any nominated person and workmen's inspectors (excluding any health record of an identifiable individual).

(3) The responsible person must send to the relevant enforcing authority such extracts from the record required to be kept under paragraph (1) as that enforcing authority may require.

(4) Any record of injuries, deaths, dangerous occurrences or diseases which the responsible person keeps for any other purpose satisfies the requirements of paragraph (1) if it covers the injuries recordable under these Regulations and includes the particulars specified in Part 2 of Schedule 1.

[3.1264]

13 Mines, quarries and offshore site disturbance

(1) Where there is a reportable incident under regulation 4, 5 or 6 at a mine, quarry or offshore workplace, or where there is a dangerous occurrence at a mine or quarry, then the place where it happened must not be disturbed, or anything at that place tampered with before—

- (a) the expiration of three clear days after the matter has been notified in accordance with these Regulations; or
- (b) the place has been visited by an inspector and, in the case of a mine or quarry, by workmen's inspectors, if that is sooner.

(2) Nothing in this regulation prohibits any person doing anything by or with the consent of an inspector, or which was necessary—

- (a) in the case of a mine or quarry, to secure the safety of the mine or quarry, or of any person; or
- (b) in the case of an offshore workplace, to secure the safety or integrity of the workplace or of any person, plant, vessel or well.

(3) In relation to a mine or quarry, this regulation does not apply if an appropriate person—

- (a) has taken adequate steps to ascertain that disturbing the site—
  - (i) is unlikely to prejudice any investigation by an inspector into the circumstances of the reportable incident; and
  - (ii) is necessary to secure the safety of any person at the mine or quarry or to avoid disrupting the normal working of the mine or quarry;
- (b) has notified any nominated person, or any person designated in writing by a nominated person to receive any such notification, of the proposed disturbance, and gives such a person a reasonable opportunity to visit the site before it is disturbed (except in the case of a non-fatal accident or dangerous occurrence where any nominated person or person designated by them cannot be contacted within a reasonable time);
- (c) has taken adequate steps to ensure that such information is obtained as will enable the preparation, without delay, of a full and accurate plan showing the position of any equipment or item relevant to the reportable incident immediately after it happened;
- (d) ensures that that plan is signed by the person who prepared it and bears the date on which it was prepared, and that a copy of that plan is supplied on request to any nominated person or any inspector; and
- (e) ensures that any equipment or item relevant to the reportable incident is kept as it was immediately after the incident, until an inspector agrees that it may be disposed of.

(4) In paragraph (3), "appropriate person" means the responsible person or—

- (a) in the case of a coal mine, a person appointed in the management structure of that mine established pursuant to regulation 10(1) of the Management and Administration of Safety and Health at Mines Regulations 1993; or

(b) in the case of a quarry, a person appointed in the management structure of that quarry established pursuant to regulation 8(1) of the 1999 Regulations.

[3.1265]

14 Restrictions on the application of regulations 4 to 10

(1) Where the injury or death of a person arises out of the conduct of any operation on, or any examination or other medical treatment of, that person (such operation, examination or other treatment being conducted by or under the supervision of a registered medical practitioner or a registered dentist), the requirements of regulations 4, 5, 6(1) and 12(1)(b) do not apply.

(2) In paragraph (1), "registered dentist" has the meaning given by section 53(1) of the Dentists Act 1984.

(3) Where the injury or death of a person arises out of or in connection with the movement of a vehicle on a road, the requirements of regulations 4, 5, 6 and 12(1)(b) do not apply, unless that person—

(a) was injured or killed by an accident involving a train;

(b) was injured or killed by exposure to a substance being conveyed by the vehicle;

(c) was engaged in work connected with the loading or unloading of any article or substance onto or off the vehicle at the time of the accident, or was injured or killed by the activities of another person who was so engaged; or

(d) was engaged in, or was injured or killed by the activities of another person who was at the time of the accident engaged in, work on or alongside a road.

(4) In paragraph (3)(d), "work on or alongside a road" means work concerned with the construction, demolition, alteration, repair or maintenance of—

(a) the road or the markings or equipment on the road;

(b) the verges, fences, hedges or other boundaries of the road;

(c) pipes or cables on, under, over or adjacent to the road; or

(d) buildings or structures adjacent to or over the road.

(5) The injury, death or diagnosis of a member of the armed forces of the Crown or of a visiting force, on duty at the time, is not subject to the requirements of regulation 4, 6, 8, 9, 10 or 12(1)(b) (and for the purposes of this paragraph a visiting force has the meaning given by section 12(1) of the Visiting Forces Act 1952).

(6) Except in relation to an offshore workplace, regulations 4 to 9 do not apply to anything which must be notified under—

(a) the Nuclear Installations Act 1965;

(b) the Merchant Shipping Act 1988;

(c) Orders and Regulations made or to be made under the enactments in (a) and (b);

(d) the Civil Aviation (Investigation of Air Accidents and Incidents) Regulations 1996;

(e) the Ionising Radiations Regulations 1999;

(f) the Electricity Safety, Quality and Continuity Regulations 2002; or

(g) the Civil Aviation (Investigation of Military Air Accidents at Civil Aerodromes) Regulations 2005.

[3.1266]

15 Restriction on parallel requirements

(1) Where the responsible person is under more than one requirement to make a notification under these Regulations, only one notification is required if the conditions in paragraph (3) are met.

(2) Where the responsible person is under more than one requirement to make a report under these Regulations, only one report is required if the conditions in paragraph (3) are met.

(3) The conditions referred to in paragraphs (1) and (2) are—

(a) the facts giving rise to each requirement are identical;

(b) the information required to be provided by each requirement is provided;

(c) where the requirements have different time limits, the shortest time limit is complied with; and

(d) in the case of a mine or quarry, all steps referred to in paragraph 4 of Part 1 of Schedule 1 are complied with.

(4) Where the responsible person is under more than one requirement to keep a record under these Regulations, only one record is required if the facts giving rise to each requirement are identical and the particulars required by each requirement are contained in the record.

[3.1267]

16 Defence

In proceedings against any person for failing to comply with a requirement of these Regulations, it is a defence for that person to prove that they were not aware of the circumstances which gave rise to that requirement, so long as that person had taken all reasonable steps to be made aware, in sufficient time, of such circumstances.

[3.1268]

17 Certificates of exemption

(1) Subject to paragraph (2)—

(a) the Executive; and

(b) in relation to activities covered by regulation 3 of the Health and Safety (Enforcing Authority for Railways and Other Guided Transport Systems) Regulations 2006, the ORR,

may exempt any person or class of persons from any requirement of these Regulations by a certificate in writing, and any such exemption may be granted subject to conditions and with or without limit of time and may be revoked by a certificate in writing at any time.

(2) Such an exemption may not be granted by the Executive or the ORR unless, having regard to the circumstances of the case and, in particular, to—

(a) the conditions, if any, which it proposes to attach to the exemption; and

(b) any other requirements imposed by or under any relevant enactments,

it is satisfied that the health and safety of persons who are likely to be affected by the exemption will not be prejudiced in consequence of it.

[3.1269]

18 Revocations, amendments and savings

(1) The revocations listed in Table 1 of Schedule 4 have effect.

(2) The amendments listed in Table 2 of Schedule 4 have effect.

(3) Any record or register required to be kept under any instrument revoked by these Regulations must be kept for the same period as if these Regulations had not been made.

[3.1270]

19 Extension outside Great Britain

These Regulations apply to premises and activities outside Great Britain to which sections 1 to 59 and 80 to 82 of the 1974 Act apply by virtue of the 2013 Order as they apply within Great Britain.

[3.1271]

20 Review

(1) The Secretary of State must from time to time—

(a) carry out a review of these Regulations;

(b) set out the conclusions of that review in a report; and

(c) publish the report.

(2) In carrying out the review the Secretary of State must, so far as is reasonable, have regard to how—

(a) Council Directive 89/391/EEC on the introduction of measures to encourage improvements in the safety and health of workers at work (which is implemented by means of regulations 4 and 6);

(b) Council Directive 92/91/EEC concerning the minimum requirements for improving the safety and health protection of workers in the mineral-extracting industries through drilling (which is implemented by means of regulations 4, 6 and 7);

(c) Council Directive 92/104/EEC on the minimum requirements for improving the safety and health protection of workers in surface and underground mineral-extracting industries (which is implemented by means of regulations 4, 6 and 7);

(d) Directive 2000/54/EC of the European Parliament and of the Council on the protection of workers from risks related to exposure to biological agents at work (which is implemented by means of regulations 7 and 9);

(e) Directive 2004/37/EC of the European Parliament and of the Council on the protection of workers from the risks related to exposure to carcinogens or mutagens at work (which is implemented by means of regulation 9);

(f) Directive 2004/49/EC of the European Parliament and of the Council on safety on the Community's railways and amending Council Directive 95/18/EC on the licensing of railway undertakings and Directive 2001/14/EC on the allocation of railway infrastructure capacity and the levying of charges for the use of railway infrastructure and safety certification (which is implemented by means of regulations 4, 6 and 7);

(g) Regulation (EC) No 1338/2008 of the European Parliament and of the Council on Community statistics on public health and health and safety at work (which is implemented by means of regulations 4, 6, 8 and 12); and

(h) Council Directive 2010/32/EU implementing the Framework Agreement on prevention from sharp injuries in the hospital and healthcare sector concluded by HOSPEEM and EPSU (which is implemented by means of regulation 7),

are implemented in other member States.

(3) The report must in particular—

(a) set out the objectives intended to be achieved by the regulatory system established by these Regulations;

(b) assess the extent to which those objectives are achieved; and

(c) assess whether those objectives remain appropriate and, if so, the extent to which they could be achieved with a system that imposes less regulation.

(4) The first report under this regulation must be published before the end of the period of five years beginning with the day on which these Regulations come into force.

(5) Reports under this regulation are afterwards to be published at intervals not exceeding five years.

SCHEDULE 1

REPORTING AND RECORDING PROCEDURES

PART 1

REPORTING PROCEDURE

Regulations 4 to 10

[3.1272]

Injuries, fatalities and dangerous occurrences

1

(1) Where required to follow the reporting procedure by regulation 4, 5, 6 or 7 (except in relation to a mine or quarry), the responsible person must—

(a) notify the relevant enforcing authority of the reportable incident by the quickest practicable means without delay; and

(b) send a report of that incident in an approved manner to the relevant enforcing authority within 10 days of the incident.

(2) Sub-paragraph (1)(a) does not apply to a self-employed person who is injured at premises owned or occupied by that self-employed person, and it is sufficient compliance with sub-paragraph (1)(b) for a self-employed person to make arrangements for the report to be sent to the relevant enforcing authority by some other person.

Diseases

2

(1) Where required to follow the reporting procedure by regulation 8 or 10, the responsible person must send a report of the diagnosis in an approved manner to the relevant enforcing authority without delay.

(2) It is sufficient compliance with sub-paragraph (1) for a self-employed person to make arrangements for the report to be sent to the relevant enforcing authority by some other person.

Carcinogens, mutagens and biological agents

3

Where required to follow the reporting procedure by regulation 9 the responsible person must notify the relevant enforcing authority in an approved manner.

Mines and quarries

4

(1) Where required to follow the reporting procedure by regulation 4, 5, 6 or 7 in the case of a mine or quarry, the responsible person must—

(a) notify the relevant enforcing authority and any nominated person of the reportable incident by the quickest practicable means without delay; and

(b) send a report of that incident in an approved manner—

(i) to any nominated person within seven days of the incident; and

(ii) to the relevant enforcing authority within 10 days of the incident.

(2) Where the responsible person becomes aware of a person subsequently dying as the result of an accident which gave rise to an injury reported in accordance with sub-paragraph (1), the responsible person must notify any nominated person of the death.

PART 2

RECORDS

Regulation 12

[3.1273]

Particulars to be kept in records of any reportable incident under regulations 4 to 7

5

The date and time of the accident or dangerous occurrence.

6

In respect of an accident injuring a person at work, that person's—

(a) full name;

(b) occupation;

(c) injury.

7

In respect of an accident injuring a person not at work, that person's—

(a) full name;

(b) status (for example "passenger", "customer", "visitor" or "bystander"); and

(c) injury,

unless these are not known and it is not reasonably practicable to ascertain them.

8

The place where the accident or dangerous occurrence happened.

9

A brief description of the circumstances in which the accident or dangerous occurrence happened.

10

The date on which the accident or dangerous occurrence was first notified or reported to the relevant enforcing authority.

11

The method by which the accident or dangerous occurrence was first notified or reported.

Particulars to be kept in records of any diagnosis reportable under regulations 8 to 10

12

The date of diagnosis of the disease.

13

The name of the person affected.

14

The occupation of the person affected.

15

The name or nature of the disease.

16

The date on which the disease was first reported to the relevant enforcing authority.

17

The method by which the disease was reported.

Particulars to be kept in records of any injuries to be recorded under regulation 12(1)(c)

18

The date and time of the accident.

19

The following particulars of the injured person—

(a) full name;

(b) occupation;

(c) injury.

20

The place where the accident happened.

21

A brief description of the circumstances in which the accident happened.

SCHEDULE 2

DANGEROUS OCCURRENCES

Regulation 7

PART 1

GENERAL

[3.1274]

Lifting equipment

1

The collapse, overturning or failure of any load-bearing part of any lifting equipment, other than an accessory for lifting.

Pressure systems

2

The failure of any closed vessel or of any associated pipework (other than a pipeline) forming part of a pressure system as defined by regulation 2(1) of the Pressure Systems Safety Regulations 2000, where that failure could cause the death of any person.

Overhead electric lines

3

Any plant or equipment unintentionally coming into—

(a) contact with an uninsulated overhead electric line in which the voltage exceeds 200 volts; or

(b) close proximity with such an electric line, such that it causes an electrical discharge.

Electrical incidents causing explosion or fire

4

Any explosion or fire caused by an electrical short circuit or overload (including those resulting from accidental damage to the electrical plant) which either—

(a) results in the stoppage of the plant involved for more than 24 hours; or

(b) causes a significant risk of death.

Explosives

5

Any unintentional—

(a) fire, explosion or ignition at a site where the manufacture or storage of explosives requires a licence or registration, as the case may be, under regulation 9, 10 or 11 of the Manufacture and Storage of Explosives Regulations 2005; or

(b) explosion or ignition of explosives (unless caused by the unintentional discharge of a weapon, where, apart from that unintentional discharge, the weapon and explosives functioned as they were designed to),

except where a fail-safe device or safe system of work prevented any person being endangered as a result of the fire, explosion or ignition.

6

The misfire of explosives (other than at a mine or quarry, inside a well or involving a weapon) except where a fail-safe device or safe system of work prevented any person being endangered as a result of the misfire.

7

Any explosion, discharge or intentional fire or ignition which causes any injury to a person requiring first-aid or medical treatment, other than at a mine or quarry.

8

(1) The projection of material beyond the boundary of the site on which the explosives are being used, or beyond the danger zone of the site, which caused or might have caused injury, except at a quarry.

(2) In this paragraph, "danger zone" means the area from which persons have been excluded or forbidden to enter to avoid being endangered by any explosion or ignition of explosives.

9

The failure of shots to cause the intended extent of collapse or direction of fall of a structure in any demolition operation.

Biological agents

10

Any accident or incident which results or could have resulted in the release or escape of a biological agent likely to cause severe human infection or illness.

Radiation generators and radiography

11

(1) The malfunction of—

(a) a radiation generator or its ancillary equipment used in fixed or mobile industrial radiography, the irradiation of food or the processing of products by irradiation, which causes it to fail to de-energise at the end of the intended exposure period; or

(b) equipment used in fixed or mobile industrial radiography or gamma irradiation, which causes a radioactive source to fail to return to its safe position by the normal means at the end of the intended exposure period.

(2) In this paragraph, "radiation generator" means any electrical equipment emitting ionising radiation and containing components operating at a potential difference of more than 5kV.

Breathing apparatus

12

The malfunction of breathing apparatus—

(a) where the malfunction causes a significant risk of personal injury to the user; or

(b) during testing immediately prior to use, where the malfunction would have caused a significant risk to the health and safety of the user had it occurred during use,

other than at a mine.

Diving operations

13

The failure, damaging or endangering of—

(a) any life support equipment, including control panels, hoses and breathing apparatus; or

(b) the dive platform, or any failure of the dive platform to remain on station,

which causes a significant risk of personal injury to a diver.

14

The failure or endangering of any lifting equipment associated with a diving operation.

15

The trapping of a diver.

16

Any explosion in the vicinity of a diver.

17

Any uncontrolled ascent or any omitted decompression which causes a significant risk of personal injury to a diver.

Collapse of scaffolding

18

The complete or partial collapse (including falling, buckling or overturning) of—

(a) a substantial part of any scaffold more than 5 metres in height;

(b) any supporting part of any slung or suspended scaffold which causes a working platform to fall (whether or not in use); or

(c) any part of any scaffold in circumstances such that there would be a significant risk of drowning to a person falling from the scaffold.

Train collisions

19

The collision of a train with any other train or vehicle, other than a collision reportable under Part 5 of this Schedule, which could have caused the death, or specified injury, of any person.

Wells

20

In relation to a well (other than a well sunk for the purpose of the abstraction of water)—

(a) a blow-out (which includes any uncontrolled flow of well-fluids from a well);

(b) the coming into operation of a blow-out prevention or diversion system to control flow of well-fluids where normal control procedures fail;

(c) the detection of hydrogen sulphide at a well or in samples of well-fluids where the responsible person did not anticipate its presence in the reservoir drawn on by the well;

(d) the taking of precautionary measures additional to any contained in the original drilling programme where a planned minimum separation distance between adjacent wells was not maintained; or

(e) the mechanical failure of any part of a well whose purpose is to prevent or limit the effect of the unintentional release of fluids from a well or a reservoir being drawn on by a well, or whose failure would cause or contribute to such a release.

Pipelines or pipeline works

21

In relation to a pipeline or pipeline works—

(a) any damage to, accidental or uncontrolled release from or inrush of anything into a pipeline;

(b) the failure of any pipeline isolation device, associated equipment or system; or

(c) the failure of equipment involved with pipeline works,

which could cause personal injury to any person, or which results in the pipeline being shut down for more than 24 hours.

22

The unintentional change in position of a pipeline, or in the subsoil or seabed in the vicinity, which requires immediate attention to safeguard the pipeline's integrity or safety.

PART 2

DANGEROUS OCCURENCES REPORTABLE EXCEPT IN RELATION TO AN OFFSHORE WORKPLACE

[3.1275]

Structural collapse

23

The unintentional collapse or partial collapse of—

(a) any structure, which involves a fall of more than 5 tonnes of material; or

(b) any floor or wall of any place of work,

arising from, or in connection with, ongoing construction work (including demolition, refurbishment and maintenance), whether above or below ground.

24

The unintentional collapse or partial collapse of any falsework.

Explosion or fire

25

Any unintentional explosion or fire in any plant or premises which results in the stoppage of that plant, or the suspension of normal work in those premises, for more than 24 hours.

Release of flammable liquids and gases

26

The sudden, unintentional and uncontrolled release—

(a) inside a building—

(i) of 100 kilograms or more of a flammable liquid;
(ii) of 10 kilograms or more of a flammable liquid at a temperature above its normal boiling point;
(iii) of 10 kilograms or more of a flammable gas; or

(b) in the open air, of 500 kilograms or more of a flammable liquid or gas.

Hazardous escapes of substances

27

The unintentional release or escape of any substance which could cause personal injury to any person other than through the combustion of flammable liquids or gases.

PART 3

DANGEROUS OCCURRENCES REPORTABLE IN RELATION TO A MINE

[3.1276]

Fires or ignition of gas

28

Any outbreak of fire below ground.

29

Any person being caused to leave any place pursuant to regulation 11(1) of the Coal and Other Mines (Fire and Rescue) Regulations 1956 or section 79 of the 1954 Act, as a result of smoke or other indication that a fire may have broken out below ground.

30

Any fire on the surface which endangers the operation of any winding or haulage apparatus installed at a shaft or unwalkable outlet or of any mechanically operated apparatus for producing ventilation below ground.

31

The ignition of any gas (other than in a safety lamp) or dust below ground.

32

The unintentional ignition of any gas in part of a firedamp drainage system on the surface or in an exhauster house.

Escapes of gas with solid matter

33

The violent unintentional escape of gas together with coal or other solid matter into the mine workings.

Failures of plant or equipment

34

The breakage or unintentional uncoupling of any belt, rope, chain, coupling, balance rope, guide rope, rope tensioning system, suspension gear or other gear used for or in connection with—

(a) carrying persons through any shaft or staple shaft;
(b) transporting persons below ground; or
(c) a belt conveyor designated by the mine manager as a man-riding conveyor.

35

The overwinding of—

(a) any conveyance being used for the carriage of persons; or
(b) any other conveyance, which becomes detached from its winding rope.

36

The bringing to rest of any conveyance operated using the friction of a rope on a winding sheave by the apparatus provided—

(a) in the headframe of the shaft; or

(b) in the part of the shaft below the lowest landing for the time being in use,

for the purpose of bringing the conveyance to rest in the event of it being overwound.

37

The stoppage of any ventilating apparatus (other than an auxiliary fan) for over 30 minutes, except for planned maintenance, which causes a reduction in mine ventilation resulting in dangerous levels of noxious or flammable gases.

38

The collapse of any headframe, winding engine house, fan house or storage bunker.

Breathing apparatus

39

The malfunction of, or development of a defect in, breathing apparatus or a smoke helmet or other apparatus serving the same purpose or a self-rescuer where—

(a) the malfunction or defect causes, or is likely to cause, a significant risk of personal injury to the user; or

(b) immediately after use and as a result of its use any person receives first-aid or medical treatment because of that person's unfitness or suspected unfitness.

Emergency escape apparatus

40

The use of any apparatus—

(a) provided at a mine in accordance with regulation 4 of the Mines (Safety of Exit) Regulations 1988; or

(b) used to leave a mine when apparatus and equipment normally so used is unavailable,

other than for the purpose of training and practice.

Inrushes of gas or flowing material

41

The inrush of noxious or flammable gas from old workings.

42

The inrush of water or material which flows when wet from any source.

Insecure tips

43

Any event (including any movement of material or any fire) which indicates that a tip to which Part 1 of the 1969 Act applies is or is likely to become insecure.

Locomotives

44

The bringing to rest of an underground locomotive by means other than its safety circuit protective devices or normal service brakes, when not used for testing purposes.

Falls of ground

45

Any fall of ground which—

(a) results from a failure of an underground support system; and

(b) prevents persons travelling through the area affected by the fall, or otherwise exposes them to danger,

other than one which is part of the normal operations at a mine.

Accidents causing specified injuries

46

Any accident in which any person suffers a specified injury.

PART 4

DANGEROUS OCCURENCES WHICH ARE REPORTABLE IN RELATION TO A QUARRY

[3.1277]

Collapse of storage bunkers

47

The collapse of any storage bunker.

Sinking of craft

48

The sinking of any water-borne craft or hovercraft.

Projection of substances outside quarry

49

(1) Following a blasting operation, the projection of any material beyond the designated danger zone or the projection of any material which caused or might have caused injury.

(2) In this paragraph, "danger zone" means the area determined for each blast under the shotfiring rules required by regulation 25(2)(a)(i) and (b) of the 1999 Regulations.

Misfires

50

Any misfire, as defined by regulation 2(1) of the 1999 Regulations.

Insecure tips

51

Any event (including any movement of material or any fire) which indicates that a tip to which the 1999 Regulations apply is or is likely to become insecure.

Movement of slopes or faces

52

Any movement or failure of an excavated slope or face which—

(a) could cause the death of any person; or

(b) adversely affects any building, contiguous land, transport system, footpath, public utility or service, watercourse, reservoir or area of public access.

Explosion or fire in vehicles or mobile plant

53

Any explosion or fire in—

(a) a dump truck with a load capacity of at least 50 tonnes; or

(b) an excavator with a bucket capacity of at least 5 cubic metres,

which results in the stoppage of that vehicle or plant for more than 24 hours, and which affects—

(i) any place where persons normally work; or

(ii) the route of egress from such a place.

PART 5

DANGEROUS OCCURRENCES WHICH ARE REPORTABLE IN RESPECT OF A RELEVANT TRANSPORT SYSTEM

[3.1278]

Collision or derailment of passenger trains

54

Any collision between a passenger train and another train.

55

The derailment of the whole or part of a passenger train.

Collision or derailment not involving passenger trains

56

Any collision between non-passenger trains—

(a) on a running line, which causes damage to a train; or

(b) in a siding, which causes damage to a train and an obstruction to a running line.

57

The derailment of a non-passenger train—

(a) on a running line, except a derailment during shunting operations which does not obstruct any other running line; or

(b) in a siding, which causes an obstruction to a running line.

Accidents involving any train

58

Any collision between a train and a buffer stop which causes damage to the train, except a collision in a siding.

59

A train striking any cattle or horse, whether or not damage is caused to the train, or striking any other animal which causes damage necessitating immediate temporary or permanent repair (including damage to the windows of the driver's cab but excluding other damage consisting solely in the breakage of glass).

60

A train on a running line striking or being struck by any object which causes damage necessitating immediate temporary or permanent repair (including damage to the windows of the driver's cab but excluding other damage consisting solely in the breakage of glass) or which might have been liable to derail the train.

61

A train, other than one on a railway, striking or being struck by a road vehicle.

62

A passenger train, or a non-passenger train not fitted with continuous self-applying brakes, becoming unintentionally divided.

Failure of train parts

63

The failure of—

(a) an axle;

(b) a wheel or tyre, including a tyre loose on its wheel;

(c) a rope or the rope's fastenings;

(d) a winding plant or equipment involved in working an incline; or

(e) any part of a train which is likely to cause an accident to that or any other train, or to cause personal injury to any person,

which occurs or is discovered whilst the train is on a running line.

Fire

64

Any fire—

(a) in or on any part of a passenger train or a train carrying dangerous goods within the meaning of the Carriage of Dangerous Goods and Use of Transportable Pressure Equipment Regulations 2009;

(b) in or on any part of a non-passenger train which was extinguished by a fire-fighting service;

(c) seriously affecting the functioning of signalling equipment;

(d) affecting the permanent way or works of a relevant transport system which necessitates the suspension of services over any line, or the closure of any part of a station or signal box or other premises, for a period—

(i) of more than 30 minutes in the case of any part of a relevant transport system below ground; and

(ii) in any other case, of more than 1 hour; or

(e) causing damage which could affect the running of a relevant transport system.

Severe electrical arcing or fusing

65

Severe electrical arcing or fusing—

(a) in or on any part of any train; or

(b) which seriously affects the functioning of signalling equipment.

Level crossings

66

Any train striking a road vehicle or gate at a level crossing.

67

Any train running onto a level crossing when not authorised to do so.

68

The failure of equipment at a level crossing which could cause a significant risk of personal injury to users of the road or path crossing the railway.

The permanent way and other works

69

The failure of a rail in a running line or of a rack rail, which results in—

(a) a complete fracture of the rail through its cross-section; or

(b) in a piece becoming detached from the rail which requires the immediate stoppage of traffic or the immediate imposition of a lower speed restriction.

70

The buckle of a running line which requires the immediate stoppage of traffic or the immediate imposition of a lower speed restriction.

71

An aircraft or vehicle of any kind either landing on, running onto or coming to rest across the line, or damaging the line, so as to cause damage—

(a) which obstructs the line; or

(b) to any railway equipment at a level crossing.

72

The runaway of an escalator, lift or passenger conveyor.

73

The following classes of accident where they are likely to cause an accident to a train or a significant risk of personal injury to any person—

(a) the failure of a tunnel, bridge, viaduct, culvert, station or other structure or any part of it including the fixed electrical equipment of an electrified relevant transport system;

(b) any failure in the signalling system which could cause a significant risk to the safe passage of trains other than a failure of a traffic light controlling the movement of vehicles on a road;

(c) a slip of a cutting or of an embankment;

(d) flooding of the permanent way;

(e) the striking of a bridge by a vessel or by a road vehicle or its load; or

(f) the failure of any other portion of the permanent way or works.

Incidents of signals passed without authority

74

Any train, travelling on a running line or entering a running line from a siding, passing a signal displaying a stop aspect without authority, unless the stop aspect was not displayed in sufficient time for the driver to stop safely at the signal.

PART 6

DANGEROUS OCCURRENCES WHICH ARE REPORTABLE IN RESPECT OF AN OFFSHORE WORKPLACE

[3.1279]

Release of petroleum hydrocarbon

75

The unintentional release of petroleum hydrocarbon on or from an offshore installation which—

(a) results in—

(i) a fire or explosion; or

(ii) the taking of action to prevent or limit the consequences of a potential fire or explosion; or

(b) could cause a specified injury to, or the death of, any person.

Fire or explosion

76

Any fire or explosion at an offshore installation, other than one caused by the release of petroleum hydrocarbon, which results in the stoppage of plant or the suspension of normal work.

Release or escape of dangerous substances

77

The unintentional or uncontrolled release or escape of any substance (other than petroleum hydrocarbon) on or from an offshore installation which could cause a significant risk of personal injury to any person.

Collapses

78

Any unintentional collapse or partial collapse of any offshore installation or of any plant on an offshore installation which jeopardises the overall structural integrity of the installation.

Equipment

79

The failure of equipment required to maintain a floating offshore installation on station which could cause a specified injury to, or the death of, any person.

Dropping objects

80

The dropping of any object on an offshore installation or on an attendant vessel or into the water adjacent to an installation or vessel which could cause a specified injury to, or the death of, any person.

Weather damage

81

Any damage to or on an offshore installation caused by adverse weather conditions and which could cause a specified injury to, or the death of, any person.

Collisions

82

Any collision between a vessel or aircraft and an offshore installation which causes damage to the installation, the vessel or the aircraft.

83

Any occurrence with the potential for a collision between a vessel and an offshore installation where, had a collision occurred, it might have jeopardised the overall structural integrity of the installation.

Subsidence or collapse of seabed

84

Any subsidence or collapse of the seabed likely to affect the foundations or the overall structural integrity of an offshore installation.

Loss of stability or buoyancy

85

Any incident which causes the loss of stability or buoyancy of a floating offshore installation.

Evacuation

86

The partial or complete evacuation of an offshore installation in the interests of safety.

Falls into water

87

Any fall of a person into water from more than 2 metres.

SCHEDULE 3

DISEASES REPORTABLE OFFSHORE

Regulation 10

[3.1280]

1

Chickenpox.

2

Cholera.

3

Diphtheria.

4

Dysentery (amoebic or bacillary).

5

Acute encephalitis.

6

Erysipelas.

7

Food poisoning.

8

Legionellosis.

9

Malaria.

10

Measles.

11

Meningitis.

12

Meningococcal septicaemia (without meningitis).

13

Mumps.

14

Paratyphoid fever.

15

Plague.

16

Acute poliomyelitis.

17

Rabies.

18

Rubella.

19

Scarlet fever.

20

Tetanus.

21

Tuberculosis.

22

Typhoid fever.

23

Typhus.

24

Viral haemorrhagic fevers.

25

Viral hepatitis.

SCHEDULE 4

REVOCATIONS AND AMENDMENTS

Regulation 18

[3.1281]–[3.1400]

Table 1
Revocations

Instrument	*Reference*	*Revocation*
The Reporting of Injuries, Diseases and Dangerous Occurrences Regulations 1995	SI 1995/3163	The whole instrument
The Diving at Work Regulations 1997	SI 1997/2776	Paragraph 5 of Schedule 2
The Quarries Regulations 1999	SI 1999/2024	The entries in Schedule 5 relating to the Reporting of Injuries, Diseases and Dangerous Occurrences Regulations 1995
The Railway Safety Regulations 1999	SI 1999/2244	Regulation 7
The Ionising Radiations Regulations 1999	SI 1999/3232	Paragraph 6 of Schedule 9
The Radiation (Emergency Preparedness and Public Information) Regulations 2001	SI 2001/2975	Paragraph 10 of Schedule 11
The Manufacture and Storage of Explosives Regulations 2005	SI 2005/1082	Paragraph 40 of Part 2 of Schedule 5
The Offshore Installations (Safety Case) Regulations 2005	SI 2005/3117	Paragraph 4 of Schedule 9
The Railways and Other Guided Transport Systems (Safety) Regulations 2006	SI 2006/599	Paragraph 1 of Schedule 6
The Reporting of Injuries, Diseases and Dangerous Occurrences (Amendment) Regulations 2012	SI 2012/199	The whole instrument

Table 2

Amendments

Instrument	*Reference*	*Amendment*
The Safety Representatives and Safety Committees Regulations 1977	SI 1977/500	For the first "a" in regulation 6(1) substitute "an over three day injury,"; and for regulation 6(3) substitute "(3) In this regulation— "notifiable accident or dangerous occurrence" and "notifiable disease" mean any accident, dangerous occurrence or disease, as the case may be, notice of which is required to be given by virtue of any of the relevant statutory provisions within the meaning of section 53(1) of the 1974 Act; and "over three day injury" means an injury required to be recorded in accordance with regulation 12(1)(b) of the Reporting of Injuries, Diseases and Dangerous Occurrences Regulations 2013.".

The Mines (Safety of Exit) Regulations 1988	SI 1988/1729	In regulation 3(4)(c) for the words "person, if any, for the time being nominated under Schedule 4 to the Reporting of Injuries, Diseases and Dangerous Occurrences Regulations 1985" substitute "nominated person, if any, within the meaning of regulation 2(1) of the Reporting of Injuries, Diseases and Dangerous Occurrences Regulations 2013"; and in regulation 7(5)(b) for the words "person, if any, nominated under Schedule 4 to the Reporting of Injuries, Diseases and Dangerous Occurrences Regulations 1985" substitute "nominated person, if any, within the meaning of regulation 2(1) of the Reporting of Injuries, Diseases and Dangerous Occurrences Regulations 2013".
The Offshore Installations (Safety Representatives and Safety Committees) Regulations 1989	SI 1989/971	After "notifiable incident" in regulation 17(3)(a) insert "or an over three day injury"; and for regulation 17(6) substitute "(6) In this regulation— "notifiable incident" means any death, injury, disease or dangerous occurrence which is required to be reported under the Reporting of Injuries, Diseases and Dangerous Occurrences Regulations 2013; and "over three day injury" means an injury required to be recorded in accordance with regulation 12(1)(b) of the Reporting of Injuries, Diseases and Dangerous Occurrences Regulations 2013.".

The Gas Safety (Management) Regulations 1996	SI 1996/551	In regulation 7(14) for the words "regulation 6(1) of the Reporting of Injuries, Diseases and Dangerous Occurrences Regulations 1995" substitute "regulation 11(1) of the Reporting of Injuries, Diseases and Dangerous Occurrences Regulations 2013".
The Offshore Installations and Wells (Design and Construction, etc) Regulations 1996	SI 1996/913	In regulation 9(2) for the words "the Reporting of Injuries, Diseases and Dangerous Occurrences Regulations 1995" substitute "the Reporting of Injuries, Diseases and Dangerous Occurrences Regulations 2013".
The Health and Safety (Consultation with Employees) Regulations 1996	SI 1996/1513	In regulation 5(2)(b) for the words "regulation 7 of the Reporting of Injuries, Diseases and Dangerous Occurrences Regulations 1995" substitute "regulation 12 of the Reporting of Injuries, Diseases and Dangerous Occurrences Regulations 2013".
The Control of Major Accident Hazards Regulations 1999	SI 1999/743	In regulation 15(4) for the words "the Reporting of Injuries, Diseases and Dangerous Occurrences Regulations 1995" substitute "the Reporting of Injuries, Diseases and Dangerous Occurrences Regulations 2013".
The Quarries Regulations 1999	SI 1999/2024	In regulation 40(3) for the words "the Reporting of Injuries, Diseases and Dangerous Occurrences Regulations 1995" substitute "the Reporting of Injuries, Diseases and Dangerous Occurrences Regulations 2013".

The Construction (Design and Management) Regulations 2007	SI 2007/320	In regulation 19(2)(d) for the words "the Reporting of Injuries, Diseases and Dangerous Occurrences Regulations 1995" substitute "the Reporting of Injuries, Diseases and Dangerous Occurrences Regulations 2013".

PART 4
CHILDREN AND YOUNG PERSONS

Part 4

Children and Young Persons Act 1963

1963 chapter 37

as amended

PART II

EMPLOYMENT OF CHILDREN AND YOUNG PERSONS

Entertainment

[4.37]–[4.38]

38 . . .

[*Repealed in relation to England and Wales by the Children and Families Act 2014, s 90 and in relation to Scotland by the Children and Young People (Scotland) Act 2014, s 94.*]

Education Act 1996

1996 chapter 56

as amended

PART X

MISCELLANEOUS AND GENERAL

CHAPTER IV

EMPLOYMENT OF CHILDREN AND YOUNG PERSONS

[4.59]

560 Work experience in last year of compulsory schooling

[(1) The enactments relating to the prohibition or regulation of the employment of children shall not apply to the employment of a child in his last two years of compulsory schooling if the employment is in pursuance of arrangements made—

(a) by a [local authority], or

(b) by the governing body of a school on behalf of such an authority,

with a view to providing him with work experience as a part of his education.

(2) For the purposes of subsection (1) a child shall be taken to be in his last two years of compulsory schooling as from the beginning of the last two school years at his school during the whole or part of which he is of compulsory school age.]

(3) Subsection (1) shall not be taken to permit the employment of a person in any way contrary to—

(a) an enactment which in terms applies to persons of less than, or not over, a specified age expressed as a number of years, or

(b) section 1(2) of the Employment of Women, Young Persons and Children Act 1920 or section 55(1) of the Merchant Shipping Act 1995 (which prohibit the employment of children in ships).

(4) No arrangements shall be made under subsection (1) for a child to be employed in any way which would be contrary to an enactment prohibiting or regulating the employment of young persons if he were a young person (within the meaning of the enactment) and not a child.

(5) Where a child is employed in pursuance of arrangements made under subsection (1), so much of any enactment as—

(a) regulates the employment of young persons (whether by excluding them from any description of work, prescribing the conditions under which they may be permitted to do it or in any other way), and

(b) would apply in relation to him if he were of an age to be treated as a young person for the purposes of that enactment,

shall apply in relation to him, in and in respect of the employment arranged for him, in all respects as if he were of an age to be so treated.

(6) Nothing in section 495 or 496 [or Chapter 2 of Part 2 of the School Standards and Organisation (Wales) Act 2013] applies in relation to any power conferred on a [local authority] . . . by subsection (1).

(7) In this section 'enactment' includes any byelaw, regulation or other provision having effect under an enactment.

[4.60]

Statutory annotations

Amendment

Sub-s (6): words 'or Chapter 2 of Part 2 of the School Standards and Organisation (Wales) Act 2013' in square brackets inserted by the School Standards and Organisation (Wales) Act 2013, s 99, Sch 5, Pt 1, para 2(1), (7).

PART 5
MACHINERY AND EQUIPMENT

Part 5A
Machinery and equipment: General

Part 5

Personal Protective Equipment at Work Regulations 1992

SI 1992/2966

as amended by SIs 1994/3017, 1999/860, 1999/3232, 2002/2174, 2005/1643, 2006/2739 and 2013/448

[5.406]

3 Disapplication of these Regulations

(1) These Regulations shall not apply to or in relation to the master or crew of a sea-going ship or to the employer of such persons in respect of the normal ship-board activities of a ship's crew under the direction of the master.

(2) Regulations 4 to 12 shall not apply in respect of personal protective equipment which is—

- (a) ordinary working clothes and uniforms which do not specifically protect the health and safety of the wearer;
- (b) an offensive weapon within the meaning of section 1(4) of the Prevention of Crime Act 1953 used as self-defence or as deterrent equipment;
- (c) portable devices for detecting and signalling risks and nuisances;
- (d) personal protective equipment used for protection while travelling on a road within the meaning (in England and Wales) of section 192(1) of the Road Traffic Act 1988, and (in Scotland) of section 151 of the Roads (Scotland) Act 1984;
- (e) equipment used during the playing of competitive sports.

(3) Regulations 4 and 6 to 12 shall not apply where any of the following Regulations apply and in respect of any risk to a person's health or safety for which any of them require the provision or use of personal protective equipment, namely—

- (a) the Control of Lead at Work Regulations 1980;
- (b) [the Ionising Radiations Regulations 1999 [SI 1999/3232]];
- (c) [the Control of Asbestos Regulations 2012];
- (d) the Control of Substances Hazardous to Health Regulations 1988;
- (e) [the Control of Noise at Work Regulations 2005]
- (f) . . .

[5.407]

Amendment

Para (3): sub-para (f) revoked by SI 2013/448, reg 2, Sch.

Part 5

Provision and Use of Work Equipment Regulations 1998

SI 1998/2306

as amended by SIs 1999/860, 1999/2001, 2001/1701, 2002/2174, 2004/129, 2005/735, 2005/1093, 2005/1643, 2006/2739, 2007/320, 2008/1597, 2013/448 and 2013/1387

PART II

GENERAL

[5.653]

12 Protection against specified hazards

(1) Every employer shall take measures to ensure that the exposure of a person using work equipment to any risk to his health or safety from any hazard specified in paragraph (3) is either prevented, or, where that is not reasonably practicable, adequately controlled.

(2) The measures required by paragraph (1) shall—

(a) be measures other than the provision of personal protective equipment or of information, instruction, training and supervision, so far as is reasonably practicable; and

(b) include, where appropriate, measures to minimise the effects of the hazard as well as to reduce the likelihood of the hazard occurring.

(3) The hazards referred to in paragraph (1) are—

(a) any article or substance falling or being ejected from work equipment;

(b) rupture or disintegration of parts of work equipment;

(c) work equipment catching fire or overheating;

(d) the unintended or premature discharge of any article or of any gas, dust, liquid, vapour or other substance which, in each case, is produced, used or stored in the work equipment;

(e) the unintended or premature explosion of the work equipment or any article or substance produced, used or stored in it.

(4) For the purposes of this regulation 'adequately' means adequately having regard only to the nature of the hazard and the nature and degree of exposure to the risk.

(5) This regulation shall not apply where any of the following Regulations apply in respect of any risk to a person's health or safety for which such Regulations require measures to be taken to prevent or control such risk, namely—

(a) the Ionising Radiations Regulations 1985;

(b) [the Control of Asbestos Regulations 2012];

(c) the Control of Substances Hazardous to Health Regulations 1994;

(d) [the Control of Noise at Work Regulations 2005];

(e) . . .

(f) the Control of Lead at Work Regulations 1998.

[(g) the Control of Vibration at Work Regulations 2005].

[5.654]

Statutory annotations

Amendment

Para (5): sub-para (e) revoked by SI 2013/448, reg 2, Sch.

SCHEDULE 1

INSTRUMENTS WHICH GIVE EFFECT TO COMMUNITY DIRECTIVES CONCERNING THE SAFETY OF PRODUCTS

Regulation 10

[5.809]

(1) ***Title***	***(2)*** ***Reference***
The Construction Plant and Equipment (Harmonisation of Noise Emission Standards) Regulations 1985	SI 1985/1968, amended by SI 1989/1127
The Construction Plant and Equipment (Harmonisation of Noise Emission Standards) Regulations 1988	SI 1988/361, amended by SI 1992/488, 1995/2357
The Electro-medical Equipment (EEC Requirements) Regulations 1988	SI 1988/1586, amended by SI 1994/3017
The Low Voltage Electrical Equipment (Safety) Regulations 1989	SI 1989/728, amended by SI 1994/3260
. . .	. . .
The Simple Pressure Vessels (Safety) Regulations 1991	SI 1991/2749, amended by SI 1994/3098
The Lawnmowers (Harmonisation of Noise Emission Standards) Regulations 1992	SI 1992/168
The Gas Appliances (Safety) Regulations 1992	SI 1992/711
The Electromagnetic Compatibility Regulations 1992	SI 1992/2372, amended by SI 1994/3080
. . .	. . .
The Personal Protective Equipment (EC Directive) Regulations 1992	SI 1992/3139, amended by SI 1993/3074, 1994/2326, 1996/3039
The Active Implantable Medical Devices Regulations 1992	SI 1992/3146, amended by SI 1995/1671
The Medical Devices Regulations 1994	SI 1994/3017
The Electrical Equipment (Safety) Regulations 1994	SI 1994/3260
The Gas Appliances (Safety) Regulations 1995	SI 1995/1629
The Equipment and Protective Systems Intended for Use in Potentially Explosive Atmospheres Regulations 1996	[SI 1996/192, amended by SI 1998/81, 2001/3766, 2005/830]
The Lifts Regulations 1997	SI 1997/831[, amended by SI 2005/831]
[The Pressure Equipment Regulations 1999	SI 1999/2001]
[The Noise Emission in the Environment by Equipment for use Outdoors Regulations 2001	SI 2001/1701]
[The Cableway Installations Regulations 2004	SI 2004/129]
[The Supply of Machinery (Safety) Regulations 2008	SI 2008/1597]

Part 5

[5.810]

Statutory annotations

Amendment

Entry relating to 'The Construction Products Regulations 1991' (omitted) revoked by SI 2013/1387, reg 28(2), Sch 5, para 1.

Part 5B
Machinery and equipment: Specific

Equipment and Protective Systems Intended for Use in Potentially Explosive Atmospheres Regulations 1996

SI 1996/192

as amended by SIs 1998/81, 2001/3766, 2005/830 and 2012/1809

SCHEDULE 5

EXCLUDED EQUIPMENT, PROTECTIVE SYSTEMS AND DEVICES

Regulation 4

[5.1745]

Medical devices intended for use in a medical environment,

Equipment and protective systems where the explosion hazard results exclusively from the presence of explosive substances or unstable chemical substances,

Equipment intended for use in domestic and non-commercial environments where potentially explosive atmospheres may only rarely be created, solely as a result of the accidental leakage of fuel gas,

Personal protective equipment covered by Directive 89/686/EEC,

Seagoing vessels and mobile offshore units together with equipment on board such vessels or units,

Means of transport, ie vehicles and their trailers intended solely for transporting passengers by air or by road, rail or water networks, as well as means of transport in so far as such means are designed for transporting goods by air, by public road or rail networks or by water. Vehicles intended for use in a potentially explosive atmosphere shall not be excluded,

The equipment covered by [Article 346(1)(b) of the Treaty on the Functioning of the European Union].

[5.1745A]

Statutory annotations

Amendment

Words 'Article 346(1)(b) of the Treaty on the Functioning of the European Union' in square brackets substituted by SI 2012/1809, art 3(1), Schedule, Pt 2.

Lifting Operations and Lifting Equipment Regulations 1998

SI 1998/2307

as amended by SI 2002/2174, 2005/831, 2007/320, 2008/1597 and SI 2013/448

[5.1825]

9 Thorough examination and inspection

(1) Every employer shall ensure that before lifting equipment is put into service for the first time by him it is thoroughly examined for any defect unless either—

- (a) the lifting equipment has not been used before; and
- (b) in the case of lifting equipment for which an EC declaration of conformity could or (in the case of a declaration under the Lifts Regulations 1997) should have been drawn up, the employer has received such declaration made not more than 12 months before the lifting equipment is put into service;

or, if obtained from the undertaking of another person, it is accompanied by physical evidence referred to in paragraph (4).

(2) Every employer shall ensure that, where the safety of lifting equipment depends on the installation conditions, it is thoroughly examined—

- (a) after installation and before being put into service for the first time; and
- (b) after assembly and before being put into service at a new site or in a new location,

to ensure that it has been installed correctly and is safe to operate.

(3) Subject to paragraph (6), every employer shall ensure that lifting equipment which is exposed to conditions causing deterioration which is liable to result in dangerous situations is—

- (a) thoroughly examined—
  - (i) in the case of lifting equipment for lifting persons or an accessory for lifting, at least every 6 months;
  - (ii) in the case of other lifting equipment, at least every 12 months; or
  - (iii) in either case, in accordance with an examination scheme; and
  - (iv) each time that exceptional circumstances which are liable to jeopardise the safety of the lifting equipment have occurred; and
- (b) if appropriate for the purpose, is inspected by a competent person at suitable intervals between thorough examinations,

to ensure that health and safety conditions are maintained and that any deterioration can be detected and remedied in good time.

(4) Every employer shall ensure that no lifting equipment—

- (a) leaves his undertaking; or
- (b) if obtained from the undertaking of another person, is used in his undertaking, unless it is accompanied by physical evidence that the last thorough examination required to be carried out under this regulation has been carried out.

(5) This regulation does not apply to winding apparatus to which the Mines (Shafts and Winding) Regulations 1993 apply.

(6) Where lifting equipment was before the coming into force of these Regulations required to be thoroughly examined by a provision specified in paragraph (7), the first thorough examination under paragraph (3) shall be made before the date by which a

thorough examination would have been required by that provision had it remained in force.

(7) The provisions referred to in paragraph (6) are—

(a) section 22(2), 25(2), 26(1)(d) and 27(2) of the Factories Act 1961;

(b) . . .

(c) regulations 28(3), 40 and 46(1) of the Construction (Lifting Operations) Regulations 1961;

(d) regulations 3(1) and (2) and 6(1) of the Offices, Shops and Railway Premises (Hoists and Lifts) Regulations 1968;

(e) regulation 6(1)(c) of and Part III of Schedule 1 to the Offshore Installations (Operational Safety, Health and Welfare) Regulations 1976;

(f) Regulation 15 of the Docks Regulations 1988.

[5.1826]

Statutory annotations

Amendment

Para (7): sub-para (b) revoked by SI 2013/448, reg 2 and Sch.

[5.1832]

13

[*Revoked by SI 2013/448, reg 2, Sch.*]

SCHEDULE 2

REVOCATION OF INSTRUMENTS

Regulation 17

[5.1839]

(1) *Title*	*(2)* *Reference*	*(3)* *Extent of revocation*
The Quarries (General) Regulations 1956	SI 1956/1780	Regulations 13 and 14.
The Quarries (Ropeways and Vehicles) Regulations 1958	SI 1958/2110	The whole Regulations.
. . .	. . .	. . .
The Shipbuilding (Particulars of Annealing) Order 1961	SI 1961/117	The whole order.
The Shipbuilding (Lifting Appliances, etc, Forms) Order 1961	SI 1961/431	The whole order.
The Construction (Lifting Operations) Regulations 1961	SI 1961/1581	The whole Regulations.
The Construction (Lifting Operations) Reports Order 1962	SI 1962/225	The whole order.
The Construction (Lifting Operations) Prescribed Particulars Order 1962	SI 1962/226	The whole order.
The Hoists Exemption Order 1962	SI 1962/715	The whole order.

(1) *Title*	*(2)* *Reference*	*(3)* *Extent of revocation*
The Hoists Exemption (Amendment) Order 1967	SI 1967/759	The whole order.
The Offices, Shops and Railway Premises (Hoists and Lifts) Regulations 1968	SI 1968/849	The whole Regulations.
The Offshore Installations (Operational Safety, Health and Welfare) Regulations 1976	SI 1976/1019	The whole Regulations.
The Hoists and Lifts (Metrication) Regulations 1983	SI 1983/1579	The whole Regulations.
The Construction (Metrication) Regulations 1984	SI 1984/1593	The whole Regulations.
The Health and Safety (Miscellaneous Modifications) Regulations 1989	SI 1989/1141	The whole Regulations.
The Lifting Plant and Equipment (Records of Test and Examination etc) Regulations 1992	SI 1992/195	The whole Regulations.
. . .	. . .	. . .

[5.1840]–[5.1843]

Statutory annotations

Amendment

Entry relating to 'The Shipbuilding and Ship-repairing Regulations 1960' (omitted) revoked by SI 2013/448, reg 2, Sch.

Control of Vibration at Work Regulations 2005

SI 2005/1093

as amended by SI 2006/557

SCHEDULE 1

HAND-ARM VIBRATION

Regulations 4(1) and 6(5)

PART I

DAILY EXPOSURE TO VIBRATION

[5.1911]

The daily exposure to vibration ($A(8)$) of a person is ascertained using the formula:

$$A(8) = a_{hv}\sqrt{\frac{T}{T_0}}$$

where:

a_{hv} is the vibration magnitude, in metres per second squared (m/s2);
T is the duration of exposure to the vibration magnitude a_{hv}; and
T_0 is the reference duration of 8 hours (28,800 seconds).

To avoid confusion between vibration magnitude and daily exposure to vibration, it is conventional to express daily exposure to vibration in m/s^2 A(8).

The vibration magnitude, a_{hv}, is ascertained using the formula:

$$a_{hv} = \sqrt{a_{hwx}^2 + a_{hwy}^2 + a_{hwz}^2}$$

where:

a_{hwx}, a_{hwy} and a_{hwz} are the root-mean-square acceleration magnitudes, in m/s^2, measured in three orthogonal directions, x, y and z, at the vibrating surface in contact with the hand, and frequency-weighted using the weighting W_h.

The definition for the frequency weighting W_h is given in British Standard BS EN ISO 5349-1:2001.

Where both hands are exposed to vibration, the greater of the two magnitudes a_{hv} is used to ascertain the daily exposure.

If the work is such that the total daily exposure consists of two or more operations with different vibration magnitudes, the daily exposure ($A(8)$) for the combination of operations is ascertained using the formula:

$$A(8) = \sqrt{\frac{1}{T_0}\sum_{i=1}^{n} a_{hvi}^2 T_i}$$

where:

n is the number of individual operations within the working day;
a_{hvi} is the vibration magnitude for operation i; and
T_i is the duration of operation i.

PART II

EXPOSURE TO VIBRATION AVERAGED OVER ONE WEEK

The exposure to vibration averaged over one week ($A(8)$week) is the total exposure occurring within a period of seven consecutive days, normalised to a reference duration of five 8-hour days (40 hours). It is ascertained using the formula:

$$A(8)_{week} = \sqrt{\frac{1}{5}\sum_{j=1}^{7} A(8)_j^2}$$

where:

$A(8)_j$ is the daily exposure for day *j*.

The exposure to vibration averaged over one week is for use only for the purposes of Regulation 6(5).

Note

These provisions are reproduced to correct typographical errors in the equations.

SCHEDULE 2

WHOLE-BODY VIBRATION

Regulations 4(2) and 6(5)

PART I

DAILY EXPOSURE TO VIBRATION

[5.1912]

The daily exposure to vibration ($A(8)$) of a person is ascertained using the formula:

$$A(8) = ka_w\sqrt{\frac{T}{T_0}}$$

where:

a_w is the vibration magnitude (root-mean-square frequency-weighted acceleration magnitude) in one of the three orthogonal directions, x, y and z, at the supporting surface;

T is the duration of exposure to the vibration magnitude a_w;

T_0 is the reference duration of 8 hours (28,800 seconds); and

k is a multiplying factor.

To avoid confusion between vibration magnitude and daily exposure to vibration, it is conventional to express daily exposure to vibration in m/s^2 A(8).

Daily exposure to vibration ($A(8)$) is evaluated separately for the x, y and z directions of vibration.

For horizontal vibration (x and y directions), $k = 1.4$ and a_w is obtained using the W_d frequency weighting. For vertical vibration (z direction), $k = 1.0$ and a_w is obtained using the W_k frequency weighting.

Definitions for the frequency weightings are given in International Standard ISO 2631-1:1997.

If the work is such that the total daily exposure consists of two or more operations with different vibration magnitudes, the daily exposure ($A(8)$) for the combination of operations is ascertained using the formula:

$$A(8) = \sqrt{\frac{1}{T_0}\sum_{i=1}^{n} a_{wi}^2 T_i}$$

where:

n is the number of individual operations within the working day;

a_{wi} is the vibration magnitude for operation i; and

T_i is the duration of operation i.

PART II

EXPOSURE TO VIBRATION AVERAGED OVER ONE WEEK

The exposure to vibration averaged over one week ($A(8)_{week}$) is the total exposure occurring within a period of seven consecutive days, normalised to a reference duration of five 8-hour days (40 hours). It is ascertained using the formula:

$$A(8)_{week} = \sqrt{\frac{1}{5}\sum_{j=1}^{7} A(8)_j^2}$$

where:

$A(8)_j$ is the daily exposure for day *j*.

The exposure to vibration averaged over one week is for use only for the purposes of Regulation 6(5).

Note

These provisions are reproduced to correct typographical errors in the equations.

Health and Safety (Sharp Instruments in Healthcare) Regulations 2013

SI 2013/645

Made 18th March 2013

The Secretary of State makes these Regulations in exercise of the powers conferred by sections 15(1) and (2) and 82(3)(a) of, and paragraphs 1(1), 8(1), 14, 15(1), 16 and 20 of Schedule 3 to, the Health and Safety at Work etc Act 1974 ("the 1974 Act").

The Regulations give effect without modifications to proposals submitted to the Secretary of State by the Health and Safety Executive under section 11(3) of the 1974 Act.

Before submitting those proposals to the Secretary of State, the Health and Safety Executive consulted the bodies that appeared to it to be appropriate as required by section 50(3) of the 1974 Act.

[5.1913]

1 Citation and commencement

These Regulations may be cited as the Health and Safety (Sharp Instruments in Healthcare) Regulations 2013 and come into force on 11th May 2013.

[5.1914]

2 Interpretation

In these Regulations—

"healthcare contractor" means an employer whose main activity is not the management, organisation or provision of healthcare, but who provides services under contract to a healthcare employer;

"healthcare employer" means an employer whose main activity is the management, organisation and provision of healthcare;

"injury" includes infection;

"medical sharp" means an object or instrument necessary for the exercise of specific healthcare activities, which is able to cut, prick or cause injury;

"safer sharp" means a medical sharp that is designed and constructed to incorporate a feature or mechanism which prevents or minimises the risk of accidental injury from cutting or pricking the skin.

[5.1915]

3 Application of requirements to employers

(1) The requirements imposed by these Regulations on an employer apply to—

(a) a healthcare employer; and

(b) a healthcare contractor whose employees, or other persons who work under the healthcare contractor's supervision and direction, are exposed to a risk of injury from medical sharps in relation to the provision of services to a healthcare employer.

(2) A requirement imposed by these Regulations on an employer that applies in relation to that employer's employees also applies, so far as is reasonably practicable, in relation to any other person who is not an employee of that employer but who works under that employer's supervision and direction.

Part 5

[5.1916]

4 Application of requirements to healthcare contractors

(1) The requirements imposed by these Regulations on a healthcare contractor apply only in relation to work—

(a) on a healthcare employer's premises; or

(b) under the authority of a healthcare employer.

(2) The requirements imposed by these Regulations on a healthcare contractor apply only to the extent that the healthcare contractor controls—

(a) a person who uses, supervises or manages the use or disposal of medical sharps; and

(b) the activities which give rise to the risk of injury from medical sharps.

[5.1917]

5 Use and disposal of medical sharps

(1) An employer must ensure that—

(a) the use of medical sharps at work is avoided so far as is reasonably practicable;

(b) when medical sharps are used at work, safer sharps are used so far as is reasonably practicable;

(c) needles that are medical sharps are not capped after use at work unless—

(i) that act is required to control a risk identified by an assessment undertaken pursuant to regulation 3 of the Management of Health and Safety at Work Regulations 1999; and

(ii) the risk of injury to employees is effectively controlled by the use of a suitable appliance, tool or other equipment;

(d) in relation to the safe disposal of medical sharps that are not designed for re-use—

(i) written instructions for employees, and

(ii) clearly marked and secure containers,

are located close to areas where medical sharps are used at work.

(2) An employer must review at suitable intervals the policies and procedures in place to meet the requirements of paragraph (1) so as to ensure that those policies and procedures remain up to date and effective.

[5.1918]

6 Information and training

(1) An employer must provide each employee of that employer who is exposed to a risk of injury at work from medical sharps with information on the matters specified in Schedule 1.

(2) In complying with paragraph (1) the employer must cooperate with worker representatives in that employer's undertaking in developing and promoting the information specified in Schedule 1.

(3) In paragraph (2), "worker representatives" means any—

(a) safety representatives within the meaning of the Safety Representatives and Safety Committees Regulations 1977; or

(b) representatives of employee safety within the meaning of the Health and Safety (Consultation with Employees) Regulations 1996.

(4) An employer must provide each employee of that employer who is exposed to a risk of injury at work from medical sharps with training on the matters specified in Schedule 2 to the extent that those matters are relevant to the type of work carried out by that employee.

[5.1919]

7 Arrangements in the event of injury

(1) Where an employer is notified of any incident at work in which an employee has suffered an injury from a medical sharp, the employer must—

(a) record the incident;

(b) investigate the circumstances and cause of the incident; and

(c) take any necessary action to prevent a recurrence.

(2) Additionally, where an employer is notified of any incident at work in which an employee has suffered an injury caused by a medical sharp that exposed, or may have exposed, the employee to a biological agent, the employer must—

(a) take immediate steps to ensure that the employee receives medical advice;

(b) ensure that any treatment advised by a registered medical practitioner, including post-exposure prophylaxis, is made available to the employee; and

(c) consider providing the employee with counselling.

(3) In this regulation—

(a) "biological agent" means a micro-organism, cell culture or human endoparasite, whether or not genetically modified, which may cause infection, allergy, toxicity or otherwise create a hazard to human health; and

(b) "post-exposure prophylaxis" means a course of treatment of medicine administered to a person after exposure, or suspected exposure, to a biological agent in order to prevent infection or development of disease caused by that biological agent.

[5.1920]

8 Notification of injuries

(1) Person "A", who is an employee or other person working under the supervision and direction of a healthcare employer or a healthcare contractor, must—

(a) as soon as practicable, notify A's employer, or any other employee of that employer with specific responsibility for the health and safety of persons at work, of any incident at work in which A has suffered an injury from a medical sharp; and

(b) provide when requested by that employer sufficient information as to the circumstances of the incident to enable the employer to comply with regulation 7.

(2) In the case of an employee or other person working under the supervision and direction of a healthcare contractor, this regulation only applies to incidents which take place—

(a) on a healthcare employer's premises; or

(b) under the authority of a healthcare employer.

[5.1921]

9 Extension outside Great Britain

These Regulations apply to and in relation to the premises and activities outside Great Britain to which sections 1 to 59 and 80 to 82 of the Health and Safety at Work 1974 apply by virtue of the Health and Safety at Work etc Act 1974 (Application Outside Great Britain) Order 2013 as they apply within Great Britain.

[5.1922]

10 Review

(1) The Secretary of State must from time to time—

(a) carry out a review of these Regulations;

(b) set out the conclusions of the review in a report; and

(c) publish the report.

(2) In carrying out the review the Secretary of State must, so far as is reasonable, have regard to how Council Directive 2010/32/EU of 10 May 2010 implementing the Framework Agreement on prevention from sharp injuries in the hospital and healthcare sector concluded by HOSPEEM and EPSU (which is implemented in part by means of these Regulations) is implemented in other Member States.

(3) The report must in particular—

(a) set out the objectives intended to be achieved by these Regulations;

(b) assess the extent to which those objectives are achieved; and

(c) assess whether those objectives remain appropriate and, if so, the extent to which they could be achieved with a system that imposes less regulation.

(4) The first report under this regulation must be published before the end of the period of five years beginning with the day on which these Regulations come into force.

(5) Reports under this regulation are afterwards to be published at intervals not exceeding five years.

SCHEDULE 1

INFORMATION TO BE PROVIDED TO EMPLOYEES

Regulation 6(1) and (2)

[5.1923]

1

The risk of injury from medical sharps.

2

Legislative requirements relating to the protection of persons at work from the risks to health and safety from medical sharps, including duties on employers and employees.

3

Good practice in preventing injury from medical sharps.

4

The benefits and drawbacks of vaccination and non-vaccination in respect of blood-borne diseases.

5

The support provided by the employer to an employee who is injured at work by a medical sharp.

SCHEDULE 2

TRAINING TO BE PROVIDED TO EMPLOYEES

Regulation 6(4)

[5.1924]–[5.2200]

1

The safe use and disposal of medical sharps.

2

The correct use of safer sharps.

3

What employees should do if they are injured at work by a medical sharp.

4

The health surveillance and other procedures to be conducted by the employer where an employee is injured by a medical sharp.

PART 7
WORKPLACES

Part 7A
Workplaces: General

Workplace (Health, Safety and Welfare) Regulations 1992

SI 1992/3004

as amended by SIs 1995/2036, 1996/1592, 1999/2024, 2002/2174, 2005/735, 2007/320 and 2013/448

[7.17]

3 Application of these Regulations

(1) These Regulations apply to every workplace but shall not apply to—

[(a) a workplace which is or is in or on a ship, save that regulations 8(2) and (3) and 12(1) and (3) apply to such a workplace where the work involves any of the relevant operations in—

(i) a shipyard, whether or not the shipyard forms part of a harbour or wet dock; or

(ii) dock premises, not being work done—

(aa) by the master or crew of a ship;

(bb) on board a ship during a trial run;

(cc) for the purposes of raising or removing a ship which is sunk or stranded; or

(dd) on a ship which is not under command, for the purposes of bringing it under command;]

[(b) a workplace which is a construction site within the meaning of the Construction (Design and Management) Regulations 2007, and in which the only activity being undertaken is construction work within the meaning of those regulations, save that—

(i) regulations 18 and 25A apply to such a workplace; and

(ii) regulations 7(1A), 12, 14, 15, 16, 18, 19 and 26(1) apply to such a workplace which is indoors] [or

(c) a workplace located below ground at a mine.]

(d) . . .

(2) In their application to temporary work sites, any requirement to ensure a workplace complies with any of regulations 20 to 25 shall have effect as a requirement to so ensure so far as is reasonably practicable.

(3) As respects any workplace which is or is in or on an aircraft, locomotive or rolling stock, trailer or semi-trailer used as a means of transport or a vehicle for which a licence is in force under the Vehicles (Excise) Act 1971 or a vehicle exempted from duty under that Act—

(a) regulations 5 to 12 and 14 to 25 shall not apply to any such workplace; and

(b) regulation 13 shall apply to any such workplace only when the aircraft, locomotive or rolling stock, trailer or semi-trailer or vehicle is stationary inside a workplace and, in the case of a vehicle for which a licence is in force under the Vehicles (Excise) Act 1971, is not on a public road.

(4) As respects any workplace which is in fields, woods or other land forming part of an agricultural or forestry undertaking but which is not inside a building and is situated away from the undertaking's main buildings—

(a) regulations 5 to 19 and 23 to 25 shall not apply to any such workplace; and

(b) any requirement to ensure that any such workplace complies with any of regulations 20 to 22 shall have effect as a requirement to so ensure so far as is reasonably practicable.

[(5) As respects any workplace which is at a quarry or above ground at a mine regulation 12 shall only apply to a floor or traffic route which is located inside a building.]

[(6) For the purposes of this regulation—

(a) 'dock premises' means any dock, wharf, quay, jetty or other place at which ships load or unload goods or embark or disembark passengers, together with neighbouring land or water which is used or occupied, or intended to be used or occupied, for those incidental activities, and any part of a ship when used for those or incidental activities;

(b) 'relevant operations' means, in relation to a ship, its repairing, refitting, painting and finishing, the scaling, scurfing or cleaning of its boilers (including combustion chambers or smoke boxes) and the cleaning of its bilges or oil-fuel tanks or any of its tanks last used for carrying oil;

(c) 'ship' includes all vessels and hovercraft which operate on water or land and water;

(d) 'shipyard' means any yard or dry dock (including the precincts thereof) in which ships or vessels are constructed, reconstructed, repaired, refitted or finished; and

(e) 'vessel' means any description of craft used for the transport of goods or passengers or the storage of goods or the accommodation of passengers on water, whether used in navigation or not.]

[7.18]

Statutory annotations

Amendment

Para (1): sub-para (a) substituted by SI 2013/448, reg 4(2).
Para (6): inserted by SI 2013/448, reg 4(3).

Work at Height Regulations 2005

SI 2005/735

as amended by SIs 2007/114, 2007/320 and SI 2013/1512

[7.105]

3 Application

(1) These Regulations shall apply—

(a) in Great Britain; and

(b) outside Great Britain as sections 1 to 59 and 80 to 82 of the 1974 Act apply by virtue of the Health and Safety at Work etc Act 1974 (Application outside Great Britain) Order 2001.

(2) The requirements imposed by these Regulations on an employer shall apply in relation to work—

(a) by an employee of his; or

(b) by any other person under his control, to the extent of his control.

(3) The requirements imposed by these Regulations on an employer shall also apply to—

(a) a self-employed person, in relation to work—

(i) by him; or

(ii) by a person under his control, to the extent of his control; and

(b) to any person other than a self-employed person, in relation to work by a person under his control, to the extent of his control.

(4) Regulations 4 to 16 of these Regulations shall not apply to or in relation to—

(a) the master and crew of a ship, or to the employer of such persons, in respect of the normal ship-board activities of a ship's crew which—

(i) are carried out solely by the crew under the direction of the master; and

(ii) are not liable to expose persons at work other than the master and crew to a risk to their safety;

(b) . . .

(c) a place specified in regulation 5(3) of the Loading and Unloading of Fishing Vessels Regulations 1988 where persons are engaged in fish loading processes; . . .

(d) . . .

(5) Regulation 11 of these Regulations shall not apply to an installation while regulation 12 of the Offshore Installations and Wells (Design and Construction, etc) Regulations 1996 apply to it.

(6) In this regulation—

(a) . . .

(b) . . .

(c) 'ship' includes every description of vessel used in navigation, other than a ship which forms part of Her Majesty's Navy.

[7.106]

Statutory annotations

Amendment

Para (4): sub-para (b) revoked by SI 2013/1512, reg 3, Schedule.

Part 7

[7.116A]

[8A Dock Operations]

[In relation to work at height where people are engaged in dock operations, every employer shall ensure that Schedule 9 is complied with.]

[7.116B]

Statutory annotations

Amendment

Inserted by SI 2013/1512, reg 5(1), (2).

SCHEDULE 9

REQUIREMENTS FOR WORK IN DOCKS

[Regulation 8a]

[7.152]

[1

In relation to work at height where people are engaged in dock operations, there shall be secure and adequate guardrails at the following places—

(a) every break, dangerous corner and other dangerous part or edge of a dock, wharf, quay or jetty; and

(b) every open side of a gangway, footway over a bridge, caisson or dock gate,

except in so far as the provision of such guardrails is impracticable because of the nature of the work carried out there and the work is in progress.

2

In this Schedule—

'dock gate' means any lock gate or other gate which can close off the entrance to the dock or part of the dock from the sea or other waterway but does not include any gate on land which controls access by vehicles or pedestrians;

'dock operations' means—

(a) the loading or unloading of goods on or from a ship at dock premises;

(b) the embarking or disembarking of passengers on or from a ship at dock premises;

(c) any activity incidental to the activities in sub-paragraph (a) or (b) which takes place on dock premises, including any of the following—

(i) the fuelling and provisioning of a ship;

(ii) the mooring of a ship;

(iii) the storing, sorting, inspecting, checking, weighing or handling of goods;

(iv) the movement of goods, passengers or vehicles;

(v) the use of welfare amenities in relation to the activities in sub-paragraph (a), (b) or (c)(i) to (iv);

(vi) attending dock premises for the purposes of the activities in sub-paragraph (a), (b) or (c)(i) to (v); or the embarking or disembarking on or from a ship of its crew at dock premises,

but does not include—

(d) the loading or unloading of goods or embarking or disembarking of persons on or from a pleasure craft or any activity incidental to those activities; or

(e) beach landing operations wholly carried out by serving members of Her Majesty's Forces or visiting forces within the meaning of the provisions of Part I of the Visiting Forces Act 1952 or a combination of both;

'dock premises' means any dock, wharf, quay, jetty or other place at which ships load or unload goods or embark or disembark passengers, together with neighbouring land or water which is used or occupied, or intended to be used or occupied, for those or incidental activities, and any part of a ship when used for those or incidental activities;

'freight container' means a container as defined in regulation 2 of the Freight Containers (Safety Convention) Regulations 1984;

'goods' includes—

(a) animals;

(b) pallets and freight containers;

(c) waste;

(d) solid ballast;

(e) vehicles which are being transported as cargo;

'pleasure craft' means any description of vessel when used solely for sport or recreation, other than for carrying fare paying passengers;

'ship' includes all vessels and hovercraft which operate on water or land and water;

'vehicle' includes all lift trucks, locomotives and rolling-stock, and trailers and semi-trailers and other mechanical plant which moves on wheels, tracks, skids or any combination thereof;

'vessel' means any description of craft used for the transport of goods or passengers or the storage of goods or the accommodation of passengers on water, whether used in navigation or not;

'welfare amenities' means—

(a) sanitary conveniences;

(b) baths and shower baths;

(c) washing facilities;

(d) a supply of wholesome drinking water;

(e) a supply of protective clothing, that is to say, clothing suitable for the protection of the wearer in refrigerated spaces or against dirt from handling dirty goods or against inclement weather;

(f) accommodation and facilities for changing into clothing worn during working hours and for storing and drying clothing;

(g) canteens, or accommodation and facilities (including facilities for heating food and boiling water) for workers employed at dock premises to eat meals provided by themselves;

(h) shelters for use during inclement weather.]

[7.153]–[7.500]

Statutory annotations

Amendment

Inserted by SI 2013/1512, reg 5.

Part 7B:
Workplaces: Specific

Factories Act 1961

1961 chapter 34

as amended

PART II

SAFETY (GENERAL PROVISIONS)

[7.1034]–[7.1035]

39

[*Repealed by SI 2014/486, reg 2(1), Schedule, Table 1.*]

PART VI

EMPLOYMENT OF WOMEN AND YOUNG PERSONS

. . .

[7.1040]–[7.1041]

117

[*Repealed by SI 2014/486, reg 2(1), Schedule, Table 1.*]

PART VII

SPECIAL APPLICATIONS AND EXTENSIONS

. . .

[7.1047]–[7.1050]

121–122

[*Repealed by SI 2014/486, reg 2(1), Schedule, Table 1.*]

PART XI

ADMINISTRATION

[7.1072]–[7.1077]

153–154A

[*Repealed by SI 2014/486, reg 2(1), Schedule, Table 1.*]

PART XII

OFFENCES, PENALTIES AND LEGAL PROCEEDINGS

[7.1085]–[7.1086]

163

[*Repealed by SI 2014/486, reg 2(1), Schedule, Table 1.*]

[7.1093]–[7.1096]

169–170

[*Repealed by SI 2014/486, reg 2(1), Schedule, Table 1.*]

PART XIV

INTERPRETATION AND GENERAL

Interpretation

[7.1107]

176 General interpretation

(1) In this Act, unless the context otherwise requires, the following expressions have the meanings hereby assigned to them respectively, that is to say:—

'bakehouse' means any place in which bread, biscuits or confectionery is or are baked by way of trade or for purposes of gain;

. . .

'bodily injury' includes injury to health;

['building operation' and 'work of engineering construction' mean 'construction work' within the meaning assigned to that phrase by regulation 2(1) of the Construction (Design and Management) Regulations [2007 (SI 2007/320)];]

'calendar year' means the period of twelve months beginning with the first day of January in any year;

. . .

['child' means any person who is not over—

(a) compulsory school age (construed in accordance with section 8 of the Education Act 1996), or

(b) school age (construed in accordance with section 31 of the Education (Scotland) Act 1980);]

'class or description', in relation to factories, includes a group of factories described by reference to locality;

'contravention' includes, in relation to any provision, a failure to comply with that provision, and the expression 'contravene' shall be construed accordingly;

'cotton cloth factory' means any room, shed or workshop, or part thereof, in which the weaving of cotton cloth is carried on;

. . .

'district council' means, as respects England and Wales, the council of a . . . district, and, as respects [Scotland, a council constituted under section 2 of the Local Government etc (Scotland) Act 1994];

'driving-belt' includes any driving strap or rope;

'fume' includes gas or vapour;

. . .

'humid factory' means a factory in which atmospheric humidity is artificially produced by steaming or other means in connection with any textile process;

['inspector'[, in relation to a factory, means an inspector appointed under section 19 of the Health and Safety at Work etc Act 1974 ('the 1974 Act')—

(a) in the case of a factory on a site in relation to which the Office for Nuclear Regulation has responsibility for the enforcement of any of the relevant statutory provisions (within the meaning of

Part 7

Part 1 of the 1974 Act) by virtue of section 18(1A) or (2) of that Act, by the Office for Nuclear Regulation;

(b) in any other case, by the Health and Safety Executive,

and references] in any provisions of this Act to the inspector for the district, the superintending inspector for the division or the chief inspector are references to an inspector so appointed for the purposes of that provision;]

'machinery' includes any driving-belt;

. . . ;

'maintained' means maintained in an efficient state, in efficient working order, and in good repair;

'the Minister' means the Minister of Labour;

'owner'—

(a) as respects England and Wales, means the person for the time being receiving the rackrent of the premises in connection with which the word is used, whether on his own account or as agent or trustee for another person, or who would so receive the rackrent if the premises were let at a rackrent; and

(b) . . .

'parent' [means a parent of a child or young person or any person who is not a parent of his but who has parental responsibility for him (within the meaning of the Children Act 1989) or who has parental responsibilities in relation to him (within the meaning of section 1(3) of the Children (Scotland) Act 1995), and includes], in relation to any child or young person, any person having direct benefit from his wages;

'period of employment' means the period (inclusive of the time allowed for meals and rest) within which persons may be employed on any day;

'prescribed' means prescribed by order of the Minister;

'prime mover' means every engine, motor or other appliance which provides mechanical energy derived from steam, water, wind, electricity, the combustion of fuel or other source;

'process' includes the use of any locomotive;

'railway' means any railway used for the purposes of public traffic whether passenger, goods, or other traffic and includes any works of the railway company connected with the railway;

'railway company' includes . . . a company or person working a railway under lease or otherwise;

'sanitary conveniences' includes urinals, water-closets, earthclosets, privies, ashpits, and any similar convenience;

['ship' and 'vessel' have the same meaning as 'ship' in the Merchant Shipping Act 1995, and 'harbour' has the same meaning as in the Merchant Shipping Act 1995;]

. . .

'tenement factory' means any premises where mechanical power from any prime mover within the close or curtilage of the premises is distributed for use in manufacturing processes to different parts of the same premises occupied by different persons in such manner that those parts constitute in law separate factories;

'tramway' means a tramway authorised by or under any Act of Parliament and used for the purpose of public traffic;

'transmission machinery' means every shaft, wheel, drum, pulley, system of fast and loose pulleys, coupling, clutch, driving-belt or other device by which the motion of a prime mover is transmitted to or received by any machine or appliance;

'week' means the period between midnight on Saturday night and midnight on the succeeding Saturday night;

'woman' means a woman who has attained the age of eighteen;

. . .

'young person' means a person who has ceased to be a child but has not attained the age of eighteen.

(2) For the purposes of this Act, machinery or plant shall be deemed to have been constructed or reconstructed, and a factory or building to have been constructed, reconstructed, extended, added to, or converted for use as a factory, before any date, if the construction, reconstruction, extension, addition, or conversion was begun before that date.

(3) For the purposes of this Act, a factory shall not be deemed to be a factory in which mechanical power is used by reason only that mechanical power is used for the purpose of heating, ventilating or lighting the workrooms or other parts of the factory.

(4) A woman, young person, or child who works in a factory, whether for wages or not, either in a process or in cleaning any part of the factory used for any process, or in cleaning or oiling any part of the machinery or plant, or in any other kind of work whatsoever incidental to or connected with the process, or connected with the article made or otherwise the subject of the process therein, shall, save as is otherwise provided by this Act, be deemed to be employed therein for the purposes of this Act or of any proceedings thereunder, . . .

(5) A young person who works in a factory, whether for wages or not, in collecting, carrying or delivering goods, carrying messages or running errands shall be deemed to be employed in the factory for the purposes of this Act or of any proceedings thereunder, but [section 119 of this Act shall not apply] to any such young person who is employed mainly outside the factory.

(6) For the purposes of this Act, employment shall be deemed to be continuous unless interrupted by an interval of at least half an hour.

(7) For the purposes of this Act, an apprentice shall be deemed to be a person employed.

(8) This Act shall in its application to London have effect as if for references to district councils there were substituted, as respects the City of London references to the common council . . .

[(8A) In the application of this Act in relation to Wales—

(a) any reference to a district council shall be construed as a reference to a county council or (as the case may be) county borough council; and

(b) any reference to the district of a district council shall be construed as a reference to a county or county borough.]

(9) References in this Act to any enactment shall be construed as references to that enactment as amended by any subsequent enactment, including this Act.

[7.1108]

Statutory annotations

Amendment

Sub-s (1): in definition 'inspector' words from ', in relation to' to 'Executive, and references' in square brackets substituted by the Energy Act 2013, s 116, Sch 12, Pt 5, para 53.

Part 7

Offices, Shops and Railway Premises Act 1963

1963 chapter 41

as amended

Scope of Act

[7.2501]

1 Premises to which this Act applies

(1) The premises to which this Act applies are office premises, shop premises and railway premises, being (in each case) premises in the case of which persons are employed to work therein.

(2) In this Act—

(a) 'office premises' means a building or part of a building, being a building or part the sole or principal use of which is as an office or for office purposes;

(b) 'office purposes' includes the purposes of administration, clerical work, handling money and telephone and telegraph operating; and

(c) 'clerical work' includes writing, book-keeping, sorting papers, filing, typing, duplicating, machine calculating, drawing and the editorial preparation of matter for publication;

and for the purposes of this Act premises occupied together with office premises for the purposes of the activities there carried on shall be treated as forming part of the office premises.

(3) In this Act—

(a) 'shop premises' means—

(i) a shop;

(ii) a building or part of a building, being a building or part which is not a shop but of which the sole or principal use is the carrying on there of retail trade or business;

(iii) a building occupied by a wholesale dealer or merchant where goods are kept for sale wholesale or a part of a building so occupied where goods are so kept, but not including a warehouse belonging to the owners, trustees or conservators of a dock, wharf or quay;

(iv) a building to which members of the public are invited to resort for the purpose of delivering there goods for repair or other treatment or of themselves there carrying out repairs to, or other treatment of, goods, or a part of a building to which members of the public are invited to resort for that purpose;

(v) any premises (in this Act referred to as 'fuel storage premises') occupied for the purpose of a trade or business which consists of, or includes, the sale of solid fuel, being premises used for the storage of such fuel intended to be sold in the course of that trade or business, but not including dock storage premises or colliery storage premises;

(b) 'retail trade or business' includes the sale to members of the public of food or drink for immediate consumption, retail sales by auction and the business of lending books or periodicals for the purpose of gain;

(c) 'solid fuel' means coal, coke and any solid fuel derived from coal or of which coal or coke is a constituent;

(d) 'dock storage premises' means fuel storage premises which constitute or are comprised in premises to which certain provisions of the Factories Act 1961 apply by virtue of section 125 (1) (docks, etc) of that Act; and

(e) 'colliery storage premises' means fuel storage premises which form part of premises which, for the purposes of the Mines and Quarries Act 1954, form part of a mine or quarry, other than premises where persons are regularly employed to work by a person other than the owner (as defined by that Act) of the mine or quarry;

and for the purposes of this Act premises occupied together with a shop or with a building or part of a building falling within sub-paragraph (ii), (iii) or (iv) of paragraph (*a*) above for the purposes of the trade or business carried on in the shop or, as the case may be, the building or part of a building, shall be treated as forming part of the shop or, as the case may be, of the building or part of the building, and premises occupied together with fuel storage premises for the purposes of the activities there carried on (not being office premises) shall be treated as forming part of the fuel storage premises, but for the purposes of this Act office premises comprised in fuel storage premises shall be deemed not to form part of the last-mentioned premises.

(4) In this Act 'railway premises' means a building occupied by railway undertakers for the purposes of the railway undertaking carried on by them and situate in the immediate vicinity of the permanent way or a part (so occupied) of a building so situate, but does not include—

(a) office or shop premises;

(b) premises used for the provision of living accommodation for persons employed in the undertaking, or hotels; or

(c) premises wherein are carried on such processes or operations as are mentioned in section 123 (1) (electrical stations) of the Factories Act 1961 and for such supply as is therein mentioned.

(5) For the purposes of this Act premises maintained in conjunction with office, shop or railway premises for the purpose of the sale or supply for immediate consumption of food or drink wholly or mainly to persons employed to work in the premises in conjunction with which they are maintained shall, if they neither form part of those premises nor are required by the foregoing provisions of this section to be treated as forming part of them, be treated for the purposes of this Act as premises of the class within which fall the premises in conjunction with which they are maintained.

[7.2502]

General note

Sections 2 and 3 exclude from the application of the Offices, Shops and Railway Premises Act (OSRPA) 1963 premises in which only employer's relatives or outworkers work, and premises where only 21 man-hours weekly are normally worked. Special provisions apply with respect to buildings of which part only fall within the definition in s 1 (see OSRPA 1963, s 42), with respect to buildings plurally owned (see OSRPA 1963, s 43) and with respect to contiguous fuel storage premises in single ownership (see OSRPA 1963, s 44). Nothing in OSRPA 1963 applies to factory premises within the meaning of the Factories Act 1961 (see OSRPA 1963, s 85(1)). With the exception of s 75(3) (which places a limitation upon the application of the Factories Act 1961 to warehouses) nothing in OSRPA 1963 applies to certain premises used for the sale of fish by wholesale (see OSRPA 1963, s 85(2)). Nothing in the Act applies to underground parts of mines within the Mines and Quarries Act 1954 (see OSRPA 1963, s 85(5)). Special provisions apply to premises used for transitory purposes (see OSRPA 1963, s 86).

Definitions

For 'office premises' see sub-ss (2) and (5); for 'Shop premises' see sub-ss (3) and (5); for 'railway premises' see sub-ss (4) and (5).

Persons

Since, by virtue of the Interpretation Act 1978, s 6(c), the plural includes the singular, premises are not outside the Act merely because only one person is employed to work therein (see *Griffith v Ferrier* 1952 JC 56, so interpreting the similar words of what is now Factories Act 1961, s 175).

Employed

For definition, see OSRPA 1963, s 90(1). For the exception of premises in which only employer's relatives or outworkers work, see OSRPA 1963, s 2; for the exception of premises where only 21 man-hours weekly are worked, see OSRPA 1963, s 3. As to the protection of this Act extending only to employed persons, see the Introduction to this Act.

Building; part of a building

By OSRPA 1963, s 90(1) 'building', except in s 1(4) (occupation by railway undertakers), includes structure; the term is not further defined. The meaning of the term as used in other legislation has been judicially considered; for example, in relation to the factory legislation. The term 'building' is used in a wide sense (per Romer LJ in *Elms v Foster Wheeler* [1954] 2 All ER 714, CA). It is quite possible for a structure to be a building notwithstanding that it is not enclosed by walls and a roof, and is not one of the more ordinary forms of buildings; see *McGuire v Power Gas Corpn Ltd* [1961] 2 All ER 544n (catalytic oil plant held to be a building); *Boyle v Kodak Ltd* [1969] 2 All ER 439, HL (oil tank 30 feet high held to be a building). See, further, *Wood v Cooper* [1894] 3 Ch 671, and *Paddington Corpn v A-G* [1906] AC 1. Each case must depend upon its own facts, bearing in mind the object of the legislation under consideration (see *Knight v Demolition and Construction Co Ltd* [1953] 2 All ER 508). In that case, Parker J held that a part of the plant in a building might be a 'building', notwithstanding that it could also be described as plant, and that gas retort blocks in a retort house were accordingly 'buildings'. On appeal ([1954] 1 All ER 711n), the point was not dealt with. In *M'Callum v Butters Bros & Co Ltd* 1954 SLT (Notes) 45, the Outer House of the Court of Session held that a crane was not a building, and in *Price v Claudgen Ltd* [1967] 1 All ER 695, HL, it was held, on the particular facts, that a neon sign fixed to the front of a cinema was not part of a building

It is thought that the principal difficulty in the interpretation of the words 'building or part of a building', in OSRPA 1963, s 1(2)(a) could arise if it became necessary to distinguish between a 'building', on the one hand, and a 'part of a building', on the other hand. It is submitted that a 'building', must have its feet upon the ground, so that in the case of a multi-storey building each storey is a part of a building and not a separate building in itself. Where, however, what outwardly appears to be a single building is divided vertically by party walls, and the various divisions are occupied as separate units without intercommunication, it is thought that each such division will constitute a separate 'building' for the purpose of the Act. The question is in any event one of fact and degree, to be decided upon an examination of the circumstances of each particular case.

Where a building as a whole does not qualify as office premises (as where it is a library) nevertheless a definable part of that building (such as a room) may constitute a 'part of a building', and if that part is solely or principally used as an office or for office purposes it is 'office premises' within OSRPA 1963, s 1(2)(a) (*Oxfordshire County Council v University of Oxford* (1980) Times, 10 December, DC).

Office purposes

See sub-s (2)(b).

Activities

It is not enough that the ancillary premises are occupied for the purposes of the office premises if they are not occupied for the purposes of the activities there carried on.

Treated as forming part of the office premises

This deeming provision only operates when the premises in question are neither a 'building . . . the sole or principal use of which is as an office or for office purposes' nor a 'part of a building the sole or principal use of which is as an office or for office purposes': see OSRPA 1963, s 1(2)(a).

Shop

This term is not defined in OSRPA 1963. In view of the extended definition of 'shop premises' contained in OSRPA 1963, s 1(3)(a), the difficulties which have arisen in interpreting the statutory definition of 'shop' contained in Shops Act 1950, s 74(1) seem unlikely to arise. Since a 'shop' is a species of the genus 'shop premises' for the purpose of OSRPA 1963, s 1(3)(a) it seems clear that nothing can be a 'shop' unless it is also 'premises'. Thus, it would seem that a mobile shop is not a 'shop' within the Act (compare the similar interpretation accorded to Shops Act

1950, s 74(1) in *Stone v Boreham* [1959] 1 QB 1, DC, and see also *Cowlairs Co-operative Society v Glasgow Corpn* 1957 JC 51; *Kahn v Newberry* [1959] 2 All ER 202, DC; *Greenwood v Whelan* [1967] 1 All ER 294, DC and *Maby v Warwick Corpn* [1972] 2 All ER 1198, DC).

In deciding what is a 'shop' for the purposes of the Act, it is submitted that once it is clear that there are 'premises' the sole remaining question is whether what is being considered is a 'shop' in the ordinary and natural sense of the word. 'Where words are . . . perfectly familiar, all one can do is to say whether or not one regards them as apt to cover or describe the circumstances in question in any particular case' (per Somervell LJ in *Bath v British Transport Commission* [1954] 2 All ER 542 at 543). It is probably true to say, however, that the carrying on of a retail trade is the hallmark of a 'shop' (see per Tindal CJ in *R v Chapman and Alderman* (1843) 7 JP 132).

Retail trade or business

'Carrying on' involves a degree of permanency (see *Golder v Thomas Johnston's (Bakers) Ltd* 1950 SLT (Sh Ct) 50, decided upon the Tenancy of Shops (Scotland) Act 1949). The definition of 'retail trade or business' in s 1(3)(b) is plainly not intended to be exhaustive (for a discussion of the meaning of the word 'includes' in statutory definitions, see *Dilworth v Stamps Comrs* [1899] AC 99, PC. It is less comprehensive than the corresponding definition in Shops Act 1950, s 74(1). A retailer is one who deals with customers, as opposed to a wholesaler, who deals only with persons who buy to sell again (per *Bacon VC in Treacher & Co Ltd v Treacher* [1874] WN 4); and see, further, *Staincross Revenue Officer v Staincross Assessment Committee and Whitehead* (1930) 143 LT 525 at 567 and *Phillips v Parnaby* [1934] 2 KB 299. The interpretation of the words, 'retail trade or business', in Shops Act 1950, s 74(1) has raised the problem whether they are apt to cover the provision of services as well as the supply of goods. In *M & F Frawley Ltd v Ve-Ri-Best Manufacturing Co Ltd* [1953] 1 QB 318, CA, Somervell LJ took the view that the word 'retail' primarily suggested the selling of goods rather than the selling of services, and whilst both he and Jenkins LJ thought that the terms of the definition suggested that a business might be a retail business although it was concerned with matters other than the sale of goods, they also took the view that to satisfy the definition where there was a sale of services the services rendered must be performed in circumstances comparable with those in which a sale of goods is carried on in a retail shop. The Court of Appeal therefore held that the business in issue, which was that of a builder and decorator, fell outside the definition. The dictum of Jenkins LJ was considered in *Ilford Corpn v Betterclean (Seven Kings) Ltd* [1965] 2 QB 222, DC, in which Parker LJ expressed the view that, within certain limits, a 'retail business' in the Shops Act 1950 covered the provision of services. These dicta, however, were not followed by the High Court of Justiciary in *Boyd v A Bell & Sons Ltd* 1969 SLT 156, holding that 'retail trade or business' in s 74(1) only applied to the selling of goods and holding, accordingly, that premises used for taking in dry cleaning were not a 'shop'. It will be noted that the expression, 'retail trade or business', in Shops Act 1950, s 74(1), is used in order to define the word 'shop' therein, so that the interpretation of that expression is necessarily coloured by its being linked with the word 'shop' (see *M & F Frawley Ltd v Ve-Ri-Best Manufacturing Co Ltd*, above). In OSRPA 1963, s 1(3)(a)(ii) the expression, 'retail trade or business', is expressly used in connection with buildings which are not shops, so that there is room for interpreting that expression in a wider sense than it is capable of bearing in the context of the Shops Act 1950.

Warehouse

Such warehouses as are here excepted fall within the provisions of Factories Act 1961, s 125, provided that mechanical power is used in or for the purposes of the warehouse; and by s 125 thereof certain provisions of the Factories Act 1961, are applied thereto. A transit shed may be a 'warehouse' (*Fisher v Port of London Authority* [1962] 1 All ER 458).

Gain

The ordinary meaning of 'gain' is acquisition. 'Gain is something obtained or acquired. It is not limited to pecuniary gain . . . and still less is it limited to commercial profits . . . ' (per Jessel MR in *ex p Hargrove & Co* (1875) 10 Ch App 542 at 546, 547n, construing the words 'not involving the acquisition of gain by the company' in Companies Act 1862, s 31); see, further, on the construction of the word 'gain' in this context, *Greenberg v Cooperstein* [1926] Ch 657. The construction of the phrase, 'for purposes of gain', in what is now Factories Act 1961, s 175(1), was considered by the Court of Appeal in *Stanger v Hendon Borough Council* [1948] 1 KB 571, to include indirect as well as direct gain.

Premises to which . . . Factories Act 1961 apply

The premises referred to in the Factories Act s 125(1) are ' . . . every dock, wharf or quay (including any warehouse belonging to the owners, trustees or conservators of the dock, wharf or quay, and any line or siding used in connection with and for the purposes of the dock, wharf or quay and not forming part of a railway or tramway) and every other warehouse (not forming part of a factory) in or for the purposes of which mechanical power is used.'

Railway premises

It is to be noted that, by virtue of OSRPA 1963, s 85(1) does not apply to premises which for the purposes of the Factories Act 1961, form part of a factory. By Factories Act 1961, s 175(2) the expression 'factory' includes the following premises in which persons are employed in manual labour:

'(f) except as provided in subsection (10) of this section, any premises in which the construction, reconstruction or repair of locomotives, vehicles or other plant for use for transport purposes is carried on as ancillary to a transport undertaking or other industrial or commercial undertaking.'

By Factories Act 1961, s 175(10):

'Premises used for the purpose of housing locomotives or vehicles where only cleaning, washing, running repairs or minor adjustments are carried out shall not be deemed to be a factory by reason only of paragraph (f) of subsection (2) of this section, unless they are premises used for the purposes of a railway undertaking where running repairs to locomotives are carried out.'

The effect of Factories Act 1961, s 175(2)(f), (10) is that railway running sheds where running repairs to locomotives are carried out are subject to the provisions of that Act. It follows that such running sheds are excepted from the definition of 'railway premises' contained in the present Act.

Railway undertakers

For definition, see OSRPA 1963, s 90(1); for 'office or shop premises' see OSRPA 1963, s 1(2), (3).

Factories Act 1961

The processes, operations and supply mentioned in Factories Act 1961, s 123(1) are 'the processes or operations of generating, transforming or converting, or of switching, controlling or otherwise regulating, electrical energy for supply by way of trade, or for supply for the purposes of any transport undertaking or other industrial or commercial undertaking or of any public building or public institution, or for supply to streets or other public places . . . '.

[7.2503]

2 Exception for premises in which only employer's relatives or outworkers work

(1) This Act shall not apply to any premises to which it would, apart from this subsection, apply, if none of the persons employed to work in the premises is other than the husband, wife[, civil partner], parent, grandparent, son, daughter, grandchild, brother or sister of the person by whom they are so employed.

(2) A dwelling shall not, for the purposes of this Act, be taken to constitute or comprise premises to which this Act applies by reason only that a person dwelling there who is employed by a person who does not so dwell does there the work that he is employed to do in compliance with a term of his contract of service that he shall do it there.

[7.2504]

Definition

For 'employed' see OSRPA 1963, s 90(1).

Statutory annotations

Amendment

Sub-s (1): words in square brackets inserted by the Civil Partnership Act 2004, s 261(1), Sch 27, para 23.

[7.2505]

3 Exception for premises where only 21 man-hours weekly normally worked

(1) This Act shall not apply to any premises to which it would, apart from this subsection, apply, if the period of time worked there during each week does not normally exceed twenty-one hours.

(2) For the purposes of this section the period of time worked in any premises shall be deemed to be—

(a) as regards a week in which one person only is employed to work in the premises, the period of time worked by him there;

(b) as regards a week in which two persons or more are so employed, the sum of the periods of time for which respectively those persons work there.

(3) . . .

[7.2506]

Worked; work

As to when persons employed by railway undertakers shall be deemed to be employed to work in premises to which the Act applies, see OSRPA 1963, s 90(3).

Definitions

For 'week' see OSRPA 1963, s 90(1); for 'employed' see OSRPA 1963, s 90(1), (4).

Statutory annotations

Amendment

Sub-s (3): repealed by SI 1974/1943, reg 2(9), Sch 1.

Special provisions with respect to buildings whereof parts are office, &c, premises and with respect to certain contiguous fuel storage premises

[7.2516]–[7.2521]

42–44

[*Repealed by SI 2014/486, reg 2(1), Schedule, Table 1.*.]

Exemptions

[7.2523]–[7.2524]

46

[*Repealed by SI 2014/486, reg 2(1), Schedule, Table 1.*.]

Enforcement

[7.2528]–[7.2531]

59–59A

[*Repealed by SI 2014/486, reg 2(1), Schedule, Table 1.*.]

Offences, penalties and legal proceedings

[7.2533]–[7.2534]

63

[*Repealed by SI 2014/486, reg 2(1), Schedule, Table 1.*.]

[7.2536]–[7.2537]

67

[Repealed by SI 2014/486, reg 2(1), Schedule, Table 1..]

[7.2539]–[7.2540]

69

[Repealed by SI 2014/486, reg 2(1), Schedule, Table 1..]

[7.2542]–[7.2545]

72

[Repealed by SI 2014/486, reg 2(1), Schedule, Table 1..]

Amendments of other Acts

[7.2551]–[7.2552]

78

[Repealed by SI 2014/486, reg 2(1), Schedule, Table 1..]

General provisions

[7.2554]–[7.2555]

80

[Repealed by SI 2014/486, reg 2(1), Schedule, Table 1..]

[7.2557]–[7.2562]

82

[Repealed by SI 2014/486, reg 2(1), Schedule, Table 1..]

[7.2565]–[7.2566]

86

[Repealed by SI 2014/486, reg 2(1), Schedule, Table 1..]

Provisions with respect to Northern Ireland, Isles of Scilly and Inner and Middle Temples

[7.2568]–[7.2569]

88

[Repealed by SI 2014/486, reg 2(1), Schedule, Table 1..]

Interpretation

[7.2571]

90 Interpretation

(1) In this Act, unless the context otherwise requires, the following expressions have the meanings hereby assigned to them respectively, that is to say:—

except in section 1(4) of this Act, 'building' includes structure;

'contravention' includes, in relation to a provision of this Act or of regulations thereunder, a failure to comply with the provision, and the expression 'contravene' shall be construed accordingly;

'employed' means employed under a contract of service or apprenticeship (whether oral or in writing, express or implied);

. . .

'fish' includes molluscs and crustaceans;

'fuel storage premises' has the meaning assigned to it by section 1 (3) (a) (v) of this Act;

. . .

. . .

. . .

'the Minister' means the [Secretary of State for Employment];

'notice' means a notice in writing;

'office premises' has the meaning assigned to it by section 1 (2) of this Act;

'owner'—

(a) as respects England and Wales, means the person for the time being receiving the rackrent of the premises, building or part of a building in connection with which the word is used, whether on his own account or as agent or trustee for another person, or who would so receive the rackrent if the premises, building or part were let at a rackrent, and

(b) as respects Scotland, means the person for the time entitled to receive or who would, if the same were let, be entitled to receive, the rents of the premises, building or part of a building in connection with which the word is used and includes a trustee, factor [or person entitled to act as legal representative of a person under disability by reason of nonage or mental or other incapacity], and in the case of public or municipal property, applies to the persons to whom the management thereof is entrusted,

and 'owned' and 'ownership' shall be construed accordingly;

. . .

. . .

. . .

'railway premises' has the meaning assigned to it by section 1 (4) of this Act;

'railway undertakers' means any persons authorised by an enactment or a provision of an order or scheme made under or confirmed by an Act to construct, work or carry on a railway;

'shop premises' has the meaning assigned to it by section 1 (3) of this Act:

'week' means the period between midnight on Saturday night and midnight on the succeeding Saturday night.

(2) . . .

(3) For the purposes of this Act—

(a) persons employed by railway undertakers to do work the general control of the doing of which is exercised at railway premises, or at office premises occupied by the undertakers for the purposes of the railway undertaking carried on by them and situate in the immediate vicinity of the permanent way, shall be deemed to be employed to work in the

Part 7

premises at which the general control of the doing of their work is exercised notwithstanding that their work is in fact done elsewhere;

(b) neither railway premises nor such office premises as aforesaid shall be taken to be premises in the case of which persons are employed to work therein by reason only of the fact that persons employed by the undertakers who occupy the premises resort to the premises for the purpose only of discharging duties whose discharge is incidental to the work that they are primarily employed to do.

(4) For the purposes of this Act, any such person as follows shall be taken to be employed, namely,—

(a) a person appointed under section 6 or 7 of the Registration Service Act 1953 who exercises and performs his powers and duties in premises provided and maintained by the council within whose area his district or sub-district is situate;

(b) . . .

(c) a member of a police force maintained by [a local policing body or] a police authority; [or]

[(d) a member of a police force seconded to the [National Crime Agency to serve as a National Crime Agency officer]].

(5) The definition of a class of premises, rooms or persons for the purposes of any regulations or order under this Act may be framed by reference to any circumstances whatever.

(6) Any reference in this Act to any other enactment shall, unless the context otherwise requires, be construed as a reference to that enactment as amended or extended by or under any subsequent enactment.

[7.2572]

Statutory annotations

Amendment

Sub-s (4): para (b) repealed by the Registration of Births, Deaths and Marriages (Scotland) Act 1965, s 58(2), Sch 2; in para (c) words 'a local policing body or' in square brackets inserted by the Police Reform and Social Responsibility Act 2011, s 99, Sch 16, Pt 3, para 83; for transitional provisions see s 98, Sch 15, Pts 3, 4 thereto; in para (d) words 'National Crime Agency to serve as a National Crime Agency officer' in square brackets substituted by the Crime and Courts Act 2013, s 15(3), Sch 8, Pt 2, para 17.

SCHEDULES 1, 2

[7.2575]–[7.3100]

[*Repealed.*]

Shipbuilding and Ship-repairing Regulations 1960

SI 1960/1932

[7.2000]

Note

These regulations were revoked by SI 2013/448.

Nuclear Industries Security Regulations 2003

SI 2003/403

as amended by SIs 2006/2815, 2007/3224, 2009/229 and 2013/190

PART 1

INTRODUCTORY

[7.3104]

2 Interpretation: general

(1) In these Regulations, unless the context otherwise requires—

'the 1974 Act' means the Health and Safety at Work etc Act 1974;

'the 2001 Act' means the Anti-terrorism, Crime and Security Act 2001;

'the 1978 Order' means the Health and Safety at Work (Northern Ireland) Order 1978;

'approved carrier' means a Class A carrier or a Class B carrier;

'approved security plan' means a security plan which has been approved by the Secretary of State under regulation 5 or 6, as amended by any amendments approved under regulation 6, and which has not been revoked (but see regulation 8 (temporary security plans));

'approved transport plan' means a transport plan which has been approved by the Secretary of State under regulation 19;

'approved transport security statement' means a transport security statement which has been approved by the Secretary of State under regulation 16, as amended by any amendments approved under that regulation, and which has not been revoked;

'carrier' means a person undertaking the transport of Category I/II nuclear material or Category III nuclear material, and includes both a carrier for hire or reward and a carrier on his own account;

'Category I/II nuclear material' has the meaning given in regulation 3(3);

'Category III nuclear material' has the meaning given in regulation 3(4);

'Class A carrier' means a carrier approved by the Secretary of State under Part 3 of these Regulations to transport Category I/II nuclear material and Category III nuclear material;

'Class B carrier' means a carrier approved by the Secretary of State under Part 3 of these Regulations to transport Category III nuclear material;

['classification policy' means the classification policy 'Information concerning the Use, Storage and Transport of Nuclear and other Radioactive Material' issued by the Secretary of State from time to time;]

'commencement date' means 22nd March 2003;

'enriched', in relation to uranium, means enriched so as to contain more than 0.711% of uranium-235;

['nuclear construction site' has the meaning given in section 77(7) of the 2001 Act;]

'nuclear material' has the meaning given in section 76(7) of the 2001 Act (as extended under regulation 3);

'nuclear premises' means—

[(a) a nuclear site, other than one in relation to which all nuclear material or other radioactive material that was used or stored has been removed as part of the decommissioning (within the meaning given in Chapter 1 of Part 1 of the Energy Act 2004) of that site;]

[(aa) a nuclear construction site on which works are being carried out—

(i) by a developer; and

(ii) pursuant to the grant or issue of a relevant consent, without which the carrying out of those works would be unlawful;]

(b) premises that form part of a nuclear site and are premises on which a person, who is not the holder of the nuclear site licence and is not acting as an officer, employee or contractor of that holder, uses or stores nuclear material or other radioactive material; or

(c) other nuclear premises on which Category I/II nuclear material or Category III nuclear material is used or stored, but excluding premises that are used solely for the purpose of the temporary storage of such material during the course of or incidental to its transport in any case where the standards, procedures and arrangements in respect of the security of the transport are contained in an approved transport security statement;

'nuclear site' has the meaning given in section 76(7) of the 2001 Act;

'nuclear site licence' has the same meaning as in section 1 of the Nuclear Installations Act 1965;

'other nuclear premises' has the meaning given in section 77(7) of the 2001 Act;

'relevant personnel', in relation to a person ('the principal') who is the responsible person in relation to any nuclear premises, a carrier or a person to whom regulation 22 applies, means—

(a) each of the principal's officers, employees, contractors and consultants, and

(b) each officer, employee, contractor or consultant of the principal's contractors and consultants;

'responsible person' has the meaning given in paragraph (2);

'security plan' must be construed in accordance with regulation 4(2) and (3);

'sensitive nuclear information' has the meaning given in section 77(7) of the 2001 Act;

'transport' means transport by any means, but excluding—

(i) transport by air, and

(ii) transport within nuclear premises or between adjacent nuclear premises;

'transport plan' must be construed in accordance with regulation 19(3);

'transport security statement' must be construed in accordance with regulation 16(2) and (3);

'United Kingdom person' has the meaning given in section 81(2) of the 2001 Act; and

'United Kingdom ship' has the meaning given in section 77(7) of the 2001 Act;

['uranium enrichment equipment' means equipment capable of being used in or in connection with the enrichment of uranium;

'uranium enrichment software' means any software capable of being used in or in connection with the enrichment or uranium].

[(1A) For the purposes of sub-paragraph (aa) of the definition of 'nuclear premises'—

'developer' means a person who is lawfully entitled to carry out works on a site with a view to its becoming a nuclear site;

'relevant consent' means—

(a) development consent within the meaning of section 31 of the Planning Act 2008;

(b) planning permission—

(i) within the meaning of section 336 of the Town and Country Planning Act 1990;

(ii) within the meaning of section 277 of the Town and Country Planning (Scotland) Act 1997;

(iii) within the meaning of article 2(2) of the Planning (Northern Ireland) Order 1991;

(c) an order under section 14 or 16 of the Harbours Act 1964;

(d) an order under section 1 of the Harbours Act (Northern Ireland) 1970;

(e) an order under section 10 of the Harbours Act (Northern Ireland) 1970.]

(2) 'Responsible person', in relation to any nuclear premises, means—

(a) in the case of a nuclear site falling within paragraph (a) of the definition of 'nuclear premises', the holder of the nuclear site licence;

[(aa) in the case of a nuclear construction site falling within sub-paragraph (aa) of the definition of 'nuclear premises', the developer;]

(b) in the case of premises falling within paragraph (b) of that definition, the person mentioned in that paragraph; and

(c) in the case of premises falling within paragraph (c) of that definition, the person who uses or stores the Category I/II nuclear material or Category III nuclear material on those premises,

but this is subject to paragraph (3).

(3) No person is the responsible person in relation to any nuclear premises falling within paragraph (b) or (c) of the definition of 'nuclear premises' by virtue of using or storing nuclear material or other radioactive material on behalf of another person if he is that other person's officer, employee or contractor.

[(4) In these Regulations and for the purposes of paragraph (b) of the definition of 'sensitive nuclear information' in section 77(7) of the 2001 Act, information which appears to the Secretary of State to be information which needs protecting in the interests of national security includes information which requires a protective marking in accordance with the classification policy.]

[7.3105]

Statutory annotations

Amendment

Para (1): definition 'nuclear construction site' inserted by SI 2013/190, reg 3(1)(a); in definition 'nuclear premises' sub-para (a) substituted by SI 2013/190, reg 3(1)(b); in definition 'nuclear premises' sub-para (aa) inserted by SI 2013/190, reg 3(1)(c).

Para (1A): inserted by SI 2013/190, reg 3(2).

Para (2): sub-para (aa) inserted by SI 2013/190, reg 3(3).

PART 2

SECURITY OF NUCLEAR PREMISES

[7.3107]

4 Requirement for approved security plan for nuclear premises

(1) There must be an approved security plan [in place at all times] for each nuclear premises (whether or not the premises form part of other premises to which this paragraph applies).

(2) A security plan must describe in writing the standards, procedures and arrangements adopted or to be adopted by the responsible person to ensure the security of—

- (a) the nuclear premises in relation to which he is the responsible person,
- (b) any Category I/II nuclear material and Category III nuclear material used or stored on the premises,
- (c) any equipment [or software] used or stored on the premises in connection with activities involving nuclear material,
- (d) any sensitive nuclear information kept on the premises, and
- (e) in the case of nuclear premises which are or form part of a nuclear site—
  - (i) any nuclear material (so far as not already mentioned in sub-paragraph (b)) and other radioactive material used or stored on the premises, and
  - (ii) any equipment [or software] used or stored on the premises in connection with activities involving other radioactive material,
- [(f) in the case of nuclear premises which are, or form part of, a nuclear construction site, any equipment used or stored on such a site].

(3) In particular, but without prejudice to the generality of paragraph (2), the plan must describe the standards, procedures and arrangements relating to—

- (a) the investigation and assessment by the Secretary of State of the suitability of relevant personnel of the responsible person with a view to ensuring the security of the premises and the material, equipment and information mentioned in paragraph (2);
- (b) the receipt and despatch of any Category I/II nuclear material and Category III nuclear material to be transported to or from the nuclear premises;
- (c) the manner in which the nuclear premises are to be policed and guarded, including the identity of the person providing any constables or persons acting as guards, the total number of constables and such persons attached to the premises and the number of such constables or other persons who will normally be present there; and
- (d) the steps to be taken by the responsible person or any person acting on his behalf if any event of a kind specified in regulation 10(5)(a), (b), (e) or (h) that requires immediate action occurs, and the regular practice of the activities required in connection with those steps.

[(3A) Further, and without prejudice to the generality of paragraph (2)—

- (a) in the case of a nuclear site (A) which is located within 5 kilometres of a nuclear site (B), the plan in relation to site A must also describe the standards, procedures and arrangements to ensure the security of site B [in relation to any] activity which is or is to be carried out at site A that may or will affect the security of site B;
- (b) in the case of a nuclear construction site falling within sub-paragraph (aa) of the definition of 'nuclear premises', which is located within 5 kilometres of a nuclear site, the plan in relation to the nuclear

Part 7

construction site must also describe the standards, procedures and arrangements to ensure the security of that nuclear site, [in relation to any] activity which is or is to be carried out at the nuclear construction site that may or will affect the security of that nuclear site.]

(4) In the case of premises which are nuclear premises on the commencement date, paragraph (1) does not apply until—

(a) if a security plan has been submitted by 22nd June 2003 to the Secretary of State for approval under regulation 5, the date on which the Secretary of State approves a security plan for the premises, or

(b) if it has not, 22nd June 2003.

[7.3108]

Statutory annotations

Amendment

Para (1): words 'in place at all times' in square brackets inserted by SI 2013/190, reg 4(1).

Para (2): sub-para (f) inserted by SI 2013/190, reg 4(2).

Para (3A): inserted by SI 2013/190, reg 4(3); in sub-paras (a), (b) words 'in relation to any' in square brackets substituted by SI 2014/526, reg 3.

[7.3112]

8 Temporary security plans during building works etc

(1) If it is proposed to carry out any work of alteration or extension to any building or other structure which is, or forms part of, nuclear premises [(other than a nuclear construction site falling within sub-paragraph (aa) of the definition of 'nuclear premises') and which is not provided for in an existing approved security plan with which the responsible person must comply]—

(a) the responsible person in relation to the premises must give notice in writing to the Secretary of State—

(i) specifying the nature of the proposed works, and

(ii) stating whether in his opinion they are likely to involve any derogation from any of the standards, procedures and arrangements described in the approved security plan for the premises, and

(b) the works may not be begun until the Secretary of State has approved a temporary security plan for them.

(2) Paragraph (1) does not apply in the case of any particular work if before the work is begun the Secretary of State has notified the responsible person in writing that that work, or any work of a description that includes that work, is in her opinion unlikely to be prejudicial to the security of the premises and the material and equipment mentioned in regulation 4(2).

(3) To obtain approval of a temporary security plan for any works, the responsible person must submit the plan in writing to the Secretary of State.

(4) The temporary security plan must describe any standards, procedures and arrangements which the responsible person proposes to adopt to ensure the security of the premises and the material and equipment mentioned in regulation 4(2) during the period whilst the works are being carried out.

(5) The Secretary of State may approve the temporary security plan as submitted or with such amendments as she may require.

(6) During the period whilst the works are being carried out, the approved security plan for the premises has effect subject to the approved temporary security plan.

(7) During that period the responsible person must comply with the standards, procedures and arrangements described in the approved temporary security plan.

(8) The responsible person may at any time submit proposals for amending the approved temporary security plan to the Secretary of State, and the Secretary of State may approve the proposals as submitted or with such amendments as she may require.

(9) In the case of premises which are nuclear premises on the commencement date, paragraphs (1) to (8) of this regulation do not apply until there is an approved security plan for the premises.

[7.3112A]

Statutory annotations

Amendment

Para (1): words from '(other than a' to 'person must comply' in square brackets inserted by SI 2013/190, reg 5.

Part 7

Part 7C
Mines and quarries

Quarries Regulations 1999

SI 1999/2024

as amended by SIs 1999/3242, 2001/2626, 2002/2099, 2002/2174, 2005/1082, 2005/1643 and SI 2013/1471

PART VII

DUTIES OF EMPLOYERS AND PARTICIPATION AND DUTIES OF PERSONS AT WORK

[7.3776]

40 Participation of persons at work

(1) It shall be the duty of the operator to make and maintain arrangements which will enable him and those persons who regularly work at the quarry to co-operate effectively in promoting and developing measures to ensure the health, safety and welfare of persons who regularly work at the quarry and in checking the effectiveness of such measures.

(2) For the purposes described in paragraph (1), a committee of persons with suitable practical experience of quarrying operations may be appointed for the quarry—

(a) in a case where there is an association or body representative of a majority of the total number of persons working at the quarry, by that association or body; or

(b) jointly by associations or bodies which are together representative of such a majority.

(3) Where an injury or dangerous occurrence which is notifiable under [the Reporting of Injuries, Diseases and Dangerous Occurrences Regulations 2013] occurs at a quarry, the operator shall permit two members of the committee appointed under paragraph (2) to inspect together the place where the injury or dangerous occurrence occurred and, so far as is necessary for ascertaining its cause, any other part of the quarry and any plant, and to take samples of the atmosphere, dust or water at that place.

(4) The operator of a quarry shall permit sufficient inspections to be carried out by members of the committee appointed under paragraph (2) to enable every part of the quarry and any plant and equipment at the quarry to be inspected once a month by two of those committee members together.

(5) The operator of a quarry shall permit members of a committee appointed under paragraph (2) who are carrying out an inspection under paragraph (4) to—

(a) scrutinise any documents which are kept at the quarry in compliance with the relevant statutory provisions;

(b) review the risk assessment referred to in regulation 7(1)(a) and the measures referred to in regulation 7(1)(b) and to suggest improvements thereto; and

(c) be accompanied by their advisers.

(6) The operator of a quarry shall ensure that any improvements suggested under paragraph (5)(b) are considered and, if they are not accepted, written reasons for this are given to the members of the committee who made the inspection.

(7) The operator of a quarry and any person nominated by him shall be entitled to accompany the two committee members appointed under paragraph (2) who are carrying out an inspection under paragraph (4) during that inspection.

(8) Where any two committee members appointed under paragraph (2) have carried out an inspection under paragraph (4), they may make a written report of the

matters ascertained as a result of the inspection and, if such a report is made, the two committee members and the operator or any person nominated by him shall sign the report.

(9) Where a written report is made, a copy of the report signed in accordance with paragraph (8) shall be posted in a conspicuous position at the quarry and kept posted there for 28 days.

[7.3776A]

Statutory annotations

Amendment

Para (3): words 'the Reporting of Injuries, Diseases and Dangerous Occurrences Regulations 2013' in square brackets substituted by SI 2013/1471, reg 18(2), Sch 4.

SCHEDULE 5

REVOCATIONS AND MODIFICATIONS OF INSTRUMENTS

Regulation 48(1)

PART II

MODIFICATIONS

Regulation 48(2)

[7.3795]

Column 1 ***Provision***	***Column 2*** ***Extent of modification***
The Mines and Quarries (Tips) Regulations 1971	
Regulations 3, 5(4), 5(5), 8, 12, 17, 18(1), 19(1), 20(1), 21(1), 22(1), 24 and 25	Omit the words 'or quarry' wherever they occur.
Regulation 2	In paragraph (1), in the definition of 'the person having responsibility for a tip', omit sub-paragraph (a).
Regulation 4	In paragraph (1), omit the words 'and quarry'.
Regulation 6	In paragraph (1), omit the words 'and of the owner of every quarry'.
	In paragraph (2), omit the words 'and quarry'.
Regulation 7	Omit the words 'mines and quarries are' and insert the words 'mine is' in their place.
Regulation 9	In paragraph (1), omit the words 'or quarry' wherever they occur and omit the words 'or the owner of the quarry'.
	In paragraphs (2) and (3), omit the words 'or quarry' wherever they occur.
Regulation 10	In paragraph (1), omit the words 'or quarry' wherever they occur and omit the words 'or, as the case may be, the owner of the quarry'.
Regulation 11	Omit the words 'or quarry' wherever they occur.
	In paragraph (a), omit the words 'or, as the case may be, the owner of the quarry' and 'or, as the case may be, by the owner of the quarry'.
Regulation 13	Omit the words 'or quarry' wherever they occur and omit the words 'or, as the case may be, of the owner of the quarry'.

Column 1 *Provision*	*Column 2* *Extent of modification*
Regulation 14	Omit the words 'or quarry' and 'or, as the case may be, of the owner of the quarry' wherever they occur.
Regulation 15	Omit the words 'or of the owner of the quarry'.
The Mines and Quarries (Tipping Plans) Rules 1971	
Regulation 3	Omit the words 'and quarries'.
The Safety Representatives and Safety Committees Regulations 1977	
Regulation 5	In paragraph (4), after the words 'Mines and Quarries Act 1954' insert the words 'or regulation 40 of the Quarries Regulations 1999'.
The Electricity at Work Regulations 1989	
Regulation 3	For sub-paragraph (b) of paragraph (1) substitute the following sub-paragraph— '(b) (i) manager, in relation to a mine within the meaning of the Mines and Quarries Act 1954, and (ii) operator, in relation to a quarry within the meaning of regulation 3 of the Quarries Regulations 1999, to ensure that all requirements or prohibitions imposed by or under these Regulations are complied with in so far as they relate to the mine of which he is the manager or quarry of which he is the operator and to matters which are within his control.'.
The Pressure Systems and Transportable Gas Containers Regulations 1989	
Regulation 2	In paragraph (1) in the definition of 'user', omit the words 'a mine or any part of a quarry (both within the meaning of section 180 of the Mines and Quarries Act 1954), it means the manager for the time being of that mine or that part of the quarry' and substitute the following words— '(a) a mine within the meaning of section 180 of the Mines and Quarries Act 1954 it means the manager for the time being of that mine; (b) a quarry within the meaning of regulation 3 of the Quarries Regulations 1999 it means the operator for the time being of that quarry.'.
The Workplace (Health, Safety and Welfare) Regulations 1992	
Regulation 2	In paragraph (1), omit the definition of 'quarry' and substitute the following definition— '"quarry" means a quarry within the meaning of the Quarries Regulations 1999;'.

Column 1 ***Provision***	***Column 2*** ***Extent of modification***
The Ionising Radiations (Outside Workers) Regulations 1993	
Regulation 2	Omit paragraph (4) and substitute the following paragraph— 'Duties under these Regulations imposed on the operator shall also be imposed upon— (a) in a mine within the meaning of section 180 of the Mines and Quarries Act 1954, the manager; and (b) in a quarry within the meaning of regulation 3 of the Quarries Regulations 1999, the operator, in so far as those duties relate to the mine or part of the mine of which he is the manager or the quarry of which he is the operator and to matters under his control.'.
The Railways (Safety Case) Regulations 1994	
Regulation 2	In paragraph (1), omit the definitions of 'mine' and 'quarry' and insert the following definitions— '"mine" has the meaning assigned to it by section 180 of the Mines and Quarries Act 1954;' and '"quarry" has the meaning assigned to it by regulation 3 of the Quarries Regulations 1999;'.
The Railways (Safety Critical Work) Regulations 1994	
Regulation 2	In paragraph (1), omit the definitions of 'mine' and 'quarry' and substitute the following definitions— '"mine" has the meaning assigned to it by section 180 of the Mines and Quarries Act 1954;' and '"quarry" has the meaning assigned to it by regulation 3 of the Quarries Regulations 1999;'.
. . .	. . .
The Carriage of Dangerous Goods by Rail Regulations 1996	
Regulation 1	In paragraph (2), omit the definitions of 'mine' and 'quarry' and substitute the following definitions— '"mine" has the meaning assigned to it by section 180 of the Mines and Quarries Act 1954;' and '"quarry" has the meaning assigned to it by regulation 3 of the Quarries Regulations 1999;'
. . .	
The Carriage of Dangerous Goods (Classification, Packaging and Labelling) and Use of Transportable Pressure Receptacles Regulations 1996	

Part 7

Column 1 ***Provision***	***Column 2*** ***Extent of modification***
Regulation 2	In subsection (1), omit the definitions of 'mine' and 'quarry' and substitute the following definitions— '"mine" has the meaning assigned to it by section 180 of the Mines and Quarries Act 1954;' and '"quarry" has the meaning assigned to it by regulation 3 of the Quarries Regulations 1999;'
The Railways Safety (Miscellaneous Provisions) Regulations 1997	
Regulation 2	In paragraph (1), omit the definitions of 'mine' and 'quarry' and substitute the following definitions— '"mine" has the meaning assigned to it by section 180 of the Mines and Quarries Act 1954;' and '"quarry" has the meaning assigned to it by regulation 3 of the Quarries Regulations 1999;'
The Health and Safety (Enforcing Authority) Regulations 1998	
Regulation 2	In paragraph (1), omit the definition of 'quarry' and substitute the following definition— '"quarry" has the meaning assigned to it by regulation 3 of the Quarries Regulations 1999;'.
Schedule 2	In paragraph 1, omit the words 'section 139(2) of the Mines and Quarries Act 1954' and substitute the following— 'regulation 45(1) of the Quarries Regulations 1999'.
The Gas Safety (Installation and Use) Regulations 1998	
Regulation 2	For sub-paragraph (a) of paragraph (4), substitute the following sub-paragraph— '(a) (i) a mine within the meaning of the Mines and Quarries Act 1954 or any place deemed to form part of a mine for the purposes of that Act, or (ii) a quarry within the meaning of the Quarries Regulations 1999 or any place deemed to form part of a quarry for the purposes of those Regulations.'
The Transport of Dangerous Goods (Safety Advisers) Regulations 1999	
Regulation 2	In paragraph (1), omit the definitions of 'mine' and 'quarry' and substitute the following definitions— '"mine" has the meaning assigned to it by section 180 of the Mines and Quarries Act 1954;' and '"quarry" has the meaning assigned to it by regulation 3 of the Quarries Regulations 1999;'
. . .	. . .

[7.3796]–[7.3800]

Statutory annotations

Amendment

Entry relating to 'Reporting of Injuries, Diseases and Dangerous Occurrences Regulations 1995' (omitted) revoked by SI 2013/1471, reg 18(1), Sch 4.

PART 8
SUBSTANCES, NOISE, RADIATION AND OTHER PHENOMENA HAZARDOUS TO HEALTH

Part 8A

Substances, noise, radiation and other phenomena hazardous to health: General

Part 8

Control of Substances Hazardous to Health Regulations 2002

SI 2002/2677

as amended by SIs 2003/978, 2004/568, 2004/3386, 2006/557, 2006/2739, 2007/1573, 2007/1894, 2008/960, 2009/716 and SI 2013/1478

SCHEDULE 2

PROHIBITION OF CERTAIN SUBSTANCES HAZARDOUS TO HEALTH FOR CERTAIN PURPOSES

Regulation 4(1)

[8.149]

	Column 1 ***Description of substance***	**Column 2** ***Purpose for which the substance is prohibited***
1	2-naphthylamine; benzidine; 4-aminodiphenyl; 4-nitrodiphenyl; their salts and any substance containing any of those compounds, in a total concentration equal to or greater than 0.1 per cent by mass.	Manufacture and use for all purposes including any manufacturing process in which a substance described in Column 1 of this item is formed.
2	Sand or other substance containing free silica.	Use as an abrasive for blasting articles in any blasting apparatus.
3	A substance— (a) containing compounds of silicon calculated as silica to the extent of more than 3 per cent by weight of dry material, other than natural sand, zirconium silicate (zircon), calcined china clay, calcined aluminous fireclay, sillimanite, calcined or fused alumina, olivine; or (b) composed of or containing dust or other matter deposited from a fettling or blasting process.	Use as a parting material in connection with the making of metal castings.
4	Carbon disulphide.	Use in the cold-cure process of vulcanising in the proofing of cloth with rubber.
5	Oils other than white oil, or oil of entirely animal or vegetable origin or entirely of mixed animal and vegetable origin.	Use for oiling the spindles of self-acting mules.

Column 1 *Description of substance*		Column 2 *Purpose for which the substance is prohibited*
6	Ground or powdered flint or quartz other than natural sand.	Use in relation to the manufacture or decoration of pottery for the following purposes: (a) the placing of ware for the biscuit fire; (b) the polishing of ware; (c) as the ingredient of a wash for saggars, trucks, bats, cranks, or other articles used in supporting ware during firing; and (d) as dusting or supporting powder in potters' shops.
7	Ground or powdered flint or quartz other than— (a) natural sand; or (b) ground or powdered flint or quartz which forms parts of a slop or paste.	Use in relation to the manufacture or decoration of pottery for any purpose except— (a) use in a separate room or building for— (i) the manufacture of powdered flint or quartz, or (ii) the making of frits or glazes or the making of colours or coloured slips for the decoration of pottery; (b) use for the incorporation of the substance into the body of ware in an enclosure in which no person is employed and which is constructed and ventilated to prevent the escape of dust.
8	Dust or powder of a refractory material containing not less than 80 per cent of silica other than natural sand.	Use for sprinkling the moulds of silica bricks, namely bricks or other articles composed of refractory material and containing not less than 80 per cent of silica.
9	White phosphorus.	Use in the manufacture of matches.
10	Hydrogen cyanide.	Use in fumigation except when— (a) released from an inert material in which hydrogen cyanide is absorbed; (b) generated from a gassing powder; or (c) applied from a cylinder through suitable piping and applicators other than for fumigation in the open air to control or kill mammal pests.
11	. . .	. . .
12	. . .	. . .
[13	. . .	. . .]

In this Schedule—

'aerosol dispenser' means an article which consists of a non-reusable receptacle containing a gas compressed, liquefied or dissolved under pressure, with or without liquid, paste or powder and fitted with a release device allowing the contents to be ejected as solid or liquid particles in suspension in a gas, as a foam, paste or powder or in a liquid state;

'blasting apparatus' means apparatus for cleaning, smoothing, roughening or removing of part of the surface of any article by the use as an abrasive of a jet of sand, metal shot or grit or other material propelled by a blast of compressed air or steam or by a wheel;

'CAS No' is the number assigned to a substance by the Chemical Abstract Service;

'cosmetic product' has the meaning assigned to it in [Article 2 of Regulation (EC) No 1223/2009 of the European Parliament and of the Council on cosmetic products (recast) as amended from time to time] (including any aerosol dispenser containing a cosmetic product);

'gassing powder' means a chemical compound in powder form which reacts with atmospheric moisture to generate hydrogen cyanide;

'medicinal product' means a substance or preparation which is—

(a) intended for use as a medicinal product within the meaning of section 130 of the Medicines Act 1968; or

(b) a substance or preparation specified in an order made under section 104 or 105 of the Medicines Act 1968 which is for the time being in force and which directs that specified provisions of that Act shall have effect in relation to that substance or preparation as such provisions have effect in relation to medicinal products within the meaning of that Act;

'use as a parting material' means the application of the material to the surface or parts of the surface of a pattern or of a mould so as to facilitate the separation of the pattern from the mould or the separation of parts of the mould;

'white oil' means a refined mineral oil conforming to a specification approved by the Executive and certified by its manufacturer as so conforming.

[8.150]

Statutory annotations

Amendment

In definition 'cosmetic product' words from 'Article 2 of' to 'time to time' in square brackets substituted by SI 2013/1478, reg 25, Sch 5, para 13.

Reach Enforcement Regulations 2008

SI 2008/2852

as amended by SIs 2010/1513, 2013/1948, 2013/2919 and 2014/2882

PART 1

INTRODUCTION

[8.254]

2 Interpretation

(1) In these Regulations—

'a listed REACH provision' means a provision of REACH listed in the REACH table;

['REACH' means Regulation (EC) No 1907/2006 of the European Parliament and of the Council concerning the Registration, Evaluation, Authorisation and Restriction of Chemicals as amended from time to time;]

'the REACH table' means the table in Schedule 1 to these Regulations.

(2) In these Regulations—

['asbestos-containing article' means an article in its entirety containing any of the asbestos fibres listed at point 6 of Annex XVII to REACH;]

['asbestos exemption certificate' has the meaning given in paragraph 2 of Schedule 5A;]

'authorised person' means a person authorised by an enforcing authority under regulation 10;

'competent authority' has the same meaning it has in REACH and in respect of the United Kingdom the authorities are—

(a) in England, the Secretary of State;

(b) in Scotland, the Scottish Ministers;

(c) in Wales, the Welsh Ministers;

(d) in Northern Ireland, the Department of Enterprise, Trade and Investment and the Department of the Environment acting alone or jointly; and

(e) in relation to matters outside the competence of a devolved administration or the Assembly, the Secretary of State;

'devolved administration or the Assembly' means the Scottish Ministers, the Welsh Ministers or the Northern Ireland Assembly;

'enforcement duty' means a duty placed on an enforcing authority under regulation 3;

'enforcing authority' means—

(a) the Department of the Environment;

(b) the Environment Agency;

(c) the Health and Safety Executive;

(d) the Health and Safety Executive for Northern Ireland;

(e) a local (consumer safety) authority;

(f) a local (health and safety) authority;

(g) the Scottish Environment Protection Agency;

(h) the Secretary of State;

['the Executive' means—

(a) in Great Britain, the Health and Safety Executive;

(b) in Northern Ireland, the Health and Safety Executive for Northern Ireland;]

'a local (consumer safety) authority' means—

(a) in Greater London, a London borough council, the Common Council of the City of London, the Sub-Treasurer of the Inner Temple and the Under-Treasurer of the Middle Temple;
(b) in England outside Greater London, a county council or, in relation to an area for which there is a district council but no county council, the district council, and the Council of the Isles of Scilly;
(c) in Northern Ireland, a district council established under section 1 of the Local Government Act (Northern Ireland) 1972;
(d) in Scotland, a council constituted under section 2 of the Local Government etc (Scotland) Act 1994;
(e) in Wales, a county council or county borough council;

'a local (health and safety) authority' means—

(a) in Greater London, a London borough council, the Common Council of the City of London, the Sub-Treasurer of the Inner Temple and the Under-Treasurer of the Middle Temple;
(b) in England, outside Greater London, a district council or, in relation to an area for which there is a county council but no district council, the county council, and the Council of the Isles of Scilly;
(c) in Northern Ireland, a district council established under section 1 of the Local Government Act (Northern Ireland) 1972;
(d) in Scotland, a council constituted under section 2 of the Local Government etc (Scotland) Act 1994;
(e) in Wales, a county council or county borough council;

['offshore installation' means—

(a) an offshore installation within the meaning of section 44(1) of the Petroleum Act 1998; or
(b) a carbon storage installation within the meaning of section 30(5) of the Energy Act 2008;]

['paint stripper' means material that, when applied to a coated substrate, softens the coating so that it can be removed easily, and 'paint stripping' is to be construed accordingly;

'paint stripper containing dichloromethane' means any paint stripper containing dichloromethane in a concentration equal to or greater than 0.1% by weight;

'professional' means any person undertaking paint stripping in the course of their professional activity other than within an industrial installation where the conditions set out in entry 59 paragraph 4 of Annex XVII to REACH are complied with.]

['railway' means—

(a) except in Northern Ireland, any system of transport the operation of which is specified in regulation 3(2) of the Health and Safety (Enforcing Authority for Railways and Other Guided Transport Systems) Regulations 2006;
(b) in relation to Northern Ireland, any railway or tramway which in either case is used for the carriage of persons or goods;]

['relevant waters', in relation to an offshore installation, has the meaning given in section 44(4) of the Petroleum Act 1998;]

['Scotland' includes Scottish controlled waters;]

['Scottish controlled waters' means—

(a) in relation to an offshore installation which is maintained (or intended to be established) for the purposes of the exploration for, or exploitation of, petroleum (within the meaning of section 1 of the Petroleum Act 1998), any waters which are controlled waters within the meaning of section 30A(1) of the Control of Pollution Act 1974; and

(b) in relation to any other offshore installation, waters within the seaward limits of the territorial sea adjacent to Scotland].

(3) Other expressions used in these Regulations which are used in REACH have the meaning they bear in REACH.

[8.254A]

Statutory annotations

Amendment

Para (1): definition 'REACH' substituted by SI 2013/2919, regs 2, 3(1).

Para (2): definition 'asbestos-containing article' inserted by SI 2013/2919, regs 2, 3(2)(a); definition 'asbestos exemption certificate' inserted by SI 2013/2919, regs 2, 3(2)(a); in definition 'enforcing authority' para (fa) inserted by SI 2013/2919, regs 2, 3(2)(b); in definition 'enforcing authority' para (i) inserted by SI 2013/755, art 4(2), Sch 4, paras 302, 303; definitions 'paint stripper, 'paint stripper containing dichloromethane' and 'professional' inserted by SI 2014/2882, reg 2(1), (2); definition 'railway' inserted by SI 2013/2919, regs 2, 3(2)(d).

PART 2

ENFORCEMENT

[8.258]

6 Health and safety enforcement

[(1) This regulation applies where the Executive is jointly under an enforcement duty with any of the following enforcing authorities, whether or not any other enforcing authority is also under an enforcement duty in respect of the same listed REACH provision—

(a) a local (health and safety) authority;

(b) the Office of Rail Regulation.]

(2) Schedule 3 (health and safety enforcement) has effect.

(3) In circumstances not provided for under Schedule 3, the Executive must perform the joint duty on behalf of a local (health and safety) authority.

(4) The arrangements under Schedule 3 and the arrangement under paragraph (3) may be varied by an enforcement agreement made in accordance with regulation 5.

(5) . . .

[8.258A]

Statutory annotations

Amendment

Para (1): substituted by SI 2013/2919, regs 2, 4(a).

Para (5): revoked by SI 2013/2919, reg 4(b).

PART 3

EXEMPTIONS

[8.260A]

[8A Placing asbestos-containing articles on the market]

[(1) A person who places an asbestos-containing article on the market does not breach the restriction on this activity provided for by Article 67 of REACH where—

(a) this is done in accordance with an asbestos exemption certificate; and

(b) the asbestos-containing article was installed or in service before 1st January 2005.

(2) Schedule 5A (placing asbestos-containing articles on the market) has effect.]

[8.260B]

Statutory annotations

Amendment

Para (1): substituted by SI 2013/2919, regs 2, 4(a).

Para (5): revoked by SI 2013/2919, reg 4(b).

[8.260C]

[8B Placing on the market and use of paint stripper containing dichloromethane

[(1) A person ('P') who places on the market paint stripper containing dichloromethane for supply to a professional or to a person for use by a professional does not breach the restriction on this activity provided for by Article 67 of REACH where P complies with the provisions of paragraph 1 of Schedule 5B.

(2) A professional who uses paint stripper containing dichloromethane does not breach the restriction on this activity provided for by Article 67 of REACH where that professional complies with the provisions of paragraphs 2 and 3 of Schedule 5B.

(3) Schedule 5B (Placing on the market and use of paint stripper containing dichloromethane) has effect.]

[8.260D]

Statutory annotations

Amendment

Inserted by SI 2014/2882, reg 2(1), (3).

[8.263]

11 Offences

(1) Except in relation to Article 67 of REACH, it is an offence for a person to contravene a listed REACH provision or cause or permit another person to do so.

(2) On and after 1st June 2009, it is an offence for a person to contravene Article 67 of REACH or cause or permit another person to do so.

(3) It is an offence for a person when subject to paragraph 6 of Schedule 4 (defence exemption certificates)—

(a) to provide a false certificate or copy; or

(b) to fail to provide when requested, as appropriate—

(i) the defence exemption certificate;

(ii) a copy of the certificate made by the Secretary of State; or

(iii) a copy of an extract of the certificate made by the Secretary of State,

or cause or permit another person to do so.

(4) On and after 1st June 2009, it is an offence for a person to contravene paragraph 5(b) or 6 of Part 1 of Schedule 5 (marketing and use of leaded paint) or cause or permit another person to do so.

[(5) Where a person ('P') is charged with an offence for a contravention of Article 67 of REACH under regulation 11(2) of these Regulations for placing on the market paint stripper containing dichloromethane for supply to—

(a) a person who is not a professional; or

(b) a person for use by a person who is not a professional,

it is a defence that P took all reasonable steps and exercised all due diligence to avoid the commission of the offence.]

[8.263A]

Statutory annotations

Amendment

Para (5) inserted by SI 2014/2882, reg 2(1), (4).

[PART 8

REVIEW

[8.275A]

[24 Review of these Regulations]

[(1) The Secretary of State must from time to time—

(a) carry out a review of these Regulations in relation to England;

(b) set out the conclusions of the review in a report; and

(c) publish the report.

(2) In carrying out the review, the Secretary of State must, so far as is reasonable, have regard to how REACH (which is enforced by means of these Regulations) is enforced in other member States.

(3) The report must, in particular—

(a) set out the objectives intended to be achieved by the regulatory system established by these Regulations;

(b) assess the extent to which the objectives are achieved; and

(c) assess whether those objectives remain appropriate and, if so, the extent to which they could be achieved with a system that imposes less regulation.

(4) The first report under this regulation must be published before the end of the period of five years beginning with 16th December 2013.

(5) Reports under this regulation are afterwards to be published at intervals not exceeding five years.]

[8.275B]

Statutory annotations

Amendment

Inserted by SI 2013/2919, regs 2, 6.

SCHEDULE 1

TABLE OF REACH PROVISIONS

Regulation 2(1)

[8.276]

Provision of REACH	***Subject Matter***	***Enforcing authority***			
		England and Wales	***Scotland***	***Northern Ireland***	***Offshore installations***
Article 5.	Prohibition on manufacture or placing on the market of substances on their own, in [mixtures] or in articles without registration.	The Health and Safety Executive.	The Health and Safety Executive.	The Health and Safety Executive for Northern Ireland.	The Health and Safety Executive. The Health and Safety Executive for Northern Ireland.
Article 7(2) (Subject to Article 7(6)).	Requirement on a producer or an importer of an article to notify the European Chemicals Agency.	The Health and Safety Executive.	The Health and Safety Executive.	The Health and Safety Executive for Northern Ireland.	The Health and Safety Executive. The Health and Safety Executive for Northern Ireland.
Article 7(3) (Subject to Article 7(6)).	Requirement on a producer or an importer to supply appropriate instructions to the recipient of the article.	The Health and Safety Executive.	The Health and Safety Executive.	The Health and Safety Executive for Northern Ireland.	The Health and Safety Executive. The Health and Safety Executive for Northern Ireland.
Article 8(2)(first sentence).	Requirement on a representative of a non-community manufacturer to comply with obligations of importers.	The Health and Safety Executive.	The Health and Safety Executive.	The Health and Safety Executive for Northern Ireland.	The Health and Safety Executive. The Health and Safety Executive for Northern Ireland.

Provision of REACH	Subject Matter	Enforcing authority			
		England and Wales	Scotland	Northern Ireland	Offshore installations
Article 9(2).	Requirement on a manufacturer, importer or producer of articles to notify the European Chemicals Agency of relevant information when that person is seeking to rely on the exemption for product and process orientated research and development.	The Health and Safety Executive.	The Health and Safety Executive.	The Health and Safety Executive for Northern Ireland.	The Health and Safety Executive. The Health and Safety Executive for Northern Ireland.
Article 9(6).	Requirement on a manufacturer, an importer or a producer of articles to comply with conditions imposed by the European Chemicals Agency.	The Health and Safety Executive. The Environment Agency. Local (health and safety) authorities. Local (consumer safety) authorities.	The Health and Safety Executive. The Scottish Environment Protection Agency Local (health and safety) authorities. Local (consumer safety) authorities	The Health and Safety Executive for Northern Ireland. The Department of the Environment. Local (health and safety) authorities. Local (consumer safety) authorities.	The Health and Safety Executive. The Health and Safety Executive for Northern Ireland. Secretary of State.

Provision of REACH	***Subject Matter***	***Enforcing authority***			
		England and Wales	***Scotland***	***Northern Ireland***	***Offshore installations***
Article 12(2).	Requirement on a manufacturer or an importer to notify the European Chemicals Agency of additional information where it reaches the next tonnage threshold.	The Health and Safety Executive.	The Health and Safety Executive.	The Health and Safety Executive for Northern Ireland.	The Health and Safety Executive. The Health and Safety Executive for Northern Ireland.
Article 14(6).	Requirement on a registrant to identify and apply the appropriate measures adequately to control the risks identified in the chemical safety assessment and where suitable to recommend them in a safety data sheet that is supplied.	The Health and Safety Executive. The Environment Agency. Local (health and safety) authorities.	The Health and Safety Executive. The Scottish Environment Protection Agency. Local (health and safety) authorities.	The Health and Safety Executive. The Department of the Environment. Local (health and safety) authorities.	The Health and Safety Executive. The Health and Safety Executive for Northern Ireland. Secretary of State.
Article 14(7).	Requirement on a registrant to keep a chemical safety report available and up to date.	The Health and Safety Executive.	The Health and Safety Executive.	The Health and Safety Executive for Northern Ireland.	The Health and Safety Executive. The Health and Safety Executive for Northern Ireland.
Article 22(1).	Requirement on a registrant to update the registrant's registration.	The Health and Safety Executive.	The Health and Safety Executive.	The Health and Safety Executive for Northern Ireland.	The Health and Safety Executive. The Health and Safety Executive for Northern Ireland.

[8.276] 8A Phenomena hazardous to health: General

Provision of REACH	***Subject Matter***	***Enforcing authority***			
		England and Wales	***Scotland***	***Northern Ireland***	***Offshore installations***
Article 22(2).	Requirement on a registrant to submit to the European Chemicals Agency an updated registration where this is required by a decision of the European Chemicals Agency.	The Health and Safety Executive.	The Health and Safety Executive.	The Health and Safety Executive for Northern Ireland.	The Health and Safety Executive. The Health and Safety Executive for Northern Ireland.
Article 24(2).	Requirement on a manufacturer or an importer to notify, in accordance with Article 10 (information to be submitted for general registration purposes) and Article 12 (information to be submitted depending on tonnage), if the quantity of a notified substance reaches the next tonnage threshold.	The Health and Safety Executive.	The Health and Safety Executive.	The Health and Safety Executive for Northern Ireland.	The Health and Safety Executive. The Health and Safety Executive for Northern Ireland.

Provision of REACH	***Subject Matter***	***Enforcing authority***			
		England and Wales	***Scotland***	***Northern Ireland***	***Offshore installations***
Article 26(1).	Requirement on a potential registrant of a non-phase in substance or a potential registrant of a phase-in substance who has not pre-registered, to inquire of the European Chemicals Agency if the same substance has been registered.	The Health and Safety Executive.	The Health and Safety Executive.	The Health and Safety Executive for Northern Ireland.	The Health and Safety Executive. The Health and Safety Executive for Northern Ireland.
Article 27(1)(a).	Requirement on a potential registrant to request information on vertebrate animals tests from a previous registrant.	The Health and Safety Executive.	The Health and Safety Executive.	The Health and Safety Executive for Northern Ireland.	The Health and Safety Executive. The Health and Safety Executive for Northern Ireland.
Article 30(6).	Subject to Article 30(5) (appeals), failure by owner of a relevant study to provide either proof of the cost of the study or the study itself.	The Health and Safety Executive.	The Health and Safety Executive.	The Health and Safety Executive for Northern Ireland.	The Health and Safety Executive. The Health and Safety Executive for Northern Ireland.

[8.276] 8A Phenomena hazardous to health: General

Provision of REACH	***Subject Matter***	***Enforcing authority***			
		England and Wales	***Scotland***	***Northern Ireland***	***Offshore installations***
Article 31(1).	Requirement on a supplier of a substance or a [mixture] to provide the recipient of that substance or [mixture] with a safety data sheet compiled in accordance with Annex II (guide to the compilation of safety data sheets).	The Health and Safety Executive.	The Health and Safety Executive.	The Health and Safety Executive for Northern Ireland.	The Health and Safety Executive. The Health and Safety Executive for Northern Ireland.
Article 31(2)-(9).	Requirement on an actor in the supply chain who has been requested to perform a chemical safety assessment to ensure that information in the safety data sheet is consistent with the information in the assessment.	The Health and Safety Executive.	The Health and Safety Executive.	The Health and Safety Executive for Northern Ireland.	The Health and Safety Executive. The Health and Safety Executive for Northern Ireland.
	Requirement on a supplier to provide a safety data sheet when requested for a [mixture] which falls within Article 31(3).	The Health and Safety Executive.	The Health and Safety Executive.	The Health and Safety Executive for Northern Ireland.	The Health and Safety Executive. The Health and Safety Executive for Northern Ireland.

Provision of REACH	***Subject Matter***	***Enforcing authority***			
		England and Wales	***Scotland***	***Northern Ireland***	***Offshore installations***
	Requirement on a supplier to provide a downstream user or a distributor with a safety data sheet when requested for a [mixture] or dangerous substance which is offered or sold to the general public.	The Health and Safety Executive.	The Health and Safety Executive.	The Health and Safety Executive for Northern Ireland.	The Health and Safety Executive. The Health and Safety Executive for Northern Ireland.
	Requirement on a supplier to provide to the recipient of a substance or a [mixture] a safety data sheet in the language of the Member State concerned.	The Health and Safety Executive.	The Health and Safety Executive.	The Health and Safety Executive for Northern Ireland.	The Health and Safety Executive. The Health and Safety Executive for Northern Ireland.
	Requirement that the safety data sheet contains the information listed in Article 31(6).	The Health and Safety Executive.	The Health and Safety Executive.	The Health and Safety Executive for Northern Ireland.	The Health and Safety Executive. The Health and Safety Executive for Northern Ireland.

[8.276] 8A Phenomena hazardous to health: General

Provision of REACH	*Subject Matter*	*Enforcing authority*			
		England and Wales	***Scotland***	***Northern Ireland***	***Offshore installations***
	Requirement on an actor in the supply chain to place the relevant exposure scenarios in an annex to the safety data sheet in accordance with section 3 of Annex XI (general rules for adaptation of the standard testing regime set out in Annexes VII to X).	The Health and Safety Executive.	The Health and Safety Executive.	The Health and Safety Executive for Northern Ireland.	The Health and Safety Executive. The Health and Safety Executive for Northern Ireland.
	Requirement on a downstream user to include the relevant exposure scenarios and use other relevant information from the safety data sheet in its own safety data sheet for identified uses.	The Health and Safety Executive.	The Health and Safety Executive.	The Health and Safety Executive for Northern Ireland.	The Health and Safety Executive. The Health and Safety Executive for Northern Ireland.

Provision of REACH	***Subject Matter***	***Enforcing authority***			
		England and Wales	***Scotland***	***Northern Ireland***	***Offshore installations***
	Requirement on a distributor to pass on relevant exposure scenarios and use other relevant information from the safety data sheet when compiling its own data sheet for identified uses.	The Health and Safety Executive.	The Health and Safety Executive.	The Health and Safety Executive for Northern Ireland.	The Health and Safety Executive. The Health and Safety Executive for Northern Ireland.
	Requirement to provide a safety data sheet free of charge either electronically or on paper [, and by no later than the date on which the substance or mixture is first supplied].	The Health and Safety Executive.	The Health and Safety Executive.	The Health and Safety Executive for Northern Ireland.	The Health and Safety Executive. The Health and Safety Executive for Northern Ireland.
	Requirement on a supplier to update a safety data sheet.	The Health and Safety Executive.	The Health and Safety Executive.	The Health and Safety Executive for Northern Ireland.	The Health and Safety Executive. The Health and Safety Executive for Northern Ireland.
[Article 31(10)	Requirement on a supplier to include the classification of substances or mixtures in a safety data sheet.	The Health and Safety Executive.	The Health and Safety Executive.	The Health and Safety Executive for Northern Ireland.	The Health and Safety Executive. The Health and Safety Executive for Northern Ireland.

Provision of REACH	*Subject Matter*	*Enforcing authority*			
		England and Wales	*Scotland*	*Northern Ireland*	*Offshore installations*
Article 32.	Duty on a supplier to provide to all recipients to whom supplies have been made within the preceding twelve months an updated safety data sheet.	The Health and Safety Executive.	The Health and Safety Executive.	The Health and Safety Executive for Northern Ireland.	The Health and Safety Executive. The Health and Safety Executive for Northern Ireland.
	Duty on a supplier who does not have to supply a safety data sheet to provide the recipient with the information in Article 32(1).	The Health and Safety Executive.	The Health and Safety Executive.	The Health and Safety Executive for Northern Ireland.	The Health and Safety Executive. The Health and Safety Executive for Northern Ireland.
Article 32(2), (3).	Duty on a supplier to provide information free of charge no later than the time of first delivery of a substance or a [mixture] after 1st June 2007.	The Health and Safety Executive.	The Health and Safety Executive.	The Health and Safety Executive for Northern Ireland.	The Health and Safety Executive. The Health and Safety Executive for Northern Ireland.
	Duty on a supplier to update the information when required by Article 32(1).	The Health and Safety Executive.	The Health and Safety Executive.	The Health and Safety Executive for Northern Ireland.	The Health and Safety Executive. The Health and Safety Executive for Northern Ireland.

Provision of REACH	***Subject Matter***	***Enforcing authority***			
		England and Wales	***Scotland***	***Northern Ireland***	***Offshore installations***
	Duty on a supplier to provide to all recipients to whom they have supplied within the preceding twelve months updated information.	The Health and Safety Executive.	The Health and Safety Executive.	The Health and Safety Executive for Northern Ireland.	The Health and Safety Executive. The Health and Safety Executive for Northern Ireland.
Article 33(1).	Duty on a supplier of an article meeting the criteria in Article 57 (substances to be included in Annex XIV) and identified in accordance with Article 59(1) to provide the recipient with sufficient information to allow safe use, including as a minimum the name of that substance.	The Health and Safety Executive. Local (health and safety) authorities.	The Health and Safety Executive. Local (health and safety) authorities.	The Health and Safety Executive for Northern Ireland. Local (health and safety) authorities.	The Health and Safety Executive. The Health and Safety Executive for Northern Ireland.

[8.276] 8A Phenomena hazardous to health: General

Provision of REACH	***Subject Matter***	***Enforcing authority***			
		England and Wales	***Scotland***	***Northern Ireland***	***Offshore installations***
Article 33(2).	Duty on a supplier of an article meeting the criteria in Article 57(substances to be included in Annex XIV) and identified in accordance with Article 59(1) to provide a consumer on request with sufficient information to allow the safe use, including as a minimum the name of that substance.	The Health and Safety Executive. [Local (consumer safety) authorities].	The Health and Safety Executive. [Local (consumer safety) authorities].	The Health and Safety Executive for Northern Ireland. [Local (consumer safety) authorities].	The Health and Safety Executive. The Health and Safety Executive for Northern Ireland.
Article 34.	Duty on an actor in the supply chain to communicate the information referred to in Article 34(a) and (b) to the next actor or distributor up the supply chain.	The Health and Safety Executive.	The Health and Safety Executive.	The Health and Safety Executive for Northern Ireland.	The Health and Safety Executive. The Health and Safety Executive for Northern Ireland.

Provision of REACH	***Subject Matter***	***Enforcing authority***			
		England and Wales	***Scotland***	***Northern Ireland***	***Offshore installations***
Article 35.	Requirement on an employer to provide workers and their representatives with access to information relating to substances or [mixtures] which they may use or be exposed to in the course of their work.	The Health and Safety Executive. Local (health and safety) authorities.	The Health and Safety Executive. Local (health and safety) authorities.	The Health and Safety Executive for Northern Ireland. Local (health and safety) authorities.	The Health and Safety Executive. The Health and Safety Executive for Northern Ireland.
Article 36(1).	Requirement on a manufacturer, importer, downstream user or distributor to keep available for at least ten years after it last manufactured, imported, supplied or used the substance or [mixture], all the information it requires to carry out its duties under REACH.	The Health and Safety Executive. The Environment Agency. Local (consumer safety) authorities. Local (health and safety) authorities.	The Health and Safety Executive. The Scottish Environment Protection Agency. Local (consumer safety) authorities. Local (health and safety) authorities.	The Health and Safety Executive for Northern Ireland. The Department of the Environment. Local (consumer safety) authorities. Local (health and safety) authorities.	The Health and Safety Executive. The Health and Safety Executive for Northern Ireland. Secretary of State.

Provision of REACH	Subject Matter	Enforcing authority			
		England and Wales	Scotland	Northern Ireland	Offshore installations
	Requirement on a manufacturer, importer, downstream user or distributor to submit or make available information to a competent authority or the European Chemicals Agency when requested to do so.	The Health and Safety Executive.	The Health and Safety Executive.	The Health and Safety Executive for Northern Ireland.	The Health and Safety Executive. The Health and Safety Executive for Northern Ireland.
Article 36(2).	Requirement on a party responsible for liquidating the registrant's, downstream user's or distributor's undertaking or assuming responsibility for the placing on the market of the substance or [mixture] concerned, to comply with Article 36(1).	The Health and Safety Executive.	The Health and Safety Executive.	The Health and Safety Executive for Northern Ireland.	The Health and Safety Executive. The Health and Safety Executive for Northern Ireland.

Provision of REACH	***Subject Matter***	***Enforcing authority***			
		England and Wales	***Scotland***	***Northern Ireland***	***Offshore installations***
Article 37(3).	Requirements providing when a manufacturer, importer or downstream user must comply with Article 14 (chemical safety report and duty to apply and recommend risk reduction measures) for registered substances.	The Health and Safety Executive.	The Health and Safety Executive.	The Health and Safety Executive for Northern Ireland.	The Health and Safety Executive. The Health and Safety Executive for Northern Ireland.
	Requirements providing when a manufacturer, importer or downstream user must comply with Article 14 for a phase-in substance.	The Health and Safety Executive.	The Health and Safety Executive.	The Health and Safety Executive for Northern Ireland.	The Health and Safety Executive. The Health and Safety Executive for Northern Ireland.

[8.276] 8A Phenomena hazardous to health: General

Provision of REACH	***Subject Matter***	***Enforcing authority***			
		England and Wales	***Scotland***	***Northern Ireland***	***Offshore installations***
	Requirement on a manufacturer, importer or downstream user to provide the European Chemicals Agency and downstream users with reasons why a use identified by a downstream user cannot be included in the chemical safety report.	The Health and Safety Executive.	The Health and Safety Executive.	The Health and Safety Executive for Northern Ireland.	The Health and Safety Executive. The Health and Safety Executive for Northern Ireland.

Part 8

Provision of REACH	***Subject Matter***	***Enforcing authority***			
		England and Wales	***Scotland***	***Northern Ireland***	***Offshore installations***
	Prohibition on a manufacturer, importer or downstream user supplying a downstream user with a substance without stating in the information referred to in Article 31 (requirements for safety data sheets) and Article 32 (duty to communicate information down the supply chain for substances on their own or in [mixtures] for which a safety data sheet is not required) the reasons why a use is not an identified-use.	The Health and Safety Executive.	The Health and Safety Executive.	The Health and Safety Executive for Northern Ireland.	The Health and Safety Executive. The Health and Safety Executive for Northern Ireland.
	Requirement on a manufacturer or an importer to include the use notified under Article 37(2) in the update of the registration under Article 22(1)(d).	The Health and Safety Executive.	The Health and Safety Executive.	The Health and Safety Executive for Northern Ireland.	The Health and Safety Executive. The Health and Safety Executive for Northern Ireland.

[8.276] 8A Phenomena hazardous to health: General

Provision of REACH	***Subject Matter***	***Enforcing authority***			
		England and Wales	***Scotland***	***Northern Ireland***	***Offshore installations***
Article 37(4).	Requirement on a downstream user to prepare a chemical safety report in accordance with Annex XII (general provisions for downstream users to assess substances and prepare chemical safety reports) for any use outside of the conditions described in an exposure scenario, a use and exposure category in a safety data sheet or for any use that a supplier advises against.	The Health and Safety Executive. The Environment Agency. Local (health and safety) authorities.	The Health and Safety Executive. The Scottish Environment Protection Agency. Local (health and safety) authorities.	The Health and Safety Executive for Northern Ireland. The Department of the Environment. Local (health and safety) authorities.	The Health and Safety Executive. The Health and Safety Executive for Northern Ireland. Secretary of State.

Provision of REACH	***Subject Matter***	***Enforcing authority***			
		England and Wales	***Scotland***	***Northern Ireland***	***Offshore installations***
Article 37(5) (read in association with Article 39(1)).	Requirement on a downstream user to identify and apply appropriate measures to adequately control risks identified in (a) a safety data sheet; (b) its own chemical safety assessment; or (c) any information received in accordance with Article 32 (duty to communicate information down the supply chain for substances on their own or in preparations for which a safety data sheet is not required).	The Health and Safety Executive. The Environment Agency. Local (health and safety) authorities.	The Health and Safety Executive. The Scottish Environment Protection Agency. Local (health and safety) authorities.	The Health and Safety Executive for Northern Ireland. The Department of the Environment. Local (health and safety) authorities.	The Health and Safety Executive. The Health and Safety Executive for Northern Ireland. Secretary of State

[8.276] 8A Phenomena hazardous to health: General

Provision of REACH	***Subject Matter***	***Enforcing authority***			
		England and Wales	***Scotland***	***Northern Ireland***	***Offshore installations***
Article 37(5) (read in association with Article 39(1)).	Requirement on a downstream user to recommend, where suitable, measures adequately to control the risks identified in (a) a safety data sheet supplied to it; (b) its own chemical safety assessment or (c) any information received in accordance with Article 32.	The Health and Safety Executive.. The Environment Agency. Local (health and safety) authorities.	The Health and Safety Executive. The Scottish Environment Protection Agency. Local (health and safety) authorities.	The Health and Safety Executive for Northern Ireland. The Department of the Environment. Local (health and safety) authorities.	The Health and Safety Executive. The Health and Safety Executive for Northern Ireland. Secretary of State.
Article 37(6).	Requirement on a downstream user to identify and apply appropriate risk management measures that are needed to ensure that the risks to human health and the environment are adequately controlled.	The Health and Safety Executive. The Environment Agency. Local (health and safety) authorities.	The Health and Safety Executive. The Scottish Environment Protection Agency. Local (health and safety) authorities.	The Health and Safety Executive for Northern Ireland. The Department of the Environment. Local (health and safety) authorities.	The Health and Safety Executive. The Health and Safety Executive for Northern Ireland. Secretary of State.
Article 37(7) (read in association with Article 39(1)).	Requirement on a downstream user to keep its chemical safety report up to date and available.	The Health and Safety Executive.	The Health and Safety Executive.	The Health and Safety Executive for Northern Ireland.	The Health and Safety Executive. The Health and Safety Executive for Northern Ireland.

Provision of REACH	***Subject Matter***	***Enforcing authority***			
		England and Wales	***Scotland***	***Northern Ireland***	***Offshore installations***
Article 38(1) (read in association with Article 39(2)).	Requirement on a downstream user to report information in Article 38(2) to the European Chemicals Agency before commencing or continuing with a particular use of a substance that has been registered by an actor up the supply chain.	The Health and Safety Executive. The Environment Agency. Local (health and safety) authorities.	The Health and Safety Executive. The Scottish Environment Protection Agency. Local (health and safety) authorities.	The Health and Safety Executive for Northern Ireland. The Department of the Environment. Local (health and safety) authorities.	The Health and Safety Executive. The Health and Safety Executive for Northern Ireland. Secretary of State.
Article 38(3).	Requirement on a downstream user to update the information provided in Article 38(2).	The Health and Safety Executive. The Environment Agency Local (health and safety) authorities.	The Health and Safety Executive. The Scottish Environment Protection Agency. Local (health and safety) authorities.	The Health and Safety Executive for Northern Ireland. The Department of the Environment. Local (health and safety) authorities.	The Health and Safety Executive. The Health and Safety Executive for Northern Ireland. Secretary of State.
Article 38(4) (read in association with Article 39(2)).	Requirement on a downstream user to report to the European Chemicals Agency if its classification of a substance is different to that of its supplier.	The Health and Safety Executive.	The Health and Safety Executive.	The Health and Safety Executive for Northern Ireland.	The Health and Safety Executive. The Health and Safety Executive for Northern Ireland.

[8.276] 8A Phenomena hazardous to health: General

Provision of REACH	*Subject Matter*	*Enforcing authority*			
		England and Wales	*Scotland*	*Northern Ireland*	*Offshore installations*
Article 41(4).	Requirement on a registrant to submit information required to the European Chemicals Agency.	The Health and Safety Executive.	The Health and Safety Executive.	The Health and Safety Executive for Northern Ireland.	The Health and Safety Executive. The Health and Safety Executive for Northern Ireland.
Article 46(2).	Requirement on a registrant to submit further information in accordance with a decision prepared by the competent authority.	The Health and Safety Executive.	The Health and Safety Executive.	The Health and Safety Executive for Northern Ireland.	The Health and Safety Executive. The Health and Safety Executive for Northern Ireland.
Article 49(a).	Requirement on a registrant to submit further information as requested by the competent authority in relation to a risk identified for on-site isolated intermediates.	The Health and Safety Executive.	The Health and Safety Executive.	The Health and Safety Executive for Northern Ireland.	The Heath and Safety Executive. The Health and Safety Executive for Northern Ireland.

Part 8

Provision of REACH	***Subject Matter***	***Enforcing authority***			
		England and Wales	***Scotland***	***Northern Ireland***	***Offshore installations***
Article 50(4).	Requirement on a registrant to provide, in accordance with Article 46 (requests for further information and check of information submitted), further information.	The Health and Safety Executive.	The Health and Safety Executive.	The Health and Safety Executive for Northern Ireland.	The Health and Safety Executive. The Health and Safety Executive for Northern Ireland.
Article 56(1) (subject to Article 56(4)-(6)).	Prohibition on a manufacturer, importer or downstream user placing a substance on the market for a use or use it itself if that substance is included in Annex XIV (list of substances subject to authorisation) unless Article 56(1)(a), (b), (c), (d) or (e) are satisfied.	The Health and Safety Executive. The Environment Agency. Local (consumer safety) authorities. Local (health and safety) authorities.	The Health and Safety Executive. The Scottish Environment Protection Agency. Local (consumer safety) authorities. Local (health and safety) authorities.	The Health and Safety Executive for Northern Ireland. The Department of the Environment. Local (consumer safety) authorities. Local (health and safety) authorities.	The Health and Safety Executive. The Health and Safety Executive for Northern Ireland. Secretary of State.

Provision of REACH	***Subject Matter***	***Enforcing authority***			
		England and Wales	***Scotland***	***Northern Ireland***	***Offshore installations***
Article 56(2) (subject to 56(4)-(6)).	Prohibition on a downstream user using a substance otherwise than in accordance with the conditions of an authorisation granted to an actor up that actor's supply chain for that use.	The Health and Safety Executive. The Environment Agency. Local (health and safety) authorities.	The Health and Safety Executive. The Scottish Environment Protection Agency. Local (health and safety) authorities.	The Health and Safety Executive for Northern Ireland. The Department of the Environment. Local (health and safety) authorities.	The Health and Safety Executive. The Health and Safety Executive for Northern Ireland. Secretary of State.
Article 60(10).	Requirement on a holder of an authorisation to ensure that the exposure is reduced to as low a level as is technically and practically possible.	The Health and Safety Executive. The Environment Agency. Local (health and safety) authorities.	The Health and Safety Executive. The Scottish Environment Protection Agency. Local (health and safety) authorities.	The Health and Safety Executive for Northern Ireland. The Department of the Environment. Local (health and safety) authorities.	The Health and Safety Executive. The Health and Safety Executive for Northern Ireland. Secretary of State.
Article 65.	Requirement on a holder of an authorisation or a downstream user to include the authorisation number on the label before placing the substance or preparation on the market for an authorised use.	The Health and Safety Executive. Local (consumer safety) authorities. Local (health and safety) authorities.	The Health and Safety Executive. Local (consumer safety) authorities. Local (health and safety) authorities.	The Health and Safety Executive for Northern Ireland. Local (consumer safety) authorities. Local (health and safety) authorities.	The Health and Safety Executive. The Health and Safety Executive for Northern Ireland.

Provision of REACH	***Subject Matter***	***Enforcing authority***			
		England and Wales	***Scotland***	***Northern Ireland***	***Offshore installations***
Article 66(1).	Requirement on a downstream user using a substance in accordance with Article 56(2) to notify the European Chemicals Agency within three months of the first supply.	The Health and Safety Executive.	The Health and Safety Executive.	The Health and Safety Executive for Northern Ireland.	The Health and Safety Executive. The Health and Safety Executive for Northern Ireland.

Provision of REACH	Subject Matter	Enforcing authority			
		England and Wales	Scotland	Northern Ireland	Offshore installations
Article 67(1) (subject to Article 67(2)).	Prohibition on the manufacture, placing on the market or use of a substance on its own, in a [mixture] or in an article for which Annex XVII (restrictions on the manufacture, placing on the market and use of certain dangerous substances, [mixtures] and articles) contains a restriction unless the manufacture, placing on the market or use complies with the conditions of that restriction.	The Health and Safety Executive. The Environment Agency. Local (consumer safety) authorities. Local (health and safety) authorities. [The Office of Rail Regulation.]	The Health and Safety Executive. The Scottish Environment Protection Agency. Local (consumer safety) authorities. Local (health and safety) authorities. [The Office of Rail Regulation.]	The Health and Safety Executive for Northern Ireland. The Department of the Environment. Local (consumer safety) authorities. Local (health and safety) authorities.	The Health and Safety Executive. The Health and Safety Executive for Northern Ireland. Secretary of State.
. . .	. . .	. . .	. . .	. . .	. . .
. . .	. . .	. . .	. . .	. . .	. . .

[8.276A]

Statutory annotations

Amendment

In entry relating to 'Article 5.' in column 2 word 'mixtures' in square brackets substituted by SI 2013/2919, regs 2, 7(1), (2)(a).

In entry relating to 'Article 9(6)' in column 3 words 'In relation to England, the Environment Agency. In relation to Wales, the Natural Resources Body for Wales.' in square brackets substituted by SI 2013/755, art 4(2), Sch 4, paras 302, 305(1), (2), (3)(a).

In entry relating to 'Article 14(6)' in column 3 words 'In relation to England, the Environment Agency. In relation to Wales, the Natural Resources Body for Wales.' in square brackets substituted by SI 2013/755, art 4(2), Sch 4, paras 302, 305(1), (2), (3)(b).

In entry relating to 'Article 31(1).' in column 2 word 'mixture' in square brackets in both places it occurs substituted by SI 2013/2919, regs 2, 7(1), (2)(b).

In entry relating to 'Article 31(2)–(9).' in column 2 word 'mixture' in square brackets in the first place it occurs substituted by SI 2013/2919, regs 2, 7(1), (2)(b).

In entry relating to 'Article 31(2)–(9).' in column 2 word 'mixture' in square brackets in the second place it occurs substituted by SI 2013/2919, regs 2, 7(1), (2)(b).

In entry relating to 'Article 31(2)–(9).' in column 2 word 'mixture' in square brackets in the third place it occurs substituted by SI 2013/2919, regs 2, 7(1), (2)(b).

In entry relating to 'Article 31(2)–(9).' in column 2 words from ', and by no' to 'is first supplied' in square brackets inserted by SI 2013/2919, regs 2, 7(1), (3).

Entry relating to 'Article 31(10)' inserted by SI 2013/2919, regs 2, 7(1), (4).

In entry relating to 'Article 32(2), (3).' in column 2 word 'mixture' in square brackets substituted by SI 2013/2919, regs 2, 7(1), (2)(b).

In entry relating to 'Article 33(2).' in column 3 words 'Local (consumer safety) authorities' in square brackets substituted by SI 2013/2919, regs 2, 7(1), (5).

In entry relating to 'Article 33(2).' in column 4 words 'Local (consumer safety) authorities' in square brackets substituted by SI 2013/2919, regs 2, 7(1), (5).

In entry relating to 'Article 33(2).' in column 5 words 'Local (consumer safety) authorities' in square brackets substituted by SI 2013/2919, regs 2, 7(1), (5).

In entry relating to 'Article 35.' in column 2 word 'mixtures' in square brackets substituted by SI 2013/2919, regs 2, 7(1), (2)(a).

In entry relating to 'Article 36(1).' in column 2 word 'mixture' in square brackets substituted by SI 2013/2919, regs 2, 7(1), (2)(b); in column 3 words 'In relation to England, the Environment Agency. In relation to Wales, the Natural Resources Body for Wales.' in square brackets substituted by SI 2013/755, art 4(2), Sch 4, paras 302, 305(1), (2), (3)(c).

In entry relating to 'Article 36(2).' in column 2 word 'mixture' in square brackets substituted by SI 2013/2919, regs 2, 7(1), (2)(b).

In entry relating to 'Article 37(3).' in column 2 word 'mixtures' in square brackets substituted by SI 2013/2919, regs 2, 7(1), (2)(a).

In entry relating to 'Article 37(4)' in column 3 words 'In relation to England, the Environment Agency. In relation to Wales, the Natural Resources Body for Wales.' in square brackets substituted by SI 2013/755, art 4(2), Sch 4, paras 302, 305(1), (2), (3)(d).

In first entry relating to 'Article 37(5)' in column 2 word 'mixtures' in square brackets substituted by SI 2013/2919, regs 2, 7(1), (2)(a).

In both entries relating to 'Article 37(5)' in column 3 words 'In relation to England, the Environment Agency. In relation to Wales, the Natural Resources Body for Wales.' in square brackets substituted by SI 2013/755, art 4(2), Sch 4, paras 302, 305(1), (2), (3)(e).

In entry relating to 'Article 37(6)' in column 3 words 'In relation to England, the Environment Agency. In relation to Wales, the Natural Resources Body for Wales.' in square brackets substituted by SI 2013/755, art 4(2), Sch 4, paras 302, 305(1), (2), (3)(f).

In entry relating to 'Article 38(1)' in column 3 words 'In relation to England, the Environment Agency. In relation to Wales, the Natural Resources Body for Wales.' in square brackets substituted by SI 2013/755, art 4(2), Sch 4, paras 302, 305(1), (2), (3)(g).

In entry relating to 'Article 38(3)' in column 3 words 'In relation to England, the Environment Agency. In relation to Wales, the Natural Resources Body for Wales.' in square brackets substituted by SI 2013/755, art 4(2), Sch 4, paras 302, 305(1), (2), (3)(h).

In entry relating to 'Article 56(1)' in column 3 words 'In relation to England, the Environment Agency. In relation to Wales, the Natural Resources Body for Wales.' in square brackets substituted by SI 2013/755, art 4(2), Sch 4, paras 302, 305(1), (2), (3)(i).

In entry relating to 'Article 56(2)' in column 3 words 'In relation to England, the Environment Agency. In relation to Wales, the Natural Resources Body for Wales.' in square brackets substituted by SI 2013/755, art 4(2), Sch 4, paras 302, 305(1), (2), (3)(j).

In entry relating to 'Article 60(10)' in column 3 words 'In relation to England, the Environment Agency. In relation to Wales, the Natural Resources Body for Wales.' in square brackets substituted by SI 2013/755, art 4(2), Sch 4, paras 302, 305(1), (2), (3)(k).

In entry relating to 'Article 67(1)' in column 2 word 'mixture' in square brackets substituted by SI 2013/2919, regs 2, 7(1), (2)(b); in column 2 word 'mixtures' in square brackets substituted by SI 2013/2919, regs 2, 7(1), (2)(a); in column 3 words 'In relation to England, the Environment Agency. In relation to Wales, the Natural Resources Body for Wales.' in square brackets substituted by SI 2013/755, art 4(2), Sch 4, paras 302, 305(1), (2), (3)(l); in column 3 words 'The

Office of Rail Regulation.' in square brackets inserted by SI 2013/2919, regs 2, 7(1), (6); in column 4 words 'The Office of Rail Regulation.' in square brackets inserted by SI 2013/2919, regs 2, 7(1), (6).

Entry relating to 'Article 113(1).' (omitted) revoked by SI 2013/2919, regs 2, 7(1), (7).

Entry relating to 'Article 113(3).' (omitted) revoked by SI 2013/2919, regs 2, 7(1), (7).

SCHEDULE 2

FUNCTIONS OF ENFORCING AUTHORITIES

Regulation 3(8)

[8.277]

1

The functions of the Environment Agency, [the Natural Resources Body for Wales,] the Scottish Environment Protection Agency and the Department of the Environment are to—

(a) prevent, minimise, remedy or mitigate the effects of pollution of the environment;

(b) ensure the safety of the environment.

2

Subject to paragraph 3, the functions of the Health and Safety Executive, the Health and Safety Executive for Northern Ireland and a local (health and safety) authority are to—

(a) secure the health, safety and welfare of persons at work;

(b) protect others against risks to health and safety in connection with activities of persons at work;

(c) control the manufacture and placing on the market of articles and substances.

3

In Northern Ireland, paragraph 2 does not apply in respect of substances for which any of the following make provision—

(a) the Explosives Acts (Northern Ireland) 1875 to 1970;

(b) the Explosives (Northern Ireland) Order 1972;

(c) the Health and Safety Quarries (Explosives) Regulations (Northern Ireland) 2006;

(d) the Manufacture and Storage of Explosives Regulations (Northern Ireland) 2006.

4

The function of a local (consumer safety) authority is to ensure that goods made available to the public for their personal consumption are safe.

5

The function of the Secretary of State is to control the use on, and discharge from, offshore installations of articles and substances in order to protect the marine environment.

[6

(1) Subject to sub-paragraph (2), the functions of the Office of Rail Regulation are to—

(a) secure the health, safety and welfare of persons at work;

(b) protect others against risks to health and safety in connection with the activities of persons at work;

(c) control the manufacture and placing on the market of articles and substances.

(2) The functions of the Office of Rail Regulation—

(a) only apply in relation to the manufacture, placing on the market and use of asbestos fibres in so far as this relates to the operation of a railway; and

(b) do not apply in relation to Northern Ireland.

(3) In sub-paragraph (2) 'operation of a railway' has the meaning given in regulation 2 of the Health and Safety (Enforcing Authority for Railways and Other Guided Transport Systems) Regulations 2006.]

[8.277A]

Statutory annotations

Amendment

Para 1: words 'the Natural Resources Body for Wales,' in square brackets inserted by SI 2013/755, art 4(2), Sch 4, paras 302, 306.

Para 6: inserted by SI 2013/2919, regs 2, 8.

SCHEDULE 3

HEALTH AND SAFETY ENFORCEMENT

Regulation 6(2)

PART 1

DEFINITIONS

[8.278]

1

For the purposes of this Schedule—

(a) 'agricultural activities'—

(i) includes horticulture, fruit growing, seed growing, dairy farming, livestock breeding and keeping, forestry, the use of land as grazing land, market gardens and nursery grounds and the preparation of land for agricultural use;

(ii) does not include such activities at a garden centre or other shop,

but 'livestock breeding and keeping' does not include activities the main purpose of which is entertainment.

(b) 'common parts' means those parts of premises used in common by, or for providing common services to or common facilities for, the occupiers of the premises;

(c) 'consumer services' means services of a type ordinarily supplied to persons who receive them otherwise than in the course of a trade, business or other undertaking carried on by them (whether for profit or not);

(d) 'dock premises' has the meaning assigned to it by—

(i) in relation to Great Britain, regulation 2(1) of the Docks Regulations 1988;

(ii) in relation to Northern Ireland, regulation 2(1) of the Docks Regulations (Northern Ireland) 1989;

(e) 'electricity system' does not include—

(i) in relation to Great Britain, the consumer's installation within the meaning of regulation 3(1) of the Electricity Safety, Quality and Continuity Regulations 2002;

(ii) in relation to Northern Ireland, the electric lines situated upon the consumer's side of the supply terminals together with any apparatus permanently connected or intended to be permanently connected thereto;

(iii) the electric lines situated upon the consumer's side of the supply terminals together with any apparatus or equipment permanently connected or intended to be permanently connected thereto;

(f) . . .

(g) 'fairground' means such part of premises as is for the time being used wholly or mainly for the operation of any fairground equipment, other than a coin-operated ride, non-powered children's playground equipment, swimming pool slide, go-kart, or plant designed to be used by members of the public for entertainment purposes for bouncing upon;

(h) 'gas' has the meaning assigned to it by—

(i) in relation to Great Britain, section 48 of the Gas Act 1986;

(ii) in relation to Northern Ireland, Part III of the Gas (Northern Ireland) Order 1996;

(i) 'gas fitting' has the meaning assigned to it by—

(i) in relation to Great Britain, section 48 of the Gas Act 1986;

(ii) in relation to Northern Ireland, regulation 2(1) of the Gas Safety (Installation and Use) Regulations (Northern Ireland) 1997;

(j) 'guided bus system' means a system of transport, used wholly or mainly for the carriage of passengers, that employs buses which for some or all of the time when they are in operation—

(i) travel along roads; and

(ii) are guided (whether while on the road or at other times) by means of—

(aa) apparatus, a structure or other device which is fixed and not part of the bus; or

(bb) a guidance system which is automatic.

(k) 'guided transport' means a system of transport, used wholly or mainly for the carriage of passengers, employing vehicles which for some or all of the time when they are in operation are guided by means of—

(i) rails, beams, slots, guides or other apparatus, structures or devices which are fixed and not part of the vehicle; or

(ii) a guidance system which is automatic,

and for this purpose 'vehicle' includes a mobile traction unit.

(l) 'livestock' means any creature kept for the production of food, wool, skins or fur or for the purpose of any agricultural activity;

(m) 'mine' has the meaning assigned to it by—

(i) in relation to Great Britain, section 180 of the Mines and Quarries Act 1954 but, notwithstanding subsection (5) of that section, does not include any railway serving the mine unless and to the extent that the railway is located within the curtilage of the mine;

(ii) in relation to Northern Ireland, section 156(1) of the Mines Act (Northern Ireland) 1969;

(n) 'office activities' includes any activity for the purposes of administration, clerical work, handling money, telephone and telegraph operating and the production of computer software by the use of computers; and for this purpose 'clerical work' includes writing, book-keeping, sorting papers, filing, typing, duplicating, machine calculating, drawing and the editorial preparation of matter for publication except where that preparation is on the premises where newspapers, magazines, periodicals or books are printed;

(o) 'pleasure craft' has the meaning assigned to it by—
   (i) in relation to Great Britain, regulation 2(1) of the Docks Regulations 1988;
   (ii) in relation to Northern Ireland, regulation 2(1) of the Docks Regulations (Northern Ireland) 1989;

(p) 'preparation dangerous for supply' means a preparation which is in one or more of the categories of danger—
   (i) in relation to Great Britain, in Schedule 1 to [the Chemicals (Hazard Information and Packaging for Supply) Regulations 2009];
   (ii) in relation to Northern Ireland, in Schedule 1 to the Chemicals (Hazard Information and Packaging for Supply) Regulations (Northern Ireland) 2002.

(q) 'quarry' has the meaning assigned to it by—
   (i) in relation to Great Britain, regulation 3 of the Quarries Regulations 1999;
   (ii) in relation to Northern Ireland, regulation 3 of the Quarries Regulations (Northern Ireland) 2006;

(r) . . .

(s) 'road'—
   (i) in England and Wales, means any length of highway or of any other road to which the public has access, and includes bridges over which a road passes;
   (ii) in Scotland, has the same meaning as in the Roads (Scotland) Act 1984;

(t) 'trolley vehicle system' means a system of transport by vehicles constructed or adapted for use on roads without rails under electric power transmitted to them by overhead wires (whether or not there is in addition a source of power on board the vehicles);

(u) 'veterinary surgery' has the meaning assigned to it by section 27 of the Veterinary Surgeons Act 1966;

(v) 'work' in relation to a gas fitting has the meaning assigned to it by—
   (i) in relation to Great Britain, regulation 2(1) of the Gas Safety (Installation and Use) Regulations 1998;
   (ii) in relation to Northern Ireland, regulation 2(1) of the Gas Safety (Installation and Use) Regulations (Northern Ireland) 1997;

(w) 'zoo' has the meaning assigned to it in relation to Great Britain by section 1(2) of the Zoo Licensing Act 1981;

(x) 'zoological establishment' has the meaning assigned to it in relation to Northern Ireland, by section 12(1) of the Welfare of Animals Act (Northern Ireland) 1972.

[8.279]

Statutory annotations

Amendment

Para 1: sub-para (f) revoked by SI 2013/2919, regs 2, 9(a); sub-para (r) revoked by SI 2013/2919, regs 2, 9(a).

PART 2

LOCAL (HEALTH AND SAFETY) AUTHORITIES

[8.280]

1

Subject to Part 4, a local (health and safety) authority must perform the joint enforcement duty on behalf of the Executive where the main activity carried on in non-domestic premises is referred to in paragraph 2.

2

The main activity referred to in paragraph 1 is—

(a) the sale of goods, or the storage of goods for retail or wholesale distribution, except—

(i) at container depots where the main activity is the storage of goods in the course of transit to or from dock premises, an airport or a railway;

(ii) where the main activity is the sale or storage for wholesale distribution of any substance or preparation dangerous for supply;

(iii) where the main activity is the sale or storage of water or sewage or their by-products or natural or town gas.

(b) the display or demonstration of goods at an exhibition for the purposes of offer or advertisement for sale;

(c) office activities;

(d) catering services;

(e) the provision of permanent or temporary residential accommodation including the provision of a site for caravans or campers;

(f) consumer services provided in a shop except dry cleaning or radio and television repairs;

(g) cleaning (wet or dry) in coin operated units in launderettes and similar premises;

(h) the use of a bath, sauna or solarium, massaging, hair transplanting, skin piercing, manicuring or other cosmetic services and therapeutic treatments, except where they are carried out under the supervision or control of a registered medical practitioner, a dentist registered under the Dentists Act 1984, a physiotherapist, an osteopath or a chiropractor;

(i) the practice or presentation of the arts, sports, games, entertainment or other cultural or recreational activities except where the main activity is the exhibition of a cave to the public;

(j) the hiring out of pleasure craft for use on inland waters;

(k) the care, treatment, accommodation or exhibition of animals, birds or other creatures, except where the main activity is horse breeding or horse training at a stable, or is an agricultural activity or veterinary surgery;

(l) the activities of an undertaker, except where the main activity is embalming or the making of coffins;

(m) church worship or religious meetings;

(n) the provision of car parking facilities within the perimeter of an airport;

(o) the provision of child care, or playgroup or nursery facilities.

3

For the purposes of paragraph 2—

(a) where a vehicle is parked in connection with the sale of—

(i) food;
(ii) drink; or
(iii) other articles,

the vehicle, including its pitch, must be regarded as separate premises for the purposes of paragraph 2;

(b) where any non-domestic premises are occupied by more than one occupier each part separately occupied must be regarded as being separate premises;

(c) where the main activity carried on in premises is the sale and fitting of motor vehicle tyres, exhausts, windscreens or sunroofs, the main activity must be regarded as the sale of goods.

PART 3

THE EXECUTIVE

[8.281]

1

Subject to Part 4, the Executive must perform the joint enforcement duty on behalf of a local (health and safety) authority in relation to—

(a) any activity in a mine or quarry other than a quarry in respect of which notice of abandonment has been given—
   (i) in relation to Great Britain, under regulation 45(1) of the Quarries Regulations 1999;
   (ii) in relation to Northern Ireland, under regulation 39(1) of the Quarries Regulations (Northern Ireland) 2006;

(b) any activity in a fairground;

(c) any activity in premises occupied by a radio, television or film undertaking in which the activity of broadcasting, recording or filming is carried on, and the activity of broadcasting, recording or filming wherever carried on, and for this purpose 'film' includes video;

(d) construction work if—
   (i) the project which includes the work is notifiable within the meaning of—
      (aa) in relation to Great Britain, regulation 2(3) of the Construction (Design and Management) Regulations 2007;
      (bb) in relation to Northern Ireland, regulation 2(3) of the Construction (Design and Management) Regulations (Northern Ireland) 2007.
   (ii) the whole or part of the work contracted to be undertaken by the contractor at the premises is to the external fabric or other external part of a building or structure;
   (iii) it is carried out in a physically segregated area of the premises, the activities normally carried out in that area have been suspended for the purpose of enabling the construction work to be carried out, the contractor has authority to exclude from that area persons who are not attending in connection with the carrying out of the work and the work is not the maintenance of insulation on pipes, boilers or other parts of heating or water systems or its removal from them.

(e) the installation, maintenance or repair of any gas system, or any work in relation to a gas fitting;

(f) the installation, maintenance or repair of electricity systems;

(g) work with ionising radiations except work in one or more of the categories set out—
  (i) in relation to Great Britain, in Schedule 1 to the Ionising Radiations Regulations 1999;
  (ii) in relation to Northern Ireland, in Schedule 1 to the Ionising Radiations Regulations (Northern Ireland) 2000;

(h) the use of ionising radiations for medical exposure;

(i) any activity in premises occupied by a radiography undertaking in which there is carried on any work with ionising radiations;

(j) in relation to Northern Ireland, any activity involving genetic modification within the meaning of regulation 2(1) of the Genetically Modified Organisms (Contained Use) Regulations (Northern Ireland) 2001;

(k) agricultural activities, and any activity at an agricultural show which involves the handling of livestock or the working of agricultural equipment;

(l) any activity on board a sea-going ship;

(m) any activity in relation to a ski slope, ski lift, ski tow or cable car;

(n) fish, maggot and game breeding except in a zoo or zoological establishment;

(o) in relation to Northern Ireland, horse breeding or horse training at a stable;

(p) any activity in relation to a pipeline within the meaning of—
  (i) in relation to Great Britain, regulation 3 of the Pipelines Safety Regulations 1996;
  (ii) in relation to Northern Ireland, regulation 3 of the Pipelines Safety Regulations (Northern Ireland) 1997.

(q) except in Northern Ireland, the operation of—
  (i) a guided bus system; or
  (ii) any other system of guided transport, other than a railway, that employs vehicles which for some or all of the time when they are in operation travel along roads.

(r) in relation to Northern Ireland, the operation of a railway;

(s) except in Northern Ireland, the operation of a trolley vehicle system;

(t) the manufacture of ammonium nitrate blasting intermediate under the Manufacture and Storage of Explosives Regulations 2005.

2

In relation to paragraph 1(d)(iii)—

(a) reference to a physically segregated area does not include an area segregated only in order to prevent the escape of asbestos; and

(b) 'asbestos' has the meaning assigned to it in relation to Great Britain, by regulation 2(1) of the [Control of Asbestos Regulations 2012] or in relation to Northern Ireland, by regulation 2(1) of the Control of Asbestos Regulations (Northern Ireland) 2007.

3

In paragraph 1, 'ionising radiations' and 'medical exposure' in relation to those radiations have the meaning assigned to them in relation to Great Britain, by regulation 2(1) of the Ionising Radiations Regulations 1999 or in relation to Northern Ireland, by regulation 2(1) of the Ionising Radiations Regulations (Northern Ireland) 2000.

4

Subject to Part 4, the Executive must perform the joint enforcement duty on behalf of a local (health and safety) authority—

(a) in relation to a body specified in paragraph 5 or 6;
(b) in relation to the officers or servants of such a body; or
(c) in relation to any part of premises occupied by such a body.

5

In Great Britain, the bodies referred to in paragraph 4 are—

(a) a local authority being—
  (i) in Greater London, a London borough council, the Common Council of the City of London, the Sub-Treasurer of the Inner Temple and the Under-Treasurer of the Middle Temple;
  (ii) in England outside Greater London, a county council or a district council, and the Council of the Isles of Scilly;
  (iii) in Scotland, a council constituted under section 2 of the Local Government etc (Scotland) Act 1994;
  (iv) in Wales, a county council or county borough council;
(b) a parish council in England or a community council in Wales or Scotland;
(c) a police authority[, local policing body] or the Receiver for the Metropolitan Police District;
(d) a fire and rescue authority under the Fire and Rescue Services Act 2004;
[(e) the Scottish Fire and Rescue Service;]
(f) a headquarters or an organisation designated for the purposes of the International Headquarters and Defence Organisation Act 1964;
(g) a service authority of a visiting force within the meaning of section 12 of the Visiting Forces Act 1952;
(h) the United Kingdom Atomic Energy Authority;
(i) the Crown, except in relation to any part of any premises occupied by the Health and Safety Executive.

6

In Northern Ireland, the bodies referred to in paragraph 4 are—

(a) a district council established under section 1 of the Local Government Act (Northern Ireland) 1972;
(b) the Northern Ireland Policing Board as defined in section 2 of the Police (Northern Ireland) Act 2000;
(c) the Northern Ireland Fire and Rescue Service Board as referred to in article 3 of the Fire and Rescue Services (Northern Ireland) Order 2006;
(d) a headquarters or an organisation designated for the purposes of the International Headquarters and Defence Organisation Act 1964;
(e) a service authority of a visiting force within the meaning of section 12 of the Visiting Forces Act 1952;
(f) the Crown.

7

Subject to Part 4, the Executive must perform the joint enforcement duty on behalf of a local (health and safety) authority in relation to—

(a) the tunnel system within the meaning it would have in section 1(7) of the Channel Tunnel Act 1987 if the words 'to be' did not appear;
(b) an offshore installation;
(c) a building or construction site, that is to say, premises where the only activities being undertaken are construction work and activities for the purposes of or in connection with such work;
(d) the campus of a university, college, school or similar educational establishment;

(e) a hospital;

(f) in relation to Northern Ireland, a railway station, railway goods yard, railway track and any part of adjacent premises, occupied in connection with them.

[8.281A]

Statutory annotations

Amendment

Para 2: in para (b) words 'the Control of Asbestos Regulations 2012' inserted by SI 2012/632, reg 33, Sch 3.

Para 5: sub-para (e) substituted by SSI 2013/119, art 4, Sch 2, para 24.

PART 4

LOCAL (HEALTH AND SAFETY) AUTHORITIES AND THE EXECUTIVE

[8.282]

1

Subject to paragraph 3, paragraph 2 applies to the common parts of any non-domestic premises where those premises are occupied by more than one occupier.

2

Where this paragraph applies—

(a) to the extent the Executive must perform a joint enforcement duty on behalf of a local (health and safety) authority under this Schedule for all other parts of the premises, it must also do so for the common parts of those premises;

(b) to the extent a local (health and safety) authority must perform a joint enforcement duty on behalf of the Executive under this Schedule for all other parts of the premises, it must also do so for the common parts of those premises.

3

In relation to land within the perimeter of an airport which consists of common parts,—

(a) for those parts not within a building or to which passengers are admitted but other members of the public are not admitted, the Executive must perform the joint enforcement duty on behalf of a local (health and safety) authority;

(b) for the remaining common parts, a local (health and safety) authority must perform the joint enforcement duty on behalf of the Executive.

[PART 5

THE OFFICE OF RAIL REGULATION]

[8.282A]

[1

The Office of Rail Regulation must perform the joint enforcement duty on behalf of the Executive in relation to Article 67 of REACH in so far as this relates to its functions in paragraph 6 of Schedule 2.]

[8.282B]

Statutory annotations

Amendment

Inserted by SI 2013/2919, regs 2, 9(b).

SCHEDULE 4

DEFENCE EXEMPTION CERTIFICATES

Regulation 7(2)

[8.283]

1

The Secretary of State may decide that it is necessary in the interests of defence for a person to be exempt from compliance with a listed REACH provision.

2

The Secretary of State may decide to apply the exemption—

(a) to a person, including the Secretary of State, or a category of persons;
(b) to one or more provision at the same time;
(c) prospectively;
(d) for a limited or unlimited period;
(e) generally or to a particular case;
(f) subject to such limitations and conditions as the Secretary of State sees fit.

3

A decision of the Secretary of State to apply the exemption must be evidenced in writing by a certificate.

4

A certificate—

(a) must contain sufficient particulars of the persons to whom, and the matters to which, it relates; and
(b) may be varied or revoked in writing.

5

The Secretary of State may provide to a person who has the benefit of a certificate—

(a) the certificate;
(b) a copy of it; or
(c) a copy of a relevant extract of the certificate.

6

A person who claims the benefit of a certificate must produce to the persons listed in paragraph 7 when reasonably requested to do so—

(a) the certificate;
(b) a copy of it made by the Secretary of State; or
(c) a copy made by the Secretary of State of a relevant extract of the certificate.

7

The persons referred to in paragraph 6—

(a) an enforcing authority;
(b) a competent authority;
(c) the equivalent of an enforcing authority of another member State;

(d) the European Chemicals Agency

8

Unless the contrary is proved—

(a) a certificate;

(b) a copy of it made by the Secretary of State; or

(c) a copy made by the Secretary of State of a relevant extract of the certificate,

is conclusive evidence of the matters to which it relates.

SCHEDULE 5

MARKETING AND USE OF LEADED PAINTS

Regulation 8

PART 1

PERMITTED MARKETING AND USE

[8.284]

1

A person may market leaded paint if it is marketed with a view to its use as set out in paragraph 2.

2

Subject to paragraph 3, a person may use leaded paint if the paint is used in the restoration or maintenance of—

(a) historic buildings or their interiors;

(b) scheduled monuments; or

(c) fine or decorative works of art,

where it is required to restore or maintain historic textures or finishes.

3

A person who intends to use leaded paint must—

(a) where that person intends to obtain the paint from a supplier of such paint, provide to that supplier a relevant declaration; or

(b) in any other case, provide to the competent body a relevant declaration and comply with paragraph 6.

4

A person may supply leaded paint if that person complies with paragraph 5.

5

A person who receives a relevant declaration pursuant to paragraph 3(a) and agrees to supply leaded paint to the intended user stated in the declaration—

(a) must send the declaration to the competent body with a notification; and

(b) must not supply the paint—

  (i) earlier than 3 weeks after providing the relevant declaration and notification to the competent body; or

  (ii) if that person receives a notice under paragraph 7(a) from the competent body.

6

A person who provides a relevant declaration under paragraph 3(b) must not use the paint—

(a) earlier than 3 weeks after providing the relevant declaration to the competent body; or

(b) if that person receives a notice under paragraph 7(a) from the competent body.

7

If a competent body is not satisfied with the content of a relevant declaration or notification, it must—

(a) within 2 weeks of receipt, give notice to that effect in writing to the person from whom it was received with reasons for its decision; and

(b) as soon as possible, provide a copy of that notice to such enforcing authorities as the competent body believes are appropriate.

8

For the purposes of this Schedule—

'competent body' means—

(a) English Heritage if the historic building, scheduled monument or work of art is in England;

(b) Cadw if the historic building, scheduled monument or work of art is in Wales;

(c) the Scottish Ministers if the historic building, scheduled monument or work of art is in Scotland;

(d) the Department of the Environment if the historic building, scheduled monument or work of art is in Northern Ireland;

'historic building' means—

(e) in relation to England and Wales, a listed building within the meaning of section 1(5) of the Planning (Listed Buildings and Conservation Areas) Act 1990 which is classified as Grade I or Grade II (starred);

(f) in relation to Scotland, a listed building within the meaning of section 1(4) of the Planning (Listed Buildings and Conservation Areas) (Scotland) Act 1997 which is classified as category A;

(g) in relation to Northern Ireland, a listed building within the meaning of article 42(7) of the Planning (Northern Ireland) Order 1991;

'leaded paint' means paint containing [lead carbonates] or lead sulphates listed at points 16 and 17 of Annex XVII of REACH;

'notification' means a written notification that contains the matters in Part 3 of this Schedule;

'relevant declaration' means a written declaration that contains the matters in Part 2 of this Schedule;

'scheduled monument' has the same meaning—

(h) in England, Wales and Scotland, as it has in section 1(11) of the Ancient Monuments and Archaeological Areas Act 1979;

(i) in Northern Ireland, as it has in article 3(2) of the Historic Monuments and Archaeological Objects (Northern Ireland) Order 1995.

[8.284A]

Statutory annotations

Amendment

Para 8: in definition 'leaded paint' words 'lead carbonates' in square brackets substituted by SI 2013/2919, regs 2, 10.

PART 2

CONTENTS OF A RELEVANT DECLARATION

[8.285]

1

The name, address and telephone number of the intended user of the paint.

2

The quantity, trade name and manufacturer of the paint intended to be used.

3

In the case of a listed building, the name, address and owner of the building and, where appropriate, its listing category.

4

In the case of a scheduled monument, the name, location and owner (if known) of the monument.

5

In the case of a listed building or scheduled monument, details of the parts of the building or monument where the paint is to be applied.

6

In the case of a work of art, its name, date, author, location, the name and address of the owner and details of how the paint is intended to be used.

7

The date the declaration is made and the signature and name of the person making it.

PART 3

CONTENTS OF A NOTIFICATION

[8.286]

1

The name, address and telephone number of the intended supplier of the paint.

2

A statement that the intended supplier agrees to supply to the intended user paint of the quantity and type stated in the relevant declaration.

3

The date the notification is made and the signature and name of the person making it.

[SCHEDULE 5A

PLACING ASBESTOS-CONTAINING ARTICLES ON THE MARKET

Regulation 8A(2)

[8.286A]

[1

For the purposes of this Schedule 'exempting authority' means—

(a) the Executive;

(b) the Office of Rail Regulation;

(c) in relation to matters outside the functions of the Executive and the Office of Rail Regulation, the competent authority.

2

An exempting authority may issue a certificate in writing exempting a person or class of persons from the prohibition on the placing on the market of asbestos-containing articles provided for by Article 67 of REACH ('asbestos exemption certificate') in relation to—

(a) an activity or a class of activities;

(b) an asbestos-containing article or a class of such articles.

3

An asbestos exemption certificate—

(a) must include conditions ensuring a high level of protection of human health;

(b) may be subject to a limit of time;

(c) may be varied or revoked by a notice in writing; and

(d) may be issued jointly by more than one exempting authority.

4

An exempting authority must not issue an asbestos exemption certificate unless it is satisfied that the health or safety of persons who are likely to be affected by the exemption will not be prejudiced in consequence of it, having regard to the circumstances of the case, and in particular to—

(a) the conditions which it proposes to attach to the exemption; and

(b) any other requirements imposed by or under any enactments which apply to the case.]

[8.286B]

Statutory annotations

Amendment

Inserted by SI 2013/2919, regs 2, 11, Schedule.

[SCHEDULE 5B

PLACING ON THE MARKET AND USE OF PAINT STRIPPER CONTAINING DICHOLOROMETHANE

Regulation 8B(3)

[8.286C]

[1

A person ('P') may place on the market paint stripper containing dichloromethane for supply to a professional or to a person for use by a professional provided the professional to whom, or for whose use, such paint stripper is supplied holds a certificate of competence in the safe use of such paint stripper issued by the Executive in accordance with paragraph 3(b).

2

A professional may use paint stripper containing dichloromethane where that professional—

(a) demonstrates proper training and competence in the safe use of paint stripper containing dichloromethane in accordance with paragraph 3; and

(b) applies in practice all appropriate safety measures, including the use of personal protective equipment that complies with the Personal Protective Equipment Regulations 2002.

3

A professional may only demonstrate proper training and competence in the safe use of paint stripper containing dichloromethane by—

(a) undertaking training in the safe use of such paint stripper which covers as a minimum—

(i) awareness, evaluation and management of risks to health, including information on existing substitutes or processes, which under their conditions of use are less hazardous to the health and safety of workers;

(ii) the use of adequate ventilation; and

(iii) the use of appropriate personal protective equipment that complies with the Personal Protective Equipment Regulations 2002; and

(b) holding a certificate of competence issued by the Executive in the safe use of paint stripper containing dichloromethane.

4

A certificate issued under paragraph 3(b) may be revoked in writing at any time by the Executive if the Executive believes that the professional holding the certificate is no longer competent in the safe use of paint stripper containing dichloromethane.]

[8.286D]

Statutory annotations

Amendment

Inserted by SI 2014/2882, reg 2(1), (5).

SCHEDULE 6

POWERS OF ENFORCEMENT

Regulation 9(1)

PART 1

THE ENVIRONMENT AGENCY, [THE NATURAL RESOURCES BODY FOR WALES,] THE SCOTTISH ENVIRONMENT PROTECTION AGENCY AND THE DEPARTMENT OF THE ENVIRONMENT

SECTION 1

POWERS OF ENTRY AND WARRANTS

[8.287]

1

The powers of any authorised person are—

(a) to enter at any reasonable time (or, in an emergency, at any time and, if need be, by force) any premises which that person has reason to believe it is necessary to enter;

(b) on entering any premises by virtue of sub-paragraph (a), to—

(i) be accompanied by any other person duly authorised by the enforcing authority and, if the authorised person has reasonable cause to apprehend any serious obstruction in the execution of the authorised person's duty, a constable;

(ii) take any equipment or materials required for any purpose for which the power of entry is being exercised;

(c) to make such examination and investigation as may in any circumstances be necessary;

(d) as regards any premises which the authorised person has power to enter, to direct that those premises or any part of them, or anything in them, shall be left undisturbed (whether generally or in particular respects) for so long as is reasonably necessary for the purpose of any examination or investigation under sub-paragraph (c);

(e) to take such measurements and photographs and make such recordings as the authorised person considers necessary for the purpose of any examination or investigation under sub-paragraph (c);

(f) to take samples, or cause samples to be taken, of any thing found in or on any premises which the authorised person has power to enter, and of the air, water or land in, on, or in the vicinity of, the premises;

(g) in the case of any thing found in or on any premises which the authorised person has power to enter and which appears to that person to be in contravention of a listed REACH provision, to cause it to be dismantled or subjected to any process or test (but not so as to destroy or damage it, unless that is necessary);

(h) in the case of any thing mentioned in sub-paragraph (g), to take possession of it and detain it for so long as is necessary for all or any of the following purposes—

(i) to examine it, or cause it to be examined, and to do, or cause to be done, to it anything which the authorised person has power to do under that sub-paragraph;

(ii) to ensure that it is not tampered with before examination of it is completed; and

(iii) to ensure that it is available for use in any proceedings for an offence under regulation 11 or 13;

(i) to require any person whom the authorised person has reasonable cause to believe to be able to give any information relevant to any examination or investigation under sub-paragraph (c) to answer (in the absence of persons other than a person nominated by that person to be present and any persons whom the authorised person may allow to be present) such questions as the authorised person thinks fit to ask and to sign a declaration of the truth of that person's answers;

(j) to require the production of, or where the information is recorded in computerised form, the furnishing of extracts from, any records which it is necessary for the authorised person to see for the purposes of an examination or investigation under sub-paragraph (c) and to inspect, and take copies of, or of any entry in, the records; and

(k) to require any person to afford the authorised person such facilities and assistance with respect to any matters or things within the other person's control or in relation to which that person has responsibilities as are necessary to enable the authorised person to exercise any of the powers conferred on the authorised person by Part 1 of this Schedule.

2

Except in an emergency, in any case where it is proposed to enter any premises used for residential purposes, or to take heavy equipment on to any premises which are to be entered, any entry by virtue of paragraph 1 must only be effected—

(a) after the expiration of at least seven days' notice of the proposed entry given to a person who appears to the authorised person in question to be in occupation of the premises in question; and

(b) either—
   (i) with the consent of the person who is in occupation of those premises; or
   (ii) under the authority of a warrant by virtue of section 2 of Part 1 of this Schedule.

3

Except in an emergency, where an authorised person proposes to enter any premises and—

(a) entry has been refused and the authorised person apprehends on reasonable grounds that the use of force may be necessary to effect entry; or

(b) the authorised person apprehends on reasonable grounds that entry is likely to be refused and that the use of force may be necessary to effect entry,

any entry on to those premises by virtue of paragraph 1 must only be effected under the authority of a warrant by virtue of section 2 of Part 1 of this Schedule.

4

In relation to any premises belonging to or used for the purposes of the United Kingdom Atomic Energy Authority, the powers under paragraph 1 have effect subject to section 6(3) of the Atomic Energy Authority Act 1954 (which restricts entry to such premises where they have been declared to be prohibited places for the purposes of the Official Secrets Act 1911).

5

Where an authorised person proposes to exercise the power conferred by paragraph 1(g), that person must, if so requested by a person who at the time is present on and has responsibilities in relation to those premises, cause anything which is to be done by virtue of that power to be done in the presence of that other person.

6

Before exercising the power conferred by paragraph 1(g), an authorised person must consult—

(a) such persons having duties on the premises where the thing is to be dismantled or subject to the process or test; and

(b) such other persons,

as appear to the authorised person to be appropriate for the purpose of ascertaining what dangers, if any, there may be in doing anything which it is proposed to do or cause to be done under the power.

7

No answer given by a person in pursuance of a requirement imposed under paragraph 1(i) is admissible in evidence in England and Wales or Northern Ireland against that person in any proceedings, or in Scotland against that person in any criminal proceedings.

8

Nothing in paragraph 1 compels the production by any person of a document which—

(a) except in relation to Scotland, that person would on grounds of legal professional privilege be entitled to withhold production on an order for discovery in an action in the County Court or High Court; or

(b) in relation to Scotland, contains confidential information made by or to an advocate or solicitor in that capacity and which that person would be entitled to withhold production on an order for the production of documents in an action in the Court of Session.

9

Section 2 of Part 1 of this Schedule has effect with respect to the powers of entry and related powers conferred by paragraph 1.

10

In this section—

'emergency' means a case in which it appears to the authorised person in question—

(a) that there is an immediate risk of serious pollution of the environment or serious harm to human health; or

(b) that circumstances exist which are likely to endanger life or health,

and that immediate entry to any premises is necessary to verify the existence of that risk or those circumstances or to ascertain the cause of that risk or those circumstances or to effect a remedy;

'premises' means any land, vehicle, vessel or plant which is designed to move or be moved whether on roads or otherwise.

SECTION 2

WARRANTS, EVIDENCE AND COMPENSATION

11

If it is shown to the satisfaction, in England and Wales of a justice of the peace, in Scotland of the sheriff or a justice of the peace, or in Northern Ireland of a lay magistrate, on sworn information in writing—

(a) that there are relevant grounds for the exercise in relation to any premises of the powers under paragraph 1; and

(b) that one or more of the conditions specified in paragraph 12 is fulfilled in relation to those premises,

the justice, sheriff or lay magistrate may by warrant authorise an enforcing authority to designate a person who is authorised to exercise the power in relation to those premises, in accordance with the warrant and, if need be, by force.

12

The conditions mentioned in paragraph 11 are—

(a) that the exercise of the power in relation to the premises has been refused;

(b) that such a refusal is reasonably apprehended;

(c) that the premises are unoccupied;

(d) that the occupier is temporarily absent from the premises and the case is one of urgency; or

(e) that an application for admission to the premises would defeat the object of the proposed entry.

13

In a case where paragraph 12 applies, a justice of the peace, sheriff or lay magistrate must not issue a warrant under paragraph 11 by virtue only of being satisfied that the exercise of the power in relation to any premises has been refused, or that a refusal is reasonably apprehended, unless the justice of the peace, sheriff or lay magistrate is also satisfied that the notice required by that paragraph has been given and that the period of that notice has expired.

14

Every warrant under paragraph 11 continues in force until the purposes for which the warrant was issued have been fulfilled.

15

An authorised person must produce evidence of that person's authorisation or designation and other authority before exercising the power.

16

Information obtained in consequence of the exercise of the powers in paragraph 1, with or without the consent of any person is admissible in evidence against that or any other person.

17

Without prejudice to the generality of paragraph 16, information obtained by means of monitoring or other apparatus installed on any premises in the exercise of the powers in paragraph 1, with or without the consent of any person in occupation of the premises, is admissible in evidence in any proceedings against that or any other person.

18

A person who, in exercise of the powers in paragraph 1, enters on any premises which are unoccupied or whose occupier is temporarily absent must leave the premises as effectively secured against trespassers as that person found them.

19

Where any person exercises any power conferred by paragraph 1(a) or (b), it is the duty of the enforcing authority under whose authorisation that person acts to make full compensation to any person who has sustained loss or damage by reason of—

(a) the exercise of a power under paragraph 1 by the authorised person; or

(b) the performance of, or failure of the authorised person to perform, the duty imposed under paragraph 18.

20

Compensation is not payable by virtue of paragraph 19 in respect of any loss or damage if—

(a) it is attributable to the default of the person who sustained it; or

(b) it is loss or damage in respect of which compensation is payable by virtue of any other enactment.

21

Any dispute as to a person's entitlement to compensation under paragraph 19, or as to the amount of any such compensation—

(a) in England and Wales, must be referred to the arbitration of a single arbitrator appointed by agreement between the enforcing authority in question and the person who claims to have sustained the loss or damage or, in default of agreement, appointed by the Secretary of State;

(b) in Scotland, must be referred to the arbitration of an arbiter, appointed by agreement between the enforcing authority in question and the person who claims to have sustained the loss or damage or, in default of agreement, appointed by the Scottish Ministers; or

(c) in Northern Ireland, must be referred to and determined by the Lands Tribunal for Northern Ireland.

22

An authorised person is not to be liable in any civil proceedings for anything done in the purported exercise of the powers under paragraph 1 if the court is satisfied that the act was done in good faith and that there were reasonable grounds for doing it.

SECTION 3

SEIZURE IN CASES OF IMMINENT DANGER OF SERIOUS POLLUTION

23

If an authorised person has entered any premises and has reasonable cause to believe that any thing there is a cause of imminent danger of serious pollution of the environment, the authorised person may seize it and cause it to be rendered harmless (whether by destruction or otherwise).

24

An authorised person who exercises the power in paragraph 23 must as soon as possible prepare and sign a written report giving particulars of the circumstances in which the thing was seized and so dealt with and must—

(a) give a signed copy of the report to a responsible person at the premises where the thing was found; and

(b) unless the person is the owner of the thing, also serve a signed copy of the report on the owner.

25

If the authorised person cannot after reasonable inquiry ascertain the name or address of the owner, the copy may be served on the owner by giving it to the person to whom a copy was given under paragraph 24(a).

26

Where the powers under paragraph 23 are exercised, the enforcing authority is entitled to recover the costs it reasonably incurs from the person who knowingly caused or permitted the thing to become a cause of imminent danger of serious pollution of the environment.

SECTION 4

NOTICES

27

An authorised person may, by a notice served on any person, require that person to furnish such information as is specified in the notice, in such form and within such period following service of the notice or at such time as is so specified.

28

If an authorised person is of the opinion that a person has contravened, is contravening or is likely to contravene a listed REACH provision, the authorised person may serve on that person an enforcement notice.

29

An enforcement notice must—

(a) state that the authorised person is of the opinion referred to in the preceding paragraph;

(b) specify the matters constituting the contravention or the matters making it likely that the contravention will arise, as the case may be;

(c) specify the steps that must be taken to remedy the contravention or to remedy the matters making it likely that the contravention will arise, as the case may be; and

(d) specify the period within which those steps must be taken.

30

An enforcement notice may be withdrawn at any time.

31

If an authorised person is of the opinion that the activities of a person, in relation to a contravention or likely contravention of a listed REACH provision, are such that they involve an imminent danger of serious pollution of the environment, the authorised person may serve on that person a prohibition notice.

32

A prohibition notice must—

(a) state that the authorised person is of the opinion referred to in the preceding paragraph;

(b) specify the danger involved in the activity; and

(c) specify the steps that must be taken to remove it and the period within which they must be taken.

33

A prohibition notice may be withdrawn at any time.

34

If a person fails to comply with an enforcement notice or prohibition notice, the enforcing authority may do what that person was required to do and may recover from that person any expenses reasonably incurred in doing so.

35

References in this Part to an 'enforcement notice' or a 'prohibition notice' have effect only for the purposes of this Part.

PART 2

THE HEALTH AND SAFETY EXECUTIVE, THE HEALTH AND SAFETY EXECUTIVE FOR NORTHERN IRELAND[, THE OFFICE OF RAIL REGULATION] AND LOCAL (HEALTH AND SAFETY) AUTHORITIES

SECTION 1

POWERS OF ENTRY

[8.288]

1

The powers of an authorised person are—

(a) at any reasonable time (or, in a situation which the authorised person believes is or may be dangerous, at any time) to enter any premises which that person has reason to believe it is necessary to enter;

(b) to be accompanied by a constable if the authorised person has reasonable cause to apprehend any serious obstruction in the execution of that person's duty;

(c) without prejudice to the preceding sub-paragraph, on entering any premises by virtue of sub-paragraph (a) to—

(i) be accompanied by any other person duly authorised by the authorised person's enforcing authority; and

(ii) take any equipment or materials required for any purpose for which the power of entry is being exercised.

(d) to make such examination and investigation as may in any circumstances be necessary for the purpose for which the power is being exercised;

(e) as regards any premises which the authorised person has power to enter, to direct that those premises or any part of them, or anything therein, must be left undisturbed (whether generally or in particular respects) for so long as is reasonably necessary for the purpose of any examination or investigation under sub-paragraph (d);

(f) to take such measurements and photographs and make such recordings as the authorised person considers necessary for the purpose of any examination or investigation under sub-paragraph (d);

(g) to take samples of any thing found in any premises which the authorised person has power to enter and of the atmosphere in or in the vicinity of any such premises;

(h) in the case of any thing found in any premises which the authorised person has power to enter and which appears to be in contravention of a listed REACH provision, to cause it to be dismantled or subjected to any process or test (but not so as to damage or destroy it unless this is in the circumstances necessary);

(i) in the case of any such thing as is mentioned in sub-paragraph (h), to take possession of it and detain it for so long as is necessary for all or any of the following purposes, namely—

   (i) to examine it and do to it anything which the authorised person has power to do under that sub-paragraph;

   (ii) to ensure that it is not tampered with before the authorised person's examination of it is completed;

   (iii) to ensure that it is available for use as evidence in any proceedings for an offence under regulation 11 or 13.

(j) to require any person whom the authorised person has reasonable cause to believe to be able to give any information relevant to any examination or investigation under sub-paragraph (d) to answer (in the absence of persons other than a person nominated by that person to be present and any persons whom the authorised person may allow to be present) such questions as the authorised person thinks fit to ask and to sign a declaration of the truth of that person's answers;

(k) to require the production of, inspect, and take copies of or of any entry in—

   (i) any books or documents which by virtue of any of the listed REACH provisions are required to be kept; and

   (ii) any other books or documents which it is necessary for the authorised person to see for the purposes of any examination or investigation under sub-paragraph (d);

(l) to require any person to afford the authorised person such facilities and assistance with respect to any matters or things within that person's control or in relation to which that person has responsibilities as are necessary to enable the authorised person to exercise any of the powers conferred upon the authorised person by this Part;

(m) to seize and cause to be rendered harmless (whether by destruction or otherwise) any thing which is found by the authorised person in any premises and which the authorised person has reasonable cause to believe that, in the circumstances in which it is found is a cause of imminent danger of serious personal injury;

(n) any other power which is necessary for the purpose of carrying out the enforcement duty.

2

Where an authorised person proposes to exercise the power conferred by paragraph 1(h) and if requested by a person who at the time is present in and has responsibilities in relation to those premises, the authorised person must cause anything which is to be done by virtue of that power to be done in the presence of that other person unless the authorised person considers that would be prejudicial to the safety of any person.

3

Before exercising the power conferred by paragraph 1(h), an authorised person must consult such persons as appear to be appropriate for the purpose of ascertaining what dangers, if any, there may be in doing anything which it is proposed to do under that power.

4

Where under the power conferred by paragraph 1(i) an authorised person takes possession of any thing found in any premises, the authorised person must leave there, either with a responsible person or, if that is impracticable, fixed in a conspicuous position, a notice giving particulars of that thing sufficient to identify it and stating that the authorised person has taken possession of it under that power; and before taking possession of any such thing under that power an authorised person must, if it is practicable to do so, take a sample and give to a responsible person at the premises a portion of the sample marked in a manner sufficient to identify it.

5

No answer given by a person in pursuance of a requirement imposed under paragraph 1(j) is admissible in evidence against that person or the spouse or civil partner of that person in any proceedings.

6

Before the power in paragraph 1(m) to render harmless any thing is exercised, an authorised person must, if it is practicable to do so, take a sample of it and give it to a responsible person at the premises a portion of the sample marked in a manner sufficient to identify it.

7

As soon as may be after any thing has been seized and rendered harmless under paragraph 1(m), the authorised person must prepare and sign a written report giving particulars of the circumstances in which it was seized and so dealt with by the authorised person, and must—

(a) give a signed copy of the report to a responsible person at the premises where the thing was found; and

(b) unless that person is the owner of it, also serve a signed copy of the report on the owner,

and if, where sub-paragraph (b) applies, the authorised person cannot after reasonable enquiry ascertain the name or address of the owner, the copy may be served on the owner by giving it to the person to whom a copy was given under paragraph 4.

8

Nothing in this Part compels the production by any person of a document which—

(a) except in relation to Scotland, that person would on grounds of legal professional privilege be entitled to withhold production on an order for discovery in an action in the County Court or High Court; or

(b) in relation to Scotland, contains confidential information made by or to an advocate or solicitor in that capacity and which that person would be entitled to withhold production on an order for the production of documents in an action in the Court of Session.

9

Where—

(a) an action has been brought against an authorised person in respect of an act done in the execution or purported enforcement of a listed REACH provision;

(b) the circumstances are such that the authorised person is not legally entitled to require an indemnity from the Executive; and

(c) the Executive is satisfied that the authorised person honestly believed that the act complained of was within that person's powers and that the duty of an authorised person required or entitled that act,

the Executive may indemnify the authorised person against the whole or part of any damages and costs which that person may be ordered to pay or may have incurred.

9A

The powers of an authorised person acting on behalf of the Office of Rail Regulation—

(a) do not include the powers in paragraph 1(m) or (n);

(b) only apply in relation to non-domestic premises.]

10

. . .

SECTION 2

NOTICES

11

If an authorised person is of the opinion that—

(a) activities are being carried on or are likely to be carried on by or under the control of a person; and

(b) the activity involves or as the case may be, will involve a risk of serious personal injury,

the authorised person may serve on that person a prohibition notice.

12

A prohibition notice must—

(a) state that the authorised person is of the opinion referred to in the preceding paragraph;

(b) specify the matters which in the authorised person's opinion give or, as the case may be, will give rise to the said risk;

(c) where in the opinion of the authorised person any of those matters involves or, as the case may be, will involve a contravention of a listed REACH provision, the authorised person must—

(i) state that opinion;

(ii) specify the provision or provisions which relate to that opinion; and

(iii) give particulars of the reasons supporting that opinion;

(d) direct that the activities to which the notice relates must not be carried on by or under the control of the person on whom the notice is served unless the matters specified in the notice in pursuance of sub-paragraph (b) and any associated contraventions of provisions so specified in pursuance of sub-paragraph (c) above have been remedied.

13

A direction contained in a prohibition notice in pursuance of paragraph 12(d) takes effect—

(a) at the end of the period specified in the notice; or

(b) if the notice so declares, immediately.

14

If an authorised person is of the opinion that a person—

(a) is contravening one or more of the listed REACH provisions; or

(b) has contravened one or more of those provisions in circumstances that make it likely that the contravention will continue or be repeated,

the authorised person may serve on that person an improvement notice.

15

An improvement notice must—

(a) state that the authorised person is of the opinion referred to in the preceding paragraph;

(b) specify the provision or provisions as to which the authorised person is of that opinion, giving particulars of the reasons supporting that opinion;

(c) require that the person to whom the improvement notice is addressed remedies the contravention or, as the case may be, the matters occasioning it within such period (ending not earlier than the period within which an appeal against the notice can be brought as provided by regulation 21) as may be specified in the notice.

16

If an authorised person is of the opinion that a person has contravened, is contravening or is likely to contravene a listed REACH provision, the authorised person may serve on that person an enforcement notice.

17

An enforcement notice must—

(a) state that the authorised person is of the opinion referred to in the preceding paragraph;

(b) specify the matters constituting the contravention or the matters making it likely that the contravention will arise, as the case may be;

(c) specify the steps that must be taken to remedy the contravention or to remedy the matters making it likely that the contravention will arise, as the case may be; and

(d) specify the period within which those steps must be taken.

18

Where a prohibition notice, an improvement notice or an enforcement notice has been served but is not to take immediate effect—

(a) the notice may be withdrawn by an authorised person at any time before the end of the period specified therein;

(b) the period so specified may be extended or further extended by an authorised person at any time when an appeal against the notice is not pending.

19

References in this Part to a 'prohibition notice', an 'improvement notice' or an 'enforcement notice' have effect only for the purposes of this Part and Part 2 of Schedule 8.

[20

This section does not apply to an authorised person acting on behalf of the Office of Rail Regulation.]

[8.288A]

Statutory annotations

Amendment

Para 9A: inserted by SI 2013/2919, regs 2, 12(b)(i).
Para 10: revoked by SI 2013/2919, regs 2, 12(b)(ii).
Para 20: inserted by SI 2013/2919, regs 2, 12(c).

PART 3

LOCAL (CONSUMER SAFETY) AUTHORITIES

SECTION 1

POWERS OF ENTRY

[8.289]

1

An authorised person may at any reasonable hour and on production, if required, of that person's credentials exercise any of the powers conferred by the following provisions of this section.

2

An authorised person may, for the purposes of ascertaining whether there has been any contravention of a listed REACH provision, inspect any goods and enter any premises other than premises occupied only as a person's residence.

3

An authorised person may, for the purpose of ascertaining whether there has been any contravention of a listed REACH provision, examine any procedure (including any arrangements for carrying out a test) connected with the production of any goods.

4

If an authorised person has reasonable grounds for suspecting that there has been a contravention of a listed REACH provisions, the authorised person may—

(a) require any person carrying on a business, or employed in connection with a business, to produce any records relating to the business;

(b) for the purpose of ascertaining (by testing or otherwise) whether there has been any such contravention, seize and detain the goods;

(c) take copies of, or of any entry in, any records produced by virtue of sub-paragraph (a).

5

An authorised person may seize and detain any goods or records which the authorised person has reasonable grounds for believing may be required as evidence in proceedings for any offence in respect of a contravention of a listed REACH provision.

6

If and to the extent that it is reasonably necessary to prevent a contravention of a listed REACH provision, the authorised person may—

(a) require any person having authority to do so to open any container or to open any vending machine; and

(b) open or break open any such container or machine where a requirement made under sub-paragraph (a) in relation to the container or machine has not been complied with.

7

An authorised person seizing any goods or records under this Part must inform the following persons that the goods or records have been so seized, that is to say—

(a) the person from whom they are seized; and

(b) in the case of imported goods seized on any premises under the control of the Commissioners for Revenue and Customs, the importer of those goods (within the meaning of the Customs and Excise Management Act 1979).

8

If a justice of the peace—

(a) is satisfied by any written information on oath that there are reasonable grounds for believing either—

(i) that any goods or records which any authorised person has power to inspect under this Part are on any premises and that their inspection is likely to disclose evidence that there has been a contravention of a listed REACH provision; or

(ii) that such a contravention has taken place, is taking place or is about to take place on any premises; and

(b) is also satisfied by any such information either—

(i) that admission to the premises has been or is likely to be refused and that notice of intention to apply for a warrant under this section has been given to the occupier; or

(ii) that an application for admission, or the giving of such a notice, would defeat the object of the entry or that the premises are unoccupied or that the occupier is temporarily absent and it might defeat the object of the entry to await the occupier's return,

the justice may by warrant, which shall continue in force for a period of one month, authorise any authorised person to enter the premises, if need be by force.

9

An authorised person entering any premises by virtue of this section may be accompanied by such other persons and may take such equipment as may appear to the authorised person to be necessary.

10

On leaving any premises which an authorised person is authorised to enter by a warrant under this section, that person must, if the premises are unoccupied or the occupier is temporarily absent, leave the premises as effectively secured against trespassers as the authorised person found them.

11

Where any goods seized by an authorised person under this Part are submitted to a test, the authorised person must inform the persons mentioned in paragraph 7 of the result of the test and, if—

(a) proceedings are brought for an offence in respect of a contravention of a listed REACH provision; and

(b) the authorised person is requested to do so and it is practicable to comply with the request,

the authorised person must allow any person who is a party to the proceedings or, as the case may be, has an interest in the goods to which the notice relates, to have the goods tested.

12

In the application of this section to Scotland, the reference in paragraph 8 to a justice of the peace includes a reference to a sheriff and the references to written information on oath include references to evidence on oath.

13

In the application of this section to Northern Ireland, the reference in paragraph 8 to a justice of the peace includes a reference to a lay magistrate and references to any information on oath include references to any complaint on oath.

SECTION 2

NOTICES

14

Where an authorised person has reasonable grounds for suspecting that a listed REACH provision has been contravened in relation to any goods, the authorised person may serve a notice (' a suspension notice').

15

A suspension notice may prohibit the person on whom it is served, for such period ending not more than six months after the date of the notice as is specified in the notice, from doing any of the following without the consent of an authorised person—

(a) supplying the goods;
(b) offering to supply them;
(c) agreeing to supply them or exposing them for supply.

16

A suspension notice served by an authorised person in respect of any goods must—

(a) describe the goods in a manner sufficient to identify them;
(b) set out the grounds on which the authorised person suspects that a listed REACH provision has been contravened in relation to the goods; and
(c) state that, and the manner in which, the person on whom the notice is served may appeal against the notice.

17

A suspension notice may require a person to keep the enforcing authority informed of the whereabouts throughout the period of the notice of any of those goods in which that person has an interest.

18

Where a suspension notice has been served on any person in respect of any goods, no further such notice must be served on that person in respect of the same goods unless—

(a) proceedings against that person for an offence in respect of a contravention, in relation to the goods, of a listed REACH provision; or
(b) proceedings for the forfeiture of the goods,

are pending at the end of the period specified in the notice.

19

A consent given by an authorised person for the purposes of paragraph 15 may impose such conditions on the doing of anything for which the consent is required as the authorised person considers appropriate.

20

Where an authorised person serves a suspension notice in respect of any goods, the authorised person is liable to pay compensation to any person having an interest in the goods in respect of any loss or damage caused by reason of the service of the notice if—

(a) there has been no contravention in relation to the goods of a listed REACH provision; and

(b) the exercise of the power is not attributable to any neglect or default by that person having an interest in the goods.

21

Any disputed question as to the right to or the amount of any compensation payable under this section must be determined by arbitration or, in Scotland, by a single arbiter appointed, failing agreement between the parties, by the sheriff.

SECTION 3

FORFEITURE

22

An authorised person in England and Wales or Northern Ireland may apply under this section for an order for the forfeiture of any goods on the grounds that there has been a contravention in relation to the goods of a listed REACH provision.

23

An application under this section may be made—

(a) where proceedings have been brought in a magistrates' court for an offence in respect of a contravention of a listed REACH provision in relation to some or all of the goods, to that court;

(b) where no application for the forfeiture of the goods has been made under sub-paragraph (a) by way of complaint to a magistrates' court.

24

On an application under this section the court may make an order for the forfeiture of any goods only if it is satisfied that there has been a contravention of a listed REACH provision in relation to the goods.

25

A court may infer for the purposes of this section that there has been a contravention of a listed REACH provision in relation to any goods if it is satisfied that any such provision has been contravened in relation to goods which are representative of those goods (whether by reason of being of the same design or part of the same consignment or batch or otherwise).

26

Any person aggrieved by an order made under this section by a magistrates' court, or by a decision of such a court not to make such an order, may appeal against that order or decision—

(a) in England and Wales, to the Crown Court;

(b) in Northern Ireland, to the county court,

and an order so made may contain such provision as appears to the court to be appropriate for delaying the coming into force of the order pending the making and determination of any appeal (including any application under section 111 of the Magistrates' Courts Act 1980 or Article 146 of the Magistrates' Courts (Northern Ireland) Order 1981(statement of case).

27

Subject to paragraph 28, where any goods are forfeited under this section they must be destroyed in accordance with such directions as the court may give.

28

On making an order under this section a magistrates' court may, if it considers it appropriate to do so, direct that the goods to which the order relates may (instead of being destroyed) be released, to such person as the court may specify, on condition that that person—

- (a) does not supply those goods to any person otherwise than as specified;
- (b) complies with any order to pay costs or expenses which has been made against that person in the proceedings for the order for forfeiture.

29

In Scotland a sheriff may make an order for forfeiture of any goods in relation to which there has been a contravention of a listed REACH provision—

- (a) on an application by the procurator-fiscal; or
- (b) where a person is convicted of any offence in respect of any such contravention, in addition to any other penalty which the sheriff may impose.

30

The procurator-fiscal making an application under paragraph 29 must serve on any person appearing to be the owner of, or otherwise to have an interest in, the goods to which the application relates a copy of the application, together with a notice giving the opportunity to appear at the hearing of the application to show cause why the goods should not be forfeited.

31

Service under paragraph 30 must be carried out, and such service may be proved, in the manner specified for citation of an accused in summary proceedings.

32

Any person upon whom notice is served under paragraph 30 and any other person claiming to be the owner of, or otherwise to have an interest in, goods to which an application under this section relates is entitled to appear at the hearing of the application to show cause why the goods should not be forfeited.

33

The sheriff must not make an order following an application under paragraph 29—

- (a) if any person on whom notice is served under paragraph 30 does not appear, unless service of the notice on that person is proved; or
- (b) if no notice under paragraph 30 has been served, unless the court is satisfied that in the circumstances it was reasonable not to serve notice on any person.

34

An order under this section may be made only if the sheriff is satisfied that there has been a contravention in relation to those goods of a listed REACH provision.

35

The sheriff may infer for the purposes of this section that there has been a contravention of a listed REACH provision in relation to any goods if the sheriff is satisfied that any such provision has been contravened in relation to any goods which are representative of those goods (whether by reason of being of the same design or part of the same consignment or batch or otherwise).

36

Where an order for the forfeiture of any goods is made following an application by the procurator-fiscal under paragraph 29, any person who appeared, or was entitled to appear, to show cause why goods should not be forfeited may, within twenty-one days of the making of the order, appeal to the High Court by Bill of Suspension on the ground of an alleged miscarriage of justice; and section 182(5)(a) to (e) of the Criminal Procedure (Scotland) Act 1995 applies to an appeal under this section as it applies to a stated case under Part X of that Act.

37

An order following an application under paragraph 29 does not take effect—

(a) until the end of the period of twenty-one days beginning with the day after the day on which the order is made; or

(b) if an appeal is made under paragraph 36 within that period, until the appeal is determined or abandoned.

38

An order under paragraph 29 does not take effect—

(a) until the end of the period within which an appeal against the order could be brought under the Criminal Procedure (Scotland) Act 1995; or

(b) if an appeal is made within that period, until the appeal is determined or abandoned.

39

Subject to paragraph 40, goods forfeited under this section must be destroyed in accordance with such directions as the sheriff may give.

40

The sheriff may direct that the goods be released, to such person as may be specified, on condition that that person does not supply those goods to any other person otherwise than as mentioned in section 46(7)(a) or (b) of the Consumer Protection Act 1987.

PART 4

THE SECRETARY OF STATE

SECTION 1

POWERS OF ENTRY

[8.290]

1

An authorised person may on producing evidence of that person's appointment—

(a) at any reasonable time (or, in a situation which in that person's opinion may give rise to a risk of significant pollution to the environment as a result of the use or discharge from an offshore installation of an article or substance, at any time) board any offshore installation;

(b) on boarding an offshore installation be accompanied by any other person authorised for those purposes by the Secretary of State and take any equipment or materials that the authorised person thinks may be required;

(c) make such examination or investigation as the authorised person considers necessary (including any examination or investigation of an offshore installation, for which purpose the authorised person may install or maintain monitoring or other apparatus on the offshore installation);

(d) give a direction requiring that any part of the offshore installation be left undisturbed (whether generally or in particular respects) for so long as reasonably necessary for the purposes of any examination or investigation under sub-paragraph (c);

(e) take such measurements and photographs and make such recordings as the authorised person considers necessary for the purpose of any examination or investigation under sub-paragraph (c);

(f) take samples of any thing found on the offshore installation or in the atmosphere or any land, seabed (including the subsoil thereof) or water in the vicinity of the offshore installation;

(g) in the case of any thing which the authorised person finds on the offshore installation, cause it to be dismantled or subjected to any process or test (but not so as to damage or destroy it unless that in the circumstances of the case is necessary);

(h) in the case of any such thing mentioned in sub-paragraph (g), take possession of it and detain it for so long as is necessary for all or any of the following purposes, namely—

  (i) to examine it and do to it anything which the authorised person has power to do under that sub-paragraph;

  (ii) to ensure that it is not tampered with before an examination of it is completed; and

  (iii) to ensure that it is available for use as evidence in any proceedings for an offence under regulation 11 or 13;

(i) require any person who the authorised person has reasonable cause to believe is able to give any information relevant to any examination or investigation under sub-paragraph (c)—

  (i) to attend at a place and time specified by the authorised person;

  (ii) to answer (in the absence of any person other than persons whom the authorised person may allow to be present and a person nominated to be present by the person on whom the requirement is imposed) such questions as the authorised person thinks fit to ask; and

  (iii) to sign a declaration of truth of that person's answers;

(j) require the production of, and inspect and take copies of or of any entry in—

  (i) any records which by virtue of any provision of any permit granted under the Offshore Chemicals Regulations 2002 are required to be kept;

  (ii) any records which the authorised person considers it necessary to see for the purposes of any examination or investigation under sub-paragraph (c);

(k) require any person to afford the authorised person such facilities and assistance with respect to any matters or things within that person's control or in relation to which that person has responsibilities as the authorised person considers are necessary to enable the authorised person to exercise any of the powers conferred on the authorised person by this Part.

SECTION 2

NOTICES

2

An authorised person may serve on a person a notice in writing ('an enforcement notice') if the authorised person is of the opinion that the person has contravened, is contravening or is likely to contravene a listed REACH provision.

3

An enforcement notice must—

(a) state that the authorised person is of the opinion referred to in the preceding paragraph;
(b) specify the matters that constitute, constituted or, as the case may be, are likely to constitute the contravention;
(c) specify the steps that must be taken to remedy or, as the case may be, prevent the contravention; and
(d) specify the period within which those steps must be taken.

4

The steps mentioned in paragraph 3(c) include steps that must be taken to remedy any pollution caused by the contravention.

5

Where a person to whom an enforcement notice is addressed has failed to take the action required by it within such time as may be specified by it and such a notice has not been revoked, the Secretary of State may undertake any action so required and the reasonable costs and expenses of the Secretary of State's so doing are recoverable as a debt from that person.

6

A person to whom an enforcement notice is addressed must afford such assistance as the Secretary of State may reasonably require for the purpose of facilitating the exercise of any powers conferred on the Secretary of State by paragraph 5.

7

An authorised person may revoke an enforcement notice.

8

An authorised person may serve on a person a notice in writing ('a prohibition notice') if the authorised person is of the opinion that where the person has contravened, is contravening or is likely to contravene a listed REACH provision, that contravention involves an imminent risk of pollution.

9

A prohibition notice must—

(a) state that the authorised person is of the opinion referred to in the preceding paragraph;
(b) specify the risk involved;
(c) specify the steps that must be taken to remove it and the period within which they must be taken.

10

An authorised person may by notice withdraw a prohibition notice wholly or in part at any time and must withdraw a notice when the authorised person is satisfied that the steps required by the notice have been taken.

11

It is the duty of the person to whom the prohibition notice is addressed to comply with its terms save to the extent that it is withdrawn wholly or in part.

SECTION 3

MISCELLANEOUS

12

An answer given by a person in compliance with a requirement imposed under paragraph 1(i) is admissible in evidence in England and Wales or Northern Ireland against that person in any proceedings or, in Scotland, against that person in criminal proceedings.

13

In criminal proceedings in which such person as is mentioned in paragraph 12 is charged with an offence to which this paragraph applies, no evidence relating to that person's answer may be adduced and no question relating to it may be asked by or on behalf of the prosecution unless evidence relating to it is adduced by or on behalf of that person.

14

Paragraph 13 applies to any offence other than one under—

- (a) regulation 18(1)(e)(ii) of the Offshore Chemicals Regulations 2002 (offences);
- (b) section 5 of the Perjury Act 1911(false statements made otherwise than on oath);
- (c) section 44(2) of the Criminal Law (Consolidation) (Scotland) Act 1995(false statements made otherwise than on oath); or
- (d) article 10 of the Perjury (Northern Ireland) Order 1979(false statutory declarations and other false unsworn statements).

15

Nothing in this Part compels the production by any person of a document which—

- (a) except in relation to Scotland, that person would on grounds of legal professional privilege be entitled to withhold production on an order for discovery in an action in the County Court or High Court; or
- (b) in relation to Scotland, contains confidential information made by or to an advocate or solicitor in that capacity and which that person would be entitled to withhold production on an order for the production of documents in an action in the Court of Session.

16

References in this Part to an 'information notice' or a 'prohibition notice' have effect only for the purposes of this Part.

SCHEDULE 7

AUTHORISATIONS

Regulation 10(2)

SECTION 1

THE ENVIRONMENT AGENCY[, THE NATURAL RESOURCES BODY FOR WALES] AND THE SCOTTISH ENVIRONMENT PROTECTION AGENCY

[8.291]

1

In respect of the Environment Agency[, the Natural Resources Body for Wales] or the Scottish Environment Protection Agency, an authorisation made under section 108 of the Environment Act 1995 (powers of enforcing authorities and persons authorised by them).

SECTION 2

THE DEPARTMENT OF THE ENVIRONMENT

2

In respect of the Department of the Environment, an authorisation made under—

(a) article 72 of the Waste and Contaminated Land (Northern Ireland) Order 1997 (powers of enforcing authorities and persons authorised by them);
(b) article 25 of the Water (Northern Ireland) Order 1999 (powers of entry and inspection); or
(c) regulation 8 of the Pollution, Prevention and Control Regulations (Northern Ireland) 2003 (appointment of chief inspector and other inspectors).

SECTION 3

THE HEALTH AND SAFETY EXECUTIVE[, THE OFFICE OF RAIL REGULATION] AND LOCAL (HEALTH AND SAFETY) AUTHORITIES IN GREAT BRITAIN

3

In respect of—
(a) the Health and Safety Executive; . . .
(b) a local (health and safety) authority in Great Britain; [or
(c) the Office of Rail Regulation],

an authorisation made under section 19 of the Health and Safety at Work etc Act 1974 (appointment of inspectors).

SECTION 4

THE HEALTH AND SAFETY EXECUTIVE FOR NORTHERN IRELAND AND LOCAL (HEALTH AND SAFETY) AUTHORITIES IN NORTHERN IRELAND

4

In respect of—
(a) the Health and Safety Executive for Northern Ireland; or
(b) a local (health and safety) authority in Northern Ireland,

an authorisation made under article 21 of the Health and Safety at Work (Northern Ireland) Order 1978 (appointment of inspectors).

SECTION 5

LOCAL (CONSUMER SAFETY) AUTHORITIES

5

In respect of—
(a) a local (consumer safety) authority in Great Britain, an authorisation made under section 72 of the Weights and Measures Act 1985 (appointment of inspectors);
(b) a local (consumer safety) authority in Northern Ireland, an authorisation made for the purposes of section 27(2) of the Consumer Protection Act 1987 (enforcement).

SECTION 6

THE SECRETARY OF STATE

6

In respect of the Secretary of State, an authorisation made under regulation 16 of the Offshore Chemicals Regulations 2002 (appointment of inspectors).

[8.291A]

Statutory annotations

Amendment

Section 1 heading: words ', the Natural Resources Body for Wales' in square brackets inserted by SI 2013/755, art 4(2), Sch 4, paras 302, 308(1), (3).

Para 1: words ', the Natural Resources Body for Wales' in square brackets inserted by SI 2013/755, art 4(2), Sch 4, paras 302, 308(1), (2).

Section 3 heading: words ', the Office of Rail Regulation' in square brackets inserted by SI 2013/2919, regs 2, 13(a).

Para 3: in sub-para (a) word omitted revoked by SI 2013/2919, regs 2, 13(b); sub-para (c) and word 'or' immediately preceding it inserted by SI 2013/2919, regs 2, 13(c).

SCHEDULE 8

APPEALS

Regulation 21(2)

PART 1

NOTICES SERVED BY THE ENVIRONMENT AGENCY, [THE NATURAL RESOURCES BODY FOR WALES,] THE SCOTTISH ENVIRONMENT PROTECTION AGENCY OR THE DEPARTMENT OF THE ENVIRONMENT

[8.292]

1

The appeal body means, where the notice is served by—

[(a) the Environment Agency, the Secretary of State;]

[(aa) the Natural Resources Body for Wales, the Welsh Ministers;]

(b) the Scottish Environment Protection Agency, the Scottish Ministers;

(c) the Department of the Environment, the Planning Appeals Commission.

2

An appeal must be received by the appeal body no later than two months from the date of the notice.

3

The appeal body may consider an appeal received after the date referred to in paragraph 2.

4

Where an appeal is brought, this does not have the effect of suspending the operation of the notice.

5

On the determination of an appeal, the appeal body may either cancel or affirm the notice and, if it affirms it, may do so either in its original form or with such modifications as the appeal body sees fit.

6

In relation to the making and determination of appeals under this section, the appeal body may adopt such procedures as it sees fit.

[7

The appeal body may—

(a) appoint a person, with or without payment, to determine the appeal;

(b) refer any matter involved in the appeal to such person as that body may appoint for that purpose, with or without payment.]

[8.292A] 8A Phenomena hazardous to health: General

[8.292A]

Statutory annotations

Amendment

Para 1: sub-para (a) substituted by SI 2013/755, art 4(2), Sch 4, paras 302, 309(1), (2)(a); sub-para (aa) inserted by SI 2013/755, art 4(2), Sch 4, paras 302, 309(1), (2)(b).

Para 7: inserted by SI 2013/2919, regs 2, 14.

PART 2

NOTICES SERVED BY THE HEALTH AND SAFETY EXECUTIVE, THE HEALTH AND SAFETY EXECUTIVE FOR NORTHERN IRELAND OR LOCAL (HEALTH AND SAFETY) AUTHORITIES

[8.293]

1

An appeal may be made to a tribunal.

2

[Schedule 1 to the Employment Tribunals (Constitution and Rules of Procedure) Regulations 2013] applies to a notice served in Great Britain by—

(a) the Health and Safety Executive;

(b) a local (health and safety) authority.

3

Under paragraph 2, an enforcement notice is to be treated in the same way as an improvement notice for the purposes of [Schedule 1] referred to.

4

Schedule 5 to the Industrial Tribunals (Constitution and Rules of Procedure) Regulations (Northern Ireland) 2005 applies to a notice served in Northern Ireland by—

(a) the Health and Safety Executive for Northern Ireland;

(b) a local (health and safety) authority.

5

Under paragraph 4, an enforcement notice is to be treated in the same way as an improvement notice for the purposes of Schedule 5 referred to.

6

The tribunal may cancel or affirm the notice and, if it affirms it, may do so in its original form or with such modifications as the tribunal may in the circumstances see fit.

7

In the case of an improvement notice, the bringing of the appeal has the effect of suspending the operation of the notice until the appeal is finally disposed of or, if the appeal is withdrawn, until the withdrawal of the appeal.

8

In the case of a prohibition notice or an enforcement notice, the bringing of the appeal has the same effect as described in paragraph 7 if, on the application of the appellant, the tribunal so directs and only from the giving of that direction.

9

One or more assessors may be appointed in respect of proceedings brought before a tribunal under this Part.

10

In this Part—

(a) 'assessor' has the same meaning as it has—

(i) in Great Britain, under section 24 of the Health and Safety at Work etc Act 1974;

(ii) in Northern Ireland, under article 26 of the Health and Safety at Work (Northern Ireland) Order 1978;

(b) 'tribunal' means—

(i) in Great Britain, an employment tribunal;

(ii) in Northern Ireland, an industrial tribunal.

Statutory annotations

AMENDMENTS

Para 2: words 'Schedule 1 to the Employment Tribunals (Constitution and Rules of Procedure) Regulations 2013' in square brackets substituted by SI 2013/1948, reg 7(1), (2).

Para 3: words 'Schedule 1' in square brackets substituted by SI 2013/1948, reg 7(1), (3).

Chemicals (Hazard Information and Packaging for Supply) Regulations 2009

SI 2009/716

as amended by SI 2011/1043, SI 2011/2131, SI 2013/1506 and SSI 2011/228

[8.352]

The Secretary of State is a Minister designated for the purpose of section 2(2) of the European Communities Act 1972 ('the 1972 Act') in relation to the regulation and control of classification, packaging and labelling of dangerous substances and preparations, and for measures relating to consumer protection.

The Secretary of State makes these Regulations:

(a) in exercise of the powers conferred upon him by section 2(2) of the 1972 Act and by sections 15(1), (2), (3)(c), (4)(b), (6)(b), (8) and (9) and 82(3)(a) of, and paragraphs 1(1)(b) and (c), (4) and (5), 2(1), 3(2), 15(1) and 16 of Schedule 3 to, the Health and Safety at Work etc Act 1974 ('the 1974 Act'); and

(b) for the purpose of giving effect without modifications to proposals submitted to him by the Health and Safety Executive under section 11(3) of the 1974 Act after the carrying out by the said Executive of consultations in accordance with section 50(3) of that Act.

These Regulations make provision for a purpose mentioned in section 2(2) of the 1972 Act and it appears to the Secretary of State that it is expedient for references in these Regulations to—

(i) the Regulation (EC) No 1272/2008 of the European Parliament and of the Council of 16 December 2008 on classification, labelling and packaging of substances and mixtures, amending and repealing Directives 67/548/EEC and 1999/45/EC, and amending Regulation (EC) No 1907/2006, be construed as including references to articles 6(5), 11(3), 12, 14, 18(3)(b), 23, 25 to 29, 35(2) second and third sub-paragraph and Annexes I to VII of that Regulation as amended from time to time; and

(ii) the *Regulation (EC) No 689/2008 of the European Parliament and of the Council of 17 June 2008 concerning the export and import of dangerous chemicals be construed as including references to Annexes I and V of that Regulation as those Annexes are amended from time to time* [Regulation (EU) No 649/2012 of the European Parliament and of the Council of 4 July 2012 concerning the export and import of hazardous chemicals, of which Annexes I, II, V and VI are as amended from time to time].

Statutory annotations

Amendment

In para (ii) words from 'Regulation (EC) No 689/2008' to 'time to time' in italics revoked and subsequent words in square brackets substituted by SI 2013/1506, reg 35, Sch 4, Pt 1, para 1.

PART 1

INTRODUCTION

Statutory annotations

Amendment

Part 1: Revoked by SI 2013/1506, reg 36(d). This revocation has effect as from 1 June 2018: see SI 2013/1506, reg 36(d).

[8.353]

1 Citation, commencement and extent

(1) These Regulations may be cited as the Chemicals (Hazard Information and Packaging for Supply) Regulations 2009 and shall come into force on 6th April 2009.

(2) These Regulations shall not extend to Northern Ireland.

[8.354]

2 Interpretation

(1) In these Regulations—

'the 1974 Act' means the Health and Safety at Work etc Act 1974;

'aerosol dispenser' means an article which consists of a non-reusable receptacle containing a gas compressed, liquefied or dissolved under pressure, with or without liquid, paste or powder and fitted with a release device allowing the contents to be ejected as solid or liquid particles in suspension in a gas, as a foam, paste or powder or in a liquid state;

'the approved classification and labelling guide' means the guide entitled 'Approved Guide to the Classification and Labelling of Dangerous Substances and Dangerous Preparations (Fifth Edition)' approved by the Health and Safety Commission on 16th April 2002, as revised or re-issued from time to time;

'category of danger' means, in relation to a dangerous substance or dangerous preparation, one of the categories of danger specified in column 1 of Schedule 1;

'the CLP Regulation' means Regulation (EC) No 1272/2008 of the European Parliament and of the Council of 16 December 2008 on classification, labelling and packaging of substances and mixtures, amending and repealing Directives 67/548/EEC and 1999/45/EC, and amending Regulation (EC) No 1907/2006, of which articles 6(5), 11(3), 12, 14, 18(3)(b), 23, 25 to 29, 35(2) second and third sub-paragraph and Annexes I to VII are as amended from time to time;

'Community workplace exposure limit' means, in respect of a substance, an exposure limit for that substance established in a [EU] instrument;

'dangerous preparation' means a preparation which is in one or more of the categories of danger specified in column 1 of Schedule 1;

'dangerous substance' means a substance—

(a) which is listed in Table 3.2 of part 3 of Annex VI of the CLP Regulation; or

(b) which, if it is not so listed, is in one or more of the categories of danger specified in column 1 of Schedule 1;

'EC number' means—

(a) in the case of a dangerous substance that appears in Table 3.2 of part 3 of Annex VI of the CLP Regulation, the EC number specified in that list;

(b) in the case of a dangerous substance that is not included in Table 3.2 of part 3 of Annex VI of the CLP Regulation or for which an EC number is not given in that list, the number for that substance specified in EINECS; or

(c) in the case of a dangerous substance that is not a phase-in substance within the meaning of REACH, the number for that substance if it is listed in ELINCS;

'EEA Agreement' means the Agreement on the European Economic Area signed at Oporto on 2nd May 1992, as adjusted by the Protocol signed at Brussels on 17th March 1993 and adopted as respects the United Kingdom by the European Economic Area Act 1993;

'EEA State' means a state which is a contracting party to the EEA Agreement;

'EINECS' means the European Inventory of Existing Commercial Chemical Substances;

'ELINCS' means the European List of Notified Chemical Substances;

'enforcing authority' shall be construed in accordance with regulation 14;

'the Executive' means the Health and Safety Executive;

'indication of danger' means, in relation to a dangerous substance or dangerous preparation, one or more of the indications of danger referred to in column 1 of Schedule 2 and—

(a) in the case of a dangerous substance listed in Table 3.2 of part 3 of Annex VI of the CLP Regulation, it is one or more of the indications of danger specified for that substance by a symbol-letter in that list;

(b) in the case of any other dangerous substance or a dangerous preparation, it is one or more indications of danger determined in accordance with the classification of that substance or preparation in accordance with regulation 4 and the approved classification and labelling guide;

['plant protection product' has the same meaning as in Article 2(1) of Regulation (EC) No 1107/2009 of the European Parliament and of the Council concerning the placing of plant protection products on the market and repealing Council Directives 79/117/EEC and 91/414/EEC;]

'the Plant Protection Products Regulations' means the Plant Protection Products Regulations 2005 and the Plant Protection Products (Scotland) Regulations 2005;

'preparation' means a mixture or solution composed of two or more substances;

'radioactive substance' means a substance which contains one or more radionuclides whose activity or concentration cannot be disregarded as far as radiation protection is concerned;

'receptacle' means a container together with any material, wrapping and component, including any closure or fastener, associated with the container which enables the container to perform its containment function;

'REACH' means Regulation (EC) No 1907/2006 of the European Parliament and of the Council concerning the Registration, Evaluation, Authorisation and Restriction of Chemicals;

'risk phrase' means a risk phrase listed in Annex III of Council Directive 67/548/EEC;

'safety phrase' means a safety phrase listed in Annex IV of Council Directive 67/548/EEC;

'substance' means a chemical element and its compounds in the natural state or obtained by any production process, including any additive necessary to preserve the stability of the product and any impurity deriving from the process used, but excluding any solvent which may be separated without affecting the stability of the substance or changing its composition;

'supply' in relation to a substance or preparation means making that substance or preparation available to another person and includes importation of the substance or preparation into Great Britain, and

'supplier' shall be construed accordingly.

(2) In these Regulations,

'package' means—

(a) subject to paragraph (3), the package in which a dangerous substance, dangerous preparation or preparation specified in regulation 11(3) is supplied, including the receptacle containing the dangerous substance or preparation in question; or

(b) a pallet or other device which enables more than one receptacle to be handled as a unit,

but does not include a container used to transport the dangerous substance or preparation unless that container is retained by the person to whom the dangerous substance or preparation is supplied for the purpose of storing that dangerous substance or preparation, and related expressions shall be construed accordingly.

(3) In the case of supply by way of retail sale, a package does not include any paper or plastic bag or other form of outer wrapping in which the package is placed when it is presented to the purchaser.

(4) In these Regulations—

(a) where reference is made to a quantity of a dangerous substance or dangerous preparation expressed in litres, that reference shall mean—

(i) in the case of a liquid, the volume in litres of that liquid;

(ii) in the case of a compressed gas, the volume in litres of the receptacle containing that gas; and

(iii) in the case of a compressed gas dissolved in a solvent, liquefied gas or solid, the same number of kilograms of that solvent, gas or solid; and

(b) for the purposes of aggregation, one kilogram of a solid shall be deemed to be equivalent to one litre of liquid or gas.

(5) In these Regulations—

(a) a risk phrase may be designated by the letter 'R' followed by a distinguishing number or combination of numbers; and

(b) a safety phrase may be designated by the letter 'S' followed by a distinguishing number or combination of numbers.

[8.354A]

Amendment

Para (1): in definition 'Community workplace exposure limit' word 'EU' in square brackets substituted by SI 2011/1043, art 6(1)(d); definition 'plant protection product' substituted by SI 2011/2131, reg 31(1), Sch 4.

[8.355]

3 Application

(1) Subject to paragraphs (2) to (6), these Regulations shall apply to any dangerous substance or dangerous preparation.

(2) These Regulations shall not apply to a substance or preparation which is—

(a) intended for use as a medicinal product within the meaning of section 130 of the Medicines Act 1968;

(b) intended for use as a veterinary medical product within the meaning of regulation 2(1) of the Veterinary Medicines Regulations 2008;

(c) intended for use as an investigational medical product within the meaning of the Medicines for Human Use (Clinical Trials) Regulations 2004;
(d) specified in an order made under section 104 or 105 of the Medicines Act 1968 which is for the time being in force and which directs that specified provisions of that Act shall have effect in relation to medicinal products within the meaning of that Act;
(e) a controlled drug within the meaning of the Misuse of Drugs Act 1971 except that these Regulations shall apply to drugs which are excepted from section 4(1)(b) of that Act (which makes it unlawful to supply a controlled drug) by Regulations made under section 7(1)(a) of that Act;
(f) a cosmetic product within the meaning of [Regulation (EC) No 1223/2009 of the European Parliament and of the Council on cosmetic products (recast), as amended from time to time];
(g) in the form of waste to which *the Waste Management Licensing Regulations 1994* [the Waste Management Licensing (Scotland) Regulations 2011], the Special Waste Regulations 1996, the Hazardous Waste (Wales) Regulations 2005 or the Hazardous Waste (England and Wales) Regulations 2005 applies;
(h) intended for use as food within the meaning of section 1 of the Food Safety Act 1990;
(i) intended for use as feeding stuff within the meaning of section 66(1) of the Agriculture Act 1970;
(j) a radioactive substance or a preparation containing radioactive substances; or
(k) a medical device within the meaning of the Medical Devices Regulations 2002 which is invasive or used in direct contact with the human body,

in the finished state, intended for the final user.

(3) These Regulations shall not apply to—
(a) a substance or preparation which is a sample taken by an authority responsible for the enforcement of any requirement imposed by or under any enactment;
(b) a substance or preparation which is under customs control; or
(c) subject to [Regulation (EU) No 649/2012 of the European Parliament and of the Council of 4 July 2012 concerning the export and import of hazardous chemicals, of which Annexes I, II, V and VI are as amended from time to time].

(4) Regulations 6 to 11 shall only apply to substances and preparations which are supplied in packages.

(5) Regulations 6 to 11 shall not apply to munitions and explosives which are supplied with a view to obtaining an explosive or pyrotechnic effect.

(6) These Regulations shall not apply to the carriage of substances or preparations by rail, road, inland waterway, sea or air.

[8.355A]

Amendment

Para (2): in sub-para (f) words from 'Regulation (EC) No 1223/2009' to 'time to time' in square brackets substituted by SI 2013/1478, reg 25, Sch 5, para 27; in sub-para (g) words 'the Waste Management Licensing Regulations 1994' in italics revoked and subsequent words in square brackets substituted, in relation to Scotland, by SSI 2011/228, reg 34(2), Sch 6, para 11.

Para (3): in para (c) words in square brackets substituted by SI 2013/1506, reg 35, Sch 4, Pt 1, para 2.

PART 2

GENERAL REQUIREMENTS

Statutory annotations

Amendment

Part 2: Revoked by SI 2013/1506, reg 36(b). This revocation has effect as from 1 June 2015: see SI 2013/1506, reg 36(b).

[8.356]

4 Classification of dangerous substances and dangerous preparations

(1) No person shall supply a dangerous substance or dangerous preparation unless it has been classified in accordance with paragraphs (2) to (7).

(2) The classification of a dangerous substance which is listed in Table 3.2 of part 3 of Annex VI of the CLP Regulation shall be the classification for that substance specified in that list.

(3) A dangerous substance which—

(a) is not a phase-in substance within the meaning of REACH; and

(b) is not listed in Table 3.2 of part 3 of Annex VI of the CLP Regulation; and

(c) has been registered in accordance with Title II of REACH,

shall be classified in conformity with that registration.

(4) Subject to paragraph (5), a dangerous substance which is not classified in accordance with paragraph (2) or (3) shall be classified—

(a) by ascertaining which of the properties specified in Column 2 of Schedule 1 applies to the dangerous substance and by placing the dangerous substance in one or more of the categories of danger specified in the corresponding entry in Column 1 of that Schedule;

(b) by assigning to the dangerous substance the appropriate risk phrases by the use of the criteria set out in the approved classification and labelling guide; and

(c) where it is proposed to classify a dangerous substance in the category of danger carcinogenic, mutagenic or toxic for reproduction, by an assessment of the evidence by a competent person.

(5) Before a dangerous substance is classified in accordance with paragraph (4), persons carrying out the classification shall make themselves aware of all relevant and accessible data which may exist in relation to the dangerous substance in question.

(6) Where a manufacturer, distributor or importer has classified a substance, in accordance with the provisions of paragraph (4), as a dangerous substance in the category of danger carcinogenic, mutagenic or toxic for reproduction, that person shall send to the Executive as soon as possible a document—

(a) summarising the information on which the classification was based; and

(b) including all relevant references and unpublished data,

unless that document has already been sent to the relevant authority in another EEA State in which the dangerous substance has been supplied.

(7) A dangerous preparation shall be classified in accordance with Schedule 3 and, where applicable, by use of the criteria contained in the approved classification and labelling guide.

[8.357]

[5 . . .

[Revoked by SI 2013/1506, reg 36(a).]

[8.357A]

[5A [Advertisements for dangerous preparations]

[(1) Subject to paragraph (2), a person who supplies a dangerous preparation shall not advertise that preparation, or arrange for the production of any such advertisement, unless mention is made in the advertisement of the type of hazard indicated on the label.

(2) Paragraph (1) shall apply only in respect of a dangerous preparation where the advertisement enables a person, otherwise than in the course of a business, to conclude a contract to purchase the dangerous preparation before that person has seen the label relating to the dangerous preparation.

(3) In this regulation, 'supply' has the same meaning as it has in section 46 of the Consumer Protection Act 1987.]]

[8.357B]

Statutory annotations

Amendments

Inserted by SI 2013/1506, reg 35, Sch 4, Pt 2, para 3.

[8.358]

6 Packaging of dangerous substances, dangerous preparations and certain specified preparations

(1) No person shall supply a dangerous substance or a dangerous preparation or a preparation specified in regulation 11(3) unless its packaging satisfies the following requirements—

- (a) subject to paragraph (2), the receptacle containing the dangerous substance or dangerous preparation is designed and constructed so that its contents cannot escape;
- (b) the materials constituting the packaging and fastenings are not susceptible to adverse attack by the contents or liable to form dangerous compounds with the contents;
- (c) the packaging and fastenings are strong and solid throughout to ensure that they will not loosen and will meet the normal stresses and strains of handling; and
- (d) any replaceable fastening fitted to the receptacle containing the dangerous substance or dangerous preparation is designed so that the receptacle can be repeatedly refastened without the contents of the receptacle escaping.

(2) Paragraph (1)(a) shall not apply where a special safety device is fitted to the receptacle.

(3) Packaging and fastenings shall be deemed to comply with the requirements of paragraphs (1)(a) to (c) if they comply with the relevant requirements of—

- (a) the Carriage of Dangerous Goods and Use of Transportable Pressure Equipment Regulations 2007;
- (b) the Merchant Shipping (Dangerous Goods and Marine Pollutants) Regulations 1997; or
- (c) the Air Navigation (Dangerous Goods) Regulations 2002.

[8.359]

7 Labelling of dangerous substances and dangerous preparations

(1) Subject to paragraphs (5) to (9) and regulation 8, no person shall supply a dangerous substance or dangerous preparation unless the particulars specified in

paragraph (2) relating to a dangerous substance or paragraph (3) relating to a dangerous preparation are clearly shown in accordance with the requirements of regulation 10—

(a) on the receptacle containing the dangerous substance or dangerous preparation; and

(b) if that receptacle is inside one or more layers of packaging, on any such layer which is likely to be the outermost layer of packaging during the supply or use of the dangerous substance or dangerous preparation, unless such packaging permits the particulars shown on the receptacle or other packaging to be clearly seen.

(2) The particulars required under paragraph (1) in relation to a dangerous substance shall be—

(a) the name, full address and telephone number of a person in an EEA State who is responsible for supplying the substance, whether the person be its manufacturer, importer or distributor;

(b) the name of the substance being—

(i) where the substance appears in Table 3.2 of part 3 of Annex VI of the CLP Regulation, the name or one of the names listed therein for that substance; or

(ii) where the substance does not appear in Table 3.2 of part 3 of Annex VI of the CLP Regulation, an internationally recognised name; and

(c) the following particulars ascertained in accordance with Part I of Schedule 4, namely—

(i) any indications of danger together with corresponding symbols;

(ii) the risk phrases, set out in full;

(iii) the safety phrases, set out in full; and

(iv) any EC number and, in the case of a substance which is listed in Table 3.2 of part 3 of Annex VI of the CLP Regulation, the words 'EC label'.

(3) The particulars required under paragraph (1) in relation to a dangerous preparation shall be—

(a) the name, full address and telephone number of a person in an EEA State who is responsible for supplying the preparation, whether that person be its manufacturer, importer or distributor;

(b) the trade name or other designation of the preparation; and

(c) the following particulars ascertained in accordance with Part I of Schedule 4, namely—

(i) identification of the constituents of the preparation which result in it being classified as a dangerous preparation,

(ii) any indications of danger together with corresponding symbols,

(iii) the risk phrases, set out in full,

(iv) the safety phrases, set out in full,

(v) in the case of a preparation intended for sale to the general public, the nominal quantity (nominal mass or nominal volume).

(4) [Indications] such as 'non-toxic', 'non-harmful', 'non-polluting', 'ecological' or any other statement indicating that the dangerous substance or preparation is not dangerous or that is likely to lead to underestimation of the dangers of the dangerous substance or dangerous preparation shall not appear on the package.

(5) Where paragraph (6) applies, the packaging of a dangerous substance or dangerous preparation classified in one or more of the categories of danger harmful, extremely flammable, highly flammable, flammable, irritant or oxidising shall not be required to be labelled in respect of that hazardous property.

(6) This paragraph shall apply where the package contains such small quantities of that substance or preparation that there is no foreseeable risk, under conditions of supply, use and disposal, arising from that hazardous property to persons handling that substance or preparation or to other persons.

(7) The packaging of a dangerous preparation classified in the category of danger dangerous for the environment (with or without the 'N' symbol) shall not be required to be labelled in respect of its environmental hazard in accordance with this regulation provided that it contains such small quantities of that preparation that there is no foreseeable risk, under conditions of supply, use and disposal, to the environment.

(8) Where the package in which a dangerous substance is supplied does not contain more than 125 millilitres of that substance the risk phrases and safety phrases need not be shown if the dangerous substance is classified only in one or more of the categories of danger—

- (a) highly flammable, flammable, oxidising or irritant; or
- (b) harmful, provided the dangerous substance is not sold to the general public.

(9) Where the package in which a dangerous preparation is supplied does not contain more than 125 millilitres of that preparation—

- (a) the risk phrases and safety phrases need not be shown if the dangerous preparation is classified only in one or more of the categories of danger—
  - (i) irritant (except those assigned the risk phrase R41);
  - (ii) dangerous for the environment and assigned the N symbol;
  - (iii) oxidising; or
  - (iv) highly flammable; and
- (b) the safety phrases need not be shown if the dangerous preparation is classified only in one or more of the categories of danger—
  - (i) flammable; or
  - (ii) dangerous for the environment and not assigned the N symbol.

(10) Where a dangerous substance or dangerous preparation is required to be labelled in accordance with these Regulations and is so labelled, that labelling shall be deemed to satisfy the requirements of—

- (a) section 5 of the Petroleum (Consolidation) Act 1928 including that section as applied to any dangerous substance by an Order in Council made under section 19 of that Act;
- (b) regulation 10 of the Dangerous Substances and Explosive Atmospheres Regulations 2002; and
- (c) regulations 3 and 4 of the Farm and Garden Chemicals Regulations 1971.

[8.359A]

Amendment

Para (4): word 'Indications' in square brackets substituted by SI 2011/2131, reg 31(1), Sch 4.

[8.360]

8 Labelling of single receptacles and receptacles in outer packagings

(1) Where except for this regulation a package would be required to show the particulars required by regulation 7 and to be labelled and marked in accordance with any of the national or international transport rules, it shall be sufficient compliance with regulation 7 if the package shows the particulars specified in paragraphs (2) or (3) of this regulation.

(2) Where the package consists of only a single receptacle, the specified particulars are—

(a) the particulars required by—
  (i) paragraph (2)(a), (b), (c)(ii), (iii) and (iv) in the case of substances; and
  (ii) paragraph (3)(a), (b), (c)(i), (iii), (iv) and (v) in the case of preparations, of regulation 7 in accordance with regulation 10;
(b) the labels and markings required by whichever of the national or international transport rules is appropriate; and
(c) where a substance or preparation has been classified as dangerous for the environment, the appropriate indication of danger and the danger symbol from Schedule 2 in accordance with regulation 10.

(3) Where the package consists of one or more receptacles in outer packagings, the particulars specified are the labels and markings required by whichever of the national or international transport rules is appropriate.

(4) For the purpose of this regulation and regulation 9(3)—
(a) the national transport rules are—
  (i) the Merchant Shipping (Dangerous Goods and Marine Pollutants) Regulations 1997;
  (ii) the Air Navigation (Dangerous Goods) Regulations 2002; and
  (iii) the Carriage of Dangerous Goods and Use of Transportable Pressure Equipment Regulations 2007; and
(b) the international transport rules are—
  (i) the European Agreement concerning the International Carriage of Dangerous Goods by Road signed at Geneva on 30th September 1957, as revised or reissued from time to time;
  (ii) the European Agreement concerning the International Carriage of Dangerous Goods by Inland Waterway, as revised or re-issued from time to time;
  (iii) the Technical Instructions for the Safe Transport of Dangerous Goods by Air, as revised or re-issued from time to time;
  (iv) the International Maritime Dangerous Goods Code, as revised or re-issued from time to time; and
  (v) the Regulation concerning the International Carriage of Dangerous Goods by Rail including its Annex which together form Appendix C to the Convention concerning the International Carriage by Rail, as revised or re-issued from time to time.

[8.361]

9 Particular labelling requirements for certain preparations

(1) In the case of preparations to which Part II of Schedule 4 applies, the appropriate provisions of that Part shall have effect to regulate the labelling of such preparations.

(2) Subject to paragraph 3, no person shall supply a preparation to which section B or C of Part II of Schedule 4 applies unless the trade name or other designation of that preparation and the name, full address and telephone number of a person in an EEA state who is responsible for supplying that preparation (whether the person is its manufacturer, importer or distributor) are clearly shown in accordance with the requirements of regulation 10—
(a) on the receptacle containing that preparation; and
(b) if that receptacle is inside one or more layers of packaging, on any such layer which is likely to be the outermost layer of packaging during the supply or use of that preparation, unless such packaging permits the particulars shown on the receptacle or other packaging to be clearly seen.

(3) Where a package would be required to be labelled and marked in accordance with any of the national or international transport rules listed in regulation 8(4) and the package consists of one or more receptacles in outer packagings, it shall be sufficient compliance with paragraph (2) if the package shows the labels and markings required by whichever of the national or international rules is appropriate.

(4) The supplier of an aerosol dispenser which contains a dangerous substance or dangerous preparation which has been classified in the category of danger 'flammable', 'highly flammable' or 'extremely flammable' may choose to omit from the label—

- (a) in the case of a dangerous substance, the particulars referred to in paragraphs (i) to (iii) of regulation 7(2)(c) which relate to that category of danger; and
- (b) in the case of a dangerous preparation, the particulars referred to in paragraphs (ii) to (iv) of regulation 7(3)(c) which relate to that category of danger,

provided the conditions specified in (5) are satisfied.

(5) The conditions referred to in paragraph (4) are that the supplier—

- (a) is in possession of evidence which shows that the contents of the aerosol dispenser do not present a risk of ignition under normal or reasonably foreseeable conditions of use; and
- (b) identifies the quantity of flammable material contained in the aerosol dispenser in the form of the following inscription on the label—
  'X% by mass of the contents are flammable'.

[(6) In the case of a plant protection product authorised or permitted, or deemed to be authorised or permitted, in accordance with Regulation (EC) No 1107/2009 of the European Parliament and of the Council concerning the placing of plant protection products on the market and repealing Council Directives 79/117/EEC and 91/414/EEC or a product approved under the Control of Pesticides Regulations 1986 which is not a plant protection product, the labelling information required by these Regulations shall be accompanied by the wording 'To avoid risks to man and the environment, comply with the instructions for use'.]

[8.361A]

Amendment

Para (6): substituted by SI 2011/2131, reg 31(1), Sch 4.

[8.362]

10 Methods of marking or labelling packages

(1) Any package which is required to be labelled in accordance with regulations 7 to 9 may carry the particulars required to be on the label clearly and indelibly marked on a part of that package reserved for that purpose and, unless the context otherwise requires, any reference in these Regulations to a label includes a reference to that part of the package so reserved.

(2) Subject to paragraph (7), any label required to be carried on a package shall be securely fixed to the package with its entire surface in contact with it and the label shall be clearly and indelibly printed.

(3) The colour and nature of the marking shall be such that any symbol and the wording stand out clearly from the background and the wording shall be of such size and spacing as to be easily read.

(4) The package shall be so labelled that the particulars can be read horizontally when the package is set down normally.

(5) Subject to paragraph (7), the dimensions of the label required by regulation 7 shall be as follows—

Capacity of package	***Dimensions of label***
(a) not exceeding 3 litres	if possible at least 52 x 74 millimetres
(b) exceeding 3 litres but not exceeding 50 litres	at least 74 x 105 millimetres
(c) exceeding 50 litres but not exceeding 500 litres	at least 105 x 148 millimetres
(d) exceeding 500 litres	at least 148 x 210 millimetres

(6) A symbol required to be shown in accordance with regulation 7(2)(c)(i) or 7(3)(c)(ii) and specified in column 3 of Schedule 2 shall be printed in black on an orange-yellow background and its size (including the orange-yellow background) shall be at least equal to an area of one-tenth of that of a label which complies with paragraph (5) and shall not in any case be less than 100 square millimetres.

(7) If the package is an awkward shape or so small that it is unsuitable to attach a label complying with paragraphs (2) and (5), the label shall be attached in some other appropriate manner.

[8.363]

11 Child resistant fastenings, tactile warning devices and other consumer protection measures

(1) The British and International Standards referred to in this regulation are further described in Schedule 5.

(2) Subject to paragraphs (4) and (5), no person shall supply to the general public a substance or a preparation specified in paragraph (3) in a receptacle of any size fitted with—

- (a) a replaceable closure unless the packaging complies with the requirements of BS EN 28317; or ISO 8317; or
- (b) a non-replaceable closure unless the packaging complies with the requirements of EN 862.

(3) The substances and preparations referred to in paragraph (2) are—

- (a) dangerous substances and dangerous preparations which are required to be labelled with the indication of danger 'very toxic', 'toxic' or 'corrosive';
- (b) preparations containing methanol in a concentration equal to or more than 3% by weight;
- (c) preparations containing dichloromethane in a concentration equal to or more than 1% by weight;
- (d) substances which are assigned the risk phrase R65 in Table 3.2 of part 3 of Annex VI of the CLP Regulation, except where such a substance is supplied in an aerosol dispenser or a container fitted with a sealed spray attachment; and
- (e) substances and preparations which are assigned the risk phrase R65 and are classified and labelled according to the approved classification and labelling guide, except where such a substance or preparation is supplied in an aerosol dispenser or a container fitted with a sealed spray attachment.

(4) Paragraph (2) shall not apply if the person supplying the substance or preparation in question can show that it is obvious that the packaging in which the substance or preparation is supplied is sufficiently safe for children because they cannot obtain access to the contents without the help of a tool.

(5) If the packaging, in which a substance or preparation specified in paragraph (3) is supplied to the general public, was approved on or before

31 May 1993 by the British Standards Institution as complying with the requirements of BS 6652, that packaging shall be deemed to comply with the requirements of BS EN 28317.

(6) No person shall supply a dangerous preparation or a preparation specified in paragraph (3) to the general public if the packaging in which that preparation is supplied has—

(a) either a shape or a designation or both likely to attract or arouse the active curiosity of children or to mislead consumers; or

(b) either a presentation or a designation or both used for—

(i) human or animal foodstuffs;

(ii) medicinal products; or

(iii) cosmetic products.

(7) Subject to paragraph (9), no person shall supply to the general public a dangerous substance or a dangerous preparation specified in paragraph (8) in a receptacle of any size, unless the packaging in which that dangerous substance or dangerous preparation is supplied carries a tactile warning of danger in accordance with EN ISO 11683.

(8) The dangerous substances and the dangerous preparations referred to in paragraph (7) are those which are required to be labelled with one or more of the following indications of danger, namely—

(a) very toxic;

(b) toxic;

(c) corrosive;

(d) harmful;

(e) extremely flammable; or

(f) highly flammable.

(9) Paragraph (7) shall not apply to an aerosol dispenser which is classified and labelled only with the indication of danger extremely flammable or highly flammable.

(10) For the purpose of ascertaining whether there has been a contravention of paragraph (2), a duly authorised officer of the enforcing authority may require the person supplying a substance or a preparation to which that paragraph applies to provide the duly authorised officer with a certificate from a qualified test house stating that—

(a) the closure is such that it is not necessary to test to BS EN 28317 or ISO 8317; or

(b) the closure has been tested and found to conform to BS EN 28317 or ISO 8317.

(11) In this regulation, 'qualified test house' means a laboratory that conforms to BS 7501 or EN 45 000.

[8.364]

12 Retention of data for dangerous preparations

(1) The person who is responsible for first supplying a dangerous preparation shall maintain a record of the information—

(a) used for the purposes of classifying that dangerous preparation in accordance with regulation 4;

(b) used for the purposes of labelling that dangerous preparation in accordance with regulation 7; and

(c) relating to any child resistant fastening or any tactile warning which forms part of the packaging in which the dangerous preparation in question is contained in accordance with regulation 11,

for at least three years after the date on which that dangerous preparation was supplied by that person for the last time.

(2) When requested by the enforcing authority to do so, a person referred to in paragraph (1) shall make the record, or a copy of the record, maintained by the person in accordance with that paragraph, available to the enforcing authority within 28 days of the date of the request.

(3) When requested to do so by the enforcing authority, a person who supplies a dangerous preparation shall provide to the enforcing authority a copy of any certificate issued by a qualified test house.

[8.365]

13 Transitional provisions for dangerous substances, dangerous preparations and certain specified preparations

(1) Regulation 4 shall not have effect on or after 1 June 2015.

(2) Where a substance has been classified, labelled and packaged in accordance with the CLP Regulation, regulations 6 to 11 shall not apply to that substance.

(3) Where a preparation has been classified, labelled and packaged in accordance with the CLP Regulation, regulations 6 to 11 shall not apply to that preparation.

(4) Insofar as they relate to substances, regulations 6 to 11 shall not have effect on or after 1 December 2010.

(5) Insofar as they relate to preparations, regulations 6 to 11 shall not have effect on or after 1 June 2015.

(6) Regulation 12 shall not apply on or after 1 June 2018.

PART 3

MISCELLANEOUS

Statutory annotations

Amendment

Part 3: Revoked by SI 2013/1506, reg 36(c). This revocation has effect as from 1 June 2015: see SI 2013/1506, reg 36(c).

[8.366]

14 Enforcement

(1) To the extent that they would not otherwise do so, sections—

(a) 16 to 28 (approval of codes of practice; enforcement; indemnification of inspectors; power to obtain information and restrictions on disclosure of information); [and]

(b) 33 to 42 (provisions as to offences)[,]

(c) . . .

of the 1974 Act shall apply to these Regulations and the CLP Regulation as if these Regulations and the CLP Regulation were health and safety Regulations for the purposes of that Act except that those sections shall not apply to duties placed by the CLP Regulation on the competent authority or the Member State.

[(1A) The maximum penalty for an offence under this regulation is—

(a) on summary conviction—

(i) in England and Wales, imprisonment for a term not exceeding three months or a fine not exceeding the statutory maximum, or both;

(ii) in Scotland, imprisonment for a term not exceeding twelve months or a fine not exceeding the statutory maximum, or both; and

(b) on conviction on indictment, imprisonment for a term not exceeding two years, or a fine, or both.]

(2) Any function of the Health and Safety Executive under any other provision of the 1974 Act under or in respect of health and safety Regulations (including their

enforcement) shall be exercisable as if these Regulations and the CLP Regulation were health and safety Regulations for the purposes of that Act to the extent that they would not otherwise do so.

(3) Notwithstanding regulation 3 of the Health and Safety (Enforcing Authority) Regulations 1998 and subject to paragraphs (4) and (5), the enforcing authority for these Regulations and the CLP Regulation shall be the Executive.

(4) Subject to paragraph (5), where a substance or preparation is supplied, or a substance, mixture or article falling within the meaning of and the provisions of the CLP Regulation is placed on the market within the meaning of the CLP Regulation in or from premises which are registered under section 75 of the Medicines Act 1968, the enforcing authority shall be the Royal Pharmaceutical Society.

(5) The enforcing authority for these Regulations and the CLP Regulation shall be the local weights and measures authority—

- (a) where a substance or preparation is supplied or a substance, mixture or article falling within the meaning of and the provisions of the CLP Regulation is placed on the market within the meaning of the CLP Regulation other than in the circumstances referred to in paragraph (4)—
  - (i) in or from any shop, mobile vehicle, market stall or other retail outlet, or
  - (ii) otherwise to members of the public, including by way of free sample, prize or mail order;
- (b) for regulation 11;
- (c) for Articles 35(2) and 48 of the CLP Regulation.

(6) In every case where, by virtue of this regulation and the CLP Regulation, these Regulations and the CLP Regulation are enforced by the Royal Pharmaceutical Society or the local weights and measures authority, they shall be enforced as if they were safety regulations made under section 11 of the Consumer Protection Act 1987 and the provisions of section 12 of that Act shall apply to these Regulations and the CLP Regulation as if they were safety regulations for the purposes of that Act and as if the maximum period of imprisonment on summary conviction specified in subsection (5) thereof were 3 months instead of 6 months.

[8.366A]

Statutory annotations

Amendment

Para (1): in sub-para (a) word 'and' in square brackets inserted by SI 2013/1666, art 5(a); in sub-para (b) comma in square brakets substituted by SI 2013/1666, art 5(b); sub-para (c) revoked by SI 2013/1666, art 5(c).

Para (1A): inserted by SI 2013/1506, reg 35, Sch 4, Pt 2, para 4.

Directive

2004/37/EC

of 29 April 2004

on the protection of workers from the risks related to exposure to carcinogens or mutagens at work (Sixth individual Directive within the meaning of Article 16(1) of Council Directive

[8.401]

THE EUROPEAN PARLIAMENT AND THE COUNCIL OF THE EUROPEAN UNION,

Having regard to the Treaty establishing the European Community, and in particular Article 137(2) thereof,

Having regard to the proposal from the Commission,

Having regard to the pinion of the European Economic and Social Committee

Having consulted the Committee of the Regions,

Acting in accordance with the procedure laid down in Article 251 of the Treaty,

Whereas:

(1) Council Directive 90/394/EEC of 28 June 1990 on the protection of workers from the risks related to exposure to carcinogens at work (Sixth individual Directive within the meaning of Article 16(1) of Directive 89/391/EEC) has been substantially amended several times. In the interests of clarity and rationality, the said Directive should therefore be codified.

(2) Compliance with the minimum requirements designed to guarantee a better standard of health and safety as regards the protection of workers from the risks related to exposure to carcinogens or mutagens at work is essential to ensure the health and safety of workers and is also intended to provide a level of minimum protection for all workers in the Community.

(3) This Directive is an individual Directive within the meaning of Article 16(1) of Council Directive 89/391/EEC of 12 June 1989 on the introduction of measures to encourage improvements in the safety and health of workers at work. Therefore the provisions of that Directive are fully applicable to the exposure of workers to carcinogens or mutagens, without prejudice to more stringent and/or specific provisions contained in this Directive.

(4) A consistent level of protection from the risks related to carcinogens or mutagens has to be established for the Community as a whole and that level of protection has to be set not by detailed prescriptive requirements but by a framework of general principles to enable Member States to apply the minimum requirements consistently.

(5) Germ cell mutagens are substances that can cause a permanent change in the amount or structure of the genetic material of a cell resulting in a change in the phenotypic characteristics of that cell, which may be transferred to descendent daughter cells.

(6) Because of their mechanism of action, germ cell mutagens are likely to have carcinogenic effects.

(7) Annex VI to Council Directive 67/548/EEC of 27 June 1967 on the approximation of laws, regulations and administrative provisions relating to the classification, packaging and labelling of dangerous substances contains the classification criteria and labelling procedures in respect of each substance.

(8) Directive 1999/45/EC of the European Parliament and of the Council of 31 May 1999 on the approximation of the laws, regulations and administrative provisions of the Member States relating to the classification, packaging and labelling of dangerous preparations contains particulars on the classification criteria and labelling procedures in respect of such preparations.

(9) In all work situations workers must be protected in respect of preparations containing one or more carcinogens or mutagens and from carcinogenic or mutagenic compounds arising at work.

(10) For some agents it is necessary to consider all absorption pathways, including the possibility of penetration through the skin, in order to ensure the best possible level of protection.

(11) Although current scientific knowledge is not such that a level can be established below which risks to health cease to exist, a reduction in exposure to carcinogens or mutagens will nonetheless reduce those risks.

(12) In order to contribute to a reduction in these risks, limit values and other directly related provisions should be established for all those carcinogens or mutagens for which the available information, including scientific and technical data, make this possible.

(13) Occupational exposure limit values must be regarded as an important component of the general arrangements for the protection of workers. Such limit values must be revised whenever this becomes necessary in the light of more recent scientific data.

(14) The precautionary principle should be applied in the protection of workers' health.

(15) Preventive measures must be taken for the protection of the health and safety of workers exposed to carcinogens or mutagens.

(16) This Directive constitutes a practical aspect of the realisation of the social dimension of the internal market.

(17) Pursuant to Council Decision 74/325/EEC, the Commission consulted the Advisory Committee on Safety and Health at Work with a view to drawing up proposals for the Directives taken over in this Directive.

(18) This Directive is without prejudice to the obligations of the Member States concerning the time limits for transposition set out in Part B of Annex IV,
HAVE ADOPTED THIS DIRECTIVE:

CHAPTER I

GENERAL PROVISIONS

[8.402]

Article 1 Objective

1. This Directive has as its aim the protection of workers against risks to their health and safety, including the prevention of such risks, arising or likely to arise from exposure to carcinogens or mutagens at work.
It lays down particular minimum requirements in this area, including limit values.

2. This Directive shall not apply to workers exposed only to radiation covered by the Treaty establishing the European Atomic Energy Community.

3. Directive 89/391/EEC shall apply fully to the whole area referred to in paragraph 1, without prejudice to more stringent and/or specific provisions contained in this Directive.

4. As regards asbestos, which is dealt with by Directive 83/477/EEC, the provisions of this Directive shall apply whenever they are more favourable to health and safety at work.

[8.403]

Article 2 Definitions

For the purposes of this Directive,

(a) 'carcinogen' means:

(i) a substance which meets the criteria for classification as a category 1 or 2 carcinogen set out in Annex VI to Directive 67/548/EEC;

(ii) a preparation composed of one or more substances referred to in point (i) where the concentration of one or more of the individual substances meets the requirements for concentration limits for the classification of a preparation as a category 1 or 2 carcinogen set out either:

— in Annex I to Directive 67/548/EEC, or

— in Part B of Annex II to Directive 1999/45/EC where the substance or substances do not appear in Annex I to Directive 67/548/EEC or appear in it without concentration limits;

(iii) a substance, preparation or process referred to in Annex I to this Directive as well as a substance or preparation released by a process referred to in that Annex;

(b) 'mutagen' means:

(i) a substance which meets the criteria for classification as a category 1 or 2 mutagen set out in Annex VI to Directive 67/548/EEC;

(ii) a preparation composed of one or more substances referred to in point (i) where the concentration of one or more of the individual substances meets the requirements for concentration limits for the classification of a preparation as a category 1 or 2 mutagen set out in either:

— Annex I to Directive 67/548/EEC, or

— Part B of Annex II to Directive 1999/45/EC where the substance or substances do not appear in Annex I to Directive 67/548/EEC or appear in it without concentration limits;

(c) 'limit value' means, unless otherwise specified, the limit of the time-weighted average of the concentration for a 'carcinogen or mutagen' in the air within the breathing zone of a worker in relation to a specified reference period as set out in Annex III to this Directive.

[8.404]

Article 3 Scope — determination and assessment of risks

1. This Directive shall apply to activities in which workers are or are likely to be exposed to carcinogens or mutagens as a result of their work.

2. In the case of any activity likely to involve a risk of exposure to carcinogens or mutagens, the nature, degree and duration of workers' exposure shall be determined in order to make it possible to assess any risk to the workers' health or safety and to lay down the measures to be taken.

The assessment shall be renewed regularly and in any event when any change occurs in the conditions which may affect workers' exposure to carcinogens or mutagens.

The employer shall supply the authorities responsible at their request with the information used for making the assessment.

3. When assessing the risk, account shall be taken of all other routes of exposure, such as absorption into and/or through the skin.

4. When the risk assessment is carried out, employers shall give particular attention to any effects concerning the health or safety of workers at particular risk and shall, inter alia, take account of the desirability of not employing such workers in areas where they may come into contact with carcinogens or mutagens.

SECTION II

EMPLOYERS' OBLIGATIONS

[8.405]

Article 4 Reduction and replacement

1. The employer shall reduce the use of a carcinogen or mutagen at the place of work, in particular by replacing it, in so far as is technically possible, by a substance, preparation or process which, under its conditions of use, is not dangerous or is less dangerous to workers' health or safety, as the case may be.

2. The employer shall, upon request, submit the findings of his investigations to the relevant authorities.

[8.406]

Article 5 Prevention and reduction of exposure

1. Where the results of the assessment referred to in Article 3(2) reveal a risk to workers' health or safety, workers' exposure must be prevented.

2. Where it is not technically possible to replace the carcinogen or mutagen by a substance, preparation or process which, under its conditions of use, is not dangerous or is less dangerous to health or safety, the employer shall ensure that the carcinogen or mutagen is, in so far as is technically possible, manufactured and used in a closed system.

3. Where a closed system is not technically possible, the employer shall ensure that the level of exposure of workers is reduced to as low a level as is technically possible.

4. Exposure shall not exceed the limit value of a carcinogen as set out in Annex III.

5. Wherever a carcinogen or mutagen is used, the employer shall apply all the following measures:

(a) limitation of the quantities of a carcinogen or mutagen at the place of work;

(b) keeping as low as possible the number of workers exposed or likely to be exposed;

(c) design of work processes and engineering control measures so as to avoid or minimise the release of carcinogens or mutagens into the place of work;

(d) evacuation of carcinogens or mutagens at source, local extraction system or general ventilation, all such methods to be appropriate and compatible with the need to protect public health and the environment;

(e) use of existing appropriate procedures for the measurement of carcinogens or mutagens, in particular for the early detection of abnormal exposures resulting from an unforeseeable event or an accident;

(f) application of suitable working procedures and methods;

(g) collective protection measures and/or, where exposure cannot be avoided by other means, individual protection measures;

(h) hygiene measures, in particular regular cleaning of floors, walls and other surfaces;

(i) information for workers;

(j) demarcation of risk areas and use of adequate warning and safety signs including 'no smoking' signs in areas where workers are exposed or likely to be exposed to carcinogens or mutagens;

(k) drawing up plans to deal with emergencies likely to result in abnormally high exposure;

(l) means for safe storage, handling and transportation, in particular by using sealed and clearly and visibly labelled containers;

(m) means for safe collection, storage and disposal of waste by workers, including the use of sealed and clearly and visibly labelled containers.

[8.407]

Article 6 Information for the competent authority

Where the results of the assessment referred to in Article 3(2) reveal a risk to workers' health or safety, employers shall, when requested, make available to the competent authority appropriate information on:

(a) the activities and/or industrial processes carried out, including the reasons for which carcinogens or mutagens are used;

(b) the quantities of substances or preparations manufactured or used which contain carcinogens or mutagens;

(c) the number of workers exposed;

(d) the preventive measures taken;

(e) the type of protective equipment used;

(f) the nature and degree of exposure;

(g) the cases of replacement.

[8.408]

Article 7 Unforeseen exposure

1. In the event of an unforeseeable event or an accident which is likely to result in an abnormal exposure of workers, the employer shall inform the workers thereof.

2. Until the situation has been restored to normal and the causes of the abnormal exposure have been eliminated:

(a) only those workers who are essential to the carrying out of repairs and other necessary work shall be permitted to work in the affected area;

(b) the workers concerned shall be provided with protective clothing and individual respiratory protection equipment which they must wear; the exposure may not be permanent and shall be kept to the strict minimum of time necessary for each worker;

(c) unprotected workers shall not be allowed to work in the affected area.

[8.409]

Article 8 Foreseeable exposure

1. For certain activities such as maintenance, in respect of which it is foreseeable that there is the potential for a significant increase in exposure of workers, and in respect of which all scope for further technical preventive measures for limiting workers' exposure has already been exhausted, the employer shall determine, after consultation of the workers and/or their representatives in the undertaking or establishment, without prejudice to the employer's responsibility, the measures necessary to reduce the duration of workers' exposure to the minimum possible and to ensure protection of workers while they are engaged in such activities.

Pursuant to the first subparagraph, the workers concerned shall be provided with protective clothing and individual respiratory protection equipment which they must wear as long as the abnormal exposure persists; that exposure may not be permanent and shall be kept to the strict minimum of time necessary for each worker.

2. Appropriate measures shall be taken to ensure that the areas in which the activities referred to in the first sub-paragraph of paragraph 1 take place are clearly demarcated and indicated or that unauthorised persons are prevented by other means from having access to such areas.

[8.410]

Article 9 Access to risk areas

Appropriate measures shall be taken by employers to ensure that access to areas in which the activities in respect of which the results of the assessment referred to in Article 3(2) reveal a risk to workers' safety or health take place are accessible solely to workers who, by reason of their work or duties, are required to enter them.

[8.411]

Article 10 Hygiene and individual protection

1. Employers shall be obliged, in the case of all activities for which there is a risk of contamination by carcinogens or mutagens, to take appropriate measures to ensure that:

- (a) workers do not eat, drink or smoke in working areas where there is a risk of contamination by carcinogens or mutagens;
- (b) workers are provided with appropriate protective clothing or other appropriate special clothing;
- (c) separate storage places are provided for working or protective clothing and for street clothes;
- (d) workers are provided with appropriate and adequate washing and toilet facilities;
- (e) protective equipment is properly stored in a well-defined place and is checked and cleaned if possible before, and in any case after, each use;
- (f) defective equipment is repaired or replaced before further use.

2. Workers may not be charged for the cost of the measures set out in paragraph 1.

[8.412]

Article 11 Information and training of workers

1. Appropriate measures shall be taken by the employer to ensure that workers and/or workers' representatives in the undertaking or establishment receive sufficient and appropriate training, on the basis of all available information, in particular in the form of information and instructions, concerning:

- (a) potential risks to health, including the additional risks due to tobacco consumption;
- (b) precautions to be taken to prevent exposure;
- (c) hygiene requirements;
- (d) wearing and use of protective equipment and clothing;
- (e) steps to be taken by workers, including rescue workers, in the case of incidents and to prevent incidents.

The training shall be:

- — adapted to take account of new or changed risk, and
- — repeated periodically if necessary.

2. Employers shall inform workers of installations and related containers containing carcinogens or mutagens, ensure that all containers, packages and installations containing carcinogens or mutagens are labelled clearly and legibly, and display clearly visible warning and hazard signs.

[8.413]

Article 12 Information for workers

Appropriate measures shall be taken to ensure that:

(a) workers and/or any workers' representatives in the undertaking or establishment can check that this Directive is applied or can be involved in its application, in particular with regard to:
  (i) the consequences for workers' safety and health of the selection, wearing and use of protective clothing and equipment, without prejudice to the employer's responsibility for determining the effectiveness of protective clothing and equipment;
  (ii) the measures determined by the employer which are referred to in the first subparagraph of Article 8(1), without prejudice to the employer's responsibility for determining such measures;
(b) workers and/or any workers' representatives in the undertaking or establishment are informed as quickly as possible of abnormal exposures, including those referred to in Article 8, of the causes thereof and of the measures taken or to be taken to rectify the situation;
(c) the employer keeps an up-to-date list of the workers engaged in the activities in respect of which the results of the assessment referred to in Article 3(2) reveal a risk to workers' health or safety, indicating, if the information is available, the exposure to which they have been subjected;
(d) the doctor and/or the competent authority as well as all other persons who have responsibility for health and safety at work have access to the list referred to in point (c);
(e) each worker has access to the information on the list which relates to him personally;
(f) workers and/or any workers' representatives in the undertaking or establishment have access to anonymous collective information.

[8.414]

Article 13 Consultation and participation of workers

Consultation and participation of workers and/or their representatives in connection with matters covered by this Directive shall take place in accordance with Article 11 of Directive 89/391/EEC.

SECTION III

MISCELLANEOUS PROVISIONS

[8.415]

Article 14 Health surveillance

1. The Member States shall establish, in accordance with national laws and/or practice, arrangements for carrying out relevant health surveillance of workers for whom the results of the assessment referred to in Article 3(2) reveal a risk to health or safety.

2. The arrangements referred to in paragraph 1 shall be such that each worker shall be able to undergo, if appropriate, relevant health surveillance:

— prior to exposure,
— at regular intervals thereafter.

Those arrangements shall be such that it is directly possible to implement individual and occupational hygiene measures.

3. If a worker is found to be suffering from an abnormality which is suspected to be the result of exposure to carcinogens or mutagens, the doctor or authority responsible for the health surveillance of workers may require other workers who have been similarly exposed to undergo health surveillance.

In that event, a reassessment of the risk of exposure shall be carried out in accordance with Article 3(2).

4. In cases where health surveillance is carried out, an individual medical record shall be kept and the doctor or authority responsible for health surveillance shall propose any protective or preventive measures to be taken in respect of any individual workers.

5. Information and advice must be given to workers regarding any health surveillance which they may undergo following the end of exposure.

6. In accordance with national laws and/or practice:

— workers shall have access to the results of the health surveillance which concern them, and

— the workers concerned or the employer may request a review of the results of the health surveillance.

7. Practical recommendations for the health surveillance of workers are given in Annex II.

8. All cases of cancer identified in accordance with national laws and/or practice as resulting from occupational exposure to a carcinogen or mutagen shall be notified to the competent authority.

[8.416]

Article 15 Record keeping

1. The list referred to in point (c) of Article 12 and the medical record referred to in Article 14(4) shall be kept for at least 40 years following the end of exposure, in accordance with national laws and/or practice.

2. Those documents shall be made available to the responsible authority in cases where the undertaking ceases activity, in accordance with national laws and/or practice.

[8.417]

Article 16 Limit values

1. The Council shall, in accordance with the procedure laid down in Article 137(2) of the Treaty, set out limit values in Directives on the basis of the available information, including scientific and technical data, in respect of all those carcinogens or mutagens for which this is possible, and, where necessary, other directly related provisions.

2. Limit values and other directly related provisions are set out in Annex III.

[8.418]

Article 17 Annexes

1. Annexes I and III may be amended in accordance only with the procedure laid down in Article 137(2) of the Treaty.

2. Purely technical adjustments to Annex II in the light of technical progress, changes in international regulations or specifications and new findings in the field of carcinogens or mutagens shall be adopted in accordance with the procedure laid down in Article 17 of Directive 89/391/EEC.

[8.419]

Article 18 Use of data

The Commission shall have access to the use made by the competent national authorities of the information referred to in Article 14(8).

[8.420]

Article 19 Notifying the Commission

Member States shall communicate to the Commission the provisions of national law which they adopt in the future in the field governed by this Directive.

[8.421]

Article 20 Repeal

Directive 90/394/EEC, as amended by the Directives referred to in Annex IV, Part A of this Directive is repealed, without prejudice to the obligations of the Member States concerning the time limits for transposition set out in Annex IV, Part B of this Directive.

References to the repealed Directive shall be construed as references to this Directive and shall be read in accordance with the correlation table in Annex V.

[8.422]

Article 21 Entry into force

This Directive shall enter into force on the twentieth day following that of its publication in the Official Journal of the European Union.

[8.421]

Article 20 Addresses

This Directive is addressed to the Member States.

ANNEX I

LIST OF SUBSTANCES, PREPARATIONS AND PROCESSES

(Article 2(a)(iii))

[8.422]

1. Manufacture of auramine.
2. Work involving exposure to polycyclic aromatic hydrocarbons present in coal soot, coal tar or coal pitch.
3. Work involving exposure to dusts, fumes and sprays produced during the roasting and electro-refining of cupro-nickel mattes.
4. Strong acid process in the manufacture of isopropyl alcohol.
5. Work involving exposure to hardwood dusts.

ANNEX II

PRACTICAL RECOMMENDATIONS FOR THE HEALTH SURVEILLANCE OF WORKERS

(Article 14(7))

[8.423]

1. The doctor and/or authority responsible for the health surveillance of workers exposed to carcinogens or mutagens must be familiar with the exposure conditions or circumstances of each worker.
2. Health surveillance of workers must be carried out in accordance with the principles and practices of occupational medicine; it must include at least the following measures:
   - keeping records of a worker's medical and occupational history,
   - a personal interview,

— where appropriate, biological surveillance, as well as detection of early and reversible effects.

Further tests may be decided upon for each worker when he is the subject of health surveillance, in the light of the most recent knowledge available to occupational medicine.

ANNEX III

LIMIT VALUES AND OTHER DIRECTLY RELATED PROVISIONS

(Article 16)

[8.424]

A. LIMIT VALUES FOR OCCUPATIONAL EXPOSURE

			Limit values			
Name of agent	EINECS [1]	CAS [2]	mg/m^3 [3]	ppm [4]	Notation	Transitional measures
Benzene	200-753-7	71-43-2	3,25 [5]	1 [5]	Skin [6]	Limit value: 3 ppm (=9,75 mg/m^3) (until 27 June 2003)
Vinyl chloride monomer	200-831	75-01-4	7,77 [5]	3 [5]	—	—
Hardwood dusts	—	—	5,00 [7]	—	—	—

1 EINECS: European Inventory of Existing Chemical Substances.

2 CAS: Chemical Abstract Service Number.

3 mg/m^3 = milligrams per cubic metre of air at 20°C and 101,3 kPa (760 mm mercury pressure).

4 ppm = parts per million by volume in air (ml/m^3).

5 Measured or calculated in relation to a reference period of eight hours.

6 Substantial contribution to the total body burden via dermal exposure possible.

7 Inhalable fraction: if hardwood dusts are mixed with other wood dusts, the limit value shall apply to all wood dusts present in that mixture.

B. OTHER DIRECTLY RELATED PROVISIONS

pm

ANNEX IV

[8.425]–[8.450]

Part A Repealed Directive and its successive amendments

Council Directive 90/394/EEC (OJ L 196, 26.7.1990, p. 1)

Council Directive 97/42/EC (OJ L 179, 8.7.1997, p. 4)

Council Directive 1999/38/EC (OJ L 138, 1.6.1999, p. 66)

Part B Deadlines for transposition into national law

90/394/EEC 31 December 1992

97/42/EC 27 June 2000

1999/38/EC 29 April 2003

Part 8B

Substances, noise, radiation and other phenomena hazardous to health: Fire and explosives

Regulatory Reform (Fire Safety) Order 2005

SI 2005/1541

as amended by SIs 2007/320 and 2008/960

PART 3

ENFORCEMENT

[8.571]

25 Enforcing authorities

[(1)] For the purposes of this Order, 'enforcing authority' means—

(a) the fire and rescue authority for the area in which premises are, or are to be, situated, in any case not falling within any of sub-paragraphs (b) to (e);

(b) the Health and Safety Executive in relation to—

(i) . . .

(ii) . . .

(iii) a ship, including a ship belonging to Her Majesty which forms part of Her Majesty's Navy, which is in the course of construction, reconstruction or conversion or repair by persons who include persons other than the master and crew of the ship;

[(iv) any workplace which is, or is on, a construction site, other than one in relation to which the Office for Nuclear Regulation is responsible for health and safety enforcement;]

[(bb) the Office for Nuclear Regulation in relation to—

(i) any premises for which a licence is required by virtue of section 1 of the Nuclear Installations Act 1965 or for which a permit is required by virtue of section 2 of that Act;

(ii) any premises for which such a licence or permit would be required but for the fact that the premises are used by, or on behalf of, the Crown;

(iii) any workplace which is, or is on, a construction site in relation to which the Office for Nuclear Regulation is responsible for health and safety enforcement;]

(c) the fire service maintained by the Secretary of State for Defence in relation to—

(i) premises, other than premises falling within paragraph (b)(iii), occupied solely for the purposes of the armed forces of the Crown;

(ii) premises occupied solely by any visiting force or an international headquarters or defence organisation designated for the purposes of the International Headquarters and Defence Organisations Act 1964;

(iii) premises, other than premises falling within paragraph (b)(iii), which are situated within premises occupied solely for the purposes of the armed forces of the Crown but which are not themselves so occupied;

(d) the relevant local authority in relation to premises which consist of—

(i) a sports ground designated as requiring a safety certificate under section 1 of the Safety of Sports Grounds Act 1975 (safety certificates for large sports stadia);

(ii) a regulated stand within the meaning of section 26(5) of the Fire Safety and Safety of Places of Sport Act 1987 (safety certificates for stands at sports grounds);

(e) a fire inspector, or any person authorised by the Secretary of State to act for the purposes of this Order, in relation to—

(i) premises owned or occupied by the Crown, other than premises falling within paragraph [(bb)(ii)] and (c);

(ii) premises in relation to which the United Kingdom Atomic Energy Authority is the responsible person, other than premises falling within paragraph [(bb)(ii)].

[(2) For the purposes of paragraph (1)—

(a) 'construction site' means a construction site, as defined in regulation 2(1) of the Construction (Design and Management) Regulations 2007, to which those Regulations apply, other than one to which regulation 46(1) of those Regulations applies;

(b) the Office for Nuclear Regulation is responsible for health and safety enforcement in relation to a construction site if, by virtue of regulations under section 18(2) of the Health and Safety at Work etc Act 1974 (enforcement), it is responsible for the enforcement of any of the relevant statutory provisions (within the meaning of Part 1 of that Act) in relation to the site.]

[8.572]

Statutory annotations

Amendment

Para (1): numbered as such by the Energy Act 2013, s 116(1), Sch 12, Pt 5, paras 87, 88(1), (2); sub-para (b)(i), (ii) revoked by the Energy Act 2013, s 116(1), Sch 12, Pt 5, paras 87, 88(1), (3)(a); sub-para (b)(iv) substituted by the Energy Act 2013, s 116(1), Sch 12, Pt 5, paras 87, 88(1), (3)(b); sub-para (bb) inserted by the Energy Act 2013, s 116(1), Sch 12, Pt 5, paras 87, 88(1), (4); in sub-para (e)(i) reference to '(bb)(ii)' in square brackets substituted by the Energy Act 2013, s 116(1), Sch 12, Pt 5, paras 87, 88(1), (5)(a); in sub-para (e)(ii) reference to '(bb)(ii)' in square brackets substituted by the Energy Act 2013, s 116(1), Sch 12, Pt 5, paras 87, 88(1), (5)(b).

Para (2): inserted by the Energy Act 2013, s 116(1), Sch 12, Pt 5, paras 87, 88(1), (6).

[8.573]

26 Enforcement of Order

(1) Every enforcing authority must enforce the provisions of this Order and any regulations made under it in relation to premises for which it is the enforcing authority and for that purpose, except where a fire inspector or other person authorised by the Secretary of State is the enforcing authority, may appoint inspectors.

(2) In performing the duty imposed by paragraph (1), the enforcing authority must have regard to such guidance as the Secretary of State may give it.

(3) A fire and rescue authority has power to arrange with [the Health and Safety Executive][, Office for Nuclear Regulation] or the Office of Rail Regulation for such of the authority's functions under this Order as may be specified in the arrangements to be performed on its behalf by the Health and Safety Executive[, Office for Nuclear Regulation] or the Office of Rail Regulation, as the case may be, (with or without payment) in relation to any particular workplace.

[8.574]

Statutory annotations

Amendment

Para (3): words 'the Health and Safety Executive' in square brackets substituted by SI 2008/960, art 22, Sch 3; words ', Office for Nuclear Regulation' in square brackets in both places they occur inserted by the Energy Act 2013, s 116(1), Sch 12, Pt 5, paras 87, 89.

Explosives Regulations 2014

SI 2014/1638

[8.655]

The Secretary of State is a Minister designated for the purposes of section 2(2) of the European Communities Act 1972 in relation to the placing on the market, transfer and safety of explosives for civil use.

It appears to the Secretary of State that it is expedient for certain references to provisions of EU instruments to be construed as references to those provisions as amended from time to time.

The Secretary of State makes—

(a) in part, regulation 8 and paragraph 15 of Schedule 11,

(b) regulations 39 to 42 and Schedules 9, 10 and 12, and

(c) to the extent that they relate to the provisions in (a) and (b) above, regulations 2(1), 3(4)(b) and (14) and (17) and 43 to 45,

under section 2(2) of that Act.

The Secretary of State makes the other provisions of these Regulations—

(a) in exercise of the powers conferred by sections 15(1), (2), (3)(a) and (c), (4), (5), (6)(b) and (9), 18(2)(za) and (a), 43(2), (4), (5) and (6), 80(1) and 82(3)(a) of, and paragraphs 1(1), (2), (3) and (4), 2, 3, 4, 5, 6(1), 7, 12, 15(1), 16, 18 and 20 of Schedule 3 to, the Health and Safety at Work etc Act 1974 ("the 1974 Act"); and

(b) apart from the modifications referred to in the next paragraph, for the purpose of giving effect without modifications to proposals submitted—

    (i) by the Health and Safety Executive under section 11(3) of the 1974 Act after carrying out consultations in accordance with section 50(3) of the 1974 Act; and

    (ii) by the Office for Nuclear Regulation under section 81(1)(a)(iv) of the Energy Act 2013 after carrying out consultations in accordance with section 81(3) of that Act.

It appears to the Secretary of State that—

(a) the modifications to the Acts referred to in paragraphs 2 to 8 of Part 1 of Schedule 13;

(b) the modifications to the instruments referred to in paragraphs 11, 12, 14, 16, 18, 20, 21, 24 and 30 of Part 2 of that Schedule; and

(c) the repeals in relation to the Acts marked with an asterisk in Part 1 of Schedule 14,

are expedient as set out in section 80(1) of the 1974 Act.

It also appears to the Secretary of State not to be appropriate to consult bodies in respect of those modifications and repeals in accordance with section 80(4) of the 1974 Act.

PART 1

INTRODUCTION

[8.656]

1 Citation and commencement

(1) These Regulations may be cited as the Explosives Regulations 2014 and, save as provided in paragraph (2), come into force on 1st October 2014.

(2) Regulations 33(7) and 36, and regulations 43 and 44 and Schedule 11 to the extent that those provisions relate to regulations 33(7) and 36, come into force on 5th April 2015.

[8.657]

2 Interpretation

(1) In these Regulations—

"the 1974 Act" means the Health and Safety at Work etc Act 1974;

"the 2005 Regulations" means the Manufacture and Storage of Explosives Regulations 2005;

"ammonium nitrate blasting intermediate" means a substance assigned in accordance with the United Nations Recommendations the U.N. no 3375;

"authorised defence site" has the meaning given in regulation 2(1) of the Health and Safety (Enforcing Authority) Regulations 1998;

"black powder" means an intimate mixture, with or without sulphur, of charcoal or other carbon with potassium nitrate or sodium nitrate, whether the mixture is in meal, granular, compressed or pelletised form, being an explosive substance assigned in accordance with the United Nations Recommendations the U.N. no 0027 or 0028;

"the CE marking" means the mark referred to in regulation 42(3);

"centre point", in relation to a store or a building, means the centre point of the store or building determined as far as is reasonably possible;

"chief officer of police"—

(a) in relation to England and Wales, has the same meaning as in section 101(1) of the Police Act 1996; and

(b) in relation to Scotland, means the person appointed to the office of chief constable under section 7(1)(a) of the Police and Fire Reform (Scotland) Act 2012; and

in relation to an area, means the chief officer of police for that area;

"civil explosive" means an explosive which has been or would be classified in accordance with the United Nations Recommendations as falling within Class 1 but it does not include—

(a) ammunition the acquisition of which is regulated or prohibited by virtue of the Firearms Acts 1968 to 1997;

(b) any explosive which it is shown is intended for lawful use by the armed forces or the police of any country;

(c) a pyrotechnic article; and

(d) an explosive which is used immediately at the place of manufacture;

"the Civil Uses Directive" means Council Directive 93/15/EEC on the harmonisation of the provisions relating to the placing on the market and supervision of explosives for civil uses as amended by Regulation (EC) No 1882/2003 of the European Parliament and of the Council of 29th September 2003, Regulation (EC) No 219/2009 of the European Parliament and of the Council of 11th March 2009 and Regulation (EU) No 1025/2012 of the European Parliament and of the Council of 25th October 2012;

"Class 1" means Class 1 in respect of explosives or the classification of dangerous goods as set out in the United Nations Recommendations;

"the Commission" means the Commission of the European Union;

"desensitised explosive" means—

(a) a solid explosive substance which has been wetted with water or alcohol or diluted with one or more other substances; or

(b) a liquid explosive substance which has been dissolved or suspended in water or one or more other substances,

to form a homogeneous mixture so as to suppress its explosive properties and which, without that treatment, would be classified in accordance with the United Nations Recommendations as falling within Class 1;

"disposes", in relation to explosives and explosive-contaminated items, means destroying the explosives or explosive-contaminated items or otherwise rendering them harmless;

"distributor" means a person in the supply chain, other than a manufacturer or an importer, who makes an explosive available on the market and "distributes" and "distribution" are to be construed accordingly;

"the Executive" means the Health and Safety Executive;

"explosive" means—

(a) any explosive article or explosive substance which would—

(i) if packaged for transport, be classified in accordance with the United Nations Recommendations as falling within Class 1; or

(ii) be classified in accordance with the United Nations Recommendations as—

(aa) being unduly sensitive or so reactive as to be subject to spontaneous reaction and accordingly too dangerous to transport, and

(bb) falling within Class 1; or

(b) a desensitised explosive,

but it does not include an explosive substance produced as part of a manufacturing process which thereafter reprocesses it in order to produce a substance or preparation which is not an explosive substance;

"explosive article" means an article containing one or more explosive substances;

"explosives certificate" has the meaning given in regulation 4(1);

"explosive substance" means a substance or preparation, not including a substance or preparation in a solely gaseous form or in the form of vapour, which is—

(a) capable by chemical reaction in itself of producing gas at such a temperature and pressure and at such a speed as could cause damage to surroundings; or

(b) designed to produce an effect by heat, light, sound, gas or smoke, or a combination of these as a result of a non-detonative, self-sustaining, exothermic chemical reaction;

"firearm" has the meaning given in section 57(1) of the Firearms Act 1968;

"fireworks" means the explosive articles assigned in accordance with the United Nations Recommendations any of the UN nos 0333 to 0337;

"GB nuclear site" has the meaning given in section 68 of the Energy Act 2013;

"harmonised standard" means a text setting out technical specifications adopted by the European Committee for Standardisation (CEN) under a mandate from the Commission in accordance with Directive 98/34 of the European Parliament and of the Council of 22nd June 1998 laying down a procedure for the provision of information in the field of technical standards and regulations, as amended by Directive 98/48/EC of the European Parliament and of the Council of 20th July 1998, Council Directive 2006/96/EC of 20th November 2006 and Regulation (EU) No 1025/2012 of the European Parliament and of the Council of 25th October 2012, and of which the reference number is published in the Official Journal of the European Communities;

"harbour" means a harbour which is within the jurisdiction of a harbour authority and includes—

(a) the areas of water within the jurisdiction of that harbour authority; and

(b) land within the jurisdiction of, or occupied by, the harbour authority and used in connection with the loading and unloading of ships,

but does not include the areas of water which are within the jurisdiction not only of the harbour authority but also of another harbour authority and which are used primarily by ships using berths within the harbour of that other harbour authority;

"harbour authority" has the meaning given in section 57 of the Harbours Act 1964;

"hazard type" means any of hazard type 1 explosive, hazard type 2 explosive, hazard type 3 explosive or hazard type 4 explosive;

"hazard type 1 explosive" means an explosive which, as a result of, or as a result of any effect of, the conditions of its storage or process of manufacture, has a mass explosion hazard;

"hazard type 2 explosive" means an explosive which, as a result of, or as a result of any effect of, the conditions of its storage or process of manufacture, has a serious projectile hazard but does not have a mass explosion hazard;

"hazard type 3 explosive" means an explosive which, as a result of, or as a result of any effect of, the conditions of its storage or process of manufacture, has a fire hazard and either a minor blast hazard or a minor projectile hazard, or both, but does not have a mass explosion hazard;

"hazard type 4 explosive" means an explosive which, as a result of, or as a result of any effect of, the conditions of its storage or process of manufacture, has a fire hazard or slight explosion hazard, or both, with only local effect;

"headquarters" means a headquarters for the time being specified in Schedule 2 to the Visiting Forces and International Headquarters (Application of Law) Order 1999;

"Her Majesty's Forces" means any of the naval, military or air forces of the Crown, whether raised inside or outside the United Kingdom and whether any such force is a regular, auxiliary or reserve force, and includes any civilian employee of the department of the Secretary of State having responsibility for defence attached to those forces;

"holder" in relation to an explosives certificate means the person named in the explosives certificate as a person fit to acquire, or to acquire and keep, explosives;

"licence", save in the definition of "prohibited person", means a licence for the manufacture or storage of explosives granted under regulation 13 and includes a varied licence;

"licensee" means a person who has been granted a licence under regulation 13 and includes a person to whom a licence is transferred and a person treated under regulation 18 as being licensed;

"licensing authority" has the meaning assigned to it by Schedule 1;

"local authority", apart from in paragraph 6(1)(c) of Schedule 11, means, in relation to—

(a) the City of London, the Common Council for the City of London;

(b) an area in the rest of London, the London borough council for that area;

(c) an area where there is a metropolitan county fire and rescue authority, that authority;

(d) the Isles of Scilly, the Council of the Isles of Scilly;

(e) an area in the rest of England, the county council for that area or where there is no county council for that area, the district council for that area;

(f) an area in Scotland, the council for the local government area; and

(g) an area in Wales, the county council or the county borough council for that area;

"manufacture" includes—

(a) in relation to explosive articles, their repair, modification, disassembly or unmaking;

(b) in relation to explosive substances, their reprocessing, modification or adaptation;

but it does not include the packing, unpacking, re-packing, labelling or testing of explosives or the division of an amount of explosives stored in bulk into smaller amounts and the placing of those smaller amounts into containers;

"mine" means an excavation or system of excavations, including all such excavations to which a common system of ventilation is provided, made for the purpose of, or in connection with, the getting, wholly or substantially by means involving the employment of persons below ground, of minerals (whether in their natural state or in solution or suspension) or products of minerals;

"new nuclear build site" has the meaning given in regulation 2A of the Health and Safety (Enforcing Authority) Regulations 1998;

"the ONR" means the Office for Nuclear Regulation;

"ONR regulated site" has the meaning given in paragraph (11);

"percussion caps" means items intended for use in small arms ammunition or small arms assigned in accordance with the United Nations Recommendations the UN no 0044 or 0378;

"police force", for the purposes of regulations 3(9)(e) and (8)(b) and 27(3)(c), includes—

(a) the police force known as the British Transport Police Force; and

(b) the constabulary known as the Civil Nuclear Constabulary by virtue of section 52(1) of the Energy Act 2004;

"preparation" means a mixture of two or more substances or a solution of any substance or substances;

"prohibited person" means a person, other than a person in respect of whom an order has been made under regulation 22(4), who—

(a) has been convicted of any offence under the Explosive Substances Act 1883, or

(b) has been sentenced to a sentence which is excluded from rehabilitation under the Rehabilitation of Offenders Act 1974 by virtue of section 5(1) and (1A) of that Act; or

(c) has been sentenced to a custodial sentence, within the meaning of section 5(7) and (8) of the Rehabilitation of Offenders Act 1974, for a term exceeding 30 months, but not exceeding 48 months, and less than the period of 7 years has elapsed, beginning with the day on which the sentence (including any licence period relating to that sentence) is completed, or less than 42 months from that day for those under the age of 18 at the time of conviction; or

(d) has been sentenced to a custodial sentence, within the meaning of section 5(7) and (8) of the Rehabilitation of Offenders Act 1974, for a term exceeding 6 months, but not exceeding 30 months, and less than the period of 48 months has elapsed, beginning with the day on which the sentence (including any licence period relating to that sentence) is completed, or less than 24 months from that day for those under the age of 18 at the time of conviction; or

(e) has been sentenced to a custodial sentence, within the meaning of section 5(7) and (8) of the Rehabilitation of Offenders Act 1974, for a term not exceeding 6 months and less than the period of 24 months has elapsed, beginning with the day on which the sentence (including any licence period relating to that sentence) is completed, or less than 18 months from that day for those under the age of 18 at the time of conviction; or

(f) has been sentenced to a sentence of service detention, within the meaning of section 5(8) of the Rehabilitation of Offenders Act 1974, for a recordable service offence and less than the period of 12 months has elapsed, beginning with the day on which the sentence is completed, or less than 6 months from that day for those under the age of 18 at the time of conviction, and, for these purposes—

(i) "conviction" includes a finding or a substituted finding that is treated as a conviction by virtue of section 376 of the Armed Forces Act 2006; and

(ii) "recordable service offence" has the meaning given by article 2(1) of the Police and Criminal Evidence Act 1984 (Armed Forces) Order 2009 and paragraph 4 of Schedule 2 to that Order;

"propellant" means a deflagrating explosive used as a propellant in firearms;

"public consultation zone" means the area around the store or proposed store, or the building where the manufacture of explosives takes place or is proposed to take place, which, from the centre point of the store or building, has a radius equivalent to double the greatest separation distance required by virtue of these Regulations to apply in the case of that store or building;

"pyrotechnic article" means any article containing explosive substances or an explosive mixture of substances designed to produce heat, light, sound, gas or smoke or a combination of such effects through self-sustained exothermic chemical reactions;

"pyrotechnic substance" means an explosive substance of a kind designed to produce an effect by heat, light, sound, gas or smoke, or a combination of any of these, as a result of non-detonative, self-sustaining, exothermic chemical reactions;

"recipient competent authority document" has the meaning in regulation 8(9);

"relevant explosive" means an explosive for which an explosives certificate is required under regulation 5 for acquiring or keeping that explosive, or would be so required were it not for regulation 3(7), and, in relation to regulations 35 and 37 and paragraph 4 of Schedule 4 it also includes—

(a) ammunition the acquisition of which is regulated or prohibited by virtue of the Firearms Acts 1968 to 1997; and

(b) smokeless powder,

even though, and to the extent that (in the case of smokeless powder), an explosives certificate is not required for their acquisition or keeping;

"relevant police force" means in the case of an explosives certificate—

(a) which will certify that the holder is a fit person to keep explosives, the police force for the police area in which the place of keeping is or is to be situated,

(b) which will certify only that the holder is a fit person to acquire explosives, the police force for the police area in which the applicant for a certificate resides or, in the case of a body corporate, in which the applicant has its registered office, or, if it has no registered office, its principal office,

(c) which has been issued, the police force stated in the certificate;

"renewal of a licence" means the granting of a licence to follow a previous licence without any amendment or gap in time;

"restricted substance" means any collection of substances which would if mixed form one or more explosive substances and which has been prepared for that purpose;

"separation distance" means the distance between the store or the building or other place in or at which explosives are, or are to be, manufactured and a building, or other place in or at which people are or are likely to be present either all the time or from time to time;

"ship" includes every description of vessel used in navigation;

"shooters' powder" means—

(a) black powder,

(b) smokeless powder which is manufactured for use in small arms, or

(c) any other substance or preparation based on potassium nitrate or nitro cellulose, whether in powder, pelletised or granular form, used, or to be used, as a propellant;

"site" means the whole area under the control of the same person and, for these purposes—

(a) all places adjoining each other under the control of the same person are to be treated as a whole area; and

(b) two or more areas under the control of the same person separated only by a road, railway or inland waterway are to be treated as a whole area;

"small arms" means any of—

(c) a firearm with a calibre not larger than 19.1 millimetres designed to fire ammunition consisting of a propelling charge and an inert projectile;

(d) a shotgun as defined by section 1(3) of the Firearms Act 1968; or

(e) a firearm intended to fire blank cartridges not more than one inch in diameter measured immediately in front of the rim or cannelure of the base of the cartridge;

"small arms ammunition" means the explosive articles assigned in accordance with the United Nations Recommendations the U.N. no 0012, 0014 or 0055 which are intended exclusively for use in small arms;

"smokeless powder" means an explosive substance assigned in accordance with the United Nations Recommendations the U.N. no 0509 or 0161 and which is intended exclusively for use in firearms;

"store" means a building, enclosed area or metal structure in which explosives are, or are to be, stored;

"substance" means any natural or artificial substance whether in solid or liquid form or in the form of a gas or vapour;

"transfer", in relation to explosives, means—

(a) in regulations 3(14) and 8, any physical movement of explosives apart from movement within one site and whether or not transferring possession of or property in the explosives is involved; and

(b) for all other provisions, transferring possession of or property in the explosives, save that, in relation to regulation 38(2) it is limited to a transfer of possession;

"UN no" means United Nations Serial Number, that is to say one of the four-digit numbers devised by the United Nations as a means of identification of types of explosives in accordance with the United Nations Recommendations;

"United Nations Recommendations" means the United Nations Recommendations on the Transport of Dangerous Goods (based on those originally prepared by the United Nations Committee of Experts on the Transport of Dangerous Goods considered by the Economic and Social Committee of Experts at its twenty-third session (Resolution 645G (XXIII) of 26 April 1957)) as revised or reissued from time to time;

"visiting force" has the same meaning as it does for the purposes of any provision of Part 1 of the Visiting Forces Act 1952;

"water-based", in relation to explosives, means explosives which are based on water and ammonium nitrate and assigned in accordance with the United Nations Recommendations the U.N. no 0241; and

"wholly-owned subsidiary" has the meaning given by section 1159 of the Companies Act 2006.

(2) For the purposes of these Regulations, the manufacture or storage of ammonium nitrate blasting intermediate is deemed to be the manufacture or storage of an explosive.

(3) For the purposes of measuring any distance required to be a separation distance by virtue of these Regulations, the distance to be measured is the horizontal distance between the outside edge of the store or the building or other place in or at in which the explosives are, or are to be, manufactured and the nearest point of the building or other place which the separation distance applies to.

(4) Any reference in these Regulations to the quantity of an explosive is to be construed as a reference to the net mass of explosive substance.

(5) For the purposes of these Regulations and subject to paragraph (6), "storage" in relation to explosives means their possession for any period after their manufacture, save for—

(a) any period during which they are being prepared at any place for use at that place; and

(b) any period during which they are being transported beyond the place where they are stored.

(6) Subject to paragraph (7), where, during any transport of any explosive beyond the place where it is stored, that explosive is, or is to be, kept at any place for more than twenty-four hours, that keeping is to be treated as storage within the meaning of these Regulations and the provisions of these Regulations apply to that keeping accordingly, notwithstanding any application of the provisions of the Carriage of Dangerous Goods and Use of Transportable Pressure Equipment Regulations 2009 to that transporting.

(7) Paragraph (6) does not apply to explosives in respect of which there is in existence an explosives licence granted under regulation 36(1) of the Dangerous Substances in Harbour Areas Regulations 1987).

(8) The performance of any function given to the chief officer of police under these Regulations may be delegated by the chief officer of police ("chief officer"), to such an extent and subject to such conditions as the chief officer may specify—

(a) to a member of the police force in respect of which the chief officer is the chief officer of police;

(b) to a person appointed to assist a police force as mentioned in paragraph 4(2) of Schedule 2 to the Police Reform and Social Responsibility Act 2011 (civilian staff); or

(c) to a person appointed to assist a police force as mentioned in section 26 of the Police and Fire Reform (Scotland) Act 2012 (police staff),

and any such delegation must be made in writing by that chief officer.

(9) Any reference in the definitions in this regulation of "desensitised explosive", "explosive substance", "pyrotechnic article", "pyrotechnic substance" or "substance" to liquid, gas, gaseous form or vapour, means, respectively, liquid, gas, gaseous form or vapour at normal atmospheric temperature and pressure.

(10) Any reference to acquiring an explosive means acquiring possession of or property in the explosive.

(11) For the purposes of these Regulations, a site is an ONR regulated site if it or any part of it is—

(a) a GB nuclear site;

(b) an authorised defence site; or

(c) a new nuclear build site.

[8.658]

3 Application and extent

(1) Except as provided by paragraphs (2) and (3), these Regulations extend to England and Wales and Scotland.

(2) In Schedule 13—

(a) the amendments made by paragraphs 16, 20 and 30 extend to Scotland only; and

(b) the amendments made by paragraphs 11 and 14 extend to England and Wales only.

(3) Regulation 29 extends to Northern Ireland in so far it applies to the importation of pyrotechnic articles and substances.

(4) These Regulations apply—

(a) within Great Britain subject to paragraphs (5) to (16); and

(b) outside Great Britain as specified in paragraphs (17) and (18).

(5) In Schedule 13—

(a) the amendments made by paragraph 14(2) and (4) apply in relation to England only; and

(b) the amendments made by paragraph 14(3) and (5) of that Schedule apply in relation to Wales only.

(6) Regulations 4, 5, 11, 19, 21, 22, 31, 32, 35 and 37 do not apply to any explosive nuclear device.

(7) Regulation 5 does not apply to the acquisition or keeping of explosives by or on behalf of—

(a) any officer referred to in section 74 of the Explosives Act 1875 exercising the power of seizure under that section;

(b) constables acting in the execution of their duties;

(c) a person—

(i) appointed to assist a police force as mentioned in paragraph 4(2) of Schedule 2 to the Police Reform and Social Responsibility Act 2011(civilian staff); or

(ii) appointed to assist a police force as mentioned in section 26 of the Police and Fire Reform (Scotland) Act 2012 (police staff),

who, in either case, is duly authorised in writing by the chief officer of police for the relevant police area;

(d) customs officers acting in the performance of their functions;

(e) the Crown, in respect of any explosive which—

(i) is in premises occupied on behalf of the Crown for, or
(ii) is intended for use for,

naval, military or air force purposes or the purposes of the department of the Secretary of State having responsibility for defence or that of the Secretary of State having responsibility for home affairs;

(f) the service authorities of visiting forces or any headquarters or organisation designated for the purposes of the International Headquarters and Defence Organisations Act 1964;
(g) a police force;
(h) the Executive; or
(i) the ONR.

(8) Regulations 6, 7, 9, 10, 12 to 18, 20, 23 and 26 to 30 do not apply to—

(a) any activity to which Part IX of the Dangerous Substances in Harbour Areas Regulations 1987 (explosives) applies;
(b) any activity to which the Carriage of Dangerous Goods and Use of Transportable Pressure Equipment Regulations 2009 apply, apart from any activity which is to be treated as storage by virtue of regulation 2(6);
(c) the master or crew of a ship or to the employer of such persons in respect of the normal shipboard activities of a ship's crew which are carried out solely by the crew under the direction of the master and in this sub-paragraph the reference to the normal shipboard activities of a ship's crew includes—
   (i) the construction, reconstruction or conversion of a ship outside, but not inside, Great Britain; and
   (ii) the repair of a ship save repair when carried out in dry dock;
(d) the transport of explosives by air;
(e) an offshore installation within the meaning of regulation 3 of the Offshore Installations and Pipeline Works (Management and Administration) Regulations 1995;
(f) the storage of explosives below ground at a mine, where the explosives are for use at that mine for the getting of minerals or ensuring the safety of the mine.

(9) Regulations 6, 7, 12 to 14, 16 to 18, 20, 23 and 27 do not apply to the manufacture or storage of explosives, at any site under the control of the Secretary of State having responsibility for defence, or held for the purposes of a visiting force or headquarters, under a scheme approved by that Secretary of State which—

(a) provides for their safe manufacture and storage; and
(b) prescribes—
   (i) separation distances, or
   (ii) a combination of separation distances and other safety measures, which are designed to ensure a standard of safety which is equivalent to that ensured by the separation distances prescribed by regulation 27 and Schedule 5.

(10) Regulations 6, 7, 12 to 14, 16 to 18, 20, 23 and 29 do not apply to explosives—

(a) seized by constables acting in the execution of their duties,
(b) received by a police force from a member of the public; or
(c) which, for reasons of public safety or protection of property, are undergoing ordnance disposal by—
   (i) persons under the direction of a member of Her Majesty's Forces or civilian employees of the department of the Secretary of State having responsibility for defence authorised in writing by that Secretary of State to carry out ordnance disposal; or

(ii) persons under the direction of a constable or authorised in writing by the chief officer of police to carry out ordnance disposal.

(11) Regulation 7 does not apply to—

(a) the Executive;

(b) the ONR;

(c) a local authority;

(d) the Commissioners for Her Majesty's Revenue and Customs;

(e) a police force;

(f) a person appointed to assist a police force as mentioned in paragraph 4(2) of Schedule 2 to the Police Reform and Social Responsibility Act 2011 (civilian staff) who is duly authorised in writing by the chief officer of police to store explosives; and

(g) a person appointed to assist a police force as mentioned in section 26 of the Police and Fire Reform (Scotland) Act 2012 (police staff) who is duly authorised in writing by the chief officer of police to store explosives.

(12) Regulations 7, 12 to 14, 16 to 18, 20, 23, 27, 29 and 30 do not apply to explosives seized by an inspector appointed under section 19 of the 1974 Act in the performance of the inspector's functions.

(13) Regulations 7 and 30 do not apply to—

(a) constables in the execution of their duties;

(b) customs officers in the performance of their functions; or

(c) inspectors appointed under section 19 of the 1974 Act in the performance of their functions.

(14) Regulations 8 and 39 to 42 do not apply in respect of the transfer of civil explosives to, by or on behalf of, or where, following the transfer, the explosives are in the possession of—

(a) a person exercising a power of seizure under section 74 of the Explosives Act 1875;

(b) constables acting in the execution of their duties;

(c) a person appointed to assist a police force as mentioned in paragraph 4(2) of Schedule 2 to the Police Reform and Social Responsibility Act 2011 (civilian staff) who is duly authorised in writing by the chief officer of police in respect of such transfer or possession;

(d) a person appointed to assist a police force as mentioned in section 26 of the Police and Fire Reform (Scotland) Act 2012 (police staff) who is duly authorised in writing by the chief officer of police in respect of such transfer or possession;

(e) customs officers acting in the performance of their functions;

(f) the Crown in respect of any civil explosives which are intended for use for the purposes of the department of the Secretary of State having responsibility for defence or that of the Secretary of State having responsibility for home affairs;

(g) inspectors appointed under section 19 of the 1974 Act exercising their power to take possession of and detain articles or substances under section 20(2)(i) of that Act as modified by Schedule 12 to these Regulations;

(h) the Executive; or

(i) the ONR.

(15) Regulations 33, 34 and 36 do not apply to—

(a) an explosive which is transported and delivered without packaging or in a mobile explosives manufacturing unit for its direct unloading into the blast-hole;

(b) fuses, which are cord-like non-detonating igniting devices;

(c) safety fuses, which consist of a core of fine grained black powder surrounded by a flexible woven fabric with one or more protective outer coverings and which, when ignited, burn at a predetermined rate without any external explosive effect; or

(d) cap-type primers, which consist of a metal or plastic cap containing a small amount of primary explosive mixture that is readily ignited by impact and which serve as igniting elements in small arms cartridges or in percussion primers for propelling charges.

(16) In paragraph (15), "mobile explosives manufacturing unit" means a moveable unit, whether mounted on a vehicle or not, for manufacturing and charging explosives from dangerous goods that are not explosives, with the unit consisting of various tanks, bulk containers and related equipment.

(17) These Regulations apply outside Great Britain as sections 1 to 59 and 80 to 82 of the 1974 Act apply by virtue of the Health and Safety at Work etc Act 1974 (Application outside Great Britain) Order 2013.

(18) Paragraph (17) does not apply to regulations 4, 5, 31 and 32.

PART 2

AUTHORISATIONS

[8.659]

4 Authorisation to acquire or acquire and keep explosives

(1) An explosives certificate is a certificate certifying that the person to whom it is issued is a fit person—

(a) to acquire explosives; or

(b) to acquire and keep explosives,

in accordance with the terms of the explosives certificate.

(2) An explosives certificate must be in a form approved for the time being for the purposes of this regulation by the Executive.

[8.660]

5

(1) Subject to paragraph (3), no person may acquire any explosives unless—

(a) that person has a valid explosives certificate certifying that person to be a fit person to acquire explosives;

(b) that person acquires no more explosives than any quantity referred to in the explosives certificate;

(c) where the explosives certificate specifies the description of explosives which that person is a fit person to acquire, that person acquires only explosives of that description; and

(d) where the explosives certificate specifies purposes for which that person is a fit person to acquire explosives, that person acquires them only for those purposes.

(2) Subject to paragraph (3), no person may keep explosives unless that person—

(a) has a valid explosives certificate certifying that person to be a fit person to keep explosives;

(b) keeps no more explosives than the quantity referred to in the explosives certificate;

(c) where the explosives certificate specifies the description of explosives which that person is a fit person to keep, keeps only explosives of that description; and

(d) keeps them at any place specified in the explosives certificate.

(3) Paragraphs (1) and (2) do not apply to—

(a) the explosives as referred to in Schedule 2;

(b) pyrotechnic articles apart from those as referred to in Schedule 3; or

(c) ammunition the acquisition of which is regulated or prohibited by virtue of the Firearms Acts 1968 to 1997.

(4) For the purposes of this regulation, where a person acts as an agent to acquire a relevant explosive for another person, the agent is to be treated as if the agent, as well as that other person, had acquired the relevant explosive and the provisions of paragraph (1) accordingly apply to the agent as well as to that other person.

[8.661]

6 Authorisation to manufacture explosives

(1) Subject to paragraph (2), no person may manufacture explosives unless that person holds a licence for that manufacture and complies with the conditions of that licence.

(2) Paragraph (1) does not apply to—

(a) the manufacture of explosives for the purpose of laboratory analysis, testing, demonstration or experimentation (but not for practical use or supply) where the total quantity of explosives being manufactured at any time does not exceed 100 grams, but nothing in this sub-paragraph is to be taken as authorising any acquisition or keeping of explosives for which an explosives certificate is required by virtue of regulation 5, without such a certificate;

(b) the making or unmaking of small arms ammunition, or ammunition with inert projectiles intended for use in recreational or occupational firearms, or the preparation of cartridges for use with firearms which are to be used at historical re-enactment events, where the total quantity of primer and propellant used at any one time does not exceed 2 kilograms and, for these purposes, the quantity of propellant used includes propellant removed from cartridges;

(c) the preparation of shot firing charges in connection with their use;

(d) the preparation, assembly, disassembly and fusing of firework displays at the place of intended use;

(e) the preparation, assembly and fusing of fireworks, in quantities of no more than 10 kilograms at a time, at a site in relation to which a person holds a licence for the storage of explosives, for the purposes of a firework display to be put on by that person;

(f) the preparation, assembly and fusing of explosives commissioned for use in theatrical, television or cinematic special effects;

(g) the reprocessing of an explosive to form a pharmaceutical product which is not in itself an explosive substance;

(h) the mixing for immediate use of—

(i) ammonium nitrate with fuel oil; or

(ii) ammonium nitrate blasting intermediate with another substance,

at a mine or quarry to produce an explosive which is not cap-sensitive;

(i) the use of desensitised explosives in the manufacture of products which are not in themselves explosives; or

(j) the manufacture of explosives by a company which is a wholly-owned subsidiary of another company at a site in relation to which that other company holds a licence to manufacture explosives and that manufacture by the wholly-owned subsidiary is in accordance with the conditions of that licence.

(3) Where any of the activities in paragraph (2) for which a licence is not required are to take place at a site in relation to which a person holds a licence, the activity may only be carried out there where to do so would not result in a breach of the conditions of that licence.

(4) In this regulation—

(a) "cap-sensitive" means an explosive which gives a positive result when tested in accordance with the Series 5(a) test of the Manual of Tests and Criteria, fifth revised edition, supporting the United Nations Recommendations;

(b) "quarry" has the meaning given in regulation 3 of the Quarries Regulations 1999;

(c) "recreational or occupational firearms" means hand-held firearms intended for the shooting of—

(i) wild game, vermin or, in the course of carrying on activities in connection with the management of an estate, wildlife; or

(ii) prepared inanimate objects;

(d) "shot firing charges" means charges used in shot firing operations; and

(e) "supply" means making available with a view to distribution or use, whether by the person making it available to another and whether for reward or free of charge.

[8.662]

7 Authorisation to store explosives

(1) Subject to paragraph (2), no person may store explosives unless that person holds a licence for their storage and complies with the conditions of that licence.

(2) Paragraph (1) does not apply to—

(a) the storage of one or more of the following—

(i) no more than 10 kilograms of shooters' powder;

(ii) no more than 5 kilograms of—

(aa) shooters' powder; or

(bb) any hazard type 3 or 4 explosive, or desensitised explosive, which is not a relevant explosive, or a combination of hazard type 3 or 4 explosives, or desensitised explosives, which are not relevant explosives; or

(cc) a combination of shooters' powder and any hazard type 3 or 4 explosives, or desensitised explosives, which are not relevant explosives;

(iii) no more than 15 kilograms of percussion caps or small arms ammunition or a mixture of them;

(b) the storage of no more than 7 kilograms of—

(i) hazard type 1 or 2 explosives, or

(ii) a combination of hazard type 1 or 2 explosives with explosives of another hazard type,

for no longer than 24 hours;

(c) the storage of hazard type 3 or 4 explosives for no longer than 24 hours;

(d) the storage of no more than 100 kilograms of—

(i) hazard type 3 explosives consisting of fireworks;

(ii) shooters' powder; or

(iii) a combination of shooters' powder and hazard type 3 and 4 explosives consisting of fireworks, provided that the explosives are stored for no longer than is necessary and in any event no more than 5 consecutive days in their place of intended use;

(e) the storage of—

(i) no more than 250 kilograms of hazard type 4 explosives provided that the explosives are stored for no longer than is necessary and in any event no more than 5 consecutive days in their place of intended use; or

(ii) no more than 50 kilograms of hazard type 4 explosives consisting solely of fireworks provided that the fireworks are stored for no longer than 21 consecutive days and are not for sale or for use at work;

(f) the storage of desensitised explosives which have been assigned in accordance with the United Nations Recommendations the U.N. no 2059, 2555, 2556 or 2557; or

(g) the storage of explosives by a company which is a wholly-owned subsidiary of another company at a site in relation to which that other company holds a licence to store explosives and that storage by the wholly-owned subsidiary is in accordance with any condition of that licence.

(3) For the purposes of paragraph (2)—

(a) no more than one of the exceptions listed in subparagraphs (a) to (g) of paragraph (2) may be relied on in relation to explosives stored at the same site at the same time, irrespective of the person who is storing them; and

(b) the quantities referred to in that paragraph are the maximum quantities of the explosives they respectively relate to which may be present at a site at any one time.

(4) Where any storage to which paragraph (2) applies for which a licence is not required is to take place at a site in relation to which a person holds a licence, that storage may only take place there where to do so would not result in a breach of the conditions of that licence.

[8.663]

8 Authorisation to transfer civil explosives

(1) Before any civil explosives are transferred, the consignee must obtain from the competent authority for the place where the transfer will terminate, a recipient competent authority document which grants approval for the transfer.

(2) No person may consign any civil explosives for carriage to a place outside Great Britain unless the approval of the relevant authority has been obtained.

(3) No person may consign any civil explosives for carriage unless that person is satisfied that the consignee has the recipient competent authority document required by paragraph (1).

(4) No person may carry civil explosives unless the civil explosives are accompanied by the recipient competent authority document required by paragraph (1) or a copy of that document certified by or on behalf of the consignee to be a true copy thereof.

(5) For the purposes of this regulation, any transfer of civil explosives to a place outside the area of the EEA States is treated as a transfer which will terminate at the place where the civil explosives are immediately before leaving the area of the EEA States.

(6) The consignee of any civil explosives must retain the recipient competent authority document or a copy thereof for a period of 3 years from the completion of the transfer.

(7) This regulation does not apply to the transfer of any component of small arms ammunition by a person for that person's own sporting or other recreational use and not for sale, or to the transfer of such component following which the component is in the possession of a person for such purposes.

(8) Except as part of a transfer carried out in accordance with this regulation, the importation of civil explosives into the United Kingdom is prohibited.

(9) In this regulation—

(a) "competent authority" means—

(i) in respect of Great Britain, the Executive, and

(ii) in respect of Northern Ireland or a place in the territory of an EEA State other than the United Kingdom, the authority whose responsibilities are specified in Article 9 of the Civil Uses Directive;

(b) "recipient competent authority document" means a document issued in accordance with Article 9.3, 9.5, or 9.6 of the Civil Uses Directive by the competent authority of the EEA State in which the transfer will terminate; and

(c) "relevant authority" means—

(i) in respect of a transfer or part of a transfer which takes place within Great Britain, the Executive, and

(ii) in respect of a transfer or part of a transfer which takes place in Northern Ireland or a place in the territory of an EEA State other than the United Kingdom, the competent authority for each place where the transfer takes or is to take place.

[8.664]

9 Prohibition concerning the acquisition and supply of fireworks

(1) No person may—

(a) acquire more than 50 kilograms of fireworks unless that person ("Person A"), or another person on behalf of Person A, holds a valid licence for the storage of those fireworks; or

(b) sell or otherwise transfer to any person ("Person B") more than 50 kilograms of fireworks unless Person B shows to the person selling or otherwise transferring the fireworks a valid licence for the storage by Person B of those fireworks.

(2) This regulation does not apply to a person who is transporting fireworks on behalf of another person.

PART 3

DEFENCES IN RELATION TO REGULATIONS 5 TO 7

[8.665]

10 Defence

(1) In proceedings against a person for a contravention of regulation 5(2), it is a defence for that person to prove that the keeping of explosives in contravention of that provision was caused by an emergency being an emergency which that person took all reasonable precautions and exercised all due diligence to avoid.

(2) In proceedings against a person for a contravention of regulation 6(1) which involves using a building or part of a building licensed for the manufacture of explosives, for another manufacturing process not specified in the licence, it is a defence for that person to prove that—

(a) that use was temporary;

(b) that other process of manufacture involved explosive of the same, or a lower, hazard type than the explosives which the conditions of the licence permit in, as the case may be, that building or part of a building;

(c) the maximum quantity of explosives in that building or part of a building at any one time permitted under the conditions of the licence was not exceeded; and

(d) that person informed—

(i) where that person's licence was granted by the ONR, the ONR; or

(ii) otherwise, the Executive,

as soon as was reasonably practicable after the start of that use.

(3) In proceedings against a person for a contravention of regulation 7(1), it is a defence for that person to prove that the storage of explosives without a licence or in breach of a condition of a licence was caused by an emergency being an emergency which that person took all reasonable precautions and exercised all due diligence to avoid.

(4) In proceedings against a person for a contravention of regulation 7(1) where it is alleged against that person that the storage concerned was for a period longer than a period ("the permitted period") referred to in regulation 7(2)(b), (c), (d) or (e)(i) or (ii), it is for that person to prove that the storage concerned was for no longer than the permitted period.

PART 4

APPLICATIONS FOR AND GRANT OF AUTHORISATIONS

[8.666]

11 Application for and issuing of an explosives certificate to acquire or acquire and keep any relevant explosive

(1) Subject to paragraphs (4) to (6), an application for an explosives certificate must be made to the chief officer of police for the relevant police force.

(2) An application for an explosives certificate must be in a form approved for the time being for the purposes of this regulation by the Executive.

(3) A chief officer of police must issue an explosives certificate to an applicant if satisfied that the applicant is a fit person to hold an explosives certificate in accordance with any terms of that certificate.

(4) This paragraph applies with respect to an explosives certificate which would, if granted, certify only that the holder is a fit person to acquire explosives and where the person who would be the holder is not resident in Great Britain or, in the case of a body corporate, does not have a registered office or any other office in Great Britain.

(5) Where paragraph (4) applies, the application for an explosives certificate may be made by a person resident in Great Britain who has knowledge of and control over any occasion when the person to whom the certificate would relate would acquire or use explosives.

(6) An application pursuant to paragraphs (4) and (5) must be made to the chief officer of police for the police area in which the person who makes the application resides or, in the case of a body corporate, has its registered office or, if it has no registered office, its principal office.

(7) The holder of an explosives certificate must inform the chief officer of police who issued it of any change in the holder's address or, where the holder is a body corporate or partnership, of its proper address for the purposes of section 46(4) of the 1974 Act, either before or immediately after any such change occurs.

[8.667]

12 Applications for licences to manufacture or store explosives

An application for a licence must be made to the licensing authority on a form approved for the time being for the purposes of this regulation by the Executive.

[8.668]

13 Grant of licences

(1) A licence or a renewal of a licence may be granted—

(a) where the licensing authority is a local authority, for such period not exceeding five years as the licensing authority determines;

(b) where the licensing authority is the chief officer of police, the Executive or the ONR, for such period not exceeding five years as that licensing authority determines, save that, where the applicant for the licence or renewal of a licence has been granted an explosives certificate, the licence or renewal, as the case may be, may only be granted for any period not exceeding the due expiry date of that explosives certificate; or

(c) for any period or without a time limit in a case—

(i) to which paragraph (3) applies; or

(ii) to which paragraph (3) does not apply by virtue of paragraph (4)(d), (e), (f), or (g); or

(iii) where the application is for a licence, or a renewal of a licence, relating only to the manufacture or storage of ammonium nitrate blasting intermediate.

(2) Subject to paragraph (3), the licensing authority must grant a licence or renewal of a licence unless any of the grounds for refusing to do so referred to in regulation 20 apply.

(3) Subject to paragraph (4), where the Executive or the ONR is the licensing authority in respect of an application for a licence the procedure set out in regulation 14 applies for obtaining the assent of—

(a) the local authority, or

(b) each local authority where the proposed site which is the subject of the application for a licence is situated partly within the area of one local authority and partly within the area of another,

and the Executive or the ONR, as the case may be, must refuse to grant the licence unless the local authority, or each local authority, as the case may be, has so assented.

(4) Paragraph (3) does not apply—

(a) where—

(i) the application is for a licence to store no more than 2000 kilograms to which paragraph (a)(i) or (b) of the definition of "explosive" in regulation 2(1) applies; and

(ii) the applicant has not notified the relevant licensing authority that the separation distances which would be required by regulation 27 and Schedule 5 could not be complied with;

(b) to an application for a licence relating to the manufacture of explosives by means of on-site mixing;

(c) to an application for a licence relating to the manufacture or storage of ammonium nitrate blasting intermediate;

(d) to an application for a licence relating to the manufacture or storage of explosives by a person who wishes to carry on such manufacture or storage within a part of a site where another person already holds a licence for the manufacture or storage of explosives; and either—

(i) the application relates to manufacturing or storage activities which would be permitted at that part of the site under the existing licence; or

(ii) in the opinion of the relevant licensing authority or a local authority whose assent would otherwise be required, no significant new health and safety issues are raised by the application;

(e) to an application for a licence relating to the manufacture of explosives by a police force maintained pursuant to section 2 of the Police Act 1996 or sections 2 and 3 of the Police and Fire Reform (Scotland) Act 2012 for their operational purposes or the training of members of that police force in relation to those purposes;

(f) to an application for a licence for the manufacture or storage of explosives at a site which, immediately before any grant of that application, is one which the disapplication in regulation 3(9) applies to and, in the opinion of the relevant licensing authority, no significant new health and safety issues are raised by the application; or

(g) to an application for a licence to follow, without a gap in time, a licence in respect of which paragraph (3) did not apply by virtue of sub-paragraph (e) or (f) and, in the opinion of the relevant licensing authority, no significant new health and safety issues are raised by the application.

(5) Every licence must include conditions which specify—

(a) the site and, within it, the places where the explosives may be stored, or, in the case of a licence to manufacture explosives, where they may be manufactured;

(b) the hazard type, if any, the description and maximum amount of explosives which may be—

(i) stored or otherwise present, or

(ii) in the case of licence to manufacture explosives, manufactured,

at any one time at any place so specified.

(6) In addition to the matters specified in paragraph (5), a licence which is granted by the relevant licensing authority in cases where the assent of the local authority was required pursuant to paragraph (3) or in cases where that assent was not required by virtue of paragraph (4)(b), (c), (d), (e), (f) or (g)—

(a) must be granted subject to such conditions as the relevant licensing authority considers appropriate which relate to separation distances,

(b) may be granted subject to such conditions as the relevant licensing authority considers appropriate which relate to—

(i) the construction, siting or orientation of any building (including any protective works around the building) where the activity will be carried on;

(ii) the activities which may be undertaken in specified buildings, rooms within those buildings, other structures or other places within the site; and

(iii) the manufacture and storage of the ingredients of explosives or articles or substances which are liable to ignite spontaneously or are flammable or otherwise dangerous in ways which could initiate or aggravate a fire or explosion,

and in this sub-paragraph—

"activity" means the manufacture or storage of explosives and includes any handling, on-site transport, testing, use and disposal of explosives and "activities" is to be construed accordingly; and

"construction" means the materials used in, and the design of, a building; and

(c) may, where both the manufacture and storage of explosives at the same site was applied for, cover both that manufacture and storage for the purposes of, respectively, regulations 6 and 7.

(7) In addition to the matters specified in paragraphs (5) and (6), where a licensing authority grants a licence which relates to the storage of pyrotechnic articles at any site where those articles are to be offered for sale, the licensing authority may attach such conditions to the licence as it considers appropriate which relate to—

(a) the storage and display of those articles in areas where they can be purchased;

(b) the prevention of risk of fire arising in respect of those articles; and

(c) the safe use of fire escapes in that area.

(8) Every person who, in a case to which paragraph (3) applies or a case to which that paragraph does not apply by virtue of paragraph (4)(c), (d), (e), (f) or (g) is granted a licence to manufacture or store explosives must ensure that the relevant licensing authority and the local planning authority in whose area the manufacture or storage takes place is, within 28 days of the licence being—

(a) granted; or

(b) varied in a way which affects the separation distances required to be maintained,

given a plan of the site and its immediate surrounding area showing the separation distances required to be maintained pursuant to the licence or varied licence.

(9) A licence granted pursuant to this regulation must be in a form approved for the time being for the purposes of this regulation by the Executive.

(10) In this regulation—

(a) "local planning authority", in relation to an area—

(i) in England and Wales has the same meaning as it has in Part I of the Town and Country Planning Act 1990 save that, where there is more than one local planning authority, it means the district planning authority for the district; and

(ii) in Scotland means the council for the local government area;

(b) "on-site mixing" means the mixing at any place of non-explosive substances or preparations to form an explosive for immediate use at that place; and

(c) "relevant licensing authority" means the Executive or the ONR (as the case may be).

[8.669]

14 Local authority assent procedure in relation to licence applications

(1) Where this regulation applies by virtue of regulation 13(3), the relevant licensing authority must, subject to regulation 20, issue the applicant with a draft licence containing the conditions which that licensing authority proposes to attach to the licence.

(2) The applicant must as soon as reasonably practicable send a copy of the application and draft licence to the local authority in whose area the manufacture or storage is proposed to take place.

(3) Within 28 days of sending to the local authority the information specified in paragraph (2), the applicant must—

(a) cause to be published in a newspaper circulating in the locality where the manufacture or storage of explosives is proposed to take place a notice which must—

(i) give details of the application;

(ii) invite representations on matters affecting the health and safety of persons other than the applicant's employees to be made in writing to the local authority within 28 days of the date that the notice is first published; and

(iii) give an address within the area of the local authority at which a copy of the application and draft licence may be inspected and the address of the local authority to which any representations must be sent; and

(b) take other reasonable steps to give that information to every person who resides or carries on a business or other undertaking within the public consultation zone.

(4) The local authority must send a copy of any representations referred to in paragraph (3)(a)(ii) to the applicant as soon as reasonably practicable after receiving them.

(5) In considering whether to assent, the local authority must have regard only to health and safety matters.

(6) Subject to paragraph (7), the local authority must, before deciding whether to assent to the application, hold a public hearing within 4 months of the date of its receipt of the copy of the application and draft licence referred to in paragraph (2).

(7) If, after the period of 28 days referred to in paragraph (3)(a)(ii) has elapsed, the local authority has received no objection to the application, or has only received objections which in its opinion are frivolous or immaterial, it may assent to the application without holding a hearing.

(8) Not less than 28 days before the hearing referred to in paragraph (6), the local authority must publish notice of the date, time and place fixed for the hearing in a newspaper circulating in the locality and send a copy of the notice to—

(a) the applicant;

(b) any person who made representations referred to in paragraph (3)(a)(ii); and

(c) the relevant licensing authority,

within 7 days from its publication.

(9) The local authority must notify the applicant and the relevant licensing authority of its decision within 7 days of making it.

(10) If the local authority fails to—

(a) send a copy of the notice referred to in paragraph (8) to the relevant licensing authority within 3 months from the date that a copy of the application and draft licence was sent to it pursuant to paragraph (2); or

(b) notify the relevant licensing authority of its decision in accordance with paragraph (9), within 2 months from the date of publication of the notice referred to in paragraph (8),

that licensing authority may make a written request to the local authority for it to state in writing whether it assents to the application.

(11) If the local authority does not respond to the written request within 28 days from the date of the request, the local authority is deemed to have assented to the application.

(12) The applicant must pay a fee to the local authority for the performance by that authority of their functions under this regulation, which fee must not exceed the sum of the costs reasonably incurred by that authority in performing those functions.

(13) In this regulation—

(a) "applicant" means the applicant for a licence or variation of a licence, as the case may be, and "application" means that person's application; and

(b) "relevant licensing authority" has the meaning given in regulation 13(10)(c).

15 Registers and retention of documents

(1) The licensing authority must—

(a) maintain a register in accordance with Schedule 4;

(b) keep a copy of any licence granted by it (together with a copy of the application for the licence) for as long as the licence remains valid; and

(c) (except where the Executive is the licensing authority), send to the Executive on request a copy of any part of the register or other document specified in this paragraph within such time as the Executive may direct.

(2) For the purposes of this regulation and Schedule 4, in the case to which regulation 3(9) applies disapplying regulations 6, 7, 12 to 14, 16 to 18, 20, 23 and 27, any reference to—

(a) the licensing authority or licensee is to be construed as a reference to the Secretary of State having responsibility for defence;

(b) any licence granted is to be construed as a reference to the scheme referred to in regulation 3(9);

(c) separation distances is to be construed as a reference to the separation distances prescribed in the scheme approved by the Secretary of State having responsibility for defence.

PART 5

VARIATION AND TRANSFER OF AUTHORISATIONS TO MANUFACTURE OR STORE EXPLOSIVES AND DEATH, BANKRUPTCY OR INCAPACITY OF A LICENSED PERSON

[8.670]

16 Variation of licences

(1) The licensing authority which grants a licence may vary it—

(a) where there has been a change in circumstances such that the separation distances can no longer be maintained and a consequent reduction in the maximum amount of explosive that may be stored is required;

(b) (where the Executive or the ONR (as the case may be) is the licensing authority in cases where the assent of the local authority was required pursuant to regulation 13(3) , or in cases where that assent was not required by virtue of regulation 13(4)(b), (c), (d), (e), (f) or (g), before the grant of the licence) where there has been a material change in circumstances so that a variation is necessary to ensure safety; or

(c) in relation to any of the matters it relates to, by agreement with the licensee.

(2) Subject to paragraphs (7) and (8), a licence may be varied on the grounds referred to in paragraph (1)(a) or (b) without the agreement of the licensee.

(3) Where the Executive or the ONR is the licensing authority in cases where the assent of the local authority—

(a) was required under regulation 13(3) before the grant of the licence; or

(b) was not required by virtue of regulation 13(4)(f) or (g); or

(c) was required under regulation 13(3) of the 2005 Regulations, before the grant of the licence; or

(d) would have been required under regulation 13(3) of the 2005 Regulations but for the operation of regulation 27(4) or (13) of those Regulations,

the provisions of regulation 14 apply in respect of a proposed variation referred to in paragraph (4) as if the reference in regulation 14(1) to regulation 13(3) were a reference to regulation 16(3).

(4) A proposed variation for the purposes of paragraph (3) is one which—

(a) relates to changes in the permitted quantities or types of explosive as a result of which the licensee could be required to maintain a separation distance greater than the separation distance required before the variation and, in the opinion of the Executive or the ONR, as the case may be, or the local authority concerned, significant new health and safety issues are raised by that proposed variation;

(b) would increase the period of the licence by more than twelve months; or

(c) would remove the period of the licence so that it would be unlimited as to time,

and the Executive or the ONR, as the case may be, must refuse to grant a varied licence unless the local authority, or each local authority in the case referred to in regulation 13(3), has so assented.

(5) Where a licensing authority proposes to vary a licence without the agreement of the licensee it must, before taking any such action, notify the licensee of its proposed course of action and afford the licensee the opportunity of making representations to the licensing authority about it, within a period of 28 days from the date of the notification.

(6) Representations made for the purpose of paragraph (1) may be made in writing, or both in writing and orally.

(7) Where the licensing authority decides to vary a licence without the agreement of the licensee it must provide the licensee with written reasons for its decision.

(8) Where the licensing authority varies a licence without the agreement of the licensee, that variation takes effect from a date to be determined by the licensing authority which must be a date after the 28 day period referred to in paragraph (5).

(9) In this regulation, any reference to varying a licence includes varying its conditions.

[8.671]

17 Transfer of licences

(1) A licence may be transferred in writing by the licensing authority which issued the licence to any other person who wishes to manufacture or store explosives in place of the licensee and who applies to the licensing authority for the transfer.

(2) A licensing authority must grant an application for a transfer of a licence unless it is of the opinion that the applicant is not a fit person

(a) to store explosives, in the case of an application to transfer a licence to store explosives; or

(b) to manufacture explosives, in the case of an application to transfer a licence to manufacture explosives.

(3) Where a licensing authority is of an opinion referred to in sub-paragraph (a) or (b) of paragraph (2), it must, subject to regulation 20, refuse the application to transfer the licence.

[8.672]

18 Death, bankruptcy or incapacity of a licensee

(1) If a licensee dies or becomes incapacitated, a person manufacturing or storing explosives in accordance with the conditions of the first-named person's licence is to be treated as being licensed in accordance with the first named person's licence until either—

(a) the expiration of a period of 60 days starting with the date of such death or incapacity;

(b) the grant or refusal of a new licence; or

(c) the transfer of, or a refusal to transfer, a licence,

whichever is the earlier.

(2) If a licensee becomes bankrupt or, in the case of a company, goes into liquidation, administration or receivership or has a receiving order made against it, any liquidator, administrator, receiver or trustee in bankruptcy is to be treated as being the licensee.

PART 6

REFUSALS OF AUTHORISATIONS

[8.673]

19 Refusal of an explosives certificate

(1) A chief officer of police must not issue an explosives certificate to a person if the chief officer of police is satisfied that the person is of unsound mind or intemperate habit.

(2) A chief officer of police must not issue an explosives certificate to a person unless the chief officer of police is satisfied that—

(a) the person has good reason for acquiring the relevant explosives;

(b) the person is a responsible person who may be permitted to acquire the relevant explosives without danger to public safety or peace;

(c) the person will take all reasonable precautions to prevent access to the relevant explosives by unauthorised persons and to prevent loss of those explosives;

(d) the person is not a prohibited person;

(e) where the person is a body corporate, no director or secretary of the body corporate, and where the person is a Scottish firm, no partner in the firm, is a prohibited person;

(f) where the application is for an explosives certificate relating only to acquisition of relevant explosives—

(i) it is not reasonably practicable for the applicant to be an occupier of a site for the storage of relevant explosives for which the person would be required to have a licence, and

(ii) either the relevant explosives will not be kept, or, if kept, the applicant will ensure that they are kept at a site where such storage is permitted pursuant to a licence or at a site occupied by the Secretary of State having responsibility for defence; and

(g) where the application is for an explosives certificate relating to the keeping of relevant explosives, either—

(i) the applicant will ensure that the relevant explosives are kept at a site where the storage is permitted pursuant to a licence;

(ii) the relevant explosives will be kept at a site occupied on behalf of the Crown, or

(iii) in the case where the applicant does not hold a licence in respect of the storage of the relevant explosives, no licence is required to be held by the applicant in respect of that storage by virtue of regulation 7(2).

[8.674]

20 Refusal of a licence and draft licence and refusal of a renewal or transfer of a licence

(1) Subject to paragraphs (3) to (5), the licensing authority must—

(a) refuse an application for a licence; and

(b) where regulation 14(1) applies, refuse to issue the draft licence referred to in regulation 14(1), where paragraph (2) applies.

(2) This paragraph applies where the licensing authority is of the opinion that—

(a) the proposed site or, within it, any place where the manufacture or storage of explosives is proposed to take place is unsuitable for that manufacture or storage; or

(b) the applicant is not a fit person—

(i) to store explosives, in the case of an application for a licence to store explosives; or

(ii) to manufacture explosives, in the case of an application for a licence to do so.

(3) Where a licensing authority proposes to refuse an application for—

(a) a licence;

(b) a renewal of a licence;

(c) a variation of a licence; or

(d) a transfer of a licence;

it must, before taking any such action, notify the applicant of its proposed course of action and afford that applicant the opportunity of making representations to the licensing authority about it, within a period of 28 days from the date of the notification.

(4) Representations made for the purpose of paragraph (3) may be made in writing, or both in writing and orally.

(5) Where the licensing authority decides to refuse an application for—

(a) a licence;

(b) a renewal of a licence;

(c) a variation of a licence; or

(d) a transfer of a licence;

it must provide the applicant with written reasons for its decision.

(6) A refusal by the licensing authority, pursuant to paragraph (1), to issue the draft licence referred to in regulation 14(1) is to be treated for the purposes of these Regulations as a refusal of an application for a licence.

PART 7

REVOCATION OF AUTHORISATIONS AND APPEALS AGAINST CERTAIN DECISIONS

[8.675]

21 Revocation and expiry of an explosives certificate

(1) A chief officer of police may revoke an explosives certificate which that chief officer of police has issued if, at any time, that chief officer of police is satisfied that the holder was not when the explosives certificate was issued, or is no longer, a fit person to hold an explosives certificate in the terms of that explosives certificate.

(2) A chief officer of police must revoke an explosives certificate which that chief officer of police has issued if, at any time, that chief officer of police is satisfied that any of the conditions in paragraph (1) or (2) of regulation 19 was not met when the explosives certificate was issued, or is no longer met.

(3) An explosives certificate ceases to be valid—

(a) 5 years after the date of issue; or

(b) after such lesser time as may be stated therein; or

(c) after notice of revocation by the chief officer of police for the relevant police force has been served on the holder of the certificate,

whichever happens first.

(4) The holder of an explosives certificate must surrender it to the chief officer of police who issued it immediately after a notice of revocation has been served on that holder under paragraph (3)(c).

[8.676]

22 Appeals and applications to the Crown Court or Sheriff

(1) A person aggrieved by any decision of a chief officer of police to refuse an explosives certificate under regulation 19 or to revoke an explosives certificate under regulation 21 may appeal within 21 days of receiving notice of that decision.

(2) An appeal under this regulation in England and Wales is to be to the Crown Court and in Scotland is to be to the sheriff by way of summary application.

(3) The court hearing an appeal under this regulation may dismiss the appeal or allow the appeal and give directions to the chief officer of police to issue or restore the explosives certificate.

(4) A prohibited person may apply to the Crown Court or, in Scotland, to the sheriff by way of summary application for an order that that person is no longer a prohibited person and the court hearing any such application may make an order exempting that person from the provisions of these Regulations relating to a prohibited person.

[8.677]

23 Revocation of a licence

(1) The licensing authority which grants a licence may, subject to paragraphs (3), (4), (5) and (6), revoke that licence—

- (a) where there has been a change in circumstances such that the site or, within it, any place in which explosives are manufactured or stored which the licence relates to is no longer suitable for that manufacture or storage;
- (b) where it appears to the licensing authority on information obtained by it after the grant of the licence that the licensee is not a fit person—
  - (i) to store explosives, in the case of a person licensed to store explosives; or
  - (ii) to manufacture explosives, in the case of a person licensed to manufacture explosives; or
- (c) by agreement with the licensee.

(2) A person whose licence is revoked must ensure that—

- (a) all explosives are removed from a site as soon as is practicable after revocation of a licence in respect of that site;
- (b) those explosives are deposited at a site which is the subject of a licence which permits any storage resulting from that depositing, or suitable arrangements are made for those explosives to be disposed of; and
- (c) the licence is returned to the licensing authority within 28 days of the date that the revocation takes effect pursuant to paragraph (6).

(3) Where a licensing authority proposes to revoke a licence, it must, before taking any such action, notify the licensee of its proposed course of action and afford that person the opportunity of making representations to the licensing authority about it, within a period of 28 days from the date of the notification.

(4) Representations made for the purpose of paragraph (3) may be made in writing, or both in writing and orally.

(5) Where the licensing authority decides to revoke a licence, it must provide in writing to the licensee the reasons for its decision.

(6) Where the licensing authority revokes a licence, that revocation takes effect from a date to be determined by the licensing authority which must be a date after the 28 day period referred to in paragraph (3).

PART 8

LICENCES FOR SITES WHICH CEASE TO BE, OR BECOME, ONR REGULATED SITES

[8.678]

24 Licences for sites which cease to be ONR regulated sites

(1) Where this regulation applies a licence granted by, or treated as granted by, the ONR is to be treated on and after the relevant date as a licence granted by the relevant licensing authority.

(2) This regulation applies where—

(a) the site in respect of which the licence was granted ceases to be an ONR regulated site; and

(b) the licence mentioned in sub-paragraph (a) remained in force immediately before the relevant date.

(3) In this regulation—

(a) "relevant date" means, in relation to a site, the date on which the site ceased to be an ONR regulated site; and

(b) "relevant licensing authority" means the licensing authority for the site determined in accordance with Schedule 1, other than paragraph 4 of that Schedule, as if an application for a licence were to be made.

[8.679]

25 Licences for sites which become ONR regulated sites

(1) Where this regulation applies a licence granted by, or treated as granted by, a licensing authority (other than the ONR) is to be treated on and after the relevant date as a licence granted by the ONR.

(2) This regulation applies where—

(a) the site in respect of which the licence was granted becomes an ONR regulated site; and

(b) the licence mentioned in sub-paragraph (a) remained in force immediately before the relevant date.

(3) In this regulation "relevant date" means, in relation to a site, the date on which the site became an ONR regulated site.

PART 9

SAFETY OF EXPLOSIVES

[8.680]

26 Fire and explosion measures

(1) Any person who manufactures or stores explosives must take appropriate measures—

(a) to prevent fire or explosion;

(b) to limit the extent of fire or explosion including measures to prevent the spreading of fires and the communication of explosions from one location to another; and

(c) to protect persons from the effects of fire or explosion.

(2) For the purposes of paragraph (1), the reference to the manufacture or storage of explosives includes a reference to any handling, on-site transport and testing of explosives which is associated with that manufacture or storage.

(3) In this regulation, "fire or explosion" means unplanned fire or explosion at the site of manufacture or storage.

[8.681]

27 Separation distances

(1) Subject to paragraphs (2) and (3), every person who stores explosives at a site must ensure that the relevant separation distance prescribed by Schedule 5 is maintained between a store and a building or other place to which that Schedule applies.

(2) Paragraph (1) does not apply to—

(a) desensitised explosives; or

(b) explosives which are stored under a licence granted by the Executive or the ONR in cases—

(i) where the assent of the local authority was required pursuant to regulation 13(3); or

(ii) where that assent was not required by virtue of regulation 13(4)(c), (d), (e), (f) or (g).

(3) Paragraph (1) does not apply to the storage of explosives where the total quantity of explosives stored at a site, excluding, in the case of sub-paragraphs (c) and (e), any amount of small arms ammunition, does not exceed—

(a) 100 grams;

(b) a combined total of 5 kilograms of shooters' powder and model rocket motors;

(c) 30 kilograms of shooters' powder and 300 grams of percussion caps;

(d) 200 detonators and—

(i) 5 kilograms of water-based explosive and detonating cord; or

(ii) 5 kilograms of water-based explosive or detonating cord; or

(e) in the case of explosives kept by a police force—

(i) 16 kilograms of stun grenades; and

(ii) 4 kilograms of explosives kept for operational purposes other than the purpose referred to in paragraph (iii) but including ordnance disposal and the training of dogs for the detection of explosives; or

(iii) 30 kilograms of explosives kept solely for the purpose of gaining entry to premises; or

(iv) 30 kilograms of explosives kept solely for the purposes of ordnance disposal;

and the explosives are stored in a safe and suitable place with all due precautions for public safety.

(4) In this regulation—

(a) "model rocket motors" means explosive articles which—

(i) are assigned in accordance with the United Nations Recommendations the U.N. no 0186, 0272, 0349, 0351 or 0471;

(ii) are intended to be used for the propulsion of model rockets or similar articles; and

(iii) in respect of each individual explosive article, contain no more than 1 kilogram of explosive; and

(b) "stun grenades" means pyrotechnic articles designed to confuse, disorientate or distract a person which are assigned in accordance with the United Nations Recommendations the U.N. no 0431 or 0432.

[8.682]

28 Discarding or disposing of explosives and decontamination of explosive-contaminated items

(1) Any person who discards or disposes of explosives or explosive-contaminated items must ensure, so far as is reasonably practicable, that they are discarded or disposed of safely.

(2) Any person who decontaminates explosive-contaminated items must ensure, so far as is reasonably practicable, that they are decontaminated safely.

[8.683]

29 Prohibitions concerning manufacture, storage and importation of certain explosives

(1) Subject to paragraph (2), no pyrotechnic substance which consists of—

(a) sulphur; or

(b) phosphorus,

mixed with chlorate of potassium or other chlorates or pyrotechnic article which contains any such mixture may be manufactured, stored or imported into the United Kingdom without the approval of the Executive.

(2) In determining whether to approve for the purposes of paragraph (1), the Executive must take into account—

(a) the risk of the mixture spontaneously combusting; and

(b) the potential for the mixture to become sensitive to ignition by friction or impact.

(3) A contravention of paragraph (1) concerning importation is punishable only under the 1974 Act.

PART 10

SECURITY OF EXPLOSIVES: PREVENTING UNAUTHORISED ACCESS OR ACQISITION

[8.684]

30 Unauthorised access

(1) Any person who manufactures, stores or keeps explosives must take all appropriate precautions for preventing unauthorised persons having access to—

(a) the places where those explosives are manufactured, stored or kept; or

(b) the explosives.

(2) No person may—

(a) without the permission of the licensee, enter—

(i) any store within a site;

(ii) any building used for the manufacture of explosives within a site, or

(iii) any site, or any place within it, with clearly marked boundaries at which explosives are stored or manufactured,

operating under a licence; or

(b) having so entered, refuse to leave that site when requested to do so by a constable or the licensee.

(3) Where following a request referred to in paragraph (2)(b) the person who has entered that place without permission refuses to leave that site, a constable or the licensee may remove that person from the site using reasonable force, if necessary.

(4) For the purposes of paragraphs (2) and (3)—

(a) "enter" includes entering onto a roof of a store or a building in which explosives are manufactured; and

(b) "licensee" includes an employee or agent of a licensee.

[8.685]

31 Prohibitions concerning the transfer of relevant explosives

(1) No person may transfer any relevant explosive to another person ("the transferee") unless—

(a) the transferee has an explosives certificate certifying that the transferee is a fit person to acquire that relevant explosive;

(b) the relevant explosive is for immediate export to a transferee resident outside the United Kingdom;

(c) the explosive is for immediate transport to Northern Ireland and the transferee has police consent under regulation 11(1) of the Explosives Regulations (Northern Ireland) 1970;

(d) the relevant explosive is for transport to an offshore installation in controlled waters as both those terms are defined in section 12 of the Mineral Workings (Offshore Installations) Act 1971; or

(e) the transferee is a person specified in regulation 3(7) and, where those provisions apply to a specified person only in particular circumstances or for particular purposes, those circumstances or purposes are satisfied in the case of the person concerned.

(2) For the purposes of this regulation, a person who acts as agent to transfer any relevant explosive for another person is to be treated as if the person acting as agent as well as that other person had transferred that relevant explosive as principal.

(3) For the purposes of this regulation, where any relevant explosive is transported (including being loaded or unloaded and during breaks which are reasonably incidental to completing the journey within a reasonable length of time) the relevant explosive is not to be treated as being transferred to or from a person who has possession of it only by reason of being—

(a) a carrier;

(b) a person engaged in the work of loading or unloading; or

(c) the occupier of a place it passes through while on the journey.

(4) Nothing in paragraph (1)(b), in relation to the transfer to which it refers, is to be taken as meaning that any requirement under other legislation applying in relation to that transfer does not apply.

[8.686]

32 Restrictions on prohibited persons

(1) Subject to paragraph (3), no employer may knowingly employ a prohibited person in a position where the employee handles or has control of any relevant explosive or any restricted substance.

(2) Subject to paragraph (3), no prohibited person, regardless of whether the person satisfies regulation 5, may acquire, handle or have control of any relevant explosive or any restricted substance.

(3) This regulation does not apply to the employment of, or the acquisition, keeping, handling or control of any relevant explosive or any restricted substance in the course of their duties by, members of Her Majesty's Forces.

PART 11

SECURITY OF EXPLOSIVES: TRACEABILITY, RECORDS AND REPORTING LOSS

[8.687]

33 Unique identification of civil explosives

(1) Subject to paragraphs (3) and (4), any person who manufactures a civil explosive must, as soon as is practicable after that manufacture and before the civil explosive may be moved away from the site where it is manufactured—

(a) mark each civil explosive item referred to in Schedule 7 relating to the civil explosive with a unique identification in accordance with that Schedule;

(b) where an associated label in respect of that marking is required by that Schedule, attach the label in accordance with those requirements; and

(c) where a passive inert electronic tag or associated tag is applied in respect of that marking, place that tag in accordance with the applicable provisions of that Schedule.

(2) The unique identification must—

(a) comprise the components described in Schedule 6; and

(b) be marked on or firmly affixed to the civil explosive item concerned in a way which ensures that it is durable and clearly legible.

(3) Paragraph (1) does not apply where the civil explosive is manufactured for export and is marked with an identification in accordance with the requirements of the importing country for allowing traceability of the civil explosive.

(4) Where a civil explosive is subject to a further manufacturing process after its original manufacture, the manufacturer must mark each civil explosive item relating to the civil explosive subjected to that further process, with a new unique identification only if the original unique identification is no longer marked in the way that paragraph (2)(b) requires and any new marking so required must be done as soon as is practicable after that further process and before the civil explosive may be moved away from the site where it is manufactured.

(5) Subject to paragraph (6), a person who imports a civil explosive into Great Britain must, as soon as is practicable after import and before acquisition of the civil explosive by another person—

(a) mark each civil explosive item referred to in Schedule 7 relating to the civil explosive with a unique identification in accordance with that Schedule;

(b) where an associated label in respect of that marking is required by that Schedule, attach the label in accordance with those requirements; and

(c) where a passive inert electronic tag or associated tag is applied in respect of that marking, place that tag in accordance with the applicable provisions of that Schedule.

(6) Paragraph (5) does not apply where the civil explosive items are marked with a unique identification before importation.

(7) Where a distributor repackages a civil explosive, the distributor must ensure that—

(a) the civil explosive items relating to the civil explosive have the unique identification marked on or affixed to them in accordance with Schedule 7;

(b) where an associated label in respect of that marking is required by that Schedule, the label is attached in accordance with those requirements; and

(c) where a passive inert electronic tag or associated tag is applied in respect of that marking, that tag is placed in accordance with the applicable provisions of that Schedule.

(8) For the purposes of this regulation and Schedule 7—

(a) "civil explosive item" means a civil explosive article, a container containing a civil explosive substance or each smallest packaging unit containing civil explosive;

(b) "civil explosive article" means an article containing one or more civil explosive substances; and

(c) "civil explosive substance" means an explosive substance in a civil explosive.

[8.688]

34 Attribution of manufacturing site codes for civil explosives

(1) This regulation applies for the purposes of the attribution of a three digit code (referred to in this regulation as the "code") to a site where civil explosives are manufactured, which is unique to that site and is a component of the unique identification described in Schedule 6.

(2) For each site within Great Britain at which civil explosives are manufactured—

(a) the manufacturer must apply to the Executive for it to attribute a code for the site; and

(b) the Executive must attribute the code and inform the manufacturer accordingly.

(3) For the purposes of the attribution of a code to a site where civil explosives are manufactured in a country that is not an EEA State—

(a) paragraph (4) applies where the manufacturer is established in an EEA State and the place of import of the civil explosives is Great Britain;

(b) paragraph (5) applies where the manufacturer is not established in an EEA State and the place of import of the civil explosives is Great Britain; and

(c) paragraph (6) applies where the manufacturer is established in Great Britain and the place of import of the civil explosives is either Northern Ireland or an EEA State other than the United Kingdom.

(4) Where this paragraph applies—

(a) in the case where the manufacturer is established in Great Britain—

(i) the manufacturer must apply to the Executive for it to attribute a code for the site where the civil explosives are manufactured; and

(ii) the Executive must attribute the code and inform the manufacturer accordingly; and

(b) in the case where the manufacturer is established in Northern Ireland or an EEA State other than the United Kingdom—

(i) the Executive must attribute a code for the site where the civil explosives are manufactured when it receives a request from the manufacturer to do so; and

(ii) the Executive must inform the manufacturer accordingly.

(5) Where this paragraph applies—

(a) the importer must apply to the Executive for it to attribute a code for the site where the civil explosives are manufactured; and

(b) the Executive must attribute the code and inform the importer accordingly.

(6) Where this paragraph applies, the manufacturer must apply to—

(a) the Secretary of State for Northern Ireland, where the place of import of the civil explosive is Northern Ireland; or

(b) the national authority of the EEA State of import of the civil explosive,

for that Secretary of State or that national authority, as the case may be, to attribute a code for the site where the civil explosives are manufactured.

[8.689]

35 Records in relation to relevant explosives

(1) Subject to paragraph (2), a person ("person A") who acquires or keeps a relevant explosive must keep a record containing the information referred to in paragraph (3).

(2) The duty imposed by paragraph (1) does not apply to—

(a) individuals who acquire any relevant explosive, otherwise than in connection with their work, solely for their own personal use; or

(b) a relevant explosive which is produced by mixing at any place non-explosive substances or preparations to form a relevant explosive for immediate use at that place.

(3) The information referred to in paragraph (1) is—

(a) the means of identifying the relevant explosive, including—

(i) its type;

(ii) its manufacturer;

(iii) a description of the relevant explosive and its name, product code or other information which enables the relevant explosive to be distinguished from every other explosive to which it is not identical;

(b) the total number of any explosive articles, the total nominal mass of explosive substance not contained in explosive articles or, in the case of such substances in cartridge form, the total number of cartridges;

(c) the location of the relevant explosive while it is in the possession of person A;

(d) the name and address of any person to whom the relevant explosive is transferred; and

(e) whether, while in the possession of person A, the relevant explosive has been—

(i) subjected to a further manufacturing process after its acquisition;

(ii) used;

(iii) transferred to another person; or

(iv) destroyed,

and the date of any such further manufacturing process, use, transfer or destruction.

(4) The record of that information must be kept up to date by person A.

(5) The system applied by person A for collecting the information must be tested by person A at regular intervals to ensure its effectiveness and the quality of the information recorded.

(6) Person A must keep the record for a period of three years from the date when the relevant explosive concerned was used, transferred to another person or destroyed.

(7) Person A must protect the record against accidental or malicious damage or destruction.

(8) Person A must provide the enforcing authority with—

(a) information as to the origin and location of each relevant explosive to which the record relates, where the enforcing authority requests it; and

(b) the name of an employee or other person who would be able to provide the enforcing authority with that information at any time and the details necessary for that authority to be able to contact that individual.

(9) Where a business of person A which acquires or keeps any relevant explosive is to cease to trade, person A must notify the enforcing authority of that fact and offer any record still required to be kept pursuant to paragraph (6) to that authority.

(10) In paragraph (3)(a)(iii), "name" means, in relation to an explosive article or explosive substance—

- (a) the name under which it is or is to be marketed; or
- (b) in the case of a military explosive (within the meaning of regulation 25(11)(a) of the Carriage of Dangerous Goods and Use of Transportable Pressure Equipment Regulations 2009), the name designated in writing for that explosive article or substance by the Secretary of State having responsibility for defence.

(11) When regulation 36 is in force, this regulation, to the extent that it concerns civil explosives to which the duty in regulation 36(1) will then apply, ceases to have effect in relation to those civil explosives.

[8.689A]

36 Records in relation to civil explosives

(1) Subject to paragraph (2), a person ("person B") who manufactures, imports, distributes, acquires or keeps any civil explosive must, in respect of any civil explosive manufactured in, or imported into, Great Britain on or after 5th April 2015, keep a record containing the information referred to in paragraph (3).

(2) The duty imposed by paragraph (1) does not apply to individuals who acquire any civil explosive, otherwise than in connection with their work, solely for their own personal use.

(3) The information referred to in paragraph (1) is—

- (a) the means of identifying and describing the civil explosive, including—
  - (i) its type; and
  - (ii) the unique identification in relation to the civil explosive;
- (b) the location of the civil explosive while it is in the possession of person B;
- (c) the name and address of any person to whom the civil explosive is transferred; and
- (d) whether, while in the possession of person B, the civil explosive has been—
  - (i) subjected to a further manufacturing process after its original manufacture;
  - (ii) used;
  - (iii) transferred to another person;
  - (iv) or destroyed,

  and the date of any such further manufacturing process, use, transfer or destruction.

(4) The record of that information must be kept up to date by person B.

(5) The system applied by person B for collecting the information must be tested by person B at regular intervals to ensure its effectiveness and the quality of the information recorded.

(6) Person B must keep the record for a period of ten years from the date when the civil explosive concerned was used, transferred to another person or destroyed.

(7) Person B must protect the record against accidental or malicious damage or destruction.

(8) Person B must provide the enforcing authority with—

- (a) information as to the origin and location of each civil explosive to which the record relates, where the enforcing authority requests it; and

(b) the name of an employee or other person who would be able to provide the enforcing authority with that information at any time and the details necessary for that authority to be able to contact that individual.

(9) Where a business of person B which manufactures, imports, distributes, acquires or keeps civil explosives is to cease to trade, person B must notify the enforcing authority of that fact and provide any record still required to be kept pursuant to paragraph (6) to that authority, who must keep that record for the remainder of the period referred to in that paragraph.

[8.689B]

37 Reporting loss

(1) Any person who acquires possession of, keeps, loads, unloads or transports any relevant explosive or is the occupier of a place where it is loaded or unloaded while on a journey must ensure that the loss of any relevant explosive is reported forthwith—

(a) to the chief officer of police for the police area in which the loss occurs; or

(b) if it is not known where the loss occurred, to the chief officer of police for the police area in which the loss is discovered, or

(c) if the loss occurs or is discovered outside Great Britain, to any chief officer of police.

(2) Where the person required by paragraph (1) to ensure the reporting of a loss is a person who had acquired possession of the relevant explosive or was keeping it, that person must also confirm the report in writing without delay including the following information (whether or not previously supplied orally)—

(a) the date and time that the loss was first discovered;

(b) the place at which that discovery was made;

(c) a description of each type of relevant explosive that has been lost sufficient to distinguish that type from other explosives which are similar but not identical;

(d) for each type lost—

(i) the number of articles lost, or

(ii) the total nominal mass of each type of explosive substance lost, except that in the case of a substance in cartridge form, the number of cartridges lost may be given.

(3) Any person who transports, loads or unloads relevant explosive or is the occupier of a place where it is loaded or unloaded must also report the loss of any relevant explosive without delay to the consignor or, if the consignor is outside the United Kingdom, to the consignee.

(4) Any person making a report under paragraph (3) must provide the consignor or consignee with any information in the possession of the person making the report which the consignor or consignee needs to comply with paragraph (5).

(5) Any consignor or consignee to whom a loss is reported under paragraph (3) must without delay notify the loss in writing to the chief officer of police for the police area in which the loss was discovered, and also (if different) the chief officer of police who issued any explosives certificate held by the consignor or consignee which relates to the relevant explosive the loss of which has been so reported, giving the information listed in paragraph (2).

(6) Employees must inform their employer without delay if they become aware of any loss of any relevant explosive which their employer must report.

(7) Where any loss of a relevant explosive occurs at a site in relation to which the Executive is the licensing authority by virtue of Schedule 1, then any requirement in this regulation to report or supply information to a chief officer of police also includes a like requirement to report or supply the same information to the Executive.

(8) Where any loss of a relevant explosive occurs at a site in relation to which the ONR is the licensing authority by virtue of paragraph 4 of Schedule 1, then any requirement in this regulation to report or supply information to a chief officer of police also includes a like requirement to report or supply the same information to the ONR.

(9) The person required by paragraph (1) to ensure that a loss is reported must maintain adequate systems for ensuring that any loss of a relevant explosive is detected.

(10) That system must be tested by that person at regular intervals to ensure its effectiveness.

(11) In determining whether any relevant explosive is lost for the purposes of this regulation, no account is to be taken of any relevant explosive in respect of which it can be shown that the cause was not theft and that the relevant explosive no longer exists.

PART 12

SECURITY PROVISIONS IN RELATION TO PLASTIC EXPLOSIVES

[8.690]

38 Prohibitions in relation to unmarked plastic explosive

(1) No person may manufacture any plastic explosive, the finished product of which is unmarked.

(2) No person may be in possession, nor transfer possession, of any unmarked plastic explosive.

(3) Paragraph (2) does not apply to a plastic explosive that is in the process of being manufactured.

(4) No person may import any unmarked plastic explosive into the United Kingdom.

(5) For the purposes of this regulation, a plastic explosive is marked if, at the time of its manufacture, it, or a sample of the plastic explosive, contains a detection agent of at least the concentration specified in the corresponding entry for that detection agent in column 2 of the Table in Part 2 of Schedule 8, whether that detection agent is introduced during the process of manufacture of the plastic explosive for the purpose of making the plastic explosive detectable or as a result of the normal formulation of that plastic explosive.

(6) In this regulation—

(a) "detection agent" means a substance named in column 1 of the Table in Part 2 of Schedule 8;

(b) "plastic explosive" has the meaning given by Part 1 of Schedule 8; and

(c) "unmarked" is to be construed in accordance with paragraph (5).

PART 13

PLACING ON THE MARKET OF CIVIL EXPLOSIVES

[8.690A]

39 Placing on the market of civil explosives

(1) No person may place any civil explosives on the market unless—

(a) the civil explosives satisfy such of the essential safety requirements as apply to those civil explosives; and

(b) the conformity of the civil explosives to the requirements of this Part has been attested in accordance with regulation 40; and

(c) the CE marking has been affixed to the civil explosives in accordance with regulation 42.

(2) For the purposes of paragraph (1), civil explosives are to be treated as satisfying the essential safety requirements if they conform to any relevant national standard.

(3) For the purposes of this regulation, "national standard" means a standard of a member State—

- (a) which transposes a relevant harmonised standard; and
- (b) the reference number of which has been published by that member State pursuant to Article 4.1 of the Civil Uses Directive.

(4) For the purpose of this regulation, civil explosives are placed on the market when they are first supplied or made available to any person with a view to their distribution or use within an EEA state or are imported, except with a view to re-export, from outside the area of the EEA states, and "place on the market" is to be construed accordingly.

(5) In this regulation, "the essential safety requirements" means the requirements listed in Schedule 9.

[8.690B]

40 Conformity attestation

(1) For the purposes of regulation 39(1)(b), the procedure for attesting conformity of civil explosives to the requirements of this Part must be—

- (a) EC type examination (Module B) referred to in Annex II(1) together with—
  - (i) type conformity (Module C) referred to in Annex II(2), or
  - (ii) production quality assurance (Module D) referred to Annex II(3), or
  - (iii) product quality assurance (Module E) referred to in Annex II(4), or
  - (iv) product verification (Module F) referred to in Annex II(5); or
- (b) unit verification (Module G) referred to in Annex II(6).

(2) In paragraph (1), a reference to a numbered Annex is a reference to the Annex to the Civil Uses Directive so numbered as amended from time to time.

[8.690C]

41 Notified bodies

(1) The Secretary of State may from time to time appoint in writing such qualified bodies as the Secretary of State thinks fit to be notified bodies for the purposes of this Part.

(2) An appointment under this regulation may relate to all or any description or class of civil explosives, may be subject to conditions or to limit of time, and may be revoked in writing at any time.

(3) For the purposes of this regulation a body is qualified if it meets the criteria set out in Annex III (minimum criteria to be taken into account by member States for the notification of bodies) or if it meets the assessment criteria laid down by a relevant harmonised standard.

(4) The Secretary of State must—

- (a) notify the Commission and the other EEA States of any body appointed under this regulation;
- (b) revoke the appointment of any such body if it appears to the Secretary of State that that body is no longer qualified; and
- (c) notify the Commission and the other EEA States of the revocation of any such appointment.

(5) A body appointed by the Secretary of State under this regulation, after agreeing with the applicant any fee in respect of the work to be undertaken by it, is to perform the functions of a notified body under the Civil Uses Directive.

(6) For the purposes of this regulation—

(a) "notified body" means a body notified to the Commission by a member State pursuant to Article 6.2 of the Civil Uses Directive; and
(b) the reference to a numbered Annex has the same meaning as in regulation 40(2).

[8.691]

42 CE marking

(1) For the purposes of regulation 39(1)(c), the CE marking is properly affixed if—
(a) it is visible, easily legible and indelible; and
(b) it is of a durable nature such that it will remain visible, easily legible and indelible during normal transport or storage; and
(c) it is affixed either—
(i) to the civil explosives themselves, or, where that is not practicable,
(ii) to an identification plate which is attached to the civil explosives and which is so designed as to make its re-use impossible, or,
(iii) where neither of the above is practicable, to the packaging of the civil explosives; and
(d) in the case of civil explosives which are subject to any EU Directive other than the Civil Uses Directive, the requirements imposed by virtue of that other Directive have also been complied with in respect of those civil explosives.

(2) No person may affix to any civil explosives the CE marking or any marking or inscription which is liable to be confused with the CE marking except by properly affixing the CE marking to civil explosives which satisfy the requirements of sub-paragraphs (a) and (b) of paragraph (1) of regulation 39.

(3) The CE marking must be in the form shown in Schedule 10.

PART 14

ENFORCEMENT OF THE REGULATIONS

[8.691A]

43 Enforcement within Great Britain

Schedule 11, which makes provision—
(a) determining the enforcing authority for these Regulations as they apply within Great Britain; and
(b) determining the enforcing authority for regulation 3 of the Management of Health and Safety at Work Regulations 1999 as it applies to the manufacture and storage of explosives,
has effect.

[8.691B]

44 Enforcement outside Great Britain

The Executive is the enforcing authority for these Regulations as they apply in any area outside Great Britain.

[8.691C]

45 Enforcement powers in relation to regulation 8 and Part 13

Schedule 12, which makes provision as to enforcement powers in relation to regulation 8 and Part 13, has effect.

PART 15

POWER TO GRANT EXEMPTIONS AND MISCELLANEOUS PROVISIONS

[8.692]

46 Exemptions

(1) Subject to paragraphs (2) and (4), the Executive may, by a certificate in writing, exempt any person or class of persons or any explosive or class of explosives from any requirement or prohibition imposed by these Regulations, and any such exemption may be granted subject to such conditions and to a limit of time and may be revoked by the Executive by a certificate in writing at any time.

(2) The Executive must not grant any such exemption unless, having regard to the circumstances of the case, and in particular to—

- (a) the conditions, if any, which it proposes to attach to the exemption; and
- (b) any other requirements imposed by or under any enactment which apply to the case,

it is satisfied that the health and safety of persons who are likely to be affected by the exemption will not be prejudiced in consequence of it and that the security of explosives will not be prejudiced.

(3) Subject to paragraph (4), the Secretary of State having responsibility for defence may, in the interests of national security, by a certificate in writing, exempt any person or class of persons, any headquarters or any explosive or class of explosives from all or any of the requirements or prohibitions imposed by these Regulations and any such exemption may be granted subject to conditions and to a limit of time and may be revoked by that Secretary of State by a certificate in writing at any time.

(4) The power to exempt in paragraph (1) or (3) does not apply to regulations 8, 33, 34, 36 and 38 to 42.

[8.692A]

47 Savings and transitional provisions

(1) A licence granted under regulation 13 of the 2005 Regulations or deemed to have been so granted by virtue of regulation 27(1) of those Regulations which was valid immediately before the relevant date is deemed to be a licence granted under regulation 13 of these Regulations and continues in force, despite the revocation of the 2005 Regulations by these Regulations, on its existing terms and conditions, subject to—

- (a) any variation under regulation 16(1)(a) to (c) of these Regulations; or
- (b) its expiry on the date it was due to expire or its revocation under regulation 23 of these Regulations, whichever is the sooner.

(2) A registration granted under regulation 11 of the 2005 Regulations which was valid immediately before the relevant date is deemed to be a licence granted under regulation 13 of these Regulations and continues in force, despite the revocation of the 2005 Regulations by these Regulations, on its existing terms until the date it was due to expire or its revocation under regulation 23, whichever is the sooner.

(3) Where an application for a licence or variation of a licence under the 2005 Regulations has been made to, and received by, the licensing authority before the relevant date and the application has not been refused nor granted by that date, the application is deemed to be an application for a licence or, as the case may be, a variation of a licence under, respectively, regulation 12 or regulation 16 of these Regulation and the provisions of these Regulations apply to the application accordingly, subject to paragraph (4) in respect of cases to which the requirements of regulations 13(3) and 14 of the 2005 Regulations applied.

(4) In relation to the application of the requirements of regulations 13(3) and 14 of these Regulations to an application referred to in paragraph (3)—

(a) a draft licence issued by the Executive or the ONR to the applicant under regulation 14(1) of the 2005 Regulations before the relevant date is deemed to be a draft licence for the purposes of regulation 14(1) of these Regulations;

(b) a notice under regulation 14(3) or (8) of the 2005 Regulations which has been published before the relevant date in respect of the application is deemed to be a notice for the purposes of, respectively, regulation 14(3) or (8) of these Regulations;

(c) if a public hearing for the purposes of regulation 14 of the 2005 Regulations would be held or continue to be held on or after the relevant date, it may continue to be so held and is deemed to be a hearing for the purposes of regulation 14 of these Regulations;

(d) the assent to the application by the local authority pursuant to regulation 14 of the 2005 Regulations, or the assent by both local authorities to the application where the assent of both is required under that regulation, given before the relevant date is deemed to be assent for the purposes of regulations 13(3) and 14 of these Regulations.

(5) An explosives certificate granted under regulation 4 of the 1991 Regulations which was valid immediately before the relevant date is deemed to be an explosives certificate granted under regulation 11 of these Regulations and continues in force, notwithstanding the revocation of the 1991 Regulations by these Regulations, on its existing terms until the date it was due to expire or its revocation under regulation 21, whichever is the sooner.

(6) Where an application for an explosives certificate under the 1991 Regulations has been made to, and received by, the chief officer of police for the relevant police force before the relevant date and the application has not been refused nor granted by that date, the application is deemed to be an application for an explosives certificate under regulation 11 of these Regulations and the provisions of these Regulations apply to the application.

(7) Despite the revocation of the 1993 Regulations by these Regulations, a recipient competent authority document issued under the 1993 Regulations which was valid immediately before the relevant date is deemed to be a valid recipient competent authority document for the purposes of regulation 8 of these Regulations.

(8) Where an application for a recipient competent authority document under the 1993 Regulations has been made to, and received by, the Executive before the relevant date and the application has not been refused nor granted by that date, the application is deemed to be an application for a recipient competent authority document under regulation 8 of these Regulations and the provisions of these Regulations apply to the application.

(9) In relation to the application of these Regulations to the storage of ammonium nitrate blasting intermediate by virtue of regulation 2(2), where a person is storing ammonium nitrate blasting intermediate on the relevant date, regulations 7, 12, 13, 14, 16 to 18, 20 and 23 do not apply to that storage by that person until 1st October 2017.

(10) Paragraph (11) applies to a person who, immediately before the relevant date, held—

(a) a licence granted under regulation 13 of the 2005 Regulations; or

(b) a registration granted under regulation 11 of the 2005 Regulations,

which, in either case, is deemed to be a licence granted under regulation 13 of these Regulations by virtue of, respectively, paragraph (1) or (2) of this regulation and, in respect of that licence or registration, was subject to the requirements of regulation 5 of the 2005 Regulations.

(11) A person to whom this paragraph applies is not subject to the requirements of regulation 27 until the later of—

(a) 1st October 2015; or

(b) the expiry of the licence or registration, as the case may be, deemed to be a licence under, respectively, paragraph (1) or (2); and

until that later date, that person must continue to comply with the requirements of regulation 5 of the 2005 Regulations.

(12) Where the storage of explosives at any place—

(a) was immediately before the relevant date exempt from the requirement for a licence under the 2005 Regulations by virtue of an exemption certificate granted under regulation 26 of those Regulations; and

(b) is not storage to which regulation 3(9) of these Regulations relates,

the person doing that storing is deemed to hold a licence granted by the Executive under regulation 13 with an expiry date of 1st October 2016.

(13) Paragraph (14) applies to a person who, immediately before the relevant date, acquires or is keeping explosives which do not require an explosives certificate under the 1991 Regulations but which do under these Regulations.

(14) A person to whom this paragraph applies is deemed to hold an explosives certificate under these Regulations which permits the acquiring and keeping of the explosives referred to in paragraph (13) until 1st October 2016.

(15) The amendments made by the 2005 Regulations to—

(a) the Stratified Ironstone, Shale and Fireclay Mines (Explosives) Regulations 1956; and

(b) the Miscellaneous Mines (Explosives) Regulations 1959,

continue to have effect despite the revocation of the 2005 Regulations by these Regulations.

(16) The amendments made by the Explosives Act 1875 and 1923 Etc (Repeals and Modifications) (Amendment) Regulations 1974 ("the Amendment Regulations") to the Explosives Act 1875 and 1923 Etc (Repeals and Modifications) Regulations 1974 continue to have effect despite the revocation of the Amendment Regulations by these Regulations.

(17) In this regulation—

(a) "the 1991 Regulations" means the Control of Explosives Regulations 1991;

(b) "the 1993 Regulations" means the Placing on the Market and Supervision of Transfers of Explosives Regulations 1993;

(c) "recipient competent authority document" has the same meaning as in regulation 8(9); and

(d) "relevant date" means 1st October 2014.

[8.692B]

48 Repeals, revocations and amendments

(1) The primary legislation specified in Part 1 of Schedule 13 and the secondary legislation specified in Part 2 of that Schedule is amended in accordance with the provisions of that Schedule.

(2) The primary legislation specified in column 1 of Part 1 of Schedule 14 is repealed to the extent specified in column 3 of that Schedule.

(3) The secondary legislation specified in column 1 of Part 2 of Schedule 14 is revoked to the extent specified in column 3 of that Schedule.

[8.692C]

49 Review

(1) The Secretary of State must from time to time—

(a) carry out a review of these Regulations,
(b) set out the conclusions of the review in a report, and
(c) publish the report.

(2) In carrying out the review the Secretary of State must, so far as is reasonable, have regard to how other member States have implemented—

(a) Council Directive 93/15/EC on the harmonization of the provisions relating to the placing on the market and supervision of explosives for civil uses, as amended by Regulation (EC) No 1882/2003 of the European Parliament and of the Council of 29th September 2003, Regulation (EC) No 219/2009 of the European Parliament and of the Council of 11th March 2009 and Regulation (EU) No 1025/2012 of the European Parliament and of the Council of 25th October 2012 which are implemented by these Regulations; and
(b) Commission Directive 2008/43/EC of 4th April 2008 setting up, pursuant to Council Directive 93/15/EEC, a system for the identification and traceability of explosives for civil uses as amended by Commission Directive 2012/4/EU, which are implemented by these Regulations.

(3) The report must in particular—

(a) set out the objectives intended to be achieved by the regulatory system established by these Regulations,
(b) assess the extent to which those objectives are achieved, and
(c) assess whether those objectives remain appropriate and, if so, the extent to which they could be achieved with a system that imposes less regulation.

(4) The first report under this regulation must be published before the end of the period of five years beginning with 1st October 2014.

(5) Reports under this regulation are afterwards to be published at intervals not exceeding five years.

SCHEDULE 1

MEANING OF 'LICENSING AUTHORITY'

Regulation 2(1)

[8.693]

1

Subject to paragraphs 2 to 5, "licensing authority" means—

(a) in relation to an application for a licence for the storage within one site of no more than 2000 kilograms of explosives to which paragraph (a)(i) or (b) of the definition of "explosive" in regulation 2(1) applies—
    (i) the chief officer of police for the area in which the storage is to take place where—
        (aa) any of the explosives are relevant explosives;
        (bb) any of the explosives are ammunition the acquisition of which is regulated or prohibited by virtue of the Firearms Acts 1968 to 1997;
        (cc) any of the explosives are smokeless powder or percussion caps; or
        (dd) the explosives are to be stored by a person who is registered as a firearms dealer under section 33 of the Firearms Act 1968;

(ii) the local authority for the area in which the storage is to take place where none of the explosives are of a type to which sub-paragraph (a)(i) applies;

(b) the Executive where the explosives are to be stored on the surface at a mine, whether in a building or not, or within a harbour;

(c) where the application for a licence relates to the manufacture or storage of ammonium nitrate blasting intermediate, the Executive;

(d) in relation to an application for a licence in any other case, the Executive.

2

In any case falling within paragraph 1(a) and subject to paragraph 5, the applicant may apply instead to the Executive for a licence, in which case the Executive is to be the licensing authority in place of the chief officer of police or local authority.

3

Where a person wishes to manufacture and store explosives at the same site, and requires a licence for that manufacture and for that storage, the Executive is to be the licensing authority in respect of any application for a licence relating to that site and the reference to "an application" in paragraph 1(d) includes any such application.

4

The ONR is the licensing authority in relation to any application for a licence for the manufacture or storage of explosives on an ONR regulated site.

5

The applicant may only apply to the Executive pursuant to paragraph 2, where—

(a) the applicant has notified the Executive that the separation distances which would be required by regulation 27 and Schedule 5 could not be complied with; or

(b) the application does not relate to the storage of pyrotechnic articles at any site where those articles are to be offered for sale and the applicant has already been granted a licence by the Executive under paragraph 1 which relates to another site.

SCHEDULE 2

EXPLOSIVES NOT REQUIRING AN EXPLOSIVES CERTIFICATE

Regulation 5(3)(a)

PART 1

LIST OF EXPLOSIVES

[8.693A]

Explosives	UN no
2-AMINO-4, 6-DINITROPHENOL, WETTED with not less than 20% water by mass	3317
CARTRIDGES, OILWELL	0277
CARTRIDGES, OIL WELL	0278
CARTRIDGES FOR TOOLS, BLANK	0014
CASES, COMBUSTIBLE, EMPTY, WITHOUT PRIMER	0446
CASES, COMBUSTIBLE, EMPTY, WITHOUT PRIMER	0447
CORD, IGNITER	0066
CUTTERS, CABLE, EXPLOSIVE	0070

FUSE, NON-DETONATING	0101
FUSE, SAFETY	0105
1-HYDROXYBENZOTRIAZOLE, ANHYDROUS	0508
1-HYDROXYBENZOTRIAZOLE, MONOHYDRATE	3474
LIGHTERS, FUSE	0131
5-MERCAPTOTETRAZOL-1-ACETIC ACID	0448
NITROCELLULOSE SOLUTION, FLAMMABLE with not more than 12.6% nitrogen, by dry mass, and not more than 55% nitrocellulose	2059
NITROCELLULOSE WITH WATER (not less than 25% water, by mass)	2555
NITROCELLULOSE WITH ALCOHOL (not less than 25% alcohol, by mass, and not more than 12.6% nitrogen, by dry mass)	2556
NITROCELLULOSE with not more than 12.6% nitrogen, by dry mass, MIXTURE WITH or WITHOUT PLASTICIZER, WITH OR WITHOUT PIGMENT	2557
4-NITROPHENYLHYDRAZINE, with not less than 30% water, by mass	3376
PRIMERS, CAP TYPE	0044
PRIMERS, CAP TYPE	0378
SODIUM PICRAMATE, dry or wetted with less than 20% water by mass	0235
SODIUM PICRAMATE, WETTED with not less than 20% water, by mass	1349
TETRAZOL-1-ACETIC ACID	0407

PART 2

FURTHER LIST OF EXPLOSIVES

[8.693B]

1

Ammunition intended for use in small arms.

2

Ammunition consisting of a propelling charge and an inert projectile intended for use in the industrial tools commonly known as kiln-guns

3

Blank ammunition intended for use in small arms.

4

Cartridges, which are empty but with a primer which—

(a) are assigned in accordance with the United Nations Recommendations the U.N. no 0055 or 0378;

(b) are intended for use in small arms; and

(c) would, if packaged for transport, be assigned in accordance with the United Nations Recommendations the UN no 0055 or 0378.

5

Cartridges power device which—

(a) is assigned in accordance with the United Nations Recommendations the UN no 0275, 0276, 0323 or 0381;
(b) is designed to produce mechanical actions such as inflation, linear or rotary motion, projection of fastening devices or extinguishing agents; and
(c) consists of a casing with a charge of deflagrating explosive and a means of ignition.

6

A desensitised explosive which is—
(a) a medicinal product as defined in regulation 2 of the Human Medicines Regulations 2012; or
(b) a veterinary medicinal product as defined in regulation 2 of the Veterinary Medicines Regulations 2013.

7

A desensitised explosive which is a substance specified in an order made under section 104 or 105 of the Medicines Act 1968 which is for the time being in force and which directs that specified provisions of the Human Medicines Regulations 2012 or the Medicines for Human Use (Clinical Trials) Regulations 2004 are to have effect in relation to that substance as such provisions have effect in relation to medicinal products as defined in regulation 2 of the Human Medicines Regulations 2012.

8

Any desensitised explosive acquired, in a quantity not exceeding 5 grams, for the purposes of—
(a) research, analysis or testing at a University;
(b) research, analysis or testing by or on behalf of the Crown; or
(c) the application of forensic science by or on behalf of a police force or the Crown.

9

Any desensitised explosive acquired, in a quantity not exceeding 1 gram, for the purposes of calibration or testing of explosives detection equipment at a port, airport or any other publicly accessible place.

10

Explosive articles which—
(a) are assigned in accordance with the United Nations Recommendations the UN no 0186, 0272, 0349, 0351 or 0471;
(b) are intended to be used for the propulsion of model rockets or similar articles; and
(c) in respect of each individual explosive article, contain no more than 1 kilogram of explosive.

11

The explosive substance Isosorbide Dinitrate which has been desensitised by mixture with not less than 60% lactose, mannose starch or calcium hydrogen phosphate and which is—
(a) assigned in accordance with the United Nations Recommendations the UN no 2907; and
(b) intended for use as an active pharmaceutical ingredient.

12

The explosive substance Nitrogen Triiodide with the chemical formulation NI3, in a quantity not exceeding 0.5 grams which has been manufactured for, and is used for, demonstration purposes as part of an educational activity within the meaning of section 1(3) of the Further Education Act 1985 and either it is wholly used up in the demonstration or any amount remaining is destroyed immediately after it.

13

A solution of the explosive substance Nitroglycerine in alcohol which is—

(a) assigned in accordance with the United Nations Recommendations the UN no 1204 or 3064; and

(b) intended for use as an active pharmaceutical ingredient.

14

Any pyrotechnic substance, in a quantity not exceeding 0.5 grams, which has been manufactured for, and is used for, demonstration purposes as part of an educational activity within the meaning of section 1(3) of the Further Education Act 1985 and either it is wholly used up in the demonstration or any amount remaining is destroyed immediately after it.

15

The explosive substance smokeless powder which is—

(a) assigned in accordance with the United Nations Recommendations the UN no 0161 or 0509 or which has been recovered from ammunition or blank ammunition intended for use in firearms; and

(b) acquired by a person who either is registered as a firearms dealer under section 33 of the Firearms Act 1968 or holds—

(i) a permit granted under section 7 of that Act;

(ii) a firearms certificate granted under section 27 of that Act;

(iii) a shotgun certificate granted under section 28 of that Act; or

(iv) a permit granted under section 17 of the Firearms (Amendment) Act 1988.

16

A solution of the explosive substance 2,4, 6-Trinitrophenol in a concentration no greater than 2% weight per volume intended for use as an analytical reagent, stain, dye or fixative.

SCHEDULE 3

PYROTECHNIC ARTICLES REQUIRING AN EXPLOSIVES CERTIFICATE

Regulation 5(3)(b)

[8.693C]

1

Pyrotechnic articles which are or would, if packaged for transport, be assigned in accordance with the United Nations Recommendations the UN no 0350, 0352, 0353, 0354, 0355, 0356, 0462, 0463, 0464, 0465, 0466, 0467, 0468, 0469, 0470 or 0472.

2

Pyrotechnic articles which—

(a) are or would, if packaged for transport, be assigned in accordance with the United Nations Recommendations the UN no 0349, 0351 or 0471; and

(b) are not intended to be used for the propulsion of model rockets or similar articles.

3

Pyrotechnic articles which—

(a) are or would, if packaged for transport, be assigned in accordance with the United Nations Recommendations the UN no 0033, 0034, 0035, 0037, 0038, 0039, 0171, 0254, 0291, 0297, 0299, 0399 or 0400; and

(b) are intended to be dropped as bombs from an aircraft.

4

Pyrotechnic articles which—

(a) are or would, if packaged for transport, be assigned in accordance with the United Nations Recommendations the UN no 0451, 0329, 0330, 0450 or 0449; and

(b) are intended to be dropped as torpedoes from an aircraft or dispatched from an installation on land or a vessel.

5

Tracers for ammunition which are or would, if packaged for transport, be assigned in accordance with the United Nations Recommendations the UN no 0212 or 0306.

SCHEDULE 4

REGISTERS

Regulation 15

[8.694]

1

The licensing authority must maintain a register ("the register") containing the information listed in paragraph 2 which relates to licences granted by it.

2

The following information must be included in the register—

(a) the name of the licensee;

(b) the licensee's permanent address, unless the licensee's home address is the licensee's only permanent address;

(c) the address of the site where the explosives are manufactured or stored (where that differs from any address included pursuant to sub paragraph (b));

(d) the hazard type if any, the description and maximum amount of explosive which may be stored or otherwise present at any one time in any place which is specified in the licence;

(e) the nature of the business of the licensee and the intended use of the explosives;

(f) the kind of explosives manufactured or stored;

(g) where separation distances are required by regulation 27 or by a condition of the licence to be maintained around the store or the building where explosives are manufactured, a plan in a suitable scale sufficient to show those separation distances;

(h) a map in a suitable scale sufficient to show the location of any stores; and

(i) the kind of store concerned, including the material out of which it is constructed.

3

Subject to paragraphs 7 and 8, where the licence only relates to explosives which are not relevant explosives, the licensing authority must—

(a) ensure that the information referred to in paragraph 2(a) to (d) in respect of that licence is available for inspection at an office of the licensing authority, at all reasonable times and free of charge, by members of the public; and
(b) provide a copy of the entry in the register relating to the information referred to in paragraph 2(a) to (d) in respect of that licence to a member of the public who requests a copy and pays a charge which must not exceed the reasonable cost of providing the copy.

4

Subject to paragraphs 7 and 8, where the licence relates to explosives which are relevant explosives, whether or not it also relates to other explosives, the licensing authority must—

(a) ensure that the information referred to in paragraph 2(a) to (d) in respect of that licence is available for inspection at an office of the licensing authority, at all reasonable times and free of charge, by a person who resides or, in the case of an undertaking, is situated within a public consultation zone concerned in relation to the licence; and
(b) provide a copy of the entry in the register relating to the information referred to in paragraph 2(a) to (d) in respect of that licence to such a person as is referred to in sub-paragraph (a) who requests a copy and pays a charge which must not exceed the reasonable cost of providing the copy.

5

The register may be kept in any form, including electronically.

6

The licensee in relation to any site in which explosives are manufactured or stored, must, where requested in writing to do so by the owner or, if not the same person, the occupier, of any premises falling within any separation distance applying in relation to that site, provide to the requesting person within 28 days of the request a scale plan of the area of land falling within that separation distance.

7

The requirements of paragraphs 3 and 4 do not apply where the information referred to in paragraph 2 is in respect of any site to which a licence relates which is used only for the storage of—

(a) less than 500 kilograms of hazard type 1 explosive or hazard type 2 explosive;
(b) less than 2 tonnes of hazard type 3 explosive or hazard type 4 explosive; or
(c) any explosives for a period of less than four weeks.

8

The requirements of paragraphs 3 and 4 do not apply where the information referred to in paragraph 2 is in respect of a licensed site in relation to which regulation 13 did not apply to the licence application for that site by virtue of regulation 13(4)(e), (f) or (g).

9

Where the licensing authority is a local authority or the ONR, that licensing authority must notify the Executive in writing, within 28 days of receipt of a written request by the Executive to do so, as to such information as is contained in the register it maintains relating to the information referred to in paragraph 2 as the Executive may require.

10

Nothing in this Schedule prevents a licensing authority from disclosing any of the information included in the register it maintains to—

(a) a fire and rescue service;
(b) a joint planning board;
(c) a local planning authority; or
(d) a police force;

for the purposes of the exercise of their respective functions.

11

For the purposes of paragraph 10, "local planning authority" and "joint planning board" have the meanings they are given in, respectively, sections 1 and 2 of the Town and Country Planning Act 1990.

SCHEDULE 5

SEPARATION DISTANCES

Regulation 27

[8.694A]

1

(1) In this Schedule—

"brick-built" means having an outer structure which is wholly or mainly of brick, concrete, stone or other similar material;

"bridleway" has the meaning given in the Highways Act 1980;

"curtain walling" means the glass, masonry or other cladding which is suspended from the structural framework of a building;

"dangerous goods" means dangerous goods to which the Carriage of Dangerous Goods and Use of Transportable Pressure Equipment Regulations 2009 apply;

"distance", save in the definition of "reference zone", means the minimum distance;

"footpath" has the meaning given in the Highways Act 1980;

"footway" has the meaning given in the Highways Act 1980;

"lightly-used road" means a road used ordinarily by more than 20 and no more than 500 vehicles every 24 hours;

"major road" means a road used ordinarily by more than 10,000 vehicles every 24 hours;

"metal-built" means built wholly or mainly of steel or other metal;

"minor road" means a road used ordinarily by more than 500 vehicles every 24 hours, other than a major road;

"mounded" means surrounded by suitable mounds;

"place of public resort" means a place where more than 100 persons are present, or are likely to be present, at any one time on a weekly or more frequent basis;

"processing" means the packing, unpacking, re-packing, labelling or testing of explosives or the division of an amount of explosives stored in bulk into smaller amounts and the placing of those smaller amounts into containers;

"protected place of Class A" means a—

(a) bridleway;
(b) footpath;
(c) footway;
(d) lightly used road; or
(e) waterway;

"protected place of Class B" means a—

(a) dock;
(b) jetty;
(c) minor road;
(d) pier;
(e) railway line;
(f) reservoir;
(g) river wall;
(h) runway for the use of aircraft; or
(i) sea wall;

"protected place of Class C" means a—

(a) major road; or
(b) place of public resort;

"protected place of Class D" means—

(a) a building, whether a dwelling or not, but not including a vulnerable building;
(b) a place, other than a building to which paragraph (a) applies, that is used for the storage in bulk of dangerous goods other than goods which are explosives; or
(c) a place, not within the site where the explosives are stored and other than a building to which paragraph (a) applies, that is used for—
    (i) the storage of explosives; or
    (ii) the manufacture or processing of explosives;

"protected place of Class E" means a vulnerable building;

"protected place of Class F" means a building within the site where the explosives are stored, other than—

(a) a building that is occupied by a person other than the licensee;
(b) a building that is a vulnerable building;
(c) a building that is normally occupied by more than 20 people; or
(d) a building that is used for the storage in bulk of dangerous goods other than goods which are explosives;

"protected place of Class G" means a building or other place within the site where the explosives are stored that is used for the storage of explosives;

"protected place of Class H" means a building or other place within the site where the explosives are stored that is used for the manufacture or processing of explosives;

"reference zone" means the area around a store having the radius from the centre point of the store specified in column 2 of the relevant Supplementary Table;

"road" means any thoroughfare on which the movement of vehicles is allowed; and

"vulnerable building" means a building or structure of vulnerable construction, that is to say—

(a) a building of more than three storeys above ground or 12m in height constructed with continuous non-load bearing curtain walling with individual glazed or frangible panels larger than 1.5 m2 and extending over more than 50% or 120 m2 of the surface of any elevation;
(b) a building of more than three storeys above ground or 12 m in height with solid walls and individual glass panes or frangible panels larger than 1.5 m2 and extending over at least 50% of any elevation;

(c) a building of more than 400 m2 plan area with continuous or individual glazing panes larger than 1.5 m2 extending over at least 50% or 120 m2 of the plan area; or

(d) any other structure that, in consequence of an event such as an explosion, may be susceptible to disproportionate damage such as progressive collapse.

(2) Any reference in this Schedule to a thoroughfare (however described), jetty, pier, waterway or railway line does not include any part of a thoroughfare, jetty, pier, waterway or railway line within the site—

(a) in which the store is situated; and

(b) which are occupied by the person storing the explosives.

(3) Subject to paragraph 8, any reference in this Schedule to "store" is to the store mentioned in regulation 27(1) in relation to which separation distances are required by that provision to be maintained.

(4) Any reference in this Schedule to a quantity of explosives shown in column 1 of a Table is to—

(a) a quantity in a store; or

(b) in cases where sub-paragraph (b) of paragraph 7 is being relied on, the sum total of the quantity of explosives in a store and the protected place of Class G to which sub-paragraph (b) of paragraph 7 applies,

which is more than the lower figure but not more than the higher figure in column 1 in the same row of the Table.

(5) Any reference in this Schedule to a building is to a building in or at which people are, or are likely to be, present either all the time or from time to time.

(6) For the purposes of this Schedule, where explosives of different hazard types are in one store, the explosives must be treated as belonging to the hazard type which would require the greatest separation distance for the total quantity of those explosives and the separation distance must be determined in relation to that total quantity.

(7) For the purposes of this Schedule, the radius for a reference zone applying in a particular case is the number in the entry in column 2 of the relevant Supplementary Table corresponding to the quantity of explosives shown in column 1 of the Supplementary Table.

(8) For the purposes of this Schedule, any reference to "bridleway", "footpath", "footway" or "waterway" does not include, respectively, a bridleway, footpath, footway used, or waterway navigated, by no more than, ordinarily, 20 persons in any 24 hour period.

(9) For the purposes of this Schedule, an area of low population density is an area where the maximum number of dwellings in a reference zone is equal to or less than the number specified in the entry in column 3 of the relevant Supplementary Table.

(10) Save as stated in paragraph (c) of the definition of "protected place of Class D" and subject to sub-paragraph (2), the references to the places referred to in the definitions from "protected place of Class A" to the definition of "protected place of Class E" are references to such places whether within a site or outside it.

2

Where the storage is—

(a) of hazard type 1 explosive in a brick-built mounded store, Table 1;

(b) of hazard type 1 explosive in a brick-built unmounded store, Table 2;

(c) of hazard type 1 explosive in a metal-built mounded store, Table 3;

(d) of hazard type 1 explosive in a metal-built unmounded store with no detonator annex attached, Table 4;

(e) of hazard type 1 explosive in a metal-built unmounded store with a detonator annex attached, Table 5;

(f) of hazard type 2 explosive, some or all items being of more than 0.7 kg net mass in a mounded store, Table 6;

(g) of hazard type 2 explosive, some or all items being of more than 0.7 kg net mass in an unmounded store, Table 7;

(h) of hazard type 2 explosive, every item being of 0.7 kg net mass or less in a mounded store, Table 8;

(i) of hazard type 2 explosive every item being of 0.7 kg net mass or less in an unmounded store, Table 9

(j) of hazard type 3 explosive, Table 10;

(k) of hazard type 4 explosive, Table 11,

applies, and any reference in this Schedule to a relevant Table is a reference to the Table which applies by virtue of this paragraph.

3

For each of Table 1, 3, 4 or 5 where that Table applies, a Supplementary Table for determining the separation distance between a store and a dwelling in an area of low population density applies as follows—

(a) for Table 1, Supplementary Table 1A;

(b) for Table 3, Supplementary Table 3A;

(c) for Table 4, Supplementary Table 4A; and

(d) for Table 5, Supplementary Table 5A,

and any reference in this Schedule to a relevant Supplementary Table is a reference to the Supplementary Table which applies by virtue of this paragraph.

4

(1) Paragraph 5 is subject to paragraph 6.

(2) Paragraphs 5 and 6 are subject to sub-paragraph (b) of paragraph 7 for determining the quantity of explosives to be applied—

(a) for determining the relevant separation distance for the purposes of paragraphs 5 and 6, and

(b) for paragraph 1(7) in relation to the radius for a reference zone,

in cases where sub-paragraph (b) of paragraph 7 is being relied on.

5

The distance between a store and any protected place of Class A, B, C, D, E, F or H is the distance specified in the entry in the column for that Class of the relevant Table corresponding to the quantity of explosives shown in column 1 of the Table.

6

Where—

(a) Table 1, 3, 4 or 5 applies; and

(b) the number of dwellings in the reference zone is equal to or less than the number specified in the entry in column 3 of the relevant Supplementary Table corresponding to the quantity of explosives shown in column 1 of the Supplementary Table,

the distance between a store and any dwelling is the distance (if any) specified in the corresponding entry in column 4 of the Supplementary Table.

7

The distance between a store and any protected place of Class G is either—

(a) the distance specified in the entry in the column for that Class of the relevant Table corresponding to the quantity of explosives shown in column 1 of the Table; or

(b) any lesser distance, provided that the quantity of explosives for determining the relevant separation distance for the purposes of paragraphs 5 and 6, and for paragraph 1(7) in relation to the radius for a reference zone, is the sum total of the quantity of explosives in the store and the protected place of Class G for which there is to be any lesser distance.

8

For cases where sub-paragraph (b) of paragraph 7 is being relied on, paragraph 1(6) has effect as if the reference to "one store" were a reference to the store and the protected place of Class G to which sub-paragraph (b) of paragraph 7 applies.

Table 1: Hazard Type 1 explosive in a brick-built mounded store

1 Quantity of explosives (kg)	*2 Class A distance (m)*	*3 Class B distance (m)*	*4 Class C distance (m)*	*5 Class D distance (m)*	*6 Class E distance (m)*	*7 Class F distance (m)*	*8 Class G distance (m)*	*9 Class H distance (m)*
0.1-25	33	50	100	100	100	50	9	18
25-30	33	50	100	100	100	50	9	18
30-40	34	51	103	103	103	51	9	18
40-50	35	53	106	106	106	53	9	18
50-60	37	55	110	110	110	55	10	22
60-70	37	56	111	114	114	57	10	22
70-80	38	57	113	118	118	59	11	22
80-90	38	58	115	121	121	61	11	22
90-100	39	59	118	125	125	63	11	22
100-150	43	64	128	142	142	71	13	27
150-200	46	70	139	156	156	78	14	27
200-250	50	75	150	169	169	85	16	30
250-300	54	80	161	170	170	85	16	30
300-350	57	86	172	172	172	86	18	33
350-400	61	91	183	183	183	92	18	33
400-450	64	97	193	193	193	97	19	36
450-500	68	102	204	204	204	102	19	36
500-550	68	102	204	204	204	102	24	56
550-600	68	102	204	204	216	102	24	56
600-650	68	102	204	227	227	113	24	56
650-700	68	102	204	231	238	116	24	56
700-750	68	102	204	235	249	118	24	56
750-800	68	102	204	238	260	119	24	56
800-850	68	102	204	242	270	121	24	56
850-900	68	102	204	245	280	123	24	56
900-950	68	102	204	248	290	124	24	56
950-1000	68	102	204	250	300	125	24	56
1000-1100	68	102	204	255	319	128	30	85
1100-1200	68	102	204	259	337	130	30	85
1200-1300	68	102	204	263	354	132	30	85
1300-1400	68	102	204	266	370	133	30	85
1400-1500	68	102	204	269	386	135	30	85
1500-1600	68	102	204	272	402	136	30	85

1600-1700	69	104	208	274	416	137	30	85
1700-1800	72	108	215	277	431	139	30	85
1800-1900	74	111	222	279	444	140	30	85
1900-2000	76	114	229	281	458	141	30	85
2000-3000	95	143	285	285	570	143	35	106
3000-4000	109	164	328	328	656	164	38	122
4000-5000	121	181	362	362	724	181	41	134
5000-10000	158	237	475	475	950	237	52	176
10000-15000	183	274	548	548	1097	274	59	204
15000-20000	202	303	606	606	1211	303	65	225
20000-25000	218	327	653	653	1306	327	70	243
25000-30000	232	347	695	695	1389	347	75	258
30000-40000	255	383	765	765	1531	384	82	275
40000-50000	275	412	825	825	1649	412	88	295
50000-60000	292	438	877	877	1753	438	94	315
60000-70000	308	461	923	923	1846	461	99	345
70000-80000	322	482	965	965	1930	482	103	345
80000-90000	335	502	1004	1004	2007	502	108	375
90000-100000	347	520	1040	1040	2079	520	111	375

Supplementary Table 1A: Hazard Type 1 explosive in a brick-built mounded store—distances for areas of low population density

1 *Quantity of explosives (kg)*	*2* *Reference zone radius (m)*	*3* *Maximum number of dwellings in the reference zone*	*4* *Distances to dwellings if the maximum number of dwellings in the reference zone is not exceeded (m)*
0.1-60	-	-	-
60-70	222	61	111
70-80	227	63	113
80-90	231	66	115
90-100	235	68	118
100-150	257	81	128
150-200	279	96	139

200-250	300	128	150
250-600	-	-	-
600-1600	408	206	204
1600-1700	416	214	208
1700-1800	431	229	215
1800-1900	444	244	222
1900-2000	458	259	229

Table 2: Hazard Type 1 explosives in a brick-built unmounded store

1 Quantity of explosives (kg)	*2 Class A distance (m)*	*3 Class B distance (m)*	*4 Class C distance (m)*	*5 Class D distance (m)*	*6 Class E distance (m)*	*7 Class F distance (m)*	*8 Class G distance (m)*	*9 Class H distance (m)*
0.1-25	47	70	141	141	141	70	141	141
25-30	48	72	144	144	144	72	144	144
30-40	50	76	151	151	151	76	151	151
40-50	53	80	159	159	159	80	159	159
50-60	56	84	168	168	168	84	168	168
60-70	59	88	176	176	176	88	176	176
70-80	61	92	184	184	184	92	184	184
80-90	64	96	191	191	191	96	191	191
90-100	66	99	199	199	199	99	199	199
100-150	77	115	230	230	230	115	230	230
150-200	85	128	256	256	256	128	256	256
200-250	92	138	276	276	276	138	276	276
250-300	98	147	293	293	293	147	293	293
300-350	103	154	308	308	308	154	308	308
350-400	107	160	320	320	320	160	320	320
400-450	110	165	331	331	331	165	331	331
450-500	113	170	340	340	340	170	340	340
500-550	116	174	348	348	348	174	348	348
550-600	118	178	355	355	355	178	355	355
600-650	120	181	361	361	361	181	361	361
650-700	122	184	367	367	367	184	367	367
700-750	124	186	372	372	372	186	372	372
750-800	126	189	377	377	377	189	377	377
800-850	127	191	381	381	381	191	381	381
850-900	128	193	385	385	385	193	385	385
900-950	130	194	389	389	389	194	389	389
950-1000	131	196	392	392	392	196	392	392
1000-1100	133	199	398	398	398	199	398	398
1100-1200	134	202	403	403	403	202	403	403
1200-1300	136	204	408	408	408	204	408	408
1300-1400	137	206	412	412	412	206	412	412

1400-1500	138	208	415	415	415	208	415	415
1500-1600	139	209	418	418	418	209	418	418
1600-1700	140	211	421	421	421	211	421	421
1700-1800	141	212	424	424	431	212	424	424
1800-1900	142	213	426	426	444	213	426	426
1900-2000	143	214	428	428	458	214	428	428
2000-3000	147	221	442	442	570	221	442	442
3000-4000	150	225	449	449	656	225	449	449
4000-5000	151	227	454	454	724	227	454	454
5000-10000	167	251	502	502	950	251	502	502
10000-15000	185	277	554	554	1097	277	554	554
15000-20000	202	303	606	606	1211	303	606	606
20000-25000	218	327	653	653	1306	327	653	653
25000-30000	232	347	695	695	1389	347	695	695
30000-40000	255	383	765	765	1531	383	765	765
40000-50000	275	412	825	825	1649	412	825	825
50000-60000	292	438	877	877	1753	438	877	877
60000-70000	308	461	923	923	1846	461	923	923
70000-80000	322	482	965	965	1930	482	965	965
80000-90000	335	502	1004	1004	2007	502	1004	1004
90000-10000	347	520	1040	1040	2079	520	1040	1040

Table 3: Hazard Type 1 explosive in a metal-built mounded store

1 Quantity of explosives (kg)	*2 Class A distance (m)*	*3 Class B distance (m)*	*4 Class C distance (m)*	*5 Class D distance (m)*	*6 Class E distance (m)*	*7 Class F distance (m)*	*8 Class G distance (m)*	*9 Class H distance (m)*
0.1-10	7	10	21	23	40	12	9	18
10-20	9	13	26	29	42	15	9	18
20-30	10	15	30	33	44	17	9	18
30-40	11	16	33	37	46	18	9	18
40-50	12	18	35	40	48	20	9	18
50-60	13	19	38	42	48	21	10	22
60-70	13	20	40	44	52	22	10	22
70-80	14	20	41	46	57	23	11	22

80-90	14	21	42	47	61	24	11	22
90-100	14	21	43	48	66	24	11	22
100-150	16	24	49	55	86	28	13	27
150-200	18	27	54	62	104	31	14	27
200-250	20	30	60	69	121	35	16	30
250-300	23	34	68	76	136	38	16	30
300-350	25	38	76	83	151	44	18	33
350-400	28	41	83	90	165	45	18	33
400-450	30	45	89	97	178	49	19	36
450-500	32	48	96	102	191	51	19	36
500-550	34	51	102	107	204	54	24	56
550-600	36	54	108	111	216	56	24	56
600-650	38	57	114	116	227	58	24	56
650-700	40	60	119	121	238	61	24	56
700-750	42	62	125	126	249	63	24	56
750-800	43	65	130	131	260	66	24	56
800-850	45	68	135	136	270	68	24	56
850-900	47	70	140	140	280	70	24	56
900-950	48	73	145	145	290	73	24	56
950-1000	50	75	150	150	300	75	24	56
1000-1100	53	80	159	159	319	80	30	85
1100-1200	56	84	168	168	337	84	30	85
1200-1300	59	88	177	177	354	89	30	85
1300-1400	62	93	185	185	370	93	30	85
1400-1500	64	97	193	193	386	97	30	85
1500-1600	67	100	201	201	402	101	30	85
1600-1700	69	104	208	208	416	104	30	85
1700-1800	72	108	215	215	431	108	30	85
1800-1900	74	111	222	222	444	111	30	85
1900-2000	76	114	229	229	458	115	30	85
2000-3000	95	143	285	285	570	143	35	106
3000-4000	109	164	328	328	656	164	38	122
4000-5000	121	181	362	362	724	181	41	134
5000-10000	158	237	475	475	950	238	52	176
10000-15000	183	274	548	548	1097	274	59	204
15000-20000	202	303	606	606	1211	303	65	225
20000-25000	218	327	653	653	1306	327	70	243
25000-30000	232	347	695	695	1389	348	75	258
30000-40000	255	383	765	765	1531	383	82	275
40000-50000	275	412	825	825	1649	412	88	295
50000-60000	292	438	877	877	1753	438	94	315

60000-70000	308		461	923	923	1846	461	99	345
70000-80000	322		482	965	965	1930	482	103	345
80000-90000	335		502	1004	1004	2007	502	108	375
90000-100000	347		520	1040	1040	2079	520	111	375

Supplementary Table 3A: Hazard type 1 explosive in a metal-built mounded store—distances for areas of low population density

1 *Quantity of explosives (kg)*	*2* *Reference zone radius (m)*	*3* *Maximum number of dwellings in the reference zone*	*4* *Distances to dwellings if the maximum number of dwellings in the reference zone is not exceeded (m)*
0.1-10	41	2	21
10-20	52	3	26
20-30	60	4	30
30-40	66	5	33
40-50	71	6	35
50-60	75	7	38
60-70	79	8	40
70-80	81	8	41
80-90	83	8	42
90-100	86	9	43
100-150	97	12	49
150-200	109	14	54
200-250	121	18	60
250-300	136	23	68
300-350	151	28	76
350-400	165	34	83
400-450	178	39	89
450-500	191	45	96
500-550	204	51	102
550-600	216	57	108
600-650	227	63	114
650-700	238	70	119
700-750	249	77	125
750-800	260	83	130
800-850	270	90	135

Table 4: Hazard Type 1 explosive in a metal built unmounded store with no detonator annex attached

1 Quantity of explosives (kg)	*2 Class A distance (m)*	*3 Class B distance (m)*	*4 Class C distance (m)*	*5 Class D distance (m)*	*6 Class E distance (m)*	*7 Class F distance (m)*	*8 Class G distance (m)*	*9 Class H distance (m)*
0.1-10	8	11	23	30	40	15	30	30
10-20	10	14	29	35	42	18	35	35
20-30	11	16	33	39	44	20	39	39
30-40	12	18	36	42	46	21	42	42
40-50	13	19	38	44	48	22	44	44
50-60	13	20	40	46	48	23	46	46
60-70	14	21	42	48	52	24	48	48
70-80	14	22	43	50	57	25	50	50
80-90	15	22	44	52	61	26	52	52
90-100	15	23	45	55	66	28	55	55
100-150	17	25	50	66	86	33	66	66
150-200	18	28	55	78	104	39	78	78
200-250	20	30	60	89	121	45	89	89
250-300	23	34	68	101	136	51	101	101
300-350	25	38	76	112	151	56	112	112
350-400	28	41	83	124	165	62	124	124
400-450	30	45	89	135	178	68	135	135
450-500	32	48	96	138	191	69	138	138
500-550	34	51	102	141	204	71	141	141
550-600	36	54	108	144	216	72	144	144
600-650	38	57	114	147	227	74	147	147
650-700	40	60	119	150	238	75	150	150
700-750	42	62	125	153	249	77	153	153
750-800	43	65	130	156	260	78	156	156
800-850	45	68	135	159	270	80	159	159
850-900	47	70	140	162	280	81	162	162
900-950	48	73	145	165	290	83	165	165
950-1000	50	75	150	168	300	84	168	168
1000-1100	53	80	159	175	319	88	175	175
1100-1200	56	84	168	181	337	91	181	181
1200-1300	59	88	177	187	354	94	187	187
1300-1400	62	93	185	193	370	97	193	193
1400-1500	64	97	193	199	386	100	199	199
1500-1600	67	100	201	205	402	103	205	205
1600-1700	69	104	208	211	416	106	211	211
1700-1800	72	108	215	217	431	108	217	217
1800-1900	74	111	222	223	444	111	223	223
1900-2000	76	114	229	229	458	114	229	229
2000-3000	95	143	285	285	570	143	285	285

3000-4000	109	164	328	328	656	164	328	328
4000-5000	121	181	362	362	724	181	362	362
5000-10000	158	237	475	475	950	237	475	475
10000-15000	183	274	548	548	1097	274	548	548
15000-20000	202	303	606	606	1211	303	606	606
20000-25000	218	327	653	653	1306	653	653	653
25000-30000	232	347	695	695	1389	347	695	695
30000-40000	255	383	765	765	1531	383	765	765
40000-50000	275	412	825	825	1649	412	825	825
50000-60000	292	438	877	877	1753	438	877	877
60000-70000	308	461	923	923	1846	461	923	923
70000-80000	322	482	965	965	1930	482	965	965
80000-90000	335	502	1004	1004	2007	502	1004	1004
90000-100000	347	520	1040	1040	2079	520	1040	1040

Supplementary Table 4A: Hazard Type 1 explosive in a metal-built unmounded store with no detonator annex attached—distances for areas of low population density

1 *Quantity of explosives (kg)*	2 *Reference zone radius (m)*	3 *Maximum number of dwellings in the reference zone*	4 *Distances to dwellings if the maximum number of dwellings in the reference zone is not exceeded (m)*
0.1-10	46	3	23
10-20	57	4	29
20-30	65	5	33
30-40	71	6	36
40-50	76	7	38
50-60	80	8	40
60-70	84	9	42
70-80	87	9	43
80-90	89	10	44
90-100	91	10	45
100-150	100	12	50
150-200	110	15	55

200-250	121	18	60
250-300	136	23	68
300-350	151	28	76
350-400	165	34	83
400-450	178	39	89
450-500	191	45	96
500-550	204	51	102
550-600	216	57	108
600-650	227	64	114
650-700	238	70	119
700-750	249	77	125
750-800	260	84	130
800-850	270	90	135
850-900	280	97	140
900-950	290	104	145
950-1000	300	111	150
1000-1100	319	125	159

Table 5: Hazard type 1 explosive in a metal-built un-mounded store with a detonator annex attached

1 *Quantity of explosives (kg)*	*2* *Class A distance (m)*	*3* *Class B distance (m)*	*4* *Class C distance (m)*	*5* *Class D distance (m)*	*6* *Class E distance (m)*	*7* *Class F distance (m)*	*8* *Class G distance (m)*	*9* *Class H distance (m)*
0.1-10	10	16	31	48	48	24	48	48
10-20	12	18	36	50	50	25	50	50
20-30	13	19	38	52	52	26	52	52
30 40	14	21	41	54	54	27	54	54
40-50	14	22	43	56	56	28	56	56
50-60	15	22	45	58	58	29	58	58
60-70	16	24	47	60	60	30	60	60
70-80	16	25	49	63	63	32	63	63
80-90	17	26	51	70	70	35	70	70
90-100	18	27	53	77	77	39	77	77
100-150	21	32	63	110	110	55	110	110
150-200	25	37	74	143	143	72	143	143
200-250	28	42	84	176	176	88	176	176
250-300	31	47	94	209	209	105	209	209
300-350	35	52	104	242	242	121	242	242
350-400	38	57	115	275	275	138	275	275
400-450	42	62	125	308	308	154	308	308
450-500	43	64	128	309	309	155	309	309
500-550	44	66	132	310	310	155	310	310
550-600	45	68	135	311	311	156	311	311

600-650	46	69	138	311	311	156	311	311
650-700	47	71	142	312	312	156	312	312
700-750	48	73	145	313	313	157	313	313
750-800	49	74	148	314	314	157	314	314
800-850	51	76	152	315	315	158	315	315
850-900	52	78	155	316	316	158	316	316
900-950	53	79	159	317	317	159	317	317
950-1000	54	81	162	318	318	159	318	318
1000-1100	56	84	169	319	319	160	319	319
1100-1200	58	88	175	321	337	161	321	321
1200-1300	61	91	182	323	354	162	323	323
1300-1400	63	94	189	325	370	163	325	325
1400-1500	65	98	195	326	386	163	326	326
1500-1600	67	101	202	328	402	164	328	328
1600-1700	70	104	209	330	416	165	330	330
1700-1800	72	108	215	332	431	166	332	332
1800-1900	74	111	222	333	444	167	333	333
1900-2000	76	114	229	335	458	168	335	335
2000-3000	95	143	285	353	570	177	353	353
3000-4000	109	164	328	370	656	185	370	370
4000-5000	121	181	362	388	724	194	388	388
5000-10000	158	237	475	475	950	237	475	475
10000-15000	183	274	548	548	1097	274	548	548
15000-20000	202	303	606	606	1211	303	606	606
20000-25000	218	327	653	653	1306	327	653	653
25000-30000	232	347	695	695	1389	347	695	695
30000-40000	255	383	765	765	1531	383	765	765
40000-50000	275	412	825	825	1649	412	825	825
50000-60000	292	438	877	877	1753	438	877	877
60000-70000	308	461	923	923	1846	461	923	923
70000-80000	322	482	965	965	1930	482	965	965
80000-90000	335	502	1004	1004	2007	502	1004	1004
90000-100000	347	520	1040	1040	2079	520	1040	1040

Supplementary Table 5A:—Hazard Type 1 explosive in a metal-built un-mounded store with a detonator annex attached—distances for areas of low population density

1 *Quantity of Explosives (kg)*	*2* *Reference zone radius (m)*	*3* *Maximum number of dwellings in the reference zone*	*4* *Distances to dwellings if the maximum number of dwellings in the reference zone is not exceeded (m)*
0.1 -10	62	5	31
10-20	71	6	36
20-30	77	7	38
30-40	82	8	41
40-50	86	9	43
50-60	90	10	45
60-70	94	11	47
70-80	98	12	49
80-90	102	13	51
90-100	106	14	53
100-150	127	20	63
150-200	147	27	74
200-250	168	35	84
250-300	188	44	94
300-350	209	54	104
350-400	229	65	115
400-450	250	77	125
450-500	257	81	128
500-550	263	85	132
550-600	270	90	135
600-650	277	95	138
650-700	284	99	142
700-750	290	104	145
750-800	297	109	148
800-850	304	114	152
850-900	310	119	155
900-950	317	124	159
950-1000	324	129	162
1000-1100	337	140	169
1100-1200	351	152	175
1200-1300	364	163	182
1300-1400	377	176	189
1400-1500	391	188	195
1500-1600	404	202	202
1600-1700	418	215	209
1700-1800	431	229	215
1800-1900	444	244	222
1900-2000	458	259	229
2000-3000	570	401	285

3000-4000	656	530	328
4000-5000	724	647	362

Table 6: Hazard Type 2 explosive, some or all items being of more than 0.7kg net mass in a mounded store

1 Quantity of explosives (kg)	*2 Class A distance (m)*	*3 Class B distance (m)*	*4 Class C distance (m)*	*5 Class D distance (m)*	*6 Class E distance (m)*	*7 Class F distance (m)*	*8 Class G distance (m)*	*9 Class H distance (m)*
0.1 - 25	20	30	60	60	120	30	9	32
25-30	20	30	60	60	120	30	9	32
30-40	25	37	74	74	148	37	9	32
40-50	29	44	88	88	176	44	9	32
50-60	33	49	99	99	198	49	9	46
60-70	36	54	108	108	216	54	9	46
70-80	39	58	116	116	232	58	9	46
80-90	41	61	123	123	246	61	9	46
90-100	43	64	129	129	258	64	9	46
100-150	51	76	152	152	304	76	9	61
150-200	56	84	168	168	336	84	9	61
200-250	60	91	181	181	362	91	9	69
250-300	64	96	191	191	382	96	9	69
300-350	67	100	200	200	400	100	10	75
350-400	69	104	207	207	414	104	10	75
400-450	71	107	213	213	426	107	10	79
450-500	73	110	219	219	438	110	10	79
500-550	75	112	224	224	448	112	10	92
550-600	76	115	229	229	458	115	10	92
600-650	78	117	233	233	466	117	10	92
650-700	79	119	237	237	474	119	10	92
700-750	80	120	241	241	482	120	10	92
750-800	81	122	244	244	488	122	10	92
800-850	83	124	248	248	496	124	10	92
850-900	84	125	251	251	502	125	10	92
900-950	84	127	253	253	506	127	10	92
950-1000	85	128	256	256	512	128	10	92
1000-1100	87	131	261	261	522	131	11	105
1100-1200	89	133	266	266	532	133	11	105
1200-1300	90	135	270	270	540	135	11	105
1300-1400	91	137	274	274	548	137	11	105
1400-1500	92	139	277	277	554	139	11	105
1500-1600	93	140	280	280	560	140	11	105
1600- 1700	95	142	284	284	568	142	11	105
1700-1800	95	143	286	286	572	143	11	105

1800-1900	96	145	289	289	578	145	11	105
1900-2000	97	146	292	292	584	146	11	105
2000-3000	104	156	312	312	624	156	12	112
3000-4000	109	163	326	326	652	163	13	117
4000-5000	112	168	337	337	674	168	14	121
5000-10000	123	185	370	370	740	185	17	133
10000-15000	129	194	388	388	776	194	20	140
15000-20000	134	201	401	401	802	201	22	144
20000-25000	137	206	411	411	822	206	24	148
25000-30000	140	210	419	419	838	210	26	151
30000-40000	144	216	431	431	862	216	29	156
40000-50000	147	220	441	441	882	220	32	159
50000-60000	150	224	449	449	898	224	35	162
60000-70000	152	227	455	455	910	227	36	166
70000-80000	153	230	460	460	920	230	38	166
80000-90000	155	233	465	465	930	233	39	170
90000-100000	157	235	470	470	940	235	40	170

Table 7: Hazard Type 2 explosive, some or all items being of more than 0.7 kg net mass in an unmounded store

1 *Quantity of explosives (kg)*	2 *Class A distance (m)*	3 *Class B distance (m)*	4 *Class C distance (m)*	5 *Class D distance (m)*	6 *Class E distance (m)*	7 *Class F distance (m)*	8 *Class G distance (m)*	9 *Class H distance (m)*
0.1-25	20	30	60	60	120	30	60	60
25-30	20	30	60	60	120	30	60	60
30-40	25	37	74	74	148	37	74	74
40-50	29	44	88	88	176	44	88	88
50-60	33	49	99	99	198	49	99	99
60-70	36	54	108	108	216	54	108	108
70-80	39	58	116	116	232	58	116	116
80-90	41	61	123	123	246	61	123	123
90-100	43	64	129	129	258	64	129	129
100-150	51	76	152	152	304	76	152	152
150-200	56	84	168	168	336	84	168	168
200-250	60	91	181	181	362	91	181	181

250-300	64	96	191	191	382	96	191	191
300-350	67	100	200	200	400	100	200	200
350-400	69	104	207	207	414	104	207	207
400-450	71	107	213	213	426	107	213	213
450-500	73	110	219	219	438	110	219	219
500-550	75	112	224	224	448	112	224	224
550-600	76	115	229	229	458	115	229	229
600-650	78	117	233	233	466	117	233	233
650-700	79	119	237	237	474	119	237	237
700-750	80	120	241	241	482	120	241	241
750-800	81	122	244	244	488	122	244	244
800-850	83	124	248	248	496	124	248	248
850-900	84	125	251	251	502	125	251	251
900-950	84	127	253	253	506	127	253	253
950-1000	85	128	256	256	512	128	256	256
1000-1100	87	131	261	261	522	131	261	261
1100-1200	89	133	266	266	532	133	266	266
1200-1300	90	135	270	270	540	135	270	270
1300-1400	91	137	274	274	548	137	274	274
1400-1500	92	139	277	277	554	139	277	277
1500-1600	93	140	280	280	560	140	280	280
1600-1700	95	142	284	284	568	142	284	284
1700-1800	95	143	286	286	572	143	286	286
1800-1900	96	145	289	289	578	145	289	289
1900-2000	97	146	292	292	584	146	292	292
2000-3000	104	156	312	312	624	156	312	312
3000-4000	109	163	326	326	652	163	326	326
4000-5000	112	168	337	337	674	168	337	337
5000-10000	123	185	370	370	740	185	370	370
10000-15000	129	194	388	388	776	194	388	388
15000-20000	134	201	401	401	802	201	401	401
20000-25000	137	206	411	411	822	206	411	411
25000-30000	140	210	419	419	838	210	419	419
30000-40000	144	216	431	431	862	216	431	431
40000-50000	147	220	441	441	882	220	441	441
50000-60000	150	224	449	449	898	224	449	449
60000-70000	152	227	455	455	910	227	455	455
70000-80000	153	230	460	460	920	230	460	460
80000-90000	155	233	465	465	930	233	465	465

90000-100000	157	235	470	470	940	235	470	470

Table 8: Hazard Type 2 explosive every item being of 0.7kg net mass or less in a mounded store

1 Quantity of explosives (kg)	*2 Class A distance (m)*	*3 Class B distance (m)*	*4 Class C distance (m)*	*5 Class D distance (m)*	*6 Class E distance (m)*	*7 Class F distance (m)*	*8 Class G distance (m)*	*9 Class H distance (m)*
0.1-25	12	18	37	37	74	18	9	15
25-30	13	19	38	38	76	19	9	15
30-40	14	20	41	41	82	20	9	15
40-50	14	22	43	43	86	22	9	15
50-60	15	22	45	45	90	22	9	18
60-70	16	23	47	47	94	23	9	18
70-80	16	24	48	48	96	24	9	18
80-90	16	25	49	49	98	25	9	18
90-100	17	25	51	51	102	25	9	18
100-150	19	28	56	56	112	28	9	22
150-200	20	30	60	60	120	30	9	22
200-250	21	32	63	63	126	32	9	24
250-300	22	33	66	66	132	33	9	24
300-350	23	34	68	68	136	34	10	25
350-400	24	35	71	71	142	35	10	25
400-450	24	36	73	73	146	36	10	27
450-500	25	37	74	74	148	37	10	27
500-550	25	38	76	76	152	38	10	31
550-600	26	39	78	78	156	39	10	31
600 650	26	39	79	79	158	39	10	31
650-700	27	40	80	80	160	40	10	31
700-750	27	41	82	82	164	41	10	31
750-800	28	41	83	83	166	41	10	31
800-850	28	42	84	84	168	42	10	31
850-900	28	43	85	85	170	43	10	31
900-950	29	43	86	86	172	43	10	31
950-1000	29	44	87	87	174	44	10	31
1000-1100	30	44	89	89	178	44	11	36
1100-1200	30	45	91	91	182	45	11	36
1200-1300	31	46	92	92	184	46	11	36
1300-1400	31	47	94	94	188	47	11	36
1400-1500	32	48	95	95	190	48	11	36
1500-1600	32	48	97	97	194	48	11	36
1600-1700	33	49	98	98	196	49	11	36
1700-1800	33	50	99	99	198	50	11	36

1800-1900	33	50	100	100	200	50	11	36
1900-2000	34	51	101	101	202	51	11	36
2000-3000	37	55	110	110	220	55	12	40
3000-4000	39	59	117	117	234	59	13	42
4000-5000	41	61	122	122	244	61	14	44
5000-10000	47	70	140	140	280	70	17	50
10000-15000	50	76	151	151	302	76	20	54
15000-20000	53	80	159	159	318	80	22	57
20000-25000	55	83	166	166	332	83	24	60
25000-30000	57	86	171	171	342	86	26	62
30000-40000	60	90	180	180	360	90	29	66
40000-50000	62	94	187	187	377	94	32	68
50000-60000	64	97	193	193	386	97	35	70
60000-70000	66	99	198	198	396	99	36	74
70000-80000	67	101	202	202	404	101	38	74
80000-90000	69	103	206	206	412	103	39	76
90000-100000	70	105	210	210	420	105	40	76

Table 9: Hazard Type 2 explosive every item being of 0.7kg net mass or less in an unmounded store

1 Quantity of explosives (kg)	*2 Class A distance (m)*	*3 Class B distance (m)*	*4 Class C distance (m)*	*5 Class D distance (m)*	*6 Class E distance (m)*	*7 Class F distance (m)*	*8 Class G distance (m)*	*9 Class H distance (m)*
0.1-25	12	18	37	37	74	18	37	37
25-30	13	19	38	38	76	19	38	38
30-40	14	20	41	41	82	20	41	41
40-50	14	22	43	43	86	22	43	43
50-60	15	22	45	45	90	22	45	45
60-70	16	23	47	47	94	23	47	47
70-80	16	24	48	48	96	24	48	48
80-90	16	25	49	49	98	25	49	49
90-100	17	25	51	51	102	25	51	51
100-150	19	28	56	56	112	28	56	56
150-200	20	30	60	60	120	30	60	60
200-250	21	32	63	63	126	32	63	63

250-300	22	33	66	66	132	33	66	66
300-350	23	34	68	68	136	34	68	68
350-400	24	35	71	71	142	35	71	71
400-450	24	36	73	73	146	36	73	73
450-500	25	37	74	74	148	37	74	74
500-550	25	38	76	76	152	38	76	76
550-600	26	39	78	78	156	39	78	78
600-650	26	39	79	79	158	39	79	79
650-700	27	40	80	80	160	40	80	80
700-750	27	41	82	82	164	41	82	82
750-800	28	41	83	83	166	41	83	83
800-850	28	42	84	84	168	42	84	84
850-900	28	43	85	85	170	43	85	85
900-950	29	43	86	86	172	43	86	86
950-1000	29	44	87	87	174	44	87	87
1000-1100	30	44	89	89	178	44	89	89
1100-1200	30	45	91	91	182	45	91	91
1200-1300	31	46	92	92	184	46	92	92
1300-1400	31	47	94	94	188	47	94	94
1400-1500	32	48	95	95	190	48	95	95
1500-1600	32	48	97	97	194	48	97	97
1600-1700	33	49	98	98	196	49	98	98
1700-1800	33	50	99	99	198	50	99	99
1800-1900	33	50	100	100	200	50	100	100
1900-2000	34	51	101	101	202	51	101	101
2000-3000	37	55	110	110	220	55	110	110
3000-4000	39	59	117	117	234	59	117	117
4000-5000	41	61	122	122	244	61	122	122
5000-10000	47	70	140	140	280	70	140	140
10000-15000	50	76	151	151	302	76	151	151
15000-20000	53	80	159	159	318	80	159	159
20000-25000	55	83	166	166	332	83	166	166
25000-30000	57	86	171	171	342	86	171	171
30000-40000	60	90	180	180	360	90	180	180
40000-50000	62	94	187	187	377	94	187	187
50000-60000	64	97	193	193	386	97	193	193
60000-70000	66	99	198	198	396	99	198	198
70000-80000	67	101	202	202	404	101	202	202
80000-90000	69	103	206	206	412	103	206	206

90000-100000	70	105	210	210	420	105	210	210

Table 10: Hazard Type 3 explosive

1 Quantity of explosives (kg)	2 Class A distance (m)	3 Class B distance (m)	4 Class C distance (m)	5 Class D distance (m)	6 Class E distance (m)	7 Class F distance (m)	8 Class G distance (m)	9 Class H distance (m)
0.1-25	0	0	0	0	0	0	9	12
25-30	7	10	20	20	20	10	9	12
30-40	7	11	22	22	22	11	9	12
40-50	8	12	23	23	23	12	9	12
50-60	8	12	25	25	25	12	9	15
60-70	9	13	26	26	26	13	9	15
70-80	9	14	27	27	27	14	9	15
80-90	9	14	28	28	28	14	9	15
90-100	10	15	29	29	29	15	9	15
100-150	11	17	34	34	34	17	9	18
150-200	12	19	37	37	37	19	9	18
200-250	13	20	40	40	40	20	10	21
250-300	14	21	43	43	43	21	10	21
300-350	15	22	45	45	45	22	11	23
350-400	16	23	47	47	47	23	11	23
400-450	16	24	49	49	49	24	12	25
450-500	17	25	50	50	50	25	12	25
500-550	17	26	52	52	52	26	14	32
550-600	18	27	54	54	54	27	14	32
600-650	18	28	55	55	55	28	14	32
650-700	19	28	56	56	56	28	14	32
700-750	19	29	58	58	58	29	14	32
750-800	20	29	59	59	59	29	14	32
800-850	20	30	60	60	60	30	14	32
850-900	20	31	61	61	61	31	14	32
900-950	21	31	62	62	62	31	14	32
950-1000	21	32	64	64	64	32	14	32
1000-1100	22	33	66	66	66	33	17	40
1100-1200	22	34	67	67	67	34	17	40
1200-1300	23	35	69	69	69	35	17	40
1300-1400	24	36	71	71	71	36	17	40
1400-1500	24	36	73	73	73	36	17	40
1500-1600	25	37	74	74	74	37	17	40
1600-1700	25	38	76	76	76	38	17	40
1700-1800	26	39	77	77	77	39	17	40
1800-1900	26	39	79	79	79	39	17	40

1900-2000	27	40	80	80	80	40	17	40
2000-3000	31	46	92	92	92	46	19	46
3000-4000	34	50	101	101	101	50	20	50
4000-5000	36	54	109	109	109	54	22	54
5000-10000	46	68	137	137	137	68	28	68
10000-15000	52	78	157	157	157	78	33	78
15000-20000	57	86	172	172	172	86	37	86
20000-25000	62	93	186	186	186	93	41	93
25000-30000	66	99	197	197	197	99	44	98
30000-40000	72	109	217	217	217	109	47	110
40000-50000	78	117	234	234	234	117	50	120
50000-60000	83	124	249	249	249	124	54	130
60000-70000	87	131	262	262	262	131	58	140
70000-80000	91	137	274	274	274	137	63	140
80000-90000	95	142	285	285	285	142	67	150
90000-100000	98	147	295	295	295	147	70	150

Table 11: Hazard Type 4 explosive

1 Quantity of explosives (kg)	*2 Class A distance (m)*	*3 Class B distance (m)*	*4 Class C distance (m)*	*5 Class D distance (m)*	*6 Class E distance (m)*	*7 Class F distance (m)*	*8 Class G distance (m)*	*9 Class H distance (m)*
0.1-250	0	0	0	0	0	0	9	11
250-300	1	1	1	1	1	1	9	11
300-340	1	1	2	2	2	1	9	11
340-370	1	2	3	3	3	2	9	11
370-400	1	2	4	4	4	2	9	11
400-450	2	3	5	5	5	3	9	11
450-500	2	3	6	6	6	3	9	11
500-550	2	4	7	7	7	4	9	12
550-650	3	5	9	9	9	5	9	12
650-700	3	5	10	10	10	5	9	12
700-750	4	6	11	11	11	6	9	12
750-800	4	6	12	12	12	6	9	12
800-900	4	7	13	13	13	7	9	12

900-950	5	7	14	14	14	7	9	12
950-1000	5	8	15	15	15	8	9	12
1000-1100	5	8	16	16	16	8	12	19
1100-1150	6	9	17	17	17	9	12	19
1150-1200	6	9	18	18	18	9	12	19
1200-1300	6	10	19	19	19	10	12	19
1300-1350	7	10	20	20	20	10	12	19
1350-1400	7	11	21	21	21	11	12	19
1400-1450	7	11	22	22	22	11	12	19
1450-1550	8	12	23	23	23	12	12	19
1550-1600	8	12	24	24	24	12	12	19
1600-1650	8	13	25	25	25	13	12	19
1650-1700	9	13	26	26	26	13	12	19
1700-1800	9	14	27	27	27	14	12	19
1800-1850	9	14	28	28	28	14	12	19
1850-1900	10	15	29	29	29	15	12	19
1900-2000	10	15	30	30	30	15	12	19
2000-3000	13	20	40	40	40	20	14	23
3000-4000	13	20	40	40	40	20	16	24
4000-5000	15	23	45	45	45	23	17	25
5000-10000	17	26	51	51	51	26	22	27
10000-15000	18	27	54	54	54	27	24	27
15000-20000	18	28	56	56	56	28	25	27
20000-25000	19	29	57	57	57	29	26	27
25000-30000	20	30	59	59	59	30	27	27
30000-40000	20	30	60	60	60	30	27	27
40000-50000	20	31	61	61	61	31	27	27
50000-60000	20	31	61	61	61	31	27	27
60000-70000	21	31	62	62	62	31	27	27
70000-80000	21	32	63	63	63	32	27	27
80000-90000	21	32	63	63	63	32	27	27
90000-100000	21	32	64	64	64	32	27	27

SCHEDULE 6

UNIQUE IDENTIFICATION FOR CIVIL EXPLOSIVES

Regulation 33(2)(a)

[8.695]

1

Subject to paragraph 2, the unique identification must comprise—

(a) a part which can be read by a human being containing the following—

(i) the name of the manufacturer;

(ii) an alphanumeric code containing—

(aa) two letters identifying the EEA State (place of production or import onto the market of the EEA States);

(bb) three digits identifying the site of manufacture; and

(cc) the unique product code and logistical information designed by the manufacturer; and

(b) a part which can be read electronically in barcode or matrix code format, or both, which relates directly to the alphanumeric identification code.

2

For articles too small to affix the unique product code and logistical information designed by the manufacturer, the information under sub-paragraphs (a)(ii)(aa) and (ii)(bb) and (b) of paragraph 1 is sufficient for the purposes of the unique identification.

SCHEDULE 7

MAKING OR AFFIXING THE UNIQUE IDENTIFICATION TO CIVIL EXPLOSIVES

Regulation 33(1), (5) and (7)

[8.695A]

Cartridged explosives and explosives in sacks

1

For a cartridged explosive and any explosive in sacks—

(a) subject to paragraph 10(1), the unique identification must be on an adhesive label attached to, or be directly printed on, each cartridge or sack;

(b) an associated label must be placed on each case of cartridges; and

(c) a passive inert electronic tag may be attached to each cartridge or sack and an associated electronic tag attached to each case of cartridges.

Packaged two-component explosives

2

Subject to paragraph 10(1), for a packaged two-component explosive, the unique identification must be on an adhesive label attached to, or be directly printed on, each smallest packaging unit containing the two components.

Plain detonators

3

For plain detonators—

(a) subject to paragraph 10(1) and (2), the unique identification must be on an adhesive label attached to, or be directly printed or stamped on, the detonator shell;

(b) an associated label must be placed on each case of detonators; and

(c) a passive inert electronic tag may be attached to each detonator and an associated tag attached to each case of detonators.

Electric, non-electric and electronic detonators

4

For electric, non-electric and electronic detonators—

(a) subject to paragraph 10(1), the unique identification must—

(i) be on an adhesive label attached to the wires or tube; or

(ii) be on an adhesive label attached to, or be directly printed or stamped on, the detonator shell;

(b) an associated label must be placed on each case of detonators; and

(c) a passive inert electronic tag may be attached to each detonator and an associated tag attached to each case of detonators.

Primers and boosters

5

For primers and boosters—

(a) subject, in the case of boosters, to paragraph 10(1) and (2), the unique identification must be on an adhesive label attached to, or be directly printed on, the primer or booster;

(b) an associated label must be placed on each case of primers or boosters; and

(c) a passive inert electronic tag may be attached to each primer or booster and an associated tag attached to each case of primers or boosters.

Detonating cords

6

For detonating cords—

(a) the unique identification must be on an adhesive label attached to, or be directly printed on, the bobbin;

(b) subject to paragraph 10(1) and (3), the unique identification must be marked every five meters on either the external envelope of the cord or the plastic extruded inner layer immediately under the exterior fibre of the cord;

(c) an associated label must be placed on each case of detonating cord; and

(d) a passive inert electronic tag may be inserted within the cord and an associated tag attached to each case of cord.

Cans, boxes and drums containing explosives

7

For cans, boxes and drums containing any explosive—

(a) subject to paragraph 10(1), the unique identification must be on an adhesive label attached to, or be directly printed on, the can, box or drum containing the explosive; and

(b) a passive inert electronic tag may be attached to each can, box and drum.

Civil explosives not referred to in paragraphs 1 to 7

8

Each civil explosive item in respect of a civil explosive not referred to in paragraphs 1 to 7 must, subject to paragraph 10(1), be marked with the unique identification.

General

9

Where adhesive detachable copies of labels showing the unique identification are attached to civil explosive articles, containers or each smallest packaging unit concerned for use by other persons, those copies must be clearly marked as copies of the original.

Small explosive articles

10

(1) For articles smaller than those to which paragraph 2 of Schedule 6 applies, which are too small to affix the information under sub-paragraphs (a)(ii)(aa) and (ii)(bb) and (b) of paragraph 1 of Schedule 6 or where it is technically impossible due to their shape or design to affix a unique identification—

(a) the unique identification specified in paragraph 1 of that Schedule must be affixed on each smallest packaging unit; and

(b) that packaging unit must be closed with a seal.

(2) In the case of each plain detonator or booster to which sub-paragraph (1) applies—

(a) the requirements of, respectively, paragraphs 3(a) and 5(a) do not apply;

(b) the information under sub-paragraph (a)(ii)(aa) and (bb) of paragraph 1 of Schedule 6 must be marked, in a durable and clearly legible way, on, as the case may be, the plain detonator or booster; and

(c) the number of plain detonators or boosters contained in each smallest packaging unit must be printed on that unit.

(3) In the case of each detonating cord to which sub-paragraph (1) applies—

(a) the requirements of paragraph 6(b) do not apply; and

(b) the unique identification referred to in paragraph 1 of Schedule 6 must be marked on the reel or spool and on any smallest packaging unit.

(4) For the purposes of this paragraph, the smallest packaging unit means the smallest packaging unit on which it is possible to affix the unique identification specified in paragraph 1 of Schedule 6.

SCHEDULE 8

MEANING OF 'PLASTIC EXPLOSIVE' AND 'DETECTION AGENT'

Regulation 38

PART 1

PLASTIC EXPLOSIVE

[8.695B]

1

For the purposes of regulation 38, "plastic explosive" means an explosive substance, commonly known as "plastic explosive", including such substance in flexible or elastic sheet form and whether or not contained in an explosive article, which is—

(a) formulated with one or more high explosives which in their pure form have a vapour pressure less than 10-4 Pa at a temperature of 25°C;

(b) formulated with a binder material; and

(c) malleable or flexible at normal room temperature.

2

The following explosives, even if meeting the description of plastic explosives in paragraph 1, are not to be regarded as plastic explosives for the purposes of regulation 38 as long as their manufacture or possession continues to be to a quantity and for a purpose specified in any of sub-paragraphs (a) to (c) of this paragraph, namely any explosive—

(a) the manufacture or possession of which is to a quantity no greater than is necessary for the purpose of, and is solely for use in, lawful research, development or testing of new or modified explosives;

(b) the manufacture or possession of which is to a quantity no greater than is necessary for the purpose of, and is solely for use in, lawful training in explosives detection or development or testing of explosives detection equipment; or

(c) the manufacture or possession of which is to a quantity no greater than is necessary for, and is solely for, the purpose of lawful forensic science.

3

In this Part "high explosives" include, but are not restricted to, cyclotetramethylenetetranitramine (HMX), pentaerythritol tetranitrate (PETN) and cyclotrimethylenetrinitramine (RDX).

PART 2

DETECTION AGENTS

Table

(1) *Name of detection agent*	*(2)* *Minimum concentration*
Ethylene glycol dinitrate (EGDN)	0.2% by mass
2,3-Dimethyl-2,3-dinitrobutane (DMNB)	1.0% by mass
para-Mononitrotoluene (p-MNT)	0.5% by mass

SCHEDULE 9

ESSENTIAL SAFETY REQUIREMENTS

Regulation 39

PART 1

GENERAL REQUIREMENTS

[8.696]

(This Schedule reproduces, with minor modifications, the provisions of Annex 1 to the Civil Uses Directive)

1

Each civil explosive must be designed, manufactured and supplied in such a way as to present a minimal risk to the safety of human life and health, and to prevent damage to property and the environment under normal, foreseeable conditions, in particular as regards the safety rules and standard practices until such time as it is used.

2

Each civil explosive must attain the performance characteristics specified by the manufacturer in order to ensure maximum safety and reliability.

3

Each civil explosive must be designed and manufactured in such a way that when appropriate techniques are employed it can be disposed of in a manner which minimises effects on the environment.

PART 2

SPECIAL REQUIREMENTS

[8.696A]

4

As a minimum, the following information and properties—where appropriate—must be considered or tested. Each civil explosive should be tested under realistic conditions. If this is not possible in a laboratory, the tests should be carried out in the conditions in which the civil explosive is to be used.

(a) Design and characteristic properties, including chemical composition, degree of homogeneity and, where appropriate, dimensions and grain size distribution.

(b) The physical and chemical stability of the civil explosive in all environmental conditions to which it may be exposed.

(c) Sensitiveness to impact and friction.

(d) Compatibility of all components as regards their physical and chemical stability.

(e) The chemical purity of the civil explosive.

(f) Resistance of the civil explosive against influence of water where it is intended to be used in humid or wet conditions and where its safety or reliability may be adversely affected by water.

(g) Resistance to low and high temperatures, where the civil explosive is intended to be kept or used at such temperatures and its safety or reliability may be adversely affected by cooling or heating of a component or of the civil explosive as a whole.

(h) The suitability of the civil explosive for use in hazardous environments (eg firedamp atmospheres, hot masses, etc) if it is intended to be used under such conditions.

(i) Safety features intended to prevent untimely or inadvertent initiation or ignition.

(j) The correct loading and functioning of the civil explosive when used for its intended purpose.

(k) Suitable instructions and, where necessary, markings in respect of safe handling, storage, use and disposal in the official language or languages of the recipient EEA state.

(l) The ability of the civil explosive, its wrapping or other components to withstand deterioration during storage until the "use by" date specified by the manufacturer.

(m) Specification of all devices and accessories needed for reliable and safe functioning of the civil explosive.

5

The various groups of civil explosives must at least also comply with the following requirements:

(a) Blasting Explosives—

(i) The proposed method of initiation must ensure safe, reliable and complete detonation or deflagration as appropriate, of the blasting explosive. In the particular case of black powder, it is the capacity as regards deflagration which is to be checked.

(ii) Blasting explosives in cartridge form must transmit the detonation safely and reliably from one end of the train of cartridges to the other.

(iii) The fumes produced by blasting explosives intended for underground use may contain carbon monoxide, nitrous gases, other gases, vapours or airborne solid residues only in quantities which do not impair health under normal operating conditions.

(b) Detonating cords, safety fuses, other fuses and shock tubes—

(i) The covering of detonating cords, safety fuses and other fuses must be of adequate mechanical strength and adequately protect the explosive filling when exposed to normal mechanical stress.

(ii) The parameters for the burning times of safety fuses must be indicated and must be reliably set.

(iii) Detonating cords must be capable of being reliably initiated, be of sufficient initiation capability and comply with requirements as regards storage even in particular climatic conditions.

(c) Detonators (including delay detonators) and relays—

(i) Detonators must reliably initiate the detonation of the blasting explosives which are intended to be used with them under all foreseeable conditions of use.

(ii) Delay connectors for detonating cords must be reliably initiated.

(iii) The initiation capability must not be adversely affected by humidity.

(iv) The delay times of delay detonators must be sufficiently uniform to ensure that the probability of overlapping of the delay times of adjacent time steps is insignificant.

(v) The electrical characteristics of electric detonators must be indicated on the packaging (eg no-fire current, resistance, etc).

(vi) The wires of electric detonators must be of sufficient insulation and mechanical strength including the solidity of the link to the detonator, taking account of their intended use.

(d) Propellants and rocket propellants—

(i) These materials must not detonate when used for their intended purpose.

(ii) Propellants where necessary (eg those based on nitrocellulose) must be stabilised against decomposition.

(iii) Solid rocket propellants, when in compressed or cast form, must not contain any unintentional fissures or gas bubbles which dangerously affect their functioning.

SCHEDULE 10

CONFORMITY MARKING

Regulation 42(3)

[8.697]

The CE conformity marking must consist of the initials "CE" taking the following form:

The marking may be reduced or enlarged but the proportions given in the above drawing must be respected.

SCHEDULE 11

ENFORCEMENT WITHIN GREAT BRITAIN

Regulation 43

PART 1

INTRODUCTORY

[8.697A]

1

(1) This Schedule makes provision for the enforcement of these Regulations within Great Britain.

(2) Despite sub-paragraph (1), nothing in paragraphs 2 to 12 or 14 applies for the purpose of making any person responsible for the enforcement of these Regulations as they apply—

(a) on sites which are GB nuclear sites; or

(b) on any part of a site which is, or forms part of, a GB nuclear site.

(3) The provisions of this Schedule apply despite the provisions of the Health and Safety (Enforcing Authority) Regulations 1998.

PART 2

ENFORCEMENT OF PROVISIONS

[8.697B]

Enforcement in respect of the manufacture and storage of explosives

2

(1) A licensing authority is the enforcing authority for the manufacture and storage provisions—

(a) for a site in relation to which it has granted a person a licence;

(b) where, in relation to a deemed licence, it would have been the licensing authority by virtue of paragraph 1 or 4 of Schedule 1 if an application for a licence had been made under these Regulations;

(c) where, in any case other than those mentioned in paragraphs (a) and (b)—

(i) it would be the licensing authority by virtue of paragraph 1 or 4 of Schedule 1 if an application for a licence is, or should have been, made under these Regulations; or

(ii) it would have been the licensing authority had the requirements of paragraph (1) of regulation 6 not been disapplied by paragraph (2) of that regulation or the requirements of paragraph (1) of regulation 7 not been disapplied by paragraph (2) of that regulation.

(2) In this paragraph—

(a) "deemed licence" means—

(i) any licence deemed by regulation 47(1) to be a licence granted under regulation 13;

(ii) any licence deemed to be held by a person pursuant to regulation 47(12); and

(iii) any registration deemed by regulation 47(2) to be a licence granted under regulation 13;

(b) "manufacture and storage provisions" means—

(i) regulations 6, 7, 23, 26, 27 and 30 as they apply otherwise than in respect of the manufacture and storage of ammonium nitrate blasting intermediate;

(ii) regulation 29 as it applies to the manufacture and storage of chlorate mixtures.

3

A licensing authority is the enforcing authority for regulation 3 of the Management of Health and Safety at Work Regulations 1999 in relation to the manufacture or storage of explosives for which it is the enforcing authority by virtue of paragraph 2.

4

The enforcing authority for regulation 13(8) as it applies in respect of the manufacture and storage of any explosive and for regulations 6, 7, 23, 26, 27 and 30 as they apply in respect of the manufacture and storage of ammonium nitrate blasting intermediate is—

(a) in relation to manufacture or storage on an authorised defence site or a new nuclear build site, the ONR;

(b) otherwise, the Executive.

Enforcement in respect of discarding, disposal and decontamination

5

The licensing authority is the enforcing authority for regulation 28 where a person discards or disposes of explosives or explosive-contaminated items, or decontaminates explosive-contaminated items, at a site in relation to which it has granted that or any other person a licence.

6

(1) The enforcing authority for regulation 28 where a person discards or disposes of explosives or explosive-contaminated items, or decontaminates explosive-contaminated items, at a place other than a site in relation to which a person holds a licence is—

(a) where the discarding, disposal or decontamination is carried out by, or on behalf of, a person who holds a licence granted by the Executive in a case in which the assent of the local authority was required under regulation 13(3) before the licence was granted, the Executive;

(b) where the discarding, disposal or decontamination is carried out by, or on behalf of, a person who holds a licence granted by the ONR in a case in which the assent of the local authority was required under regulation 13(3) before the licence was granted, the ONR;

(c) where neither paragraph (a) or (b) applies—

(i) in any case where the local authority is by virtue of the 1998 Regulations the enforcing authority for the premises, or part of premises, at which the disposal or decontamination is carried out, the local authority;

(ii) otherwise, the Executive.

(2) In this paragraph, "local authority" in sub-paragraph 1(c) has the meaning given in regulation 2(1) of the Health and Safety (Enforcing Authority) Regulations 1998.

Enforcement in respect of import of chlorate mixtures

7

The Executive is the enforcing authority for regulation 29 as it applies to the import of pyrotechnic substances or pyrotechnic articles which consist of sulphur or phosphorous mixed with chlorate of potassium or other chlorates.

Enforcement in respect of acquisition and supply of fireworks

8

The enforcing authority for regulation 9 is—

(a) for a site in relation to which it has granted a person a licence, the Executive;

(b) subject to sub-paragraph (a), a local authority which is a licensing authority in the area of that local authority.

Enforcement in respect of explosives certificates and access to relevant explosives

9

(1) The enforcing authority for regulations 5, 11(5), 21(4), 31, 32 and 37 in any area of Great Britain is the chief officer of police for that area.

(2) Despite sub-paragraph (1), the enforcing authority for regulations 31, 32 and 37 against a police force, any member of a police force or any relevant person is—

(a) so far as the relevant regulation applies in relation to activities carried out on, or in relation to, any authorised defence site or a new nuclear build site, the ONR;

(b) otherwise, the Executive.

(3) For the purposes of this paragraph "relevant person" means a person—

(a) appointed to assist a police force as mentioned in paragraph 4(2) of Schedule 2 to the Police Reform and Social Responsibility Act 2011 (civilian staff); or

(b) appointed to assist a police force as mentioned in section 26 of the Police and Fire Reform (Scotland) Act 2012 (police staff).

Enforcement in respect of identification and traceability

10

The enforcing authority for regulation 33—

(a) at a site in relation to which a person holds a licence granted by the chief officer of police for the area in which the site is situated, is that chief officer of police;

(b) at an authorised defence site or a new nuclear build site, is the ONR;

(c) otherwise, is the Executive.

11

(1) The enforcing authority for regulation 34 is the Executive.

(2) Despite sub-paragraph (1), the ONR is the enforcing authority for regulation 34 to the extent that it imposes requirements on manufacturers who—

(a) are established in Great Britain; and

(b) manufacture explosives on an authorised defence site or a new nuclear build site.

Enforcement in respect of record keeping

12

(1) The enforcing authority for regulation 35 is—

(a) the Executive—

(i) in respect of any place within a site in relation to which it is the licensing authority;

(ii) subject to paragraph (b)(ii), in respect of any place occupied on behalf of the Crown for naval, military or air force purposes or for the purposes of the department of the Secretary of State having responsibility for defence;

(iii) in respect of any place below ground in any mine;

(iv) subject to paragraph (b)(iii), against a police force, any member of a police force or relevant person;

(b) the ONR—

(i) in respect of any place within a site in relation to which it is the licensing authority;

(ii) in respect of any place within an authorised defence site or a new nuclear build site;

(iii) against a police force, any member of a police force or any relevant person, so far as regulation 35 applies in relation to activities carried out on, or in relation to, any authorised defence site or new nuclear build site;

(c) in any other case, the chief officer of police for the area.

(2) In this paragraph "relevant person" has the meaning given in paragraph 9.

13

The enforcing authority for regulation 36—

(a) at a site in relation to which a person holds a licence granted by the chief officer of police for the area in which the site is situated, is, subject to sub-paragraph (b), that chief officer of police;

(b) at a site which is, or forms part of, an authorised defence site or a new nuclear build site, is the ONR;

(c) in any other case, is the Executive.

Enforcement in respect of unmarked plastic explosives

14

(1) The enforcing authority for regulation 38 is the Executive.

(2) Despite sub-paragraph (1), the ONR is the enforcing authority for paragraphs (1) and (2) of regulation 38 as they apply on, or in relation to, an authorised defence site or a new nuclear build site.

Enforcement in respect of transfers, and the placing on the market, of civil explosives

15

(1) The Executive is the enforcing authority for regulation 8 and Part 13.

(2) Despite sub-paragraph (1), the ONR is the enforcing authority for regulation 8 as it applies on, or in relation to, any activity carried out wholly or mainly on an authorised defence site or a new nuclear build site.

SCHEDULE 12

ENFORCEMENT POWERS IN RESPECT OF TRANSFERS, AND THE PLACING ON THE MARKET, OF CIVIL EXPLOSIVES

Regulation 45

[8.697C]

1

This Schedule applies in relation to the enforcement of—

(a) regulation 8, to the extent that it is made under section 2(2) of the European Communities Act 1972, by the Executive or the ONR; and

(b) Part 13 by the Executive.

2

In relation to the enforcement of the provisions referred to in paragraph 1—

(a) sections 19 to 28, 33 to 35, 38, 39, 41 and 42 of, and Schedule 3A to, the 1974 Act apply as provided in paragraph 3; and

(b) sections 36(1) and (2) and 37 of the 1974 Act apply in relation to offences under section 33 of the 1974 Act as applied by paragraph (a) and the modifications specified in paragraph 3.

3

For the purposes of the enforcement of the provisions referred to in paragraph 1, and in respect of any related proceedings for a contravention of any of those provisions, the provisions of the 1974 Act mentioned in paragraph 2 apply as if—

(a) any reference to the relevant statutory provisions were a reference to—
  (i) those provisions modified by this paragraph; and
  (ii) regulation 8 and Part 13 of these Regulations;

(b) any reference to an enforcing authority or a responsible enforcing authority were a reference to the Executive or the ONR, as the case may be;

(c) any reference to the field of responsibility, however expressed were omitted;

(d) in section 20—
  (i) subsection 2(h), the reference to any article or substance which appears to an inspector to have caused or to be likely to cause danger to health or safety included a reference to any civil explosive item which an inspector has reasonable cause to believe will be unlawfully acquired, used or dealt in;
  (ii) the reference in subsection (2)(i) to "the preceding paragraph" included also a reference to subsection (2)(h) as modified by this paragraph;
  (iii) subsection (3) were omitted;
  (iv) the reference to subsection (2)(h) in subsections (4) and (5) included also a reference to subsection (2)(h) as modified by this paragraph; and
  (v) the reference to subsection (2)(i) in subsection (6) included also a reference to subsection (2)(i) as modified by this paragraph;

(e) section 22 permitted an inspector to serve a prohibition notice, in addition to the circumstances specified in that section, in any case where—
  (i) a manufacturer, or any person acting as agent for the manufacturer, or, failing them, the person responsible for placing the explosives on the market, has failed to comply with the requirements of regulation 42 in relation to the CE marking; and
  (ii) the manufacturer, agent or the person responsible for placing the explosives on the market, as the case may be, has been served with a notice under paragraph 4 of this Schedule or an improvement notice under section 21 of the 1974 Act in respect of that failure and has continued to fail to comply after the period for remedying the contravention specified in the respective notice;

(f) subsections (3), (4) and (6) of section 23 were omitted;

(g) in section 33—
  (i) in subsection (1)—
    (aa) paragraphs (a), (b) and (d) were omitted;
    (bb) in paragraph (c), any reference to health and safety

regulations were a reference to regulation 8 and Part 13; and

(ii) subsection (3) were omitted; and

(h) in section 34—

(i) paragraphs (a) and (b) of subsection (1) were omitted; and

(ii) in subsection (3), the reference to six months were a reference to twelve months.

Civil explosives which do not satisfy the requirements of regulation 39(1)(a) and (b)

4

Where it is satisfied that the CE marking has been affixed to civil explosives but that the civil explosives do not satisfy the requirements of sub-paragraphs (a) and (b) of paragraph (1) of regulation 39, the Executive may serve notice in writing requiring the person on whom it is served to take such measures as are necessary to ensure that the civil explosives do satisfy the requirements of those sub-paragraphs.

5

Where it is satisfied that civil explosives in respect of which a notice has been served in accordance with paragraph 4 continue not to satisfy the requirements of sub-paragraphs (a) and (b) of paragraph (1) of regulation 39, the Executive may serve notice in writing requiring the person on whom it is served to take such measures as are necessary to ensure the civil explosives are withdrawn from the market.

Explosives which may compromise safety when used for their intended purpose

6

Where it is satisfied that civil explosives to which the CE marking has been affixed may compromise safety when being used for their intended purpose, the Executive may serve notice in writing requiring the person on whom it is served to take such measures as are necessary to ensure the civil explosives are withdrawn from the market.

Supplementary provisions

7

A notice referred to in paragraph 4 or 5 may be served on—

(a) the manufacturer of the civil explosives or any person acting as agent for the manufacturer; or, failing them,

(b) the person responsible for placing the civil explosives on the market.

8

A notice referred to in paragraph 6 may be served on any person the Executive has reasonable grounds for believing is in a position to take the measures specified in the notice.

9

A notice referred to in paragraphs 4 to 6—

(a) must specify measures to be taken under the notice,

(b) may be subject to conditions, and

(c) must—

(i) have immediate effect; or

(ii) allow such time for compliance as the Executive specifies in the notice.

SCHEDULE 13

AMENDMENTS

Regulation 48(1)

[8.698]

PART 1

AMENDMENTS TO PRIMARY LEGISLATION

Explosives Act 1875

1

(1) The Explosives Act 1875 is amended as follows.

(2) Omit sections 23 and 61.

(3) In section 74 (seizure and detention of explosives liable to forfeiture)—

(a) in the opening words, omit "any inspector appointed by the Health and Safety Executive under section 19 of the 1974 Act, or";

(b) in subsection (1)—

(i) for the words from "he is an inspector" to "a justice", substitute "he is authorised by an order from a justice"; and

(ii) omit "or of an inspector appointed by the Health and Safety Executive under section 19 of the 1974 Act,"; and

(c) after subsection (6), omit the words from "In this section" to the end of the section.

London Building Act 1930

2

In section 143 of the London Building Act 1930 (regulations for building near dangerous business), in subsection (4)(a), for "the Manufacture and Storage of Explosives Regulations 2005" substitute "the Explosives Regulations 2014".

Fireworks Act 1951

3

The Fireworks Act 1951 is repealed.

Customs and Excise Management Act 1979

4

In section 75(1) of the Customs and Excise Management Act 1979 (explosives), for "the Manufacture and Storage of Explosives Regulations 2005" substitute "the Explosives Regulations 2014".

Isle of Man Act 1979

5

In section 8(2) of the Isle of Man Act 1979 (removal of goods from Isle of Man to United Kingdom)—

(a) at the end of paragraph (b) insert "or"; and

(b) for paragraphs (c) and (d) substitute—

"(c) any explosives the importation of which into the United Kingdom is prohibited by regulation 29 of the Explosives Regulations 2014.".

Environmental Protection Act 1990

6

In section 142(7) of the Environmental Protection Act 1990 (powers to obtain information about potentially hazardous substances), for "the Manufacture and Storage of Explosives Regulations 2005" substitute "the Explosives Regulations 2014".

Fireworks Act 2003

7

In section 14(2) of the Fireworks Act 2003 (prohibition of supply etc of other explosives), for "the Explosives Act 1875 (c 17)" substitute "the Explosives Regulations 2014".

Energy Act 2013

8

(1) The Energy Act 2013 is amended as follows.

(2) In Schedule 12 (minor and consequential amendments relating to Part 3), omit paragraphs 50 to 52.

[8.698A]

PART 2

AMENDMENTS TO SECONDARY LEGISLATION

The Clean Air (Emission of Dark Smoke) (Exemption) Regulations 1969

9

In Schedule 1 to the Clean Air (Emission of Dark Smoke) (Exemption) Regulations 1969 (exempted matter), for "the Manufacture and Storage of Explosives Regulations 2005" substitute "the Explosives Regulations 2014".

The Explosives Acts 1875 and 1923 etc (Repeals and Modifications) Regulations 1974

10

In Schedule 2 to the Explosives Acts 1875 and 1923 etc (Repeals and Modifications) Regulations 1974 (the Explosives Act 1875: modifications), omit paragraph 5.

The Rehabilitation of Offenders Act 1974 (Exceptions) Order 1975

11

The Rehabilitation of Offenders Act 1974 (Exceptions) Order 1975 is amended as follows—

(a) in paragraph 8 of Part III of Schedule 1 (excepted professions, offices, employments, work and occupations), for "regulations 4 and 7 of the Control of Explosives Regulations 1991", substitute "regulations 4, 5 and 11 of the Explosives Regulations 2014";

(b) in paragraph 3 of Schedule 2 (excepted licences, certificates and permits), for "regulations 4 and 7 of the Control of Explosives Regulations 1991", substitute "regulations 4, 5 and 11 of the Explosives Regulations 2014"; and

(c) for paragraph 13 of Schedule 3 (excepted proceedings) substitute—

"13

Proceedings in respect of—

(a) an application to the chief officer of police for an explosives certificate pursuant to regulations 4, 5 and 11 of the Explosives Regulations 2014 ("the 2014 Regulations") as to the fitness of the applicant to acquire or acquire and keep explosives, including consideration as to whether to refuse the application on any of the grounds specified in regulation 19 of the 2014 Regulations;

(b) the revocation of such certificates pursuant to regulation 21 of the 2014 Regulations;

(c) an appeal or application pursuant to regulation 22 of the 2014 Regulations against a decision taken under regulation 19 or 21.".

The Isles of Scilly (Functions) Order 1979

12

In article 3(e) of the Isles of Scilly (Functions) Order 1979—

(a) omit "the Explosives Act 1875;"; and

(b) for "the Manufacture and Storage of Explosives Regulations 2005" substitute "the Explosives Regulations 2014".

The Dangerous Substances in Harbour Areas Regulations 1987

13

Regulation 33 of the Dangerous Substances in Harbour Areas Regulations 1987 (Application of Part IX) is amended as follows—

(a) in sub-paragraph (b) of paragraph 1, for the words from "article 8" to "Order 2001" substitute "article 11 of the Health and Safety at Work etc Act 1974 (Application Outside Great Britain) Order 2013"; and

(b) in sub-paragraph (e) of paragraph 2—

(i) in paragraph (i), for " licensed under the Manufacture and Storage of Explosives Regulations 2005" substitute "in relation to which a person holds a licence under the Explosives Regulations 2014 ("the 2014 Regulations")"; and

(ii) for paragraph (ii), substitute—

"(ii) in relation to which a person is deemed to be licensed under the 2014 Regulations by virtue of regulation 47(1) of those Regulations in cases where, in relation to that deemed licence, the assent of the local authority would have been required pursuant to regulation 13(3) of the 2014 Regulations had a licence been applied for under those Regulations;".

The Planning (Hazardous Substances) Regulations 1992

14

(1) Schedule 1 to the Planning (Hazardous Substances) Regulations 1992 (hazardous substances and controlled quantities) is amended as follows.

(2) In column 1 of entry number 54 in Part A as it applies in England, for paragraph (1) substitute—

"(1) cellulose nitrate—

(a) for which a licence is required and has been granted under the Explosives Regulations 2014 by the Health and Safety Executive where it is the licensing authority by virtue of—

(i) paragraph 1(b) of Schedule 1 to those Regulations in cases where the assent of the local authority was required pursuant to regulation 13(3) of those Regulations or would have been so required but for regulation 13(4)(b), (c), (d), (e), (f) or (g) of those Regulations; or

(ii) paragraph 1(d) of Schedule 1 to those Regulations; or

(b) for which a licence is required and has been granted under the Explosives Regulations 2014 by the Office for Nuclear Regulation in cases where the assent of the local authority was required pursuant to regulation 13(3) of those Regulations or would have been so required but for regulation 13(4)(b), (c), (d), (e), (f) or (g) of those Regulations; or".

(3) In column 1 of entry number 54 in Part A as it applies in Wales, for paragraph (1) substitute—

"(1) cellulose nitrate—

(a) for which a licence is required and has been granted under the Explosives Regulations 2014 by the Health and Safety Executive where it is the licensing authority by virtue of—

(i) paragraph 1(b) of Schedule 1 to those Regulations in cases where the assent of the local authority was required pursuant to regulation 13(3) of those Regulations or would have been so required but for regulation 13(4)(b), (c), (d), (e), (f) or (g) of those Regulations; or

(ii) paragraph 1(d) of Schedule 1 to those Regulations; or

(b) for which a licence is required and has been granted under the Explosives Regulations 2014 by the Office for Nuclear Regulation in cases where the assent of the local authority was required pursuant to regulation 13(3) of those Regulations or would have been so required but for regulation 13(4)(b), (c), (d), (e), (f) or (g) of those Regulations; or".

(4) In Part B as it applies in England—

(a) in column 1 of entry 4, for the words from "EXPLOSIVE" to "Regulations 1987" substitute—

"EXPLOSIVE (see Note 2 to this Part) where the substance, preparation or article falls under UN/ADR Division 1.4, excluding those—

(a) for which a licence is required and has been granted under the Explosives Regulations 2014 by the Health and Safety Executive where it is the licensing authority by virtue of—

(i) paragraph 1(b) of Schedule 1 to those Regulations in cases where the assent of the local authority was required pursuant to regulation 13(3) of those Regulations or would have been so required but for regulation 13(4)(b), (c), (d), (e), (f) or (g) of those Regulations; or

(ii) paragraph 1(d) of Schedule 1 to those Regulations, or

(b) for which a licence is required and has been granted under the Explosives Regulations 2014 by the Office for Nuclear Regulation in cases where the assent of the local authority was required pursuant to regulation 13(3) of those Regulations or would have been so required but for regulation 13(4)(b), (c), (d), (e), (f) or (g) of those Regulations; or

(c) licensed under the Dangerous Substances in Harbour Areas Regulations 1987"; and

(b) in column 1 of entry 5, for the words from "EXPLOSIVE" to "Regulations 1987" substitute—

"EXPLOSIVE (see Note 2 to this Part) where the substance, preparation or article falls under any of: UN/ADR Divisions 1.1, 1.2, 1.3, 1.5 or 1.6 or risk phrase R2 or R3, excluding those—

(a) for which a licence is required and has been granted under the Explosives Regulations 2014 by the Health and Safety Executive where it is the licensing authority by virtue of—

(i) paragraph 1(b) of Schedule 1 to those Regulations in cases where the assent of the local authority was required pursuant to regulation 13(3) of those Regulations or would have been so required but for regulation 13(4)(b), (c), (d), (e), (f) or (g) of those Regulations; or

(ii) paragraph 1(d) of Schedule 1 to those Regulations; or

(b) for which a licence is required and has been granted under the Explosives Regulations 2014 by the Office for Nuclear Regulation in cases where the assent of the local authority was required pursuant to regulation 13(3) of those Regulations or would have been so required but for regulation 13(4)(b), (c), (d), (e), (f) or (g) of those Regulations; or

(c) licensed under the Dangerous Substances in Harbour Areas Regulations 1987".

(5) In Part B as it applies in Wales—

(a) in column 1 of entry 4, for the words from "EXPLOSIVE" to "Regulations 1987" substitute—

"EXPLOSIVE (see Note 2 to this Part) where the substance, preparation or article falls under UN/ADR Division 1.4, excluding those—

(a) for which a licence is required and has been granted under the Explosives Regulations 2014 by the Health and Safety Executive where it is the licensing authority by virtue of—

(i) paragraph 1(b) of Schedule 1 to those Regulations in cases where the assent of the local authority was required pursuant to regulation 13(3) of those Regulations or would have been so required but for regulation 13(4)(b), (c), (d), (e), (f) or (g) of those Regulations; or

(ii) paragraph 1(d) of Schedule 1 to those Regulations; or

(b) for which a licence is required and has been granted under the Explosives Regulations 2014 by the Office for Nuclear Regulation in cases where the assent of the local authority was required pursuant to regulation 13(3) of those Regulations or would have been so required but for regulation 13(4)(b), (c), (d), (e), (f) or (g) of those Regulations; or

(c) licensed under the Dangerous Substances in Harbour Areas Regulations 1987"; and

(b) in column 1 of entry 5, for the words from "EXPLOSIVE" to "Regulations 1987 substitute—

"EXPLOSIVE (see Note 2 to this Part) where the substance, preparation or article falls under any of: UN/ADR Divisions 1.1, 1.2, 1.3, 1.5 or 1.6 or risk phrase R2 or R3, excluding those—

(a) for which a licence is required and has been granted under the Explosives Regulations 2014 by the Health and Safety Executive where it is the licensing authority by virtue of—

(i) paragraph 1(b) of Schedule 1 to those Regulations in cases where the assent of the local authority was required pursuant to regulation 13(3) of those Regulations or would have been so required but for regulation 13(4)(b), (c), (d), (e), (f) or (g) of those Regulations; or

(ii) paragraph 1(d) of Schedule 1 to those Regulations; or

(b) for which a licence is required and has been granted under the Explosives Regulations 2014 by the Office for Nuclear Regulation in cases where the assent of the local authority was required pursuant to regulation 13(3) of those Regulations or would have been so required but for regulation 13(4)(b), (c), (d), (e), (f) or (g) of those Regulations; or

(c) licensed under the Dangerous Substances in Harbour Areas Regulations 1987".

Coal and Other Safety-Lamp Mines (Explosives) Regulations 1993

15

In regulation 2(1) of the Coal and Other Safety-Lamp Mines (Explosives) Regulations 1993 (interpretation), in the definition of "explosives store" for "or registration made under the Manufacture and Storage of Explosives Regulations 2005" substitute "under the Explosives Regulations 2014".

The Town and Country Planning (Hazardous Substances) (Scotland) Regulations 1993

16

Schedule 1 to the Town and Country Planning (Hazardous Substances) (Scotland) Regulations 1993 (hazardous substances and controlled quantities) is amended as follows—

(a) in Part A, in column 1 of entry number 54, for paragraph (1), substitute—

"(1) cellulose nitrate—

(a) for which a licence is required and has been granted under the Explosives Regulations 2014 by the Health and Safety Executive where it is the licensing authority by virtue of—

(i) paragraph 1(b) of Schedule 1 to those Regulations in cases where the assent of the local authority was required pursuant to regulation 13(3) of those Regulations or would have been so required but for regulation 13(4)(b), (c), (d), (e), (f) or (g) of those Regulations; or

(ii) paragraph 1(d) of Schedule 1 to those Regulations; or

(b) for which a licence is required and has been granted under the Explosives Regulations 2014 by the Office for Nuclear Regulation in cases where the assent of the local authority was required pursuant to regulation 13(3) of those Regulations or would have been so required but for regulation 13(4)(b), (c), (d), (e), (f) or (g) of those Regulations; or"; and

(b) in Part B—

(i) in column 1 of entry number 4, for the words from "EXPLOSIVE" to "Regulations 1987" substitute—

"EXPLOSIVE (see Note 2 to Part B) where the substance, preparation or article falls under UN/ADR Division 1.4, excluding those—

(a) for which a licence is required and has been granted under the Explosives Regulations 2014 by the Health and Safety Executive where it is the licensing authority by virtue of—

(i) paragraph 1(b) of Schedule 1 to those Regulations in cases where the assent of the local authority was required pursuant to regulation 13(3) of those Regulations or would have been so required but for regulation 13(4)(b), (c), (d), (e), (f) or (g) of those Regulations; or

(ii) paragraph 1(d) of Schedule 1 to those Regulations; or

(b) for which a licence is required and has been granted under the Explosives Regulations 2014 by the Office for Nuclear Regulation in cases where the assent of the local authority was required pursuant to regulation 13(3) of those Regulations or would have been so required but for regulation 13(4)(b), (c), (d), (e), (f) or (g) of those Regulations; or

(c) licensed under the Dangerous Substances in Harbour Areas Regulations 1987"; and

(ii) in column 1 of entry number 5, for the words from "EXPLOSIVE" to "Regulations 1987", substitute—

"EXPLOSIVE (see Note 2 to Part B) where the substance, preparation or article falls under any of: UN/ADR Divisions 1.1, 1.2, 1.3, 1.5 or 1.6 or risk phrase R2 or R3, excluding those—

(a) for which a licence is required and has been granted under the Explosives Regulations 2014 by the Health and Safety Executive where it is the licensing authority by virtue of—

(i) paragraph 1(b) of Schedule 1 to those Regulations in cases where the assent of the local authority was required pursuant to regulation 13(3) of those Regulations or would have been so required but for regulation 13(4)(b), (c), (d), (e), (f) or (g) of those Regulations; or

(ii) paragraph 1(d) of Schedule 1 to those Regulations; or

(b) for which a licence is required and has been granted under the Explosives Regulations 2014 by the Office for Nuclear Regulation in cases where the assent of the local authority was required pursuant to regulation 13(3) of those Regulations or would have been so required but for regulation 13(4)(b), (c), (d), (e), (f) or (g) of those Regulations; or

(c) licensed under the Dangerous Substances in Harbour Areas Regulations 1987".

The Health and Safety (Enforcing Authority) Regulations 1998

17

In regulation 4 of the Health and Safety (Enforcing Authority) Regulations 1998 (exceptions), omit paragraphs (7) to (12).

The Visiting Forces and International Headquarters (Application of Law) Order 1999

18

In Schedule 6 to the Visiting Forces and International Headquarters (Application of Law) Order 1999 (enactments conferring specific exemptions, privileges etc), omit the entry relating to the Fireworks Act 1951.

The Quarries Regulations 1999

19

In regulation 2(1) of the Quarries Regulations 1999 (interpretation), in the definition of "explosives store", for "or registration made under the Manufacture and Storage of Explosives Regulations 2005" substitute "under the Explosives Regulations 2014".

Building (Scotland) Regulations 2004

20

For paragraph 1 of Schedule 1 to the Building (Scotland) Regulations 2004 (exempted buildings and services, fittings and equipment), substitute—

"1

(1) Any building in which explosives are manufactured or stored under a licence granted under the Explosives Regulations 2014 where the whole building is used for that manufacture or storage.

(2) Where only a part of a building is used for the manufacture or storage of explosives under a licence granted under the Explosives Regulations 2014, that part of the building where the licence specifies that that manufacture or storage may take place.

(3) Except—sub-paragraphs (1) and (2) do not include any building or, as the case may be, any part of a building as is referred to in, respectively, sub-paragraph (1) or (2) in relation to which—

(a) no minimum separation distance is required to be maintained by virtue of regulation 27(2)(a) or (3) of the Explosives Regulations 2014; or

(b) a minimum separation distance of 0 metres is prescribed by virtue of regulation 27(1) of, and Schedule 5 to, the Explosives Regulations 2014 and the requirement for the assent of the local authority under regulation 13(3) of those Regulations did not apply by virtue of regulation 13(4)(a) of those Regulations.".

The REACH Enforcement Regulations 2008

21

In paragraph 1 of Part 3 of Schedule 3 to the REACH Enforcement Regulations 2008 (health and safety enforcement) for sub-paragraph (t), substitute—

"(t) in relation to Great Britain, the manufacture or storage of ammonium nitrate blasting intermediate under the Explosives Regulations 2014.".

The Health and Safety (Miscellaneous Amendments and Revocations) Regulations 2009

22

In regulation 2 of the Health and Safety (Miscellaneous Amendments and Revocations) Regulations 2009 (extension outside Great Britain)—

(a) omit paragraphs (a) and (b); and

(b) omit ", respectively, regulation 14 of the Control of Explosives Regulations 1991, regulation 3(1)(b) of the Manufacture and Storage of Explosives Regulations 2005 and".

The Carriage of Dangerous Goods and Use of Transportable Pressure Equipment Regulations 2009

23

In regulation 7 of the Carriage of Dangerous Goods and Use of Transportable Pressure Equipment Regulations 2009 (additional security requirements for carriage by road), for paragraph (5)(c) substitute—

"(c) "a safe and secure place" means a safe and secure place—

(i) within a site in relation to which a person is licensed to manufacture or store explosives under regulation 13 of the Explosives Regulations 2014; or

(ii) at which the manufacture or storage of explosives may lawfully take place by virtue of a certificate of exemption granted under those Regulations.".

The Building Regulations 2010

24

For paragraph 1 of Class 1 of Schedule 2 to the Building Regulations 2010 (exempt buildings and work), substitute—

"1

(1) Any building in which explosives are manufactured or stored under a licence granted under the Explosives Regulations 2014 where—

(a) the whole building is used for that manufacture or storage, and either

(b) a minimum separation distance of greater than 0 metres is prescribed by virtue of regulation 27(1) of, and Schedule 5 to, those Regulations; or

(c) a minimum separation distance of 0 metres is prescribed by virtue of the provisions referred to in paragraph (b) and the assent of the local authority was required by regulation 13(3) of those Regulations or would have been so required but for regulation 13(4)(b), (c), (d), (e), (f) or (g) of those Regulations.

(2) Where only a part of a building is used for the manufacture or storage of explosives under a licence granted under the Explosives Regulations 2014 and—

(a) a minimum separation distance of greater than 0 metres is prescribed by virtue of regulation 27(1) of, and Schedule 5 to, those Regulations; or

(b) a minimum separation distance of 0 metres is prescribed by virtue of the provisions referred to in paragraph (a) and the assent of the local authority was required by regulation 13(3) of those Regulations or would have been so required but for regulation 13(4)(b), (c), (d), (e), (f) or (g) of those Regulations,

that part of the building where the licence specifies that that manufacture or storage may take place.".

The Carriage of Dangerous Goods and Use of Transportable Pressure Equipment (Amendment) Regulations 2011

25

The Carriage of Dangerous Goods and Use of Transportable Pressure Equipment (Amendment) Regulations 2011 are amended as follows—

(a) in regulation 17(2) (consequential amendments), omit sub-paragraphs (c), (d) and (e); and

(b) in the Schedule omit paragraphs 3, 4 and 5.

The Health and Safety (Fees) Regulations 2012

26

The Health and Safety (Fees) Regulations 2012 are amended as follows.

27

In regulation 1(4) (citation, commencement and interpretation), for "explosives certificate, licence or registration" substitute "explosive certificate or licence".

28

(1) In regulation 9 (fees payable under the Manufacture and Storage of Explosives Regulations 2005 and certain other provisions concerning explosives, including acetylene, and under the Petroleum (Consolidation) Act 1928 and the Petroleum (Transfer of Licences) Act 1936)—

(a) for paragraph (1), substitute—

"(1) Where an application in relation to a provision specified in column 1 of Part 1 of Schedule 8, for a purpose specified in column 2 of that Part, is made to a licensing authority which is the licensing authority by virtue of—

(a) paragraph 1(c) or (d) of Schedule 1 to the 2014 Regulations, or

(b) paragraphs 1(b), 2 or 4 of that Schedule in cases where the assent of the local authority is required under regulation 13(3) of those Regulations or is not required by virtue of regulation 13(4)(b) to (g) of those Regulations,

the fee specified in the corresponding entry in column 3 of that Part is payable by the applicant to that licensing authority.";

(b) in paragraph (2), after "manufacture" insert "or store";

(c) for paragraph (3), substitute—

"(3) Where an application in relation to a provision specified in column 1 of Part 2 of Schedule 8, for a purpose specified in column 2 of that Part, is made to a licensing authority, which is the licensing authority by virtue of—

(a) paragraph 1(a) of Schedule 1 to the 2014 Regulations, or

(b) paragraphs 1(b), 2 or 4 of Schedule 1 to those Regulations in cases where the requirement for assent of the local authority under regulation 13(3) of those Regulations is disapplied by regulation 13(4)(a) of those Regulations,

the fee specified in the corresponding entry in column 3 of that Part is payable by the applicant to that licensing authority.";

(d) in paragraph (11)—

(i) for "1991" substitute "2014"; and

(ii) for "4(6)(d)" substitute "19(2)(d)"; and

(e) in paragraph (13)—

(i) omit the definition of "the 1991 Regulations";
(ii) for the definition of "the 2005 Regulations" substitute—
""the 2014 Regulations" means the Explosives Regulations 2014;";
(iii) for the definition beginning ""ammonium nitrate blasting intermediate"" substitute ""ammonium nitrate blasting intermediate", "chief officer of police", "explosives certificate", "licence", "licensing authority", "manufacture", "on-site mixing", "prohibited person", "shooters' powder" and "site" have the same meanings as in the 2014 Regulations;"; and
(iv) omit the definition beginning ""chief officer of police"".

29

In Schedule 8 (fees payable under the Manufacture and Storage of Explosives Regulations 2005 and certain other provisions concerning explosives, including acetylene, and under the Petroleum (Consolidation) Act 1928 and the Petroleum (Transfer of Licences) Act 1936)—

(a) for the heading and Part 1, substitute—

"FEES PAYABLE IN RELATION TO THE EXPLOSIVES REGULATIONS 2014, THE ACETYLENE SAFETY (ENGLAND, WALES AND SCOTLAND) REGULATIONS 2014 AND THE PETROLEUM (CONSOLIDATION) REGULATIONS 2014

PART 1

FEES FOR APPLICATIONS FOR LICENCES, OR VARIATIONS TO, OR TRANSFER OF, LICENCES, TO MANUFACTURE OR TO STORE EXPLOSIVES MADE TO LICENSING AUTHORITIES WHICH ARE LICENSING AUTHORITIES BY VIRTUE OF PARAGRAPHS 1(C) OR (D) OF SCHEDULE 1 TO THE EXPLOSIVES REGULATIONS 2014 OR PARAGRAPHS 1(B), 2 OR 4 OF THAT SCHEDULE IN LOCAL AUTHORITY ASSENT CASES OR WHERE NO ASSENT REQUIRED BY VIRTUE OF REGULATION 13(4)(B) TO (G)

Table 1

1 ***Provision under which a licence is granted***	2 ***Purpose of application***	3 ***Fee***	4 ***Fee for work by Specialist Inspector***
The 2014 Regulations			
Regulation 13, as extended by regulation 2(2) of those Regulations to the manufacture and storage of ammonium nitrate blasting intermediate	Licence to manufacture explosives not being ammonium nitrate blasting intermediate nor relating to on-site mixing	£631	£127 per hour worked

	Licence to manufacture or store only ammonium nitrate blasting intermediate	£156 per hour worked	
	Licence to manufacture explosives by means of on-site mixing	£234	£127 per hour worked
	Licence to store explosives:	£631	£127 per hour worked
	Renewal of any of the above licences	£83	£127 per hour worked
Regulation 16	Varying a licence to manufacture or store explosives, not being ammonium nitrate blasting intermediate:	£432	£127 per hour worked
	Varying a licence to manufacture or store ammonium nitrate blasting intermediate	£156 per hour worked	
Regulation 17	Transfer of any of the above licences	£52	
	Replacement of any licences referred to in this Part if lost	£52"	

(b) for Part 2 substitute—

"PART 2

FEES FOR APPLICATIONS FOR LICENCES, OR VARIATIONS TO, OR TRANSFER OF, LICENSES TO STORE EXPLOSIVES MADE TO LICENSING AUTHORITIES WHICH ARE LICENSING AUTHORITIES BY VIRTUE OF PARAGRAPH 1(A) OF SCHEDULE 1 TO THE 2014 REGULATIONS OR PARAGRAPH 1(B), 2 OR 4 OF THAT SCHEDULE IN CASES WHERE LOCAL AUTHORITY ASSENT IS NOT REQUIRED BY VIRTUE OF REGULATION 13(4)(A)

Table 2

1 ***Provision under which a licence is granted***	*2* ***Purpose of application***	*3* ***Fee***
The 2014 Regulations		
Regulation 13	Licence to store explosives where, by virtue of regulation 27 of, and Schedule 5 to, the 2014 Regulations, a minimum separation distance of greater than 0 metres is prescribed:	
	(a) one year's duration	£178
	(b) two years' duration	£234
	(c) three years' duration	£292
	(d) four years' duration	£360
	(e) five years' duration	£407
	Licence to store explosives where, by virtue of regulation 27 of, and Schedule 5 to, the 2014 Regulations, no minimum separation distance or a 0 metres minimum separation distance is prescribed:	
	(a) one year's duration	£105
	(b) two years' duration	£136
	(c) three years' duration	£166
	(d) four years' duration	£198
	(e) five years' duration	£229

	Renewal of licence to store explosives where, by virtue of regulation 27 of, and Schedule 5 to, the 2014 Regulations, a minimum separation distance of greater than 0 metres is prescribed:	
	(a) one year's duration	£83
	(b) two years' duration	£141
	(c) three years' duration	£198
	(d) four years' duration	£256
	(e) five years' duration	£313
	Renewal of licence to store explosives where, by virtue of regulation 27 of, and Schedule 5 to, the 2014 Regulations, no minimum separation distance or a 0 metres minimum separation distance is prescribed:	
	(a) one year's duration	£52
	(b) two years' duration	£83
	(c) three years' duration	£115
	(d) four years' duration	£146
	(e) five years' duration	£178
Regulation 16	Varying a licence:	
	(a) varying name of licensee or address of site	£35
	(b) any other kind of variation	The reasonable cost to the licensing authority of having the work carried out
Regulation 17	Transfer of licence	£35
	Replacement of licence	£35

Note: The fee payable for a licence or renewal of a licence—

(a) of less than one year's duration is, respectively, the fee set out above for a licence, or renewal of a licence of one year's duration decreased proportionately according to the duration of the period for which the licence renewal is granted;

(b) of more than one but less than two years' duration is, respectively, the fee set out above for a licence, or a renewal of a licence of one year's duration increased proportionately according to the duration of the period for which the licence or renewal is granted;

(c) of more than two but less than three years' duration is, respectively, the fee set out above for a licence or renewal of a licence of two years' duration increased proportionately according to the duration of the period for which the licence or renewal is granted;

(d) of more than three but less than four years' duration is, respectively, the fee set above for a licence or renewal of a licence of three years' duration increased proportionately according to the duration of the period for which the licence or renewal is granted;

(e) of more than four but less than five years' duration is, respectively, the fee set out above for a licence or renewal of a licence of four years' duration increased proportionately according to the duration of the period for which the licence or renewal is granted."; and

(c) for Part 8 substitute—

"PART 8

FEES FOR EXPLOSIVES CERTIFICATES UNDER THE EXPLOSIVES REGULATIONS 2014

Table 1

1 ***Provisions under which a fee is payable***	2 ***Purpose of application***	3 ***Fee***
The 2014 Regulations		
Regulation 11 (see Note)	(a) Explosives certificate for acquiring and keeping explosives, not including an application for an explosives certificate referred to in entries (c), (g) or (i), at a site in relation to which a person holds a licence to store explosives and, by virtue of regulation 27 of, and Schedule 5 to, the 2014 Regulations no minimum separation distance is prescribed or a 0 metres minimum separation distance is prescribed:	
	(i) one year's duration	£125
	(ii) two years' duration	£156
	(iii) three years' duration	£188
	(iv) four years' duration	£219
	(v) five years' duration	£251

(b) Renewal of the explosives certificate referred to in (a):	
(i) one year's duration	£110
(ii) two years' duration	£130
(iii) three years' duration	£151
(iv) four years' duration	£173
(v) five years' duration	£193
(c) Explosives certificate for acquiring and keeping explosives, not including an application for an explosives certificate referred to in entries (a), (g) or (i), at a site in relation to which a person holds a licence for the storage of no more than 2000 kilograms of explosives:	
(i) one year's duration	£136
(ii) two years' duration	£166
(iii) three years' duration	£198
(iv) four years' duration	£229
(v) five years' duration	£261
(d) Renewal of the explosives certificate referred to in (c):	
(i) one year's duration	£130
(ii) two years' duration	£156
(iii) three years' duration	£183
(iv) four years' duration	£209
(v) five years' duration	£234
(e) Explosives certificate for acquiring and keeping explosives, not including an application for an explosives certificate referred to in entries (g) or (i), at a site in relation to which a person holds a licence for the storage of more than 2000 kilograms of explosives:	
(i) one year's duration	£183

(ii) two years' duration	£219
(iii) three years' duration	£256
(iv) four years' duration	£292
(v) five years' duration	£329
(f) Renewal of the explosives certificate referred to in (e):	
(i) one year's duration	£161
(ii) two years' duration	£193
(iii) three years' duration	£224
(iv) four years' duration	£256
(v) five years' duration	£287
(g) Explosives certificate for acquiring and keeping only shooters' powder at a site in relation to which the applicant holds a licence, where the applicant also makes a relevant application under the 1968 Act to the chief officer of police determining the explosives certificate application which is to be determined at the same time	£24
(h) Renewal of the explosives certificate referred to in (g) where the applicant also makes a relevant application under the 1968 Act to the chief officer of police determining the renewal application which is to be determined at the same time	£15

(i) Explosives certificate for the acquiring and keeping only shooters' powder at a site in relation to which the applicant holds a licence and a relevant certificate, where no relevant application under the 1968 Act by the applicant is to be determined at the same time	£44
(j) Renewal of the explosives certificate referred to in (i) where no relevant application under the 1968 Act by the applicant is to be determined at the same time	£18
(k) Explosives certificate for acquiring more than 15 kilograms of explosives, not including an application for an explosives certificate referred to in entries (m) or (o):	
(i) one year's duration	£125
(ii) two years duration	£156
(iii) three years duration	£188
(iv) four years duration	£219
(v) five years duration	£251
(l) Renewal of the explosive certificate referred to in (k):	
(i) one year's duration	£110
(ii) two years duration	£130
(iii) three years duration	£151
(iv) four years duration	£173
(v) five years duration	£193

(m) Explosives certificate for acquiring more than 15 kilograms of shooters' powder only, where the applicant also makes a relevant application under the 1968 Act to the chief officer of police determining the explosives certificate application which is to be determined at the same time	£24
(n) Renewal of the explosives certificate referred to in (m) where the applicant also makes a relevant application under the 1968 Act to the chief officer of police determining the renewal application which is to be determined at the same time	£15
(o) Explosives certificate for acquiring more than 15 kilograms of shooters' powder only, where the applicant holds a relevant certificate and no relevant application under the 1968 Act is to be determined at the same time	£44
(p) Renewal of the explosives certificate referred to in (o) where no relevant application under the 1968 Act by the applicant is to be determined at the same time	£18
(q) Replacement of any explosives certificate referred to in (a) to (f) and (k) and (l) if lost	£35
(r) Replacement of any explosives certificate referred to in (g) to (j) and (m) to (p) if lost	£10

Table 2

The fee for a check carried out for the purposes of regulation 19(2)(d) of the 2014 Regulations is £5

Note: The fee payable for an explosives certificate or renewal of an explosives certificate ("renewal")—

(a) of less than one year's duration is, respectively, the fee set out above for a certificate, or renewal of an explosives certificate of one year's duration decreased proportionately according to the duration of the period for which the explosives certificate or renewal is granted;

(b) of more than one but less than two years' duration is, respectively, the fee set out above for an explosives certificate, or a renewal of an explosives certificate of one year's duration increased proportionately according to the duration of the period for which the explosives certificate or renewal is granted;

(c) of more than two but less than three years' duration is, respectively, the fee set out above for an explosives certificate or renewal of an explosives certificate of two years' duration increased proportionately according to the duration of the period for which the explosives certificate or renewal is granted;

(d) of more than three but less than four years' duration is, respectively, the fee set above for an explosives certificate or renewal of an explosives certificate of three years' duration increased proportionately according to the duration of the period for which the explosives certificate or renewal is granted;

(e) of more than four but less than five years' duration is, respectively, the fee set out above for an explosives certificate or renewal of an explosives certificate of four years' duration increased proportionately according to the duration of the period for which the explosives certificate or renewal is granted.".

The Rehabilitation of Offenders Act 1974 (Exclusions and Exceptions) (Scotland) Order 2013

30

The Rehabilitation of Offenders Act 1974 (Exclusions and Exemptions) (Scotland) Order 2013 is amended as follows—

(a) for paragraph 10 of Schedule 1 (proceedings) substitute—

"10

Proceedings under the Explosives Regulations 2014 ("the 2014 Regulations") in respect of—

(a) the application to the chief officer of police for an explosives certificate pursuant to regulations 4, 5 and 11 of the 2014 Regulations certifying a person to be a fit and proper person to acquire or acquire and keep explosives;

(b) the revocation of such certificates pursuant to regulation 21 of the 2014 Regulations;

(c) an appeal or application to the Sheriff under regulation 22 of the 2014 Regulations against a decision taken under regulation 19 or 21.";

(b) in paragraph 3(3)(c) of Schedule 3 (exclusions of section 4(2)(a) and (b) of the Act), for "regulation 4 of the Control of Explosives Regulations 1991" substitute "regulation 11 of the Explosives Regulations 2014"; and

(c) in paragraph 4 of Part 3 of Schedule 4 (excepted professions, offices, employments and occupations) for "regulation 4 of the Control of Explosives Regulations 1991" substitute "regulations 4, 5 and 11 of the Explosives Regulations 2014".

The Reporting of Injuries, Diseases and Dangerous Occurrences Regulations 2013

31

The Reporting of Injuries, Diseases and Dangerous Occurrences Regulations 2013 are amended as follows—

(a) in regulation 2(1) (interpretation), in the definition of "explosives" for "the Manufacture and Storage of Explosives Regulations 2005" substitute "the Explosives Regulations 2014"; and

(b) in paragraph 5(a) of Part 1 of Schedule 2 (dangerous occurrences) for "or registration, as the case may be, under regulation 9, 10 or 11 of the Manufacture and Storage of Explosives Regulations 2005" substitute "under regulations 6 or 7 of the Explosives Regulations 2014".

SCHEDULE 14

REPEALS AND REVOCATIONS

Regulation 48(2) and (3)

[8.699]

PART 1

REPEALS

1 *Title*	2 *Reference*	3 *Extent of repeal*
Explosives Act 1875	c 17 (38 & 39 Vict)	Sections 23 and 61; in section 74, in the opening words, "any inspector appointed by the Health and Safety Executive under section 19 of the 1974 Act, or"; in subsection (1), "or of an inspector appointed by the Health and Safety Executive under section 19 of the 1974 Act,"; after subsection (6), the words from "In this section" to the end of the section.
Fireworks Act 1951*	c 58	The whole Act.
Energy Act 2013*	c 32	In Schedule 12, paragraphs 50 to 52.

[8.700]

PART 2

REVOCATIONS

1 *Title*	*2* *Reference*	*3* *Extent of revocation*
Order of Secretary of State (No 11), dated September 20, 1924, making Byelaws as to the Conveyance of Explosives on Roads, and in certain special cases.	SR & O 1924/1129	The whole instrument.
Order in Council (No 26) Relating to Picric Acid, Picrates and Mixtures of Picric Acid with other Substances.	SR & O 1926/823	The whole instrument.
The Explosives Acts 1875 and 1923 etc (Repeals and Modifications) Regulations 1974.	SI 1974/1885	Regulation 2(a) and Schedule 1, except to the extent that they relate to sections 73, 75 and 89 of the Explosives Act 1875 as amended by SI 1974/2166; regulations 3 to 7, paragraphs 1 to 4 and 6 to 21 of Schedule 2, and Schedule 3.
The Explosives Acts 1875 and 1923 etc (Repeals and Modifications) (Amendment) Regulations 1974.	SI 1974/2166	The whole instrument.
The Control of Explosives Regulations 1991.	SI 1991/1531	The whole instrument.
The Placing on the Market and Supervision of Transfers of Explosives Regulations 1993.	SI 1993/2714	The whole instrument.
The Marking of Plastic Explosives for Detection Regulations 1996	SI 1996/890	The whole instrument
The Health and Safety (Enforcing Authority) Regulations 1998	SI 1998/494	In regulation 4, paragraphs (7) to (12).

[8.700] 8B Phenomena hazardous to health: Fire and explosives

The Manufacture and Storage of Explosives Regulations 2005.	SI 2005/1082	The whole instrument.
The Manufacture and Storage of Explosives and the Health and Safety (Enforcing Authority) (Amendment and Supplementary Provisions) Regulations 2007.	SI 2007/2598	The whole instrument.
The Health and Safety (Miscellaneous Amendments and Revocations) Regulations 2009.	SI 2009/693	In regulation 2, paragraphs (a) and (b); paragraphs 1, 2 and 4 of Schedule 1.
The Carriage of Dangerous Goods and Use of Transportable Pressure Equipment (Amendment) Regulations 2011.	SI 2011/1885	In regulation 17(2), sub-paragraphs (c), (d) and (e); paragraphs 3, 4 and 5 of the Schedule.
The Identification and Traceability of Explosives Regulations 2013.	SI 2013/449	The whole instrument.
The Energy Act 2013 (Office for Nuclear Regulation) (Consequential Amendments, Transitional Provisions and Savings) Order 2014.	SI 2014/469	In Schedule 2, paragraph 1. In Schedule 3, paragraphs 46 to 51, 53 to 55, 110 to 122 and 146.

Part 8E

Substances, noise, radiation and other phenomena hazardous to health: Asbestos, electricity, flammable liquids, gas, gmos, lead and radiation

Control of Asbestos in the Air Regulations 1990

SI 1990/556

as amended by SI 2013/390

Made 6th March 1990

[8.2363]

The Secretary of State for the Environment, as respects England, the Secretary of State for Wales, as respects Wales, and the Secretary of State for Scotland, as respects Scotland, in exercise of their powers (as respects regulations 1 to 3) under sections 1(1)(b) and (d), 15(1) and (3) and 53(1) of the Health and Safety at Work etc Act 1974, after consulting, in accordance with section 50(1) of that Act, the Health and Safety Commission and such other bodies as it appeared to them to be appropriate to consult, and (as respects regulations 1, 4 and 5) under section 2(2) of the European Communities Act 1972 being the Ministers designated for the purposes of that sub-section in relation to the control of air pollution, and of all other powers enabling them in that behalf, hereby make the following Regulations

[8.2364]

1 ***Citation, commencement and interpretation***

(1) These Regulations may be cited as the Control of Asbestos in the Air Regulations 1990 and subject to paragraph (2) shall come into force on 5th April 1990.

(2) In the case of any premises to which section 5 of the Health and Safety at Work etc Act 1974 applied before 31st December 1988, regulations 2 and 3 shall come into force on 30th June 1991.

(3) In these Regulations 'asbestos' means the following fibrous silicates—

actinolite, amosite, anthophyllite, chrysotile, crocidolite or tremolite.

[8.2364A]

Statutory annotations

Amendment

Revoked, in relation to England and Wales, by SI 2013/390, reg 58(1)(c).

[8.2365]

2 ***Limit value for the discharge of asbestos into the air during the use of asbestos***

(1) Any person having control of any premises to which section 5 of the Health and Safety at Work etc Act 1974 applies and from which asbestos is emitted through discharge outlets into the air during the use of asbestos shall ensure that—

- *(a) the concentration of asbestos so discharged, measured by the sampling and analysis procedures described in the Annex to Council Directive 87/217/EEC on the prevention and reduction of environmental pollution by asbestos, or any other method which gives equivalent results, does not exceed 0.1 milligram of asbestos per cubic metre of air, and*
- *(b) the concentration of asbestos so emitted is measured as mentioned in paragraph (a) at regular intervals of not more than 6 months.*

(2) In this regulation—

'raw asbestos' means the product resulting from the primary crushing of asbestos ore; and

'use of asbestos' means activities which involve—

(a) the production of raw asbestos from ore, excluding any process directly associated with the mining of the ore, or

(b) the manufacturing and industrial finishing of any of the following products using raw asbestos—

asbestos cement, asbestos cement products, asbestos fillers, asbestos filters, asbestos floor coverings, asbestos friction products, asbestos insulating board, asbestos jointing, packaging and reinforcement materials, asbestos packing, asbestos paper and card, and asbestos textiles.

[8.2365A]

Statutory annotations

Amendment

Revoked, in relation to England and Wales, by SI 2013/390, reg 58(1)(c).

[8.2366]

3 ***Enforcing authority***

The Secretary of State shall be the authority responsible for enforcing regulation 2.

[8.2366A]

Statutory annotations

Amendment

Revoked, in relation to England and Wales, by SI 2013/390, reg 58(1)(c).

[8.2367]

4 ***Control of environmental pollution by asbestos resulting from the working of products containing asbestos***

(1) Any person undertaking activities involving the working of products containing asbestos shall ensure that those activities do not cause significant environmental pollution by asbestos fibres or dust emitted into the air; and

(2) Any person undertaking the demolition of buildings, structures and installations containing asbestos and the removal from them of asbestos or materials containing asbestos involving the release of asbestos fibres or dust into the air shall ensure that significant environmental pollution is not caused thereby.

(3) In this regulation—

'fibre' means a fibre within the meaning of Part B.II of the Annex to Council Directive 87/217/EEC; and

'the working of products containing asbestos' means activities other than the use of asbestos (within the meaning of regulation 2) which are liable to release asbestos into the environment.

[8.2367A]

Statutory annotations

Amendment

Revoked, in relation to England and Wales, by SI 2013/390, reg 58(1)(c).

[8.2368]

5

(1) Any person who contravenes the provisions of regulation 4 commits an offence and a person guilty of such an offence shall be liable on conviction on indictment to a fine, or on summary conviction to a fine not exceeding level 4 on the standard scale.

(2) Section 37 (offences by bodies corporate) of the Health and Safety at Work etc Act 1974 shall apply to an offence under this regulation as it applies to an offence under any of the relevant statutory provisions of that Act.

[8.2368A]

Statutory annotations

Amendment

Revoked, in relation to England and Wales, by SI 2013/390, reg 58(1)(c).

Gas Safety (Management) Regulations 1996

SI 1996/551

as amended by SI 2013/1471

[8.2376]

7 Gas escapes and investigations

(1) It shall be the duty of British Gas plc to provide a continuously manned telephone service (which shall be contactable within Great Britain by the use of one telephone number) for enabling persons to report an escape of gas from a network or from a gas fitting supplied with gas from a network.

(2) Where British Gas plc is notified of such an escape of gas it shall, if it is not responsible for preventing the escape under paragraphs (4) or (5), report it forthwith to the person who is.

(3) A person referred to in regulation 6(2)(a) to (c) and the holder of a licence issued under section 7A(1) of the Gas Act 1986 who discovers or is notified of any such escape of gas (other than by virtue of a report made to him pursuant to paragraph (2)) shall report it forthwith to British Gas plc.

(4) Where any gas escapes from a network the person conveying the gas in the part of the network from which the gas escapes shall, as soon as is reasonably practicable after being so informed of the escape, attend the place where the gas is escaping, and within 12 hours of being so informed of the escape, he shall prevent the gas escaping.

(5) Where any gas escapes from a gas fitting supplied with gas from a network, the person conveying the gas in the part of the network immediately upstream of the emergency control for the supply of gas to that fitting shall, as soon as is reasonably practicable after being so informed of the escape, attend the place where the gas is escaping, and within 12 hours of being so informed of the escape, he shall prevent the gas escaping.

(6) Where a person conveying gas in a network has reasonable cause to suspect that gas conveyed by him which has escaped has entered, or may enter, any premises, he shall, so far as is reasonably practicable, take all the steps necessary to avert danger to persons from such entry.

(7) If the responsible person for any premises knows or has reason to suspect that gas is escaping from a gas fitting in those premises supplied with gas from a network he shall immediately take all reasonable steps to cause the supply of gas to be shut off at such place as may be necessary to prevent further escape of gas.

(8) If gas continues to escape into those premises after the supply of gas has been shut off or when a smell of gas persists, the responsible person for the premises discovering such escape or smell shall immediately give notice of the escape or smell to British Gas plc.

(9) Where an escape of gas has been stopped by shutting off the supply, no person shall cause or permit the supply to be re-opened (other than in the course of repair) until all necessary steps have been taken to prevent a recurrence of such escape.

(10) In any proceedings against a person for an offence consisting of a contravention of paragraphs (4) or (5) above it shall, in so far as the contravention is not preventing the escape of gas within the period of 12 hours referred to in those paragraphs, be a defence for the person to prove that it was not reasonably practicable for him effectually to prevent the gas from escaping within that period, and that he did effectually prevent the escape of gas as soon as it was reasonably practicable for him to do so.

(11) A person conveying gas may appoint another person to act on his behalf to prevent an escape of gas, and where he does so in advance of discovering or being notified of such an escape—

- (a) he shall notify British Gas plc of the name of the person appointed;
- (b) the appointee shall in relation to the escape be responsible for complying with paragraphs (4) to (6) in substitution for the person conveying the gas, and paragraph (6) shall have effect as if the reference to the person conveying gas having reasonable cause to suspect that the gas has entered or may enter premises were a reference to the appointee having such cause.

(12) Where an escape of gas from a gas fitting on domestic premises has resulted in a fire or explosion, the person conveying the gas in the part of the network immediately upstream of the emergency control for the supply of gas to that fitting shall, as soon as is reasonably practicable after receiving notice of the fire or explosion, cause an investigation to be carried out so as to establish, so far as is reasonably practicable, whether the escape was from installation pipework or from an, and if so which, appliance.

(13) Where an escape of gas from a network has or was likely to have resulted in a fire or explosion, the person conveying the gas in the part of the network where the gas escaped shall, as soon as is reasonably practicable after receiving notice of the escape, cause an investigation to be carried out so as to establish the source of the escape and, so far as is reasonably practicable, the reason for it.

(14) Where an incident notifiable under [regulation 11(1) of the Reporting of Injuries, Diseases and Dangerous Occurrences Regulations 2013] has arisen as a result of an escape of carbon monoxide gas from incomplete combustion of gas in a gas fitting, the person who supplied the gas shall, as soon as is reasonably practicable after receiving notice of the incident, cause an investigation to be carried out so as to establish, so far as is reasonably practicable, the cause of the escape and accumulation of the carbon monoxide gas.

(15) Where a person who conveys gas receives notice of an incident referred to in paragraph (14), he shall, as soon as is reasonably practicable, inform the relevant gas supplier of that fact.

(16) A person who causes an investigation to be carried out pursuant to paragraphs (12), (13) or (14) shall—

- (a) ensure that the individuals who carry it out are competent;
- (b) notify the Executive before the investigation begins of the intention to carry it out;
- (c) ensure that a report of the investigation is prepared and a copy of it is sent to the Executive as soon as is reasonably practicable after the investigation has been completed.

(17) In this Regulation—

- (a) 'the responsible person' has the same meaning as in regulation 2(1) of the 1994 Regulations;
- (b) any reference to an escape of gas from a gas fitting includes a reference to an escape or emission of carbon monoxide gas resulting from incomplete combustion of gas in such a fitting;
- (c) any reference to a fire or explosion of gas is a reference to an unintended fire or explosion of gas;
- (d) any reference to a person supplying gas does not include a reference to a person to whom the gas is supplied and who provides it for use in a flat or part of premises let by him.

[8.2376A]

Statutory annotations

Amendment

Para (14): words 'regulation 11(1) of' to 'Reporting of Injuries, Diseases and Dangerous Occurrences Regulations 2013' in square brackets substituted by SI 2013/1471, reg 18(2), Sch 4.

Ionising Radiations Regulations 1999

SI 1999/3232

SCHEDULE 8

QUANTITIES AND CONCENTRATIONS OF RADIONUCLIDES

Regulation 2(4) and 30(1) and (2) and Schedule 1

PART II

QUANTITY RATIOS FOR MORE THAN ONE RADIONUCLIDE

[8.3068]–[8.3100]

1

For the purpose of Regulation 2(4), the quantity ratio for more than one radionuclide is the sum of the quotients of the quantity of a radionuclide present Q_p divided by the quantity of that radionuclide specified in the appropriate column of Part I of this Schedule Q_{lim}, namely—

$$\Sigma(Qp / Q_{lim})$$

2

In any case where the isotopic composition of a radioactive substance is not known or is only partially known, the quantity ratio for that substance shall be calculated by using the values specified in the appropriate column in Part I for 'other radionuclides not listed above' for any radionuclide that has not been identified or where the quantity of a radionuclide is uncertain, unless the employer can show that the use of some other value is appropriate in the circumstances of a particular case, when he may use that value.

Note

This provision is reproduced to correct typographical errors in the equation.

Genetically Modified Organisms (Contained Use) Regulations 2000

SI 2000/2831

as amended by SIs 2001/2626, 2002/63, 2002/2443, 2002/3188, 2005/2466, 2005/2759, 2008/960, 2009/693, 2009/1892, 2010/2840, 2011/1043, 2013/1666 and SSI 2002/541

PART V

MISCELLANEOUS AND GENERAL

[8.3135]

26 Enforcement . . .

(1) Subject to paragraph (2) and to the extent they would not otherwise do so, the provisions of—

(a) sections 16 to 26 (approved codes of practice and enforcement) [and] sections 33 to 42 (provisions as to offences) . . . of the 1974 Act; and

(b) the Health and Safety (Training for Employment) Regulations 1990,

shall apply to these Regulations as if they were health and safety regulations for the purposes of that Act, and [any function of the Health and Safety Executive under any other provision of the 1974 Act] under or in respect of health and safety regulations (including their enforcement) shall be exercisable as if these Regulations were, to the extent they would not otherwise be so, health and safety regulations for the purposes of that Act.

(2) A failure to discharge a duty—

(a) placed on the competent authority or the Executive by these Regulations; or

(b) placed on any other person by Schedule 11,

shall not be an offence, and section 33(1)(c) of the 1974 Act shall have effect accordingly.

(3) Notwithstanding regulation 3 of the Health and Safety (Enforcing Authority) Regulations 1998, the enforcing authority for these Regulations shall be the Executive.

[8.3136]

Statutory annotations

Amendment

Provision heading: words omitted revoked by SI 2013/1666, art 3(a).

Para (1): in sub-para (a) word 'and' in square brackets substituted by SI 2013/1666, art 3(b)(i); in sub-para (a) words omitted revoked by SI 2013/1666, art 3(b)(ii); words from 'any function of' to 'of the 1974 Act' in square brackets substituted by SI 2008/960, art 22, Sch 3.

Radiation (Emergency Preparedness and Public Information) Regulations 2001

SI 2001/2975

as amended by SIs 2002/2099, 2002/2469, 2004/568, 2006/557, 2007/1573 and SI 2013/235

[8.3203]

2 Interpretation

(1) In these Regulations, unless the context otherwise requires—

. . .

'the 1999 Regulations' means the Ionising Radiations Regulations 1999;

. . .

'the Agency' in relation to premises or transport or a plan relating to premises or transport—

(a) in England and Wales, means the Environment Agency,

(b) in Scotland, means the Scottish Environment Protection Agency;

'approved dosimetry service' means an approved dosimetry service within the meaning of the 1999 Regulations and which is approved for the purpose of regulation 14 of these Regulations;

'carrier' shall be construed in accordance with paragraph (2);

'carrier's emergency plan' shall be construed in accordance with regulation 8;

. . .

'dose assessment' means the dose assessment made and recorded by an approved dosimetry service in accordance with regulation 21 of the 1999 Regulations;

'dose record' means the record made and maintained in respect of an employee by the approved dosimetry service in accordance with regulation 21 of the 1999 Regulations;

'emergency exposure' means an exposure of an employee engaged in an activity of or associated with the response to a radiation emergency or potential radiation emergency in order to bring help to endangered persons, prevent exposure of a large number of persons or save a valuable installation or goods, whereby one of the individual dose limits referred to in paragraphs 1 or 2 of Part I of Schedule 4 to the 1999 Regulations could be exceeded;

'emergency services' means—

(a) those police, fire and ambulance services who are likely to be required to respond to a radiation emergency which has occurred at the premises of an operator or at the location of a radiation emergency during the course of the transport of a radioactive substance, and

(b) where appropriate, Her Majesty's Coastguard;

'the Executive' means the Health and Safety Executive;

['health authority' means—

(a) in relation to England, [a clinical commissioning group established under section 14D of the National Health Service Act 2006];

(b) in relation to Wales, a Health Authority established under section 8 of that Act; and

(c) in relation to Scotland, a Health Board established under section 2 of the National Health Service (Scotland) Act 1978;]

'installation' means a unit in which the radioactive substances present are, or are intended to be, produced, used, handled or stored, and it includes—

(a) equipment, structures, pipework, machinery and tools,

(b) railway sidings, docks and unloading quays serving the unit, and

(c) jetties, warehouses or similar structures, whether floating or not,

which are necessary for the operation of the unit;

'intervention' means a human activity that prevents or decreases the exposure of persons to radiation from a radiation emergency or from an event which could lead to a radiation emergency, by acting on the sources of radiation, the paths by which such radiation may be transmitted to persons and on persons themselves;

'ionising radiation' means the transfer of energy in the form of particles or electromagnetic waves of a wavelength of 100 nanometres or less or a frequency of 3 x 1015 hertz or more capable of producing ions directly or indirectly;

'licensed site' means a site in respect of which a nuclear site licence has been granted and is in force;

'licensee' means the person to whom a nuclear site licence has been granted;

'local authority' means—

(a) subject to sub-paragraphs (b) and (c) below, in relation to—

  (i) London, the London Fire and Emergency Planning Authority,

  (ii) an area where there is a fire and civil defence authority, that authority,

  (iii) the Isles of Scilly, the Council of the Isles of Scilly,

  (iv) an area in the rest of England, the county council for that area or, where there is no county council for that area, the district council for that area,

  (v) an area in Scotland, the council for the local government area, and

  (vi) an area in Wales, the county council or the county borough council for that area;

(b) for the purposes of regulation 16(2), in relation to—

  (i) London, the London Fire and Emergency Planning Authority, and, in the City of London, the Common Council for the City of London, or, in an area in the rest of London, the London Borough Council for that area,

  (ii) any other area where there is a fire and civil defence authority, that authority and the district council for that area,

  (iii) the Isles of Scilly, the Council of the Isles of Scilly,

  (iv) an area in the rest of England, the county council, if any, for that area and the district council, if any, for that area,

  (v) an area in Scotland, the council for the local government area, and

  (vi) an area in Wales, the county council or the county borough council for that area;

(c) for the purposes of regulation 16(3), in relation to—

(i) the City of London, the Common Council for the City of London,

(ii) an area in the rest of London, the London Borough Council for that area,

(iii) the Isles of Scilly, the Council of the Isle of Scilly,

(iv) an area in the rest of England, the district council for that area or, where there is no district council for that area, the county council for that area,

(v) an area in Scotland, the council for the local government area, and

(vi) an area in Wales, the county council or the county borough council for that area;

'medical surveillance' means medical surveillance carried out in accordance with regulation 24 of the 1999 Regulations;

'member of the public' means any person not being—

(a) a person for the time being present upon premises where a radiation emergency is reasonably foreseeable or where a radiation emergency has actually occurred, or

(b) a person engaged in an activity of or associated with the response to a radiation emergency;

'non-dispersible source' means a sealed source or a radioactive substance which in either case by virtue of its physical and chemical form cannot cause a radiation emergency in any reasonably foreseeable event but it does not include any radioactive substance inside a nuclear reactor or any nuclear fuel element;

'nuclear site licence' has the meaning assigned to it by section 1(1) of the Nuclear Installations Act 1965;

'off-site emergency plan' shall be construed in accordance with regulation 9;

'operator' shall be construed in accordance with paragraph (3);

'operator's emergency plan' shall be construed in accordance with regulation 7;

'premises' means—

(a) the whole area under the control of the same person where radioactive substances are present in one or more installations, and for this purpose two or more areas under the control of the same person and separated only by a road, railway or inland waterway shall be treated as one whole area, or

(b) where radioactive substances are present on a licensed site, that licensed site,

and, where a radioactive substance forms an integral part of a vessel and is used in connection with the operation of that vessel, it includes that vessel when it is at fixed point moorings or alongside berths, save that such vessel shall be deemed to be separate premises only where such moorings or berths do not form part of a licensed site or part of premises under the control of the Secretary of State for Defence;

'radiation accident' means an accident where immediate action would be required to prevent or reduce the exposure to ionising radiation of employees or any other persons and includes a radiation emergency;

'radiation emergency' means any event (other than a pre-existing situation) which is likely to result in any member of the public being exposed to ionising radiation arising from that event in excess of any of the doses set out in Schedule 1 and for this purpose any health protection measure to be taken during the 24 hours immediately following the event shall be disregarded;

'radioactive substance' means any substance which contains one or more radionuclides whose activity cannot be disregarded for the purposes of radiation protection;

'sealed source' means a source containing any radioactive substance whose structure is such as to prevent, under normal conditions of use, any dispersion of radioactive substances into the environment;

'transport' means—

(a) . . .

(b) transferring or conveying a radioactive substance through any public place otherwise than—

(i) by rail, road, inland waterway, sea or air; or

(ii) by means of a pipeline or similar means;

. . .

'work with ionising radiation' means work involving the production, processing, handling, use, holding, storage . . . or disposal of radioactive substances which can increase the exposure of persons to radiation from an artificial source, or from a radioactive substance containing naturally occurring radionuclides which are processed for their radioactive, fissile or fertile properties.

(2) In these Regulations, any reference to a carrier is a reference to—

(a) . . .

(b) an employer transferring or conveying a radioactive substance through any public place otherwise than by rail, road, inland waterway, sea or air or by means of a pipeline or similar means.

(3) In these Regulations, any reference to an operator is a reference to—

(a) in relation to any premises other than a licensed site, the person who is, in the course of a trade, business or other undertaking carried on by him, in control of the operation of premises, and

(b) in relation to a licensed site, the licensee,

and any duty imposed by these Regulations on the operator shall extend only in relation to those premises.

(4) In these Regulations, unless the context otherwise requires, any reference to—

(a) an employer includes a reference to a self-employed person and any duty imposed by these Regulations on an employer in respect of his employee shall extend to a self-employed person in respect of himself;

(b) exposure to ionising radiation is a reference to exposure to ionising radiation arising from work with ionising radiation.

(5) Any reference in these Regulations to—

(a) a numbered regulation or Schedule is a reference to the regulation in or Schedule to these Regulations so numbered; and

(b) a numbered paragraph is a reference to the paragraph so numbered in the regulation or Schedule in which the reference appears.

[8.3204]

Statutory annotations

Amendment

Para (1): in definition 'health authority' in para (a) words from 'a clinical commissioning group' to 'National Health Service Act 2006' in square brackets substituted by SI 2013/235, art 11, Sch 2, Pt 1, para 47(1), (2).

[8.3210]

7 Operator's emergency plan

(1) Where the assessment made by an operator in accordance with regulation 4(1) or regulation 5 shows that it is reasonably foreseeable that a radiation emergency might arise (having regard to the steps taken by the operator under regulation 4(2)), the operator shall prepare an adequate emergency plan (in these Regulations referred to as an 'operator's emergency plan') designed to secure, so far as is reasonably practicable, the restriction of exposure to ionising radiation and the health and safety of persons who may be affected by such reasonably foreseeable emergencies as are identified by the said assessment.

(2) Without prejudice to paragraph (1), the operator's emergency plan shall contain the information specified in Part I of Schedule 7.

(3) No person shall carry out work with ionising radiation to which the assessment made in accordance with regulation 4(1) applies unless—

(a) the operator has complied with the requirements of paragraphs (1) and (2); and

(b) the local authority has complied with the requirements of regulation 9(1), (2), (8) and (9) and has provided confirmation of this to the operator in accordance with regulation 9(13).

(4) The operator's emergency plan shall be drawn up having regard to the principles set out in Part I of Schedule 8.

(5) Without prejudice to the generality of paragraph (1), the operator's emergency plan shall secure, where appropriate, intervention for the purposes set out in Part II of Schedule 8.

(6) For the purpose of preparing the operator's emergency plan pursuant to paragraph (1) or of reviewing the plan pursuant to regulation 10(1), the operator shall consult—

(a) his employees, any person carrying out work on behalf of the operator, the Executive, the local authority in whose area the premises of the operator are situated, the emergency services, the health authority for the area in which the premises of the operator are situated and the Agency; and

[(aa) if the premises of the operator are situated in England, the National Health Service Commissioning Board;

(ab) if the premises of the operator are situated in England or Wales, Public Health England, an executive agency of the Department of Health; and]

(b) such other persons, bodies and authorities as the operator considers appropriate,

and, in a case where the emergency services form part of the plan, shall give such information to those services as will enable them to perform their functions in accordance with the plan.

(7) The operator shall ensure that any employee who may be involved with or may be affected by arrangements in the operator's emergency plan is or has been provided with—

(a) suitable and sufficient information, instruction and training; and

(b) the equipment necessary to restrict that employee's exposure to ionising radiation including, where appropriate, the issue of suitable dosemeters or other devices obtained in either case from the approved dosimetry service with which the operator has entered into an arrangement under regulation 21 of the 1999 Regulations.

(8) The operator shall provide to the Executive upon request and within such reasonable time as the Executive may specify a copy of the operator's emergency plan or such parts of that plan as the Executive may require.

[8.3210A]

Statutory annotations

Amendment

Para (6): sub-paras (aa), (ab) substituted by SI 2013/235, art 11, Sch 2, Pt 1, para 47(1), (3).

[8.3211]

8 Carrier's emergency plan

(1) Where the assessment made in accordance with regulation 4(1) or regulation 5 shows that it is reasonably foreseeable that a radiation emergency might arise in respect of the transport of a radioactive substance (having regard to the steps taken by the carrier under regulation 4(2)), the carrier shall prepare or ensure that there has been prepared an adequate emergency plan in respect of the transport of such substances (in these Regulations referred to as a 'carrier's emergency plan') designed to secure, so far as is reasonably practicable, the restriction of exposure to ionising radiation and the health and safety of persons who may be affected by such reasonably foreseeable emergencies as are identified by the said assessment.

(2) Without prejudice to paragraph (1), the carrier's emergency plan shall contain the information specified in Part II of Schedule 7.

(3) A carrier shall not undertake the transport of any radioactive substance to which the assessment made in accordance with regulation 4(1) applies unless he has complied with the requirements of paragraph (1) and (2).

(4) . . .

(5) The carrier's emergency plan shall be drawn up having regard to the principles set out in Part I of Schedule 8.

(6) Without prejudice to the generality of paragraph (1), the carrier's emergency plan shall secure, where appropriate, intervention for the purposes set out in Part II of Schedule 8.

(7) For the purpose of preparing a carrier's emergency plan pursuant to paragraph (1) or of reviewing the plan pursuant to regulation 10(1), the carrier shall ensure that consultation is carried out with—

- (a) the Executive . . . and the Agency; and
- [(aa) the National Health Service Commissioning Board and Public Health England, an executive agency of the Department of Health; and]
- (b) such local authorities, emergency services, health authorities and other persons, bodies or authorities (or in each case representatives thereof) as the carrier considers appropriate.

(8) The carrier shall ensure that any employee under his control who may be involved with or may be affected by arrangements in the carrier's emergency plan is or has been provided with—

- (a) suitable and sufficient information, instruction and training; and
- (b) the equipment necessary to restrict that employee's exposure to ionising radiation including, where appropriate, suitable dosemeters or other devices obtained in either case from the approved dosimetry service with which the carrier has entered into an arrangement under regulation 21 of the 1999 Regulations.

(9) Where requested by the Executive, the carrier shall provide to the Executive within such reasonable time as may be specified a copy of the carrier's emergency plan or such parts of the plan as the Executive may require.

[8.3212]–[8.3220]

Statutory annotations

Amendment

Para (7): sub-para (aa) substituted by SI 2013/235, art 11, Sch 2, Pt 1, para 47(1), (4).

[8.3221]

9 Off-site emergency plan

(1) The local authority in whose area there is situated premises at which there is carried out work with ionising radiation to which these Regulations apply and in respect of which an assessment made by the operator pursuant to regulation 4(1) or regulation 5 shows that it is reasonably foreseeable that a radiation emergency might arise (having regard to the steps taken by the operator under regulation 4(2)) shall prepare an adequate emergency plan (in these Regulations referred to as an 'off-site emergency plan') designed to secure, so far as is reasonably practicable, the restriction of exposure to ionising radiation and the health and safety of persons who may be affected by such reasonably foreseeable emergencies as are identified in that assessment and the plan shall be prepared in respect of such area as in the opinion of the Executive any member of the public is likely to be affected by such radiation emergencies.

(2) Without prejudice to paragraph (1), the off-site emergency plan shall contain the information specified in Part III of Schedule 7.

(3) The off-site emergency plan prepared pursuant to paragraph (1) shall address each reasonably foreseeable radiation emergency that has been identified by the operator for the purposes of regulation 7(1).

(4) Where an assessment has been made pursuant to regulation 4(1), within 28 days of sending the report of the assessment to the Executive in accordance with regulation 6(1), the operator shall supply to the local authority such information as is necessary for the purpose of enabling the authority to prepare the off-site emergency plan required by paragraph (1).

(5) Without prejudice to paragraph (4), the operator shall further supply to the local authority—

- (a) any additional information the local authority may reasonably request to enable the off-site emergency plan to be prepared; and
- (b) details of any material change to the information provided under paragraph (4) resulting from—
  - (i) a further assessment made pursuant to regulation 5(1) or (2); or
  - (ii) a revision of the operator's emergency plan pursuant to regulation 10(1).

(6) The information provided to a local authority pursuant to paragraphs (4) or (5)(a) shall be reviewed and where necessary revised by the operator at suitable intervals not exceeding 3 years from the date at which information was last supplied to the local authority under those paragraphs and the operator shall within 28 days inform the local authority of the outcome of that review.

(7) The operator shall, within 28 days of any further assessment or revision referred to in paragraph (5)(b) inform the local authority of any material change to the information supplied arising from that assessment or review.

(8) Subject to paragraph (9), the off-site emergency plan shall be prepared no later than 6 months (or such longer period as the Executive may agree in writing) after whichever is the later of—

- (a) the receipt by the local authority of a notice from the Executive informing the local authority of the need to prepare an off-site emergency plan in respect of the area; or
- (b) the receipt by the local authority of the information referred to in paragraph (4).

(9) The off-site emergency plan shall be prepared before the operator carries out work with ionising radiation to which the assessment made in accordance with regulation 4(1) applies.

(10) The off-site emergency plan shall be drawn up having regard to the principles set out in Part I of Schedule 8.

(11) Without prejudice to the generality of paragraph (1), the off-site emergency plan shall secure, where appropriate, intervention for the purposes set out in Part II of Schedule 8.

(12) For the purpose of preparing an off-site emergency plan pursuant to paragraph (1) or of reviewing the plan pursuant to regulation 10(1), the local authority shall consult—

- (a) the operator carrying out the work with ionising radiation to which the plan relates, the Executive, the emergency services, each health authority in the vicinity of the premises of the operator and the Agency; and
- [(aa) if the premises of the operator are situated in England, the National Health Service Commissioning Board;
- (ab) if the premises of the operator are situated in England or Wales, Public Health England, an executive agency of the Department of Health; and]
- (b) such other persons, bodies and authorities and members of the public as the local authority considers appropriate.

(13) Once it has prepared the off-site emergency plan the local authority shall confirm in writing to the operator that it has done so.

(14) The employer of any employee who may be required to participate in the implementation of an off-site emergency plan shall ensure that such employees of his are or have been provided with—

- (a) suitable and sufficient information, instruction and training; and
- (b) the equipment necessary to restrict that employee's exposure to ionising radiation including, where appropriate, the issue of suitable dosemeters or other devices.

(15) The local authority shall provide to the Executive upon request and within such reasonable time as may be specified a copy of the off-site emergency plan or such parts of that plan as the Executive may require.

[8.3221A]

Statutory annotations

Amendment

Para (12): sub-paras (aa), (ab) substituted by SI 2013/235, art 11, Sch 2, Pt 1, para 47(1), (5).

[8.3225]

13 Implementation of emergency plans

(1) An operator or carrier who has prepared (or, in relation to a carrier, has ensured that there has been prepared) an emergency plan pursuant to regulation 7 or 8, as the case may be, shall take reasonable steps to put it, or such parts of it as are necessary, into effect without delay when—

- (a) a radiation emergency occurs, or
- (b) an event occurs which could reasonably be expected to lead to a radiation emergency,

and shall notify such occurrence to the Executive without delay.

(2) A local authority which has prepared an emergency plan pursuant to regulation 9 shall take reasonable steps to ensure that it, or such parts of it as are necessary, is put into effect without delay when informed by the operator that—

- (a) a radiation emergency has occurred; or
- (b) an event has occurred which could reasonably be expected to lead to a radiation emergency.

(3) In the event of a radiation emergency resulting from his work with ionising radiation, the operator or carrier shall—

(a) as soon as is reasonably practicable, make (or, in relation to a carrier, ensure that there has been made) a provisional assessment of the circumstances and consequences of such an emergency and for this purpose shall consult—

(i) in the case of the operator, the emergency services, the local authority, the health authority, [the National Health Service Commissioning Board (if the premises of the operator are situated in England), Public Health England, an executive agency of the Department of Health, (if the premises of the operator are situated in England or Wales),] the Agency and such other persons, bodies or authorities as have functions under the operator's emergency plan or the off-site emergency plan; and

(ii) in the case of the carrier, . . . [the Agency, the National Health Service Commissioning Board and Public Health England, an executive agency of the Department of Health,] and any emergency services, local authority and health authority who were involved in the response to the emergency and such other persons, bodies or authorities as have functions under the carrier's emergency plan;

(b) as soon as is practicable and in any event within 12 months or such longer time as the Executive may agree, make (or in relation to a carrier ensure that there has been made) a full assessment of the consequences of that emergency and the effectiveness of the emergency plans put into effect as a result of that emergency; and

(c) within 28 days of the completion of the assessment under sub-paragraph (b) make a report of the findings of the assessment and retain that report or a copy thereof for at least 50 years from the date upon which the report was completed.

(4) The operator or carrier shall provide to the Executive within 28 days of the making of the report under paragraph (3)(c) above a copy of that report.

[8.3226]

Statutory annotations

Amendment

Para (3): in sub-para (a)(i) words from 'the National Health Service Commissioning Board' to 'England or Wales),' in square brackets inserted by SI 2013/235, art 11, Sch 2, Pt 1, para 47(1), (6)(a); in sub-para (a)(ii) words from 'the Agency, the' to 'Department of Health,' in square brackets substituted by SI 2013/235, art 11, Sch 2, Pt 1, para 47(1), (6)(b).

Justification of Practices Involving Ionising Radiation Regulations 2004

SI 2004/1769

as amended by SIs 2005/525, 2013/235 and 2013/1821

PART 5

INFORMATION, CONSULTATION AND PUBLICITY

[8.3419]

18 Consultation

(1) Before making a justification decision, a determination under regulation 12 or serving a contravention notice under regulation 22, the Justifying Authority or Secretary of State (as the case may be)—

(a) shall consult—

(i) the Health and Safety Executive;

(ii) the Food Standards Agency;

[(iii) Public Health England, an executive agency of the Department of Health; and]

(iv) where the class or type of practice involves a radioactive substance, the Environment Agency, [the Natural Resources Body for Wales,] the Scottish Environment Protection Agency and the Department of the Environment for Northern Ireland;

(b) may consult any other person whom he considers it appropriate to consult;

(c) shall take such steps as he considers appropriate to bring the proposed decision, determination or contravention notice to the attention of any person likely to be affected by it.

(2) Before exercising any function under Part 3 or under regulation 16, 17, 22, 26 or 27, the person proposing to exercise the function shall consult the other persons listed in regulation 6(1).

[8.3420]

Statutory annotations

Amendment

Para (1): sub-para (a)(iii) substituted by SI 2013/235, art 11, Sch 2, Pt 1, para 68(1), (2); in sub-para (a)(iv) words 'the Natural Resources Body for Wales,' in square brackets inserted by SI 2013/1821, art 32(1), (2).

PART 8

ENFORCEMENT AND OFFENCES

[8.3424]

22 Contravention notices

(1) For the purposes of this regulation and of regulation 23, a 'relevant breach' means a contravention of regulation 4(5), 5(3) or 20; failure to comply with any condition imposed pursuant to regulation 11(1); or failure to comply with any requirement imposed pursuant to regulation 11(2)(b) or 16(1).

(2) If the Justifying Authority is of the opinion that a person has committed a relevant breach he shall serve a notice (in these Regulations referred to as a 'contravention notice') on that person.

(3) The Justifying Authority shall give notice of any contravention notice served pursuant to paragraph (2) to—

- (a) such of the persons listed in regulation 6(1) as are not the person serving the notice;
- (b) the Health and Safety Executive;
- (c) the Food Standards Agency;
- [(d) Public Health England, an executive agency of the Department of Health;]
- (e) where the class or type of practice involves a radioactive substance, the Environment Agency[, the Natural Resources Body for Wales] and the Scottish Environment Protection Agency; and
- (f) any other person he considers appropriate.

(4) The contravention notice shall—

- (a) state that the Justifying Authority is of the opinion that the person has committed a relevant breach;
- (b) specify the matters constituting the contravention or failure to comply;
- (c) specify the steps that must be taken to remedy the contravention or failure to comply;
- (d) specify the period within which those steps must be taken;
- (e) specify the date on which the notice is to take effect;
- (f) invite the person upon whom the notice is served to notify the Justifying Authority within 14 days of any grounds upon which the notice ought to be varied or withdrawn under paragraph (5); and
- (g) not take effect before 28 days after the date of service of the notice, unless in the opinion of the Justifying Authority it is necessary for the notice to have immediate effect or otherwise to be expedited.

(5) The Justifying Authority may vary or withdraw any contravention notice by serving a further notice; and paragraph (3) and, in the case of a variation, paragraph (4) shall apply to that notice.

[8.3425]

Statutory annotations

Amendment

Para (3): sub-para (d) substituted by SI 2013/235, art 11, Sch 2, Pt 1, para 68(1), (3); in sub-para (e) words ', the Natural Resources Body for Wales' in square brackets inserted by SI 2013/1821, art 32(1), (3).

Petroleum (Consolidation) Regulations 2014

SI 2014/1637

[8.3741]

The Secretary of State makes these Regulations in exercise of the powers conferred upon him by sections 15(1), (2), (3)(a) and (c), (4), (6)(b) and (8), 18(2)(a), 43(2) and (4), 80(1) and 82(3)(a) of, and paragraphs 1(1), (2), (3) and (4), 3(1), 4, 9, 12 and 15(1) of Schedule 3 to, the Health and Safety at Work etc Act 1974 ("the 1974 Act")—

(a) for the purpose of giving effect without modifications to proposals submitted to him by the Health and Safety Executive under section 11(3) of the 1974 Act, the Executive having consulted in accordance with section 50(3) of that Act, and

(b) it appearing to him that the repeal of the Acts and the provisions of the Act mentioned in Part 1 of Schedule 4, and modification of the provisions of the Acts and instrument mentioned in paragraphs 1 to 5 of Part 2 of Schedule 4, are expedient as set out in section 80(1) of the 1974 Act and after consulting such bodies which appeared to him to be appropriate in accordance with section 80(4) of that Act.

PART 1

INTRODUCTION AND GENERAL PROHIBITION ON THE KEEPING OF PETROL

[8.3742]

1 Citation and commencement

(1) These Regulations may be cited as the Petroleum (Consolidation) Regulations 2014.

(2) They come into force on 1st October 2014.

[8.3743]

2 Interpretation

In these Regulations—

"the 1974 Act" means the Health and Safety at Work etc Act 1974;

"dispenser" means a mechanical or electrical device for the dispensing of petrol;

"dispensing of petrol" means the manual or electrical pumping of petrol from a storage tank into—

(a) the fuel tank for an internal combustion engine; or

(b) any container, including a demountable fuel tank;

"dispensing premises" means a workplace for the dispensing of petrol that is not within the curtilage of a domestic premises and does not include a road tanker or similar vehicle;

"demountable fuel tank" means the fuel tank for the internal combustion engine of a motor boat or similar vessel, motor vehicle, hovercraft or aircraft that has been designed so that it can be safely removed and re-affixed to the engine without leaking fuel;

"dock" includes any pier, jetty or other place, in or at which ships can ship or unship goods or passengers;

"domestic or other relevant premises" means—

(a) domestic premises; or

(b) non-domestic premises that are not a workplace;

"domestic or other relevant user" means a person who stores petrol in domestic or other relevant premises;

"harbour" means any harbour, whether natural or artificial, and any port, haven, estuary, tidal or other river, canal or inland navigation navigated by sea-going ships and any dock;

"motor vehicle" means all mechanically propelled vehicles intended or adapted for use on land;

"petroleum enforcement authority" means—

(a) in England—

(i) in Greater London or a metropolitan county, the fire and rescue authority;

(ii) in a non-metropolitan county, the county council, and may include district borough, city or unitary councils;

(iii) in the Isles of Scilly, the Council for the Isles of Scilly;

(b) in Wales, the county council or county borough council;

(c) in Scotland, a council constituted under section 2 of the Local Government etc (Scotland) Act 1994;

"petrol" means petroleum or a mixture of petroleum with one or more substances which—

(a) is liquid or viscous at a temperature of 15°C and a pressure of 101.325 kPa (1013.25 mb); and

(b) when tested in accordance with Part A.9 of the Annex to Council Regulation (EC) No 440/2008 (laying down the test methods pursuant to Regulation (EC) No 1907/2006 of the European Parliament and of the Council on the Registration, Evaluation, Authorisation and Restriction of Chemicals REACH), has a flash point (as defined in that Part) of less than 21°C;

"portable petrol storage container" means any container designed, manufactured or sold for storing petrol, other than a demountable fuel tank or a fuel tank attached to an internal combustion engine, that is designed to be carried by hand;

"relevant date" means the date on which these Regulations come into force;

"ship", except in regulation 18, includes every description of vessel used in navigation, whatever its method of propulsion, including hovercraft;

"workplace" means a workplace within the meaning of the Dangerous Substances and Explosive Atmospheres Regulations 2002.

[8.3744]

3 Application

(1) These Regulations do not apply to—

(a) a workplace except one—

(i) used, or intended for use, for the dispensing of petrol; or

(ii) within the curtilage of domestic premises; or

(b) any establishment to which the Control of Major Accident Hazards Regulations 1999 apply by virtue of regulation 3 of those Regulations.

(2) Regulations 12(2), (3) and (4), 18(1)(b) and 21 do not apply to Her Majesty's Forces.

(3) In this regulation "Her Majesty's Forces" means any of the naval, military or air forces of the Crown, whether raised inside or outside the United Kingdom and whether any such force is a regular, auxiliary or reserve force, and includes any civilian employed by those forces.

[8.3745]

4 General prohibition on the keeping of petrol

No person is to keep petrol except in compliance with regulation 5 or regulation 13.

PART 2

STORAGE OF PETROL ON DISPENSING PREMISES

[8.3746]

5 Storage under certificate

A person keeps petrol in compliance with this regulation if the petrol is kept on dispensing premises in respect of which there is—

(a) a valid storage certificate; and

(b) compliance with the notification requirements set out in regulations 9 and 10, as applicable.

[8.3747]

6 Application and grant of storage certificates

(1) A petroleum enforcement authority must grant a certificate in respect of dispensing premises (a "storage certificate") where—

(a) an application has been made in accordance with this regulation; and

(b) it is satisfied that the containment system for petrol at the dispensing premises, including storage tanks, pipework and dispensers, may reasonably be used to store petrol and would not create an unacceptable risk to the health or safety of any person.

(2) An application under this regulation may be made by any person.

(3) An application under this regulation must include the following information—

(a) the name and address of the applicant;

(b) the address of the dispensing premises;

(c) a drawing of the layout of the dispensing premises;

(d) a drawing of the containment system for petrol at the dispensing premises, including storage tanks and pipework;

(e) a drawing of the drainage system for petrol at the dispensing premises.

(4) A storage certificate must contain the following information—

(a) the name of the petroleum enforcement authority granting the certificate;

(b) the address of the dispensing premises in respect of which the certificate is granted;

(c) a drawing of the layout of the dispensing premises;

(d) a drawing of the containment system for petrol at the dispensing premises, including storage tanks and pipe work;

(e) a drawing of the drainage system for petrol at the dispensing premises.

(5) An application for a storage certificate must be made within—

(a) a maximum of six months; and

(b) a minimum of 28 days,

before the day on which the storage of petrol is to be undertaken at the dispensing premises.

(6) A storage certificate is not personal to the person to whom it is granted and remains valid on transfer of ownership, operation or management of the dispensing premises.

[8.3748]

7 Validity of storage certificates

A storage certificate remains valid for so long as—

(a) there is no prescribed material change (see regulation 8); and

(b) the dispensing premises do not cease to be used for the keeping of petrol for a continuous period exceeding 12 months.

[8.3749]

8 Prescribed material changes

(1) For the purposes of regulation 7 a "prescribed material change" is a change to dispensing premises of a description falling within Schedule 1.

(2) A prescribed material change must be notified to the petroleum enforcement authority at least 28 days before the commencement of any works to effect the change.

(3) Where a prescribed material change has been effected to dispensing premises, the petroleum enforcement authority must issue, subject to the requirements of regulation 6(1)(b) being met, a new storage certificate containing the information specified in regulation 6(4) to—

(a) the person to whom the certificate which it replaces was granted; or

(b) if that petroleum enforcement authority has subsequently been notified of the keeping of petrol at the premises by another person under regulation 10, the last such person to notify the petroleum enforcement authority.

[8.3750]

9 Notification before ceasing to occupy dispensing premises in respect of which there is a storage certificate

(1) A person keeping petrol on dispensing premises in respect of which there is a storage certificate must notify the petroleum enforcement authority in accordance with this regulation before ceasing to occupy the dispensing premises.

(2) The notification must be made in writing and must state—

(a) the person's name and address;

(b) the address of the dispensing premises; and

(c) the date on which the person proposes to cease occupation of the dispensing premises.

(3) The notification must be given—

(a) a maximum of six months; and

(b) a minimum of 28 days,

before the day on which the person ceases to occupy the dispensing premises.

(4) A contravention of paragraph (3)(b) does not relieve a person from the duty to notify the petroleum enforcement authority under paragraphs (1) and (2).

[8.3751]

10 Notification before keeping petrol on dispensing premises in respect of which there is an existing storage certificate

(1) This regulation applies where a person proposes to keep petrol on dispensing premises in respect of which there is a valid storage certificate, and that person was not the person to whom the storage certificate was granted.

(2) A person proposing to keep petrol where this regulation applies must notify the petroleum enforcement authority in accordance with this regulation before keeping petrol on the dispensing premises.

(3) The notification must be made in writing and must state—

(a) the person's name and address;
(b) the address of the dispensing premises; and
(c) the date on which the person proposes to commence keeping petrol at the dispensing premises.

(4) The notification must be given—
(a) a maximum of six months; and
(b) a minimum of 28 days,

before the day on which the person proposes to keep petrol on the dispensing premises.

(5) A contravention of paragraph (4)(b) does not relieve a person from the duty to notify the petroleum enforcement authority under paragraphs (2) and (3).

[8.3752]

11 Appeals from refusals by petroleum enforcement authorities to grant storage certificates

(1) A person who is aggrieved by a decision of a petroleum enforcement authority not to grant a storage certificate may appeal to the Secretary of State.

(2) Subsections (2) to (6) of section 44 of the 1974 Act apply for the purposes of this regulation as they apply to an appeal under section 44(1) of that Act.

(3) The Health and Safety Licensing Appeals (Hearings Procedure) Rules 1974 apply for the purposes of this regulation, but as if any reference in those Rules—
(a) to an appeal were a reference to an appeal under paragraph (1); and
(b) to a licensing authority were a reference to a petroleum enforcement authority.

(4) Paragraph (3) applies in Scotland as if the reference to the Health and Safety Licensing Appeals (Hearings Procedure) Rules 1974 were a reference to the Health and Safety Licensing Appeals (Hearings Procedure) (Scotland) Rules 1974.

(5) The Secretary of State, when determining the appeal, may direct the petroleum enforcement authority to grant a storage certificate.

(6) In paragraph (5) "Secretary of State" includes a person who under section 44(2) of the 1974 Act determines the appeal on behalf of the Secretary of State.

[8.3753]

12 General prohibitions on dispensing and supply of petrol

(1) No person is to dispense, or allow the dispensing of, petrol from a dispenser on dispensing premises except—
(a) for use in a motor vehicle or motor boat;
(b) to fill a suitable portable container or a demountable fuel tank; or
(c) for the purpose of maintenance or calibration of the dispenser.

(2) No person under the age of sixteen years is to operate a dispenser on dispensing premises.

(3) No person is to supply, or allow the supply of, petrol to a person under the age of sixteen years.

(4) No person is to dispense petrol from a dispenser otherwise than on dispensing premises except to fill a suitable portable container.

(5) Paragraph (4) does not apply in relation to the re-fuelling of an aircraft by a bowser or vehicle specifically designed for the purpose of re-fuelling aircraft.

(6) In this regulation "suitable portable container" has the meaning given by paragraph 6 of Schedule 2.

PART 3

STORAGE OF PETROL IN DOMESTIC OR OTHER RELEVANT PREMISES

[8.3754]

13 Storage of petrol in domestic or other relevant premises

(1) A person keeps petrol in compliance with this regulation if the petrol—

(a) is kept in domestic or other relevant premises;

(b) is kept otherwise than for sale; and

(c) is kept—

(i) in accordance with paragraph 1 or 2 of Schedule 2; or

(ii) where there is a valid licence authorising the keeping of petrol granted by the petroleum enforcement authority under regulation 14, in accordance with that licence.

(2) A person keeps petrol in compliance with this regulation if the person keeps no more than 500 millilitres of petrol in a suitable container that is securely stopped, in the circumstances described in paragraph (1)(a) and (b), whether or not the person also keeps petrol pursuant to paragraph (1)(c).

(3) Petrol kept pursuant to paragraph (2) is not to be counted towards the amount of petrol kept in a storage place in accordance with Schedule 2.

[8.3755]

14 Licences for storage in domestic or other relevant premises

(1) A petroleum enforcement authority may grant a person a licence authorising the keeping of petrol in excess of 275 litres in domestic or other relevant premises.

(2) A licence may only authorise the keeping of petrol for private use.

(3) "Private use" does not include keeping petrol for the purpose of sale.

(4) A petroleum enforcement authority may only grant a licence if it is satisfied that the keeping of petrol in accordance with the licence, including any proposed conditions of the licence, would not create an unacceptable risk to the health or safety of any person.

(5) A petroleum enforcement authority may attach conditions to a licence concerning any or all of the following matters—

(a) the nature and location of the premises in which petrol is to be kept;

(b) the amount of petrol kept;

(c) the mode of storage of petrol;

(d) the dispensing of petrol;

(e) the nature of any goods with which petrol is to be stored;

(f) the systems, if any, for fire detection;

(g) the equipment, if any, for the extinguishing of fires;

(h) the security of the site and petrol stored on the premises;

(i) the safe-keeping of petrol, generally.

(6) A licence must specify the premises, and the places within the premises where petrol is to be kept, and the name of the licensee.

(7) A licence may be granted for a maximum of three years.

(8) A licence is personal to the licensee and is not transferrable.

[8.3756]

15 Appeals from refusals by petroleum enforcement authorities to grant licences

(1) A person who is aggrieved by a decision of a petroleum enforcement authority not to grant a licence under regulation 14 may appeal to the Secretary of State.

(2) Subsections (2) to (6) of section 44 of the 1974 Act apply for the purposes of this regulation as they apply to an appeal under section 44(1) of that Act.

(3) The Health and Safety Licensing Appeals (Hearings Procedure) Rules 1974 apply for the purposes of this regulation, but as if any reference in those Rules—

(a) to an appeal were a reference to an appeal under paragraph (1); and

(b) to a licensing authority were a reference to a petroleum enforcement authority.

(4) Paragraph (3) applies in Scotland as if the reference to the Health and Safety Licensing Appeals (Hearings Procedure) Rules 1974 were a reference to the Health and Safety Licensing Appeals (Hearings Procedure) (Scotland) Rules 1974.

(5) The Secretary of State, when determining the appeal, may direct the petroleum enforcement authority to grant a licence.

(6) In paragraph (5) "Secretary of State" includes a person who under section 44(2) of the 1974 Act determines the appeal on behalf of the Secretary of State.

[8.3757]

16 Renewal and variation of licences

(1) A petroleum enforcement authority may renew a licence, on the application of the licensee, for a period of up to three years if it is satisfied that the keeping of petrol in accordance with the conditions of the licence would not create an unacceptable risk to the health or safety of any person.

(2) A petroleum enforcement authority may vary the conditions of a licence, on the application of the licensee, if the petroleum enforcement authority is satisfied that the storage of petrol in accordance with the conditions of the licence, as it is proposed they be varied, would not create an unacceptable risk to the health or safety of any person.

(3) In varying the conditions of a licence a petroleum enforcement authority may attach only conditions that it would be able to attach pursuant to regulation 14 on the grant of a licence.

[8.3758]

17 Revocation of licences

(1) A petroleum enforcement authority may revoke a licence where—

(a) there has been a change in circumstances so that the continued keeping of petrol in accordance with the conditions of the licence would create an unacceptable risk to the health or safety of any person; or

(b) a condition of the licence has been contravened.

(2) Where a licence is revoked, the licensee must return the licence to the petroleum enforcement authority within 28 days of the date of the revocation.

PART 4

MISCELLANEOUS PROVISIONS

[8.3759]

18 Enforcing authorities

(1) A petroleum enforcement authority is the enforcing authority within its area for—

(a) these Regulations;

(b) sections 2 to 4 and 7 and 8 of the 1974 Act for the purposes of regulation 21(1), even if the road tanker in question is on a road at the time of that filling or replenishing; and

(c) the Dangerous Substances and Explosive Atmospheres Regulations 2002 in so far as they apply to any activity—
(i) at a retail petrol filling station relating to fuelling motor vehicles, ships with vehicle fuel or aircraft with petrol; or
(ii) at a non-retail petrol filling station relating to fuelling motor vehicles, ships or aircraft with petrol.

(2) Paragraph (1)(c) does not apply to—
(a) any activity at a retail petrol filling station connected with—
(i) repairing motor vehicles, ships or aircraft; or
(ii) retailing goods other than vehicle fuel; or
(b) any activity at a non-retail petrol filling station connected with repairing motor vehicles, ships or aircraft.

(3) In this regulation—

"non-retail petrol filling station" means premises used, or intended for use, for dispensing petrol for use in motor vehicles, ships or aircraft and includes any petrol dispenser, other apparatus or storage tank for storing petrol used there in connection with that fuelling, but it does not include any retail petrol filling station;

"retail petrol filling station" means premises used, or intended for use, for dispensing petrol to the public for use in motor vehicles, ships or aircraft by way of sale and includes any vehicle fuel dispenser, other apparatus or storage tank for storing vehicle fuel used there in connection with that fuelling;

"ship" includes every description of vessel used in navigation propelled by means of an internal combustion engine and includes hovercraft;

"vehicle fuel" means petrol or any other substance which provides the power in an internal-combustion engine in a motor vehicle, ship or aircraft;

"vehicle fuel dispenser" means a mechanical or electrical device for the dispensing of vehicle fuel.

[8.3760]

19 Manufacture etc of portable petrol storage containers

(1) Any person who manufactures, imports or supplies any container for use as a portable petrol storage container must ensure that the container meets the standards for such containers specified in Schedule 3.

(2) In any proceedings for an offence under paragraph (1) it is a defence for any person who supplied the container, but did not manufacture it, to prove that the person took all reasonable precautions and exercised all due diligence to avoid the commission of the offence.

[8.3761]

20 Repairs of vessels or containers that have contained petrol

A domestic or other relevant user must not repair, or allow the repair of, any container or other vessel, which to the person's knowledge has contained petrol, or still contains petrol, without first taking all reasonable precautions to ensure that the container or other vessel has been rendered free from petrol and any flammable petrol vapours.

[8.3762]

21 Prohibition on direct filling of fuel tanks from road tankers

(1) No person is to fill or replenish a fuel tank for an internal combustion engine or a portable container with petrol directly from a road tanker conveying petrol.

(2) A road tanker is conveying petrol where it is conveying petrol by road in circumstances where the Carriage of Dangerous Goods and Use of Transportable Pressure Equipment Regulations 2009 apply to that tanker.

[8.3763]

22 Limited continuation of activities that were formerly compliant

(1) A petrol storage container that either—

(a) is made of metal and immediately before the relevant date complied with the requirements of regulations 2(1) and 4 of the Petroleum-Spirit (Motor Vehicles, etc) Regulations 1929; or

(b) immediately before the relevant date complied with sub-paragraphs (a) to (d) of regulation 3(1) of, and the Schedule to, the Petroleum-Spirit (Plastic Containers) Regulations 1982,

may be manufactured and sold for use in Great Britain for up to two years after the relevant date.

(2) A person storing petrol in domestic or other relevant premises in accordance with the storage requirements in the Petroleum-Spirit (Motor Vehicles, etc) Regulations 1929 immediately before the relevant date is to be treated as complying with the storage requirements in Schedule 2 for up to two years after the relevant date.

(3) A person storing up to 15 litres of petrol in domestic or other relevant premises in accordance with the storage requirements in section 1 of the Petroleum (Consolidation) Act 1928 immediately before the relevant date is to be treated as complying with the storage requirements in Schedule 2 for up to two years after the relevant date.

(4) For petrol stored domestically in accordance with the storage requirements in the Petroleum-Spirit (Motor Vehicles, etc) Regulations 1929 any valid notice already given under those Regulations is deemed to be notice given in accordance with paragraph 4 of Schedule 2 to these Regulations, and remains valid for up to one year after the relevant date, or until the user gives or renews notice under these Regulations, whichever occurs first.

(5) An application for a licence under section 2 of the Petroleum (Consolidation) Act 1928 received, but not determined, by a petroleum licensing authority before the relevant date, is to be considered under regulation 6, 14 or 16 of these Regulations, as applicable, and the provisions of these Regulations apply.

(6) An application for the transfer of a licence under section 1 of the Petroleum (Transfer of Licences) Act 1936 received, but not determined, by the petroleum licensing authority before the relevant date, is to be considered under regulation 6 or 14 of these Regulations, as applicable, and the provisions of these Regulations apply.

(7) Fees paid for the grant or transfer of a licence in the circumstances described in paragraphs (5) and (6) are deemed to be fees paid in full for the determination of the application under these Regulations.

[8.3764]

23 Transitional provisions in relation to licences granted under the Petroleum (Consolidation) Act 1928

(1) This regulation applies to the keeping of petrol that immediately before the relevant date was authorised by a licence granted under section 2(2) of the Petroleum (Consolidation) Act 1928.

(2) Where the keeping of petrol to which this regulation applies—

(a) occurs on dispensing premises; and

(b) fulfils the requirements in paragraph (4),

it is deemed to be kept in compliance with a valid certificate granted under regulation 6.

(3) Where the keeping of petrol to which this regulation applies—

(a) occurs on domestic or other relevant premises;

(b) is for private use; and

(c) fulfils the requirements in paragraph (4),

it is deemed to be kept in compliance with a valid licence granted under regulation 14.

(4) The requirements are the conditions attached to the licence referred to in paragraph (1), but excluding any provision for renewal of the licence.

(5) "Private use" does not include keeping petrol for the purpose of sale.

(6) Except for paragraph (7), this regulation ceases to have effect at the end of the period of three years beginning with the relevant date.

(7) A licence deemed to be tantamount to a valid storage certificate under paragraph (2), where the storage arrangements remain unaltered, is to be converted, by the petroleum enforcement authority, to a storage certificate on its expiry without having to fulfil the requirements set out in regulation 6.

[8.3765]

24 Savings in relation to byelaws

Despite the revocation of regulation 17(5) of the Dangerous Substances and Explosive Atmospheres Regulations 2002 (which saves byelaws made or having effect under section 9 of the Petroleum (Consolidation) Act 1928), byelaws made or having effect under section 9 of the Petroleum (Consolidation) Act 1928 which are in force immediately before the relevant date are to continue.

[8.3766]

25 Repeals, revocations and consequential amendments

(1) The Acts and instruments referred to in column 1 of Part 1 of Schedule 4 are repealed or revoked to the extent specified in the corresponding entry in column 3 of that Part.

(2) Part 2 of Schedule 4 (which makes consequential amendments) has effect.

[8.3767]

26 Review

(1) Before the end of the review period, the Secretary of State must—

(a) carry out a review of these Regulations;

(b) set out the conclusions of the review in a report; and

(c) publish the report.

(2) The report must in particular—

(a) set out the objectives intended to be achieved by the regulatory system established by these Regulations;

(b) assess the extent to which those objectives are achieved; and

(c) assess whether those objectives remain appropriate and, if so, the extent to which they could be achieved with a system that imposes less regulation.

(3) "Review period" means the period of five years beginning with the relevant date.

SCHEDULE 1

PRESCRIBED MATERIAL CHANGES TO DISPENSING PREMISES

Regulation 8

[8.3768]

1

The cessation of use of one or more of the petrol storage tanks.

2

The removal or permanent decommissioning of one or more of the petrol storage tanks.

3

The installation of any tank, pipework or vapour pipework associated with the storage and dispensing of petrol.

4

The installation of any petrol pump, any other automotive pump, or dispenser in a new location.

SCHEDULE 2

PETROL STORAGE IN DOMESTIC OR OTHER RELEVANT PREMISES

Regulation 13(1)(c)

[8.3769]

Requirements for keeping up to 30 litres of petrol in suitable portable containers or in a single demountable fuel tank

1

A person keeps petrol in accordance with this paragraph if—

(a) no more than a total of 30 litres in suitable portable containers or in a single demountable fuel tank is kept, of which—
  (i) no more than 30 litres is kept in one or more suitable portable containers;
  (ii) no more than 30 litres is kept in one demountable fuel tank; or
  (iii) no more than 30 litres is kept in no more than two suitable portable containers in any motor vehicle, motor boat, hovercraft or aircraft;

(b) the storage place is—
  (i) within, above, below, attached to, or within the curtilage of a building (but, subject to paragraphs (ii) and (iii), not a flat or public building);
  (ii) below a flat;
  (iii) attached to a public building; or
  (iv) in a vehicle for the purpose of using it as fuel for any internal combustion engine;

(c) the storage place, where it is—
  (i) within a building, is fire-separated from the rest of the building and any exit route from the building; and
  (ii) above, below or attached to a building, is fire-separated from the building; and

(d) the common storage requirements of paragraph 3 are met.

Requirements for keeping of up to 275 litres of petrol

2

A person keeps petrol in accordance with this paragraph if—

(a) petrol is kept in accordance with the common storage requirements of paragraph 3;

(b) notice has been given to the petroleum enforcement authority in accordance with paragraph 4;

(c)

(i) more petrol is kept than may be kept in accordance with paragraph 1; and

(ii) within the entirety of the domestic or other relevant premises no more than 275 litres is kept; and either

(d) no more than so much petrol as may be kept in accordance with paragraph 1 is kept in accordance with that paragraph and the remainder is kept in accordance with (e); or

(e) all petrol is kept—

(i) in suitable portable containers, or demountable fuel tanks, or both of those types of container;

(ii) in a storage place that is—

(aa) a detached building; or

(bb) attached to a building (but not domestic premises), and is fire-separated from the building; and

(iii) where fire extinguishing and spillage containment apparatus is available either within the storage place, or as close as is reasonably practicable to the storage place.

Common storage requirements for the keeping of intermediate quantities of petrol

3

The requirements of this paragraph are met if—

(a) the storage place is not living accommodation;

(b) no dispensing of petrol occurs at the storage place;

(c) where the storage place is not in the open air, it has a direct exit to the open air and is ventilated to the open air;

(d) all reasonable precautions are taken in the storage place to prevent any sources of ignition or heat that would be liable to ignite petrol or its vapour; and

(e) petrol is not used in the storage place otherwise than—

(i) in the fuel tank of any internal combustion engine; or

(ii) in quantities, not exceeding at any one time 150 millilitres, for cleaning or as a solvent for repair purposes.

Notification of petrol storage

4

A petroleum enforcement authority is given notice in accordance with this paragraph where notice—

(a) is made in writing, stating—

(i) the name of the occupier of the storage place;

(ii) the address of the domestic or other relevant premises where the storage place is located;

(iii) that petrol is kept, or to be kept, on those domestic or other relevant premises in accordance with paragraph 2; and

(b) is given by the occupier of the storage place—

(i) before keeping, or permitting the keeping, of petrol in accordance with paragraph 2; and

(ii) after that, annually in the month of January.

Calculating the amount of petrol kept in a storage place

5

For the purpose of calculating the amount of petrol kept in domestic or other relevant premises, any petrol contained in a fuel tank that is connected to an internal combustion engine in the way it would ordinarily be connected when the engine is running, is not to be counted towards the total amount of petrol kept.

Suitable portable containers

6

For the purposes of this Schedule (and regulation 12) a container is a "suitable portable container" if it—

- (a) is in a good state of repair, secure from leaking, has not degraded from exposure to chemicals or light and is not unsteady when placed on a level surface; and
- (b) fulfils one of the following requirements—
  - (i) it meets the requirements of Schedule 3;
  - (ii) it is made of metal and immediately before the relevant date complied (or would have complied if it were used for the purpose of keeping petrol) with the requirements of regulations 2(1) and 4 of the Petroleum-Spirits (Motor Vehicles, etc) Regulations 1929; or
  - (iii) immediately before the relevant date it had the characteristics described in sub-paragraphs (a) to (d) of regulation 3(1) of, and the Schedule to, the Petroleum-Sprits (Plastic Containers) Regulations 1982.

Interpretation

7

In this Schedule—

"fire-resistant internal linings" means materials or products used in lining any partition, wall, ceiling or other internal structure that—

- (a) adequately resist the spread of flame over their surfaces; and
- (b) have, if ignited, either a rate of heat release or a rate of fire growth, that is reasonable in the circumstances;

"fire-separated" means—

- (a) separated by an external wall of a building which is able to resist adequately the spread of fire over the wall and from one building to another, having regard to the height, use and position of the building;
- (b) separated by a partition, wall, ceiling or other internal structure, containing fire-resistant internal linings; or
- (c) separated by the means described in sub-paragraph (a) or (b), other than a floor or ceiling, where there is an opening fitted with a self-closing door constructed of hard wood or other fire-resisting material;

"flat" means a separate and self-contained premises constructed or adapted for use for residential purposes and forming part of a building, from some other part of which it is divided horizontally;

"public building" means a building consisting of or containing a place of public resort but a building is not to be treated as a place of public resort because it is, or it contains, a dwelling to which the public are occasionally admitted;

"storage place" in domestic or other relevant premises means the place where petrol is stored.

SCHEDULE 3

PORTABLE PETROL STORAGE CONTAINERS

Regulation 19

[8.3770]

1

A portable petrol storage container must—

(a) have a nominal capacity—
  (i) no greater than 10 litres if made of plastic; and
  (ii) no greater than 20 litres if made of metal;

(b) have a total capacity between 10% and 15% more than the nominal capacity;

(c) be made of either metal or plastic that is suitable and safe for the purpose and will not significantly degrade due to exposure to petrol or naturally occurring ultra-violet radiation;

(d) be designed and constructed so that—
  (i) it is reasonably robust and not liable to break under the normal conditions of use;
  (ii) the escape of liquid or vapour is prevented;
  (iii) petrol can be poured safely from it; and
  (iv) it is not unsteady when placed on a flat surface;

(e) be marked or labelled in a legible and indelible form with—
  (i) the words "PETROL" and "HIGHLY FLAMMABLE";
  (ii) an appropriate hazard warning sign;
  (iii) the nominal capacity in litres; and
  (iv) the manufacturer's name and the date and month of manufacture.

2

In paragraph 1—

"nominal capacity" means the maximum volume of liquid that the container is intended to hold at 20°C;

"total capacity" means the maximum volume of liquid that the container would hold if filled to the brim.

SCHEDULE 4

Regulation 25

PART 1

REPEALS AND REVOCATIONS

[8.3771]

Column 1 *Title*	*Column 2* *Reference*	*Column 3* *Extent of repeal or revocation*
London County Council (General Powers) Act 1912	c civ	The whole Act.
Petroleum (Consolidation) Act 1928	c 32	The whole Act.

Petroleum (Transfer of Licences) Act 1936	c 27	The whole Act.
Regulatory Enforcement and Sanctions Act 2008	c 13	The entries in Schedule 3 relating to the Petroleum (Consolidation) Act 1928 and the Petroleum (Transfer of Licences) Act 1936, and the entry in Schedule 6 relating to the Petroleum (Consolidation) Act 1928, sections 1, 5 and 18.
Petroleum-Spirit (Motor Vehicles, etc) Regulations 1929	SR & O 1929/952	The whole Regulations.
Petroleum (Mixtures) Order 1929	SR & O 1929/993	The whole Order.
Petroleum (Liquid Methane) Order 1957	SR & O 1957/859	The whole Order.
Petroleum (Regulation) Acts 1928 and 1936 (Repeals and Modifications) Regulations 1974	SI 1974/1942	The whole Regulations.
Petroleum (Consolidation) Act 1928 (Enforcement) Regulations 1979	SI 1979/427	The whole Regulations.
Petroleum-Spirit (Plastic Containers) Regulations 1982	SI 1982/630	The whole Regulations.
Dangerous Substances in Harbour Areas Regulations 1987	SI 1987/37	Regulation 47(3).
Health and Safety (Explosives and Petroleum Fees) (Modification) Regulations 1987	SI 1987/52	The whole Regulations.
Health and Safety (Miscellaneous Provisions) (Metrication etc) Regulations 1992	SI 1992/1811	The entry in the first column in the table in Part 1 of Schedule 1 referring to the Petroleum (Consolidation) Act 1928 and the corresponding entries below in the first, second, third and fourth columns in that table.

Health and Safety Information for Employees (Modifications and Repeals) Regulations 1995	SI 1995/2923	The entry in the first column in the table in Part 1 of the Schedule referring to the Petroleum (Consolidation) Act 1928 and the corresponding entry in the second column in that table.
Carriage of Dangerous Goods by Rail Regulations 1996	SI 1996/2089	The whole Regulations.
Carriage of Dangerous Goods by Road Regulations 1996	SI 1996/2095	The whole Regulations.
Quarries Regulations 1999	SI 1999/2024	In column 1 of Part II of Schedule 5, the reference to the Carriage of Dangerous Goods by Rail Regulations 1996 and the corresponding entries in column 2 of that Part.
Dangerous Substances and Explosive Atmospheres Regulations 2002	SI 2002/2776	Regulation 17(4) and (5). Paragraphs 2, 3, 4, 5, and 6 of Schedule 6. The entry in the first column in the table in Part 1 of Schedule 7 referring to the Petroleum (Consolidation) Act 1928 and the corresponding entries in the second and third columns in that table.

PART 2

[8.3772]–[8.3780]

1

In section 143(4)(a) (regulations for building near dangerous businesses) of the London Building Act 1930 for "licence is in force under the provisions of the Petroleum (Consolidation) Act 1928" substitute "storage certificate or licence is in force under the provisions of the Petroleum (Consolidation) Regulations 2014".

2

For section 69(8) (control of petroleum-filling stations) of the London County Council (General Powers) Act 1933 substitute—

"(8) In this section the expression "petroleum-filling station" means any premises or place used or intended to be used by way of trade or for purposes of gain for fuelling motor vehicles with petrol by means of a pump or standpipe or other like apparatus and includes any building advertisement pump or other apparatus used in connection with such premises or place and the expression "petrol" has the meaning given by regulation 2 of the Petroleum (Consolidation) Regulations 2014.".

3

In section 73 (derelict petrol tanks) of the Public Health Act 1961—

(1) in subsection (1), for "petroleum spirit" substitute "petrol";

(2) in subsection (5), for "section twenty-three of the Petroleum (Consolidation) Act 1928" substitute "section 57 of the Harbours Act 1964"; and

(3) for subsection (6) substitute—

"(6) In this section "petrol" has the meaning given by regulation 2 of the Petroleum (Consolidation) Regulations 2014.".

4

In section 94 (disused petrol containers) of the Civic Government (Scotland) Act 1982—

(1) in subsection (1), for "petroleum spirit" substitute "petrol";

(2) in subsection (4), for "section 23 of the Petroleum (Consolidation) Act 1928" substitute "section 57 of the Harbours Act 1964"; and

(3) for subsection (5) substitute—

"(5) In this section "petrol" has the meaning given by regulation 2 of the Petroleum (Consolidation) Regulations 2014.".

5

In article 3(e) of the Isles of Scilly (Functions) Order 1979 (functions), for "the Petroleum (Consolidation) Act 1928" substitute "the Petroleum (Consolidation) Regulations 2014".

6

In the Dangerous Substances in Harbour Areas Regulations 1987—

(a) in regulation 2(1) (interpretation), for the definition beginning with "petroleum-spirit" substitute ""petroleum-spirit" means petrol within the meaning of regulation 2 of the Petroleum (Consolidation) Regulations 2014";

(b) for regulation 5(2)(a) (application of these Regulations) substitute—

"(a) petroleum-spirit, intended for use in any internal combustion engine and not wholly or partly for the purpose of sale, kept in a container which meets the requirements of the Petroleum (Consolidation) Regulations 2014.";

(c) for regulation 29(b) (application of this Part) substitute—

"(b) to petroleum-spirit kept in accordance with the requirements of the Petroleum (Consolidation) Regulations 2014.".

7

In the Dangerous Substances (notification and marking of sites) Regulations 1990, in Schedule 1 (Exceptions)—

(a) for paragraph 2(c) substitute—

"(c) sites in respect of which a storage certificate or licence is in force for the keeping of petrol under the Petroleum (Consolidation) Regulations 2014";

(b) for paragraph 4 substitute—

"4

Regulations 5 and 6 (which relate to access and location marking respectively) do not apply to retail or non-retail petrol filling stations within the meaning of regulation 18 of the Petroleum (Consolidation) Regulations 2014.".

8

In paragraph (10) of regulation 7 (labelling of dangerous substances and dangerous preparations) of the Chemicals (Hazard Information and Packaging for Supply) Regulations 2009, for sub-paragraph (a) substitute—

"(a) regulation 19 of and paragraph 1(e) of Schedule 3 to the Petroleum (Consolidation) Regulations 2014;".

9

In the Health and Safety (Fees) Regulations 2012—

(a) in the heading to regulation 9 for "Petroleum (Consolidation) Act 1928 and the Petroleum (Transfer of Licences) Act 1936" substitute "Petroleum (Consolidation) Regulations 2014";

(b) for paragraph (5) of regulation 9 substitute—

"(5) Where an application in relation to the provision specified in column 1 of Part 4 of Schedule 8 is made for the purpose described in the corresponding entry in column 2 of that Part, the fee specified in the corresponding entry in column 3 of that Part is payable by the applicant to the petroleum enforcement authority.";

(c) in paragraph (12) of regulation 9 omit ", 4" in each place where it appears;

(d) for regulation 10 substitute—

"Date from which fees are payable under Petroleum (Consolidation) Regulations 2014

10

(1) The fee prescribed by these Regulations in respect of an application for a storage certificate under regulation 6 of the Petroleum (Consolidation) Regulations 2014 is payable on the certificate first having effect on or after the date the Petroleum (Consolidation) Regulations 2014 come into force, irrespective of the date of the application for that storage certificate.

(2) The fee prescribed by these Regulations in respect of an application for a licence under regulation 14 of the Petroleum (Consolidation) Regulations 2014 is payable on the licence first having effect or any renewal of a licence first taking effect, on or after the date the Petroleum (Consolidation) Regulations 2014 come into force, irrespective of the date of the application for that licence or renewal.";

(e) for Part 4 of Schedule 8 substitute—

"PART 4

FEE PAYABLE IN RESPECT OF APPLICATIONS FOR THE GRANTING OF STORAGE CERTIFICATES AND PETROL LICENCES

1 *Provision under which a fee is payable*	2 *Purpose of application*	3 *Fee*
Petroleum (Consolidation) Regulations 2014		
Regulation 6	Storage certificate, where premises to be used to store petrol of a quantity—	
	not exceeding 2,500 litres	£42 for each year of certificate

	exceeding 2,500 litres but not exceeding 50,000 litres	£58 for each year of certificate
	exceeding 50,000 litres	£120 for each year of certificate
Regulation 14	Licence to keep petrol of a quantity—	
	not exceeding 2,500 litres	£42 for each year of licence
	exceeding 2,500 litres but not exceeding 50,000 litres	£58 for each year of licence
	exceeding 50,000 litres	£120 for each year of licence".

Acetylene Safety (England and Wales and Scotland) Regulations 2014

SI 2014/1639

[8.3781]

The Secretary of State makes these Regulations—

(a) in exercise of the powers conferred by sections 15(1), (2), (3)(a) and (c), (4)(a), (5) and (9), 43(2), (4), (5) and (6), 80(1) and 82(3)(a) of, and paragraphs 1(1), (2), (3), (5), 2(1), 3, 4, 9 and 12 of Schedule 3 to, the Health and Safety at Work etc Act 1974 ("the 1974 Act");

(b) for the purposes of giving effect without modifications to proposals submitted to him by—

(i) the Health and Safety Executive under section 11(3) of the 1974 Act; and

(ii) the Office for Nuclear Regulation under section 81(1)(a)(iv) of the Energy Act 2013 ('the 2013 Act'); and

(c) it appearing to him that the repeals of the provisions of the Acts marked with an asterisk in Part 1 of Schedule 3 are expedient as set out in section 80(1) of the 1974 Act and that it was not appropriate to consult bodies in respect of those repeals in accordance with section 80(4) of that Act.

Before submitting proposals for these Regulations to the Secretary of State, the Health and Safety Executive has carried out consultations in accordance with section 50(3) of the 1974 Act, and the Office for Nuclear Regulation has carried out consultations in accordance with section 81(3) of the 2013 Act.

[8.3782]

1 Citation, commencement, application and extent

(1) These Regulations may be cited as the Acetylene Safety (England and Wales and Scotland) Regulations 2014 and come into force on 1st October 2014.

(2) These Regulations apply—

(a) within Great Britain; and

(b) apart from paragraph 8 of Schedule 1 (cylinder colouring), outside Great Britain as sections 1 to 59 and 80 to 82 of the 1974 Act apply by virtue of the Health and Safety at Work etc Act 1974 (Application outside Great Britain) Order 2013.

(3) These Regulations extend to England and Wales and Scotland only.

[8.3783]

2 Interpretation

In these Regulations—

"the 1974 Act" means the Health and Safety at Work etc Act 1974;

"the 2009 Regulations" means the Carriage of Dangerous Goods and Use of Transportable Pressure Equipment Regulations 2009;

"acetylene" means the hydrocarbon chemical listed under that name in the European inventory of existing commercial chemical substances, published in OJ C 146A on 15th June 1990, under entry number 200-816-9;

"acetylene equipment" means any equipment containing compressed acetylene gas, excluding a cylinder and acetylene manifold but including a pipe, hose, valve, or device that is—

(a) designed or manufactured for use with compressed acetylene gas, or

(b) used with compressed acetylene gas;

"acetylene gas" means acetylene gas—

(a) in a free gaseous state,

(b) dissolved in a solution, or

(c) contained in a homogeneous porous material, with or without a solvent;

"acetylene manifold" means a system of interconnected pipework in which compressed acetylene gas is contained, and which connects to, but excludes, a cylinder;

"bar(g)" means pressure, above atmospheric pressure, measured in the unit bar;

"compressed acetylene gas" means acetylene gas at pressure equal to or greater than 0.62 bar(g);

"cylinder" means—

(a) transportable pressure equipment within the meaning of article 2(1)(a) of the Transportable Pressure Equipment Directive; or

(b) an old pressure receptacle which—

(i) satisfies the requirements of regulation 14(4) of the 2009 Regulations, and

(ii) is to be treated as an old pressure receptacle under regulation 15(9) of these Regulations;

"licence" means a licence granted under regulation 9 and includes a renewed, varied or transferred licence;

"licensee" means a person who has been granted a licence under regulation 9 and includes—

(c) a person to whom a licence is transferred;

(d) a person treated as being the licensee under regulation 13; and

(e) a person deemed to hold a licence under regulation 15.

"licensing authority" means—

(a) in relation to a relevant activity carried out on ONR regulated premises, the ONR;

(b) otherwise, the Executive;

"old pressure receptacle" means an old pressure receptacle within the meaning of regulation 14(6) of the 2009 Regulations;

"ONR" means the Office for Nuclear Regulation;

"ONR regulated premises" means any premises which are, or are on—

(i) a GB nuclear site (within the meaning of section 68 of the Energy Act 2013),

(ii) an authorised defence site (within the meaning of regulation 2(1) of the Health and Safety (Enforcing Authority) Regulations 1998), or

(iii) a new nuclear build site (within the meaning of regulation 2A of those Regulations).

"relevant activity" means any of the following or a combination of any of the following—

(a) the manufacture of compressed acetylene gas,

(b) the compression of acetylene gas equal to or greater than 0.62 bar(g), or

(c) the filling of a cylinder with compressed acetylene gas;

"relevant date" means the date when these Regulations come into force;

"risk" means the likelihood of a person's safety being affected by any uncontrolled release of energy caused by—

(a) the decomposition of acetylene gas or acetylene—derived compounds, or

(b) any uncontrolled combustion of acetylene gas or acetylene—derived compounds; and

"the Transportable Pressure Equipment Directive" means Directive 2010/35/EU of the European Parliament and Council repealing Council Directives 76/767/EEC, 84/525/EEC, 84/526/EEC, 84/527/EEC and 1999/36/EC.

[8.3784]

3 Prohibition concerning liquid and solid acetylene

No person is to manufacture, use, store, transport, import or sell liquid or solid acetylene, whether on its own or in any mixture.

[8.3785]

4 Prohibition concerning compressed acetylene gas

(1) Subject to paragraph (2), no person is to use, store, transport, import or sell compressed acetylene gas.

(2) Paragraph (1) does not apply to—

(a) the use of compressed acetylene gas at equal to or lower than 1.5 bar(g) provided that—

(i) the requirements in Schedule 1 are complied with; and

(ii) risk is reduced so far as is reasonably practicable;

(b) the storage of compressed acetylene gas in an acetylene manifold, provided that—

(i) the requirements in Schedule 1 are complied with; and

(ii) risk is reduced so far as is reasonably practicable;

(c) the storage of compressed acetylene gas in a cylinder which would, if used for carriage, meet the requirements of—

(i) the 2009 Regulations; and

(ii) paragraph 8 of Schedule 1 (cylinder colouring) to these Regulations;

(d) the transportation, importation and sale of compressed acetylene gas in a cylinder, provided that—

(i) the relevant requirements of the 2009 Regulations are complied with; and

(ii) the cylinder complies with paragraph 8 of Schedule 1 (cylinder colouring) to these Regulations;

(e) the use or storage of the requisite amount of compressed acetylene gas, in conditions other than those specified in sub-paragraphs (a) to (d), where the use or storage is required to assess the compliance of acetylene equipment, an acetylene manifold or cylinder with any relevant standards of a prescribed standards body or with relevant legal requirements, provided the use or storage is undertaken in such a way that risk is reduced so far as is reasonably practicable; and

(f) the use or storage of compressed acetylene gas in conditions other than those specified in sub-paragraphs (a) to (d), during the process of testing equipment, including devices and valves, where the testing occurs for the purpose of regulatory investigations, enforcement actions or other legal proceedings initiated by the licensing authority.

(3) In this regulation—

"prescribed standards body" means any full member from time to time of the International Organization for Standardization;

"requisite amount" means—

(a) where there is a minimum amount required for the assessment in the relevant standard or in the relevant legal requirements, that minimum amount; or

(b) where there is not a minimum amount required for the assessment in the relevant standard or the relevant legal requirements, the lowest amount at which the assessment can be performed effectively;

"relevant legal requirements" means any enactment of national, European or international origin imposing, from time to time, safety or technical requirements on acetylene equipment, an acetylene manifold or cylinder, that is applicable in Great Britain or in any other jurisdiction.

[8.3786]

5 Prohibition of a relevant activity without a licence

(1) Subject to paragraph (2), no person is to conduct a relevant activity unless the person—

(a) holds a licence for that activity;

(b) complies with the conditions of that licence; and

(c) complies with the requirements of Schedule 1.

(2) Paragraph (1) does not apply to the compression of naturally occurring, by-product, or traces of, compressed acetylene gas that is mixed with other gases excluding oxygen in a proportion insufficient to sustain decomposition.

[8.3787]

6 Applications in relation to licences

(1) An application for a licence or a renewal, variation or transfer of a licence must be made to the licensing authority in a form approved for the time being by the Executive for the purpose of these Regulations.

(2) The licensing authority may request additional information from the applicant where the licensing authority considers it necessary to determine whether to grant, renew, vary or transfer a licence.

(3) Where the licensing authority considers it necessary, when determining whether to grant, renew, vary or transfer a licence, to obtain evidence that the equipment, including devices and valves, proposed to be used for a relevant activity is suitable for that activity, then it may—

(a) test the equipment, including devices and valves; or

(b) require the applicant to provide evidence that such equipment, including devices and valves, is suitable.

[8.3788]

7 Premises which cease to be ONR regulated premises: effect on licences

(1) Where this regulation applies any licence granted by, or treated as granted by, the ONR is to be treated on and after the relevant date applicable to the premises as a licence granted by the Executive.

(2) This regulation applies where—

(a) the premises in connection with which the licence was granted cease to be ONR regulated premises; and

(b) the licence mentioned in sub-paragraph (a) remained in force (with or without variations) immediately before the date on which the premises ceased to be ONR regulated premises.

(3) In this regulation "relevant date applicable to the premises" means the date on which the premises ceased to be ONR regulated premises.

[8.3789]

8 Premises which become ONR regulated premises: effect on licences

(1) Where this regulation applies any licence granted by, or treated as granted by, the Executive is to be treated on and after the relevant date applicable to the premises as a licence granted by the ONR.

(2) This regulation applies where—

(a) premises in connection with which the licence was granted become ONR regulated premises; and

(b) the licence mentioned in sub-paragraph (a) remained in force (with or without variations) immediately before the date on which the premises became ONR regulated premises.

(3) In this regulation "relevant date applicable to the premises" means the date on which the premises became ONR regulated premises.

[8.3790]

9 Grants, renewals, variations and transfers of licences

(1) The licensing authority may grant a licence—

(a) for any relevant activity; and

(b) with or without a time limit as to its duration.

(2) The licensing authority may grant a licence only if it is of the opinion that—

(a) the premises where the applicant proposes to carry out a relevant activity are suitable for that activity;

(b) the construction, site and orientation of any building, including protective works around the building, where the applicant proposes to carry out a relevant activity are suitable for that activity;

(c) the equipment, including devices and valves, which the applicant proposes to use for any relevant activity are suitable for that activity; and

(d) there are no other reasons to conclude that permitting the relevant activity would result in significant risk that may not be adequately reduced by the imposition of conditions under regulation 9(3).

(3) The licensing authority may make a licence subject to any of the following conditions that it considers appropriate to reduce risk—

(a) the maximum amount of compressed acetylene gas which may be manufactured on the premises;

(b) the maximum amount of acetylene gas which may be compressed at equal to or greater than 0.62 bar(g) on the premises;

(c) the maximum number of cylinders which may be filled with compressed acetylene gas on the premises;

(d) conditions as to the construction, location or orientation of any building, including protective works around the building, where the relevant activity is to take place;

(e) conditions as to the activities which may be undertaken in a building or part of a building within the premises specified in the licence;

(f) conditions as to the equipment, including devices and valves, that is to be used for any relevant activity;

(g) conditions as to the operational requirements for equipment, including devices and valves; and

(h) any other condition the licensing authority considers appropriate to reduce risk.

(4) The licensing authority may, following an application by the licensee, renew a time-limited licence where it is satisfied that—

(a) the licensee has complied with the licence conditions, if any;

(b) the requirements in regulation 9(2) are still satisfied; and

(c) the conditions of the licence, if any, are still sufficient adequately to reduce risk.

(5) The licensing authority may vary a licence, including the conditions attached to a licence—

(a) without the consent of the licensee, if there has been a change in circumstances such that a variation is necessary adequately to reduce risk; or

(b) on the application of the licensee, if the licensing authority is satisfied that after the variation the—

(i) requirements in regulation 9(2) would still be satisfied; and

(ii) conditions of the licence, if any, are still sufficient adequately to reduce risk.

(6) Following an application by the licensee, the licensing authority may transfer a licence, subject to the conditions of that licence, to any other person who wishes to conduct the activity permitted in the licence in place of the licensee, if the licensing authority is satisfied that after the transfer the requirements in regulation 9(2) would still be satisfied.

[8.3791]

10 Form of licences

(1) A licence must be in a form approved for the time being by the Executive for the purposes of this regulation.

(2) A licence must specify—

(a) the name and address of the licensee;

(b) the relevant activity that is permitted;

(c) the premises, and the places within the premises, where any relevant activity may take place;

(d) any conditions attached to that licence under regulation 9(3); and

(e) the duration for which it is granted, if it is of limited duration.

[8.3792]

11 Revocation of licences

The licensing authority may revoke a licence if there has been a change in circumstances leading the licensing authority to conclude that the requirements in regulation 9(2) are no longer satisfied.

[8.3793]

12 Further provisions concerning refusals, variations and revocations

(1) Where the licensing authority proposes to—

(a) refuse an application for—

(i) a licence,

(ii) a renewal of a licence,

(iii) a transfer of a licence, or

(iv) a variation of a licence;

(b) vary a licence without the agreement of the licensee; or

(c) revoke a licence,

it must notify the applicant or the licensee in writing, as applicable, of its proposed course of action and afford them a period of 28 days beginning with the date of the notification to make written representations to the licensing authority about the proposed course of action.

(2) After the 28 day period for making representations in paragraph (1) has elapsed, the licensing authority must consider any written representations received in that 28 day period from the applicant or the licensee, as applicable, prior to making a final decision.

(3) Where the licensing authority decides to—

(a) refuse an application for—

(i) a licence,

(ii) a renewal of a licence,

(iii) a transfer of a licence, or

(iv) a variation of a licence;

(b) vary a licence without the agreement of the licensee; or

(c) revoke a licence,

it must provide the applicant or licensee, as applicable, with written reasons.

[8.3794]

13 Death, incapacity, bankruptcy and insolvency

(1) If a licensee dies or becomes incapacitated, a person carrying on a relevant activity in accordance with the conditions of the licensee's licence is to be treated as being the licensee until the first to occur of—

(a) the expiration of a period of 60 days beginning with the date of the death or incapacity;

(b) the grant or refusal of a new licence; or

(c) the transfer of, or refusal to transfer, the licence.

(2) If a licensee becomes bankrupt or, in the case of a company, goes into liquidation, administration or receivership or has a receiving order made against it, any liquidator, administrator, trustee in bankruptcy or receiver is to be treated as being the licensee.

[8.3795]

14 Power to grant exemptions

(1) Subject to paragraph (3), the Executive may, by a certificate in writing, exempt any person or class of persons from all or any of the requirements or prohibitions imposed by or under these Regulations, save where such an exemption would contravene rules imposed by other enactments.

(2) An exemption granted under paragraph (1) may be—

(a) subject to conditions,

(b) subject to a time limit, and

(c) revoked at any time by a certificate in writing.

(3) The Executive must not grant an exemption under paragraph (1) unless, having regard to the circumstances of the case, and in particular to—

(a) the conditions, if any, which it proposes to attach to the exemption under paragraph (2); and

(b) any requirements imposed by or under any enactments which apply to the case,

it is satisfied that the safety of persons who are likely to be affected by the exemption will not be prejudiced in consequence of it.

[8.3796]

15 Transitional provisions

(1) Where, immediately before the relevant date—

(a) the manufacture of acetylene gas was permitted at premises specified in an approval by the Executive under the Order in Council (No 30) prohibiting the Manufacturing, Importation, Keeping, Conveyance or Sale of Acetylene when an Explosive as defined by the Order of 2nd February 1937;

(b) the compression of acetylene gas was permitted at premises specified in an approval by the Executive under the 1919 Order; or

(c) any person or class of persons was exempt from the general prohibition on compression or manufacture of acetylene gas by virtue of an exemption certificate granted under the 1979 Regulations,

the person carrying on that manufacture, compression, or combination of these activities is deemed to hold a licence granted under regulation 9 permitting such activities as constitute a relevant activity subject to any conditions in the original approval or exemption.

(2) Where the compression of acetylene gas was permitted at premises specified in an approval by the Executive under the 1919 Order the deemed licence is to be taken to allow the filling of any cylinder with compressed acetylene gas as well as the compression of acetylene gas at equal to or greater than 0.62 bar(g).

(3) Nothing in paragraph (1) or (2) may be taken to exempt the holder of a deemed licence from any of the requirements in Schedule 1, except insofar as the previous approval expressly allowed a specified activity or practice that would contravene any such requirement.

(4) Where an application for an approval for any one or more of the manufacture or compression of, or filling of any cylinder with, acetylene gas was made to and received by the Executive before the relevant date and the application has not been refused nor the approval granted by the relevant date, the application is to be treated as an application for a licence to the licensing authority under these Regulations and the provisions of these Regulations are to apply to that application.

(5) Where an application to amend, renew or transfer an approval was made to and received by the Executive before the relevant date and the application has not been refused nor the approval or transfer granted by the relevant date, the application is to be treated as an application for a variation, renewal or transfer of a deemed licence to the licensing authority under these Regulations and the provisions of these Regulations are to apply to that application.

(6) Where fees have already been paid in relation to an application for an approval, amendment, renewal or transfer of an approval which was made to and received by the Executive before the relevant date where the approval, amendment, renewal or transfer was not refused or granted before the relevant date, no additional fees will be payable for the determination of that application.

(7) Where acetylene gas is kept in an acetylene manifold in conditions which, immediately before the relevant date would have been permissible under the Certificate of Exemption No 4 of 1984 made under the 1979 Regulations, keeping acetylene gas in an acetylene manifold in this manner is to be treated as falling within the exception in regulation 4(2)(b) until midnight on 31st December 2015.

(8) Where acetylene gas is kept in conditions which, immediately before the relevant date, would have been permissible under Certificate of Exemption No 2 of 1989 made under the 1979 Regulations, keeping acetylene gas in this manner is to be treated as falling within the exception in regulation 4(2)(a) until midnight on 31st December 2015, and for these purposes keeping includes use.

(9) Where, immediately before the relevant date—

(a) a gas container was permitted for the keeping, importing, conveyance or sale of acetylene under Certificate of Exemption Number 1 of 1995 made under the 1979 Regulations; or

(b) a transportable pressure receptacle was permitted for the keeping, importing, conveyance or sale of acetylene under Certificate of Exemption Number 1 of 2001 made under the 1979 Regulations,

such gas container or transportable pressure receptacle is to be treated as an old pressure receptacle for the purposes of these Regulations.

(10) Where, immediately before the relevant date, a transportable pressure receptacle complies with the requirement to be painted maroon, under paragraph 2(3) of the Certificate of Exemption No 1 of 2001 made under the 1979 Regulations, that transportable pressure receptacle is to be treated as complying with the requirement under paragraph 8 of Schedule 1 (cylinder colouring), until such time as the colour is no longer visibly maroon.

(11) In this regulation—

"the 1919 Order" means the Order of Secretary of State (No 9), dated June 23, 1919, relating to Compressed Acetylene contained in a Porous Substance (1919);

"the 1979 Regulations" means the Explosives Act 1875 (Exemptions) Regulations 1979; and

"deemed licence" means a licence deemed to have been granted under regulation 9 by virtue of paragraph (1).

[8.3797]

16 Amendments, repeals and revocations

(1) The secondary legislation specified in Schedule 2 is amended in accordance with the provisions of that Schedule.

(2) The primary legislation specified in column 1 of Part 1 of Schedule 3 is repealed to the extent specified in column 3 of that Schedule.

(3) The secondary legislation specified in column 1 of Part 2 of Schedule 3 is revoked to the extent specified in column 3 of that Schedule.

[8.3798]

17 Review

(1) Before the end of the review period, the Secretary of State must—

(a) carry out a review of these Regulations;

(b) set out the conclusions of the review in a report; and

(c) publish the report.

(2) The report must in particular—

(a) set out the objectives intended to be achieved by the regulatory system established by these Regulations;

(b) assess the extent to which those objectives are achieved; and

(c) assess whether those objectives remain appropriate and, if so, the extent to which they could be achieved with a system that imposes less regulation.

(3) In this regulation "review period" means the period of five years beginning with the relevant date.

SCHEDULE 1

SAFETY REQUIREMENTS

Regulations 4(2) and 5(1)

Requirements for any acetylene equipment or acetylene manifold

[8.3799]

1

Any acetylene equipment or acetylene manifold may only be used if designed, manufactured and operated to prevent, so far as is reasonably practicable—

(a) the uncontrolled combustion of acetylene gas;

(b) the decomposition of acetylene gas; and

(c) the formation of acetylene—derived compounds that pose a risk or are liable to initiate decomposition of acetylene gas.

2

Any acetylene equipment or acetylene manifold may only be used if designed and manufactured to—

(a) withstand the thermal and mechanical stresses of any decomposition of the acetylene gas that it contains; or

(b) dissipate or direct the thermal and mechanical stresses of any decomposition of the acetylene gas that it contains.

Further requirements for an acetylene manifold

3

An acetylene manifold may only be used if it is—

(a) designed to prevent the mixture of air, or oxygen, with acetylene gas within the acetylene manifold;

(b) fitted with rigid pipework with an internal diameter equal to or less than 25mm;

(c) fitted with a flexible hose where rigid pipework is not practicable and the flexible hose—

   (i) is kept at the minimum practicable length; and

   (ii) has an internal diameter equal to or less than 25mm; and

(d) not subjected to pressure greater than that within any attached cylinder.

Requirements of a system of connected acetylene equipment

4

A system of connected acetylene equipment may only be used if it is fitted—

(a) with a pressure regulation device that is—

   (i) designed and constructed for use with compressed acetylene gas; and

   (ii) positioned as close as is reasonably practicable to the acetylene manifold or, where no acetylene manifold is used, to the cylinder;

(b) within one metre of the pressure regulation device, with a flame arrestor; and

(c) with a non-return device, effective against the return of gas towards the cylinder, and a quick-acting shut-off device, and both devices are positioned as close as is reasonably practicable to the acetylene manifold or, where no acetylene manifold is used, to the cylinder.

5

A system of connected acetylene equipment may only be used if the internal diameter of the pipework does not exceed the maximum diameter specified in the Table for the respective pressure of the compressed acetylene gas in use.

Requirements for a flame arrestor

6

A flame arrestor that is put into service after the relevant date may only be used if, at the time it is put into service, it complies with the requirements of BSEN 730-1:2002 or any other relevant international standard recognised for use in any EEA State at the relevant date.

Requirement for an isolation valve

7

A manually operated isolation valve must be fitted as close as is reasonably practicable upstream of each acetylene burner or acetylene-consuming equipment.

Cylinder colouring

8

A cylinder must be painted, on both its body and shoulder, in the colour matching number 3007 of the Classic RAL system.

Interpretation

9

In this Schedule—

"acetylene burner" means any equipment, including devices and valves, designed for use with ignited acetylene gas;

"acetylene-consuming equipment" means any equipment designed to convert acetylene gas into another chemical form;

"flame arrestor" means a device designed and constructed to arrest the progression of any flame resulting from the decomposition or uncontrolled combustion of acetylene gas;

"system of connected acetylene equipment" means any connected equipment, whether permanently or temporarily connected, which is intended for use with compressed acetylene gas, or which is used with compressed acetylene gas, including acetylene equipment, a cylinder, an acetylene manifold and other accessories; and

"quick-acting shut-off device" means a safety device effective against the continued release of—

(a) acetylene gas; and

(b) products of decomposition caused by—

(i) the decomposition of acetylene gas; or

(ii) any uncontrolled combustion of acetylene gas.

Table

Pipework maximum internal diameter (mm)	Maximum pressure of contained compressed acetylene gas (bar(g))
23	1.5

25	1.3
35	1.0
42	0.8

SCHEDULE 2

AMENDMENTS

Regulation 16(1)

The Health and Safety (Enforcing Authority) Regulations 1998

[8.3800]

1

In Schedule 2 to the Health and Safety (Enforcing Authority) Regulations 1998 after paragraph 13 insert—

"14

Any activity regulated by the Acetylene Safety (England and Wales and Scotland) Regulations 2014.".

The Carriage of Dangerous Goods and Use of Transportable Pressure Equipment Regulations 2009

2

In regulation 14(5) of the 2009 Regulations after "undertaken" insert ", after these Regulations have come into force,".

The Health and Safety (Fees) Regulations 2012

3

The Health and Safety (Fees) Regulations 2012 are amended as follows—

(a) in regulation 9 (Fees payable etc)—

(i) after paragraph (4) insert—

"(4A) Where an application in relation to the 2014 Acetylene Regulations is made for a purpose specified in column 1 of the Table in Part 3A of Schedule 8, the fee specified in the corresponding entry in column 2 of that Table is payable by the applicant to the Executive, or to the ONR where it is the licensing authority by virtue of the 2014 Acetylene Regulations.

(4B) An additional fee is payable by the applicant to the Executive, or to the ONR where it is the licensing authority by virtue of the 2014 Acetylene Regulations, where work is required to be carried out by its specialist inspectors in connection with any application in relation to the 2014 Acetylene Regulations that is made for a purpose specified in column 1 of the Table in Part 3A of Schedule 8 for which there is a corresponding entry in column 3 of that Table, and the fee for work in connection with each such purpose is that specified in the corresponding entry in column 3 of that Table for each hour worked, adjusted pro rata for a period worked of less than one hour, and such fee is payable prior to the notification of the result of the application."

(ii) in paragraph (13) before the definition of the "2014 Regulations" insert—

""the 2014 Acetylene Regulations" means the Acetylene Safety (England and Wales and Scotland) Regulations 2014;".

(b) in Schedule 8 (Fees payable etc) after Part 3 insert—

"PART 3A

APPLICATIONS UNDER, AND REPLACEMENT OF LICENCE UNDER, THE ACETYLENE SAFETY (ENGLAND AND WALES AND SCOTLAND) REGULATIONS 2014

1 **Purpose of Application**	2 **Fee**	3 **Fee for work by specialist inspector**
Original application for a licence for the manufacture of compressed acetylene gas, the compression of acetylene gas, the filling of a cylinder with compressed acetylene gas or any combination of those activities under regulation 6	£39	£127 per hour worked
Application to renew a licence under regulation 6	£39	£127 per hour worked
Application to vary a licence under regulation 6	£39	£127 per hour worked
Application to transfer a licence under regulation 6	£39	£127 per hour worked
Replacement of any of the licences referred to in this part if lost	£39"	

SCHEDULE 3

REPEALS AND REVOCATIONS

Regulation 16(2) and (3)

PART 1

REPEALS

[8.3801]

1 Title	2 Reference	3 Extent of repeal
Explosives Act 1875	c 17	Sections 43 and 104
Regulatory Enforcement and Sanctions Act 2008*	c 13	In Schedule 6, in the entry relating to the Explosives Act 1875, "43"
Health and Safety (Offences) Act 2008*	c 20	Paragraph 1 of Schedule 3

PART 2

REVOCATIONS

[1.3802]–[1.3810]

Instrument	Reference	Extent of Revocation
The Order of Secretary of State (No 5) relating to Compressed Acetylene in Admixture with Oil-Gas, 1898	S R & O 1898/248	The whole instrument
The Order of Secretary of State (No 5A) relating to Compressed Acetylene in Admixture with Oil-Gas (1905)	S R & O 1905/1128	The whole instrument
The Order of Secretary of State (No 9), dated June 23, 1919, relating to Compressed Acetylene contained in a Porous Substance (1919)	S R& O 1919/809	The whole instrument
The Order in Council (No 30) prohibiting the Manufacture, Importation, Keeping, Conveyance or Sale of Acetylene when an Explosive as defined by the Order, 1937	S R & O 1937/54	The whole instrument
The Compressed Acetylene Order 1947	S R & O 1947/805	The whole instrument
The Compressed Acetylene (Importation) Regulations 1978	SI 1978/1723	The whole instrument
The Explosives Act 1875 (Exemptions) Regulations 1979	SI 1979/1378	The whole instrument
The Explosives Act 1875 etc (Metrication and Miscellaneous Amendment) Regulations 1984	SI 1984/510	The whole instrument except for regulations 1(1) and (2)(a), 2 and Schedule 1 to the extent that they relate to section 32 of the Explosives Act 1875
The Health and Safety (Fees) Regulations 2012	SI 2012/1652	Regulation 9(4) and (7) to (9) and Parts 3 and 5 to 7 of Schedule 8

Genetically Modified Organisms (Contained Use) Regulations 2014

SI 2014/1663

[8.3811]

The Secretary of State has been designated for the purposes of section 2(2) of the European Communities Act 1972 in relation to the control and regulation of genetically modified organisms.

The Secretary of State makes these Regulations in exercise of the powers conferred by section 2(2) of the European Communities Act 1972 and by sections 15(1), (2), (3)(b) and (5)(b), 52(2) and (3) and 82(3)(a) of, and paragraphs 1(1)(b) and (c), (2), (4) and (5), 3, 4(1), 5, 6, 8(2), 9, 11, 13(1) and (3), 14, 15(1), 16, 17 and 20 of Schedule 3 to, the Health and Safety at Work etc Act 1974 ("the 1974 Act").

These Regulations give effect without modification to proposals submitted to the Secretary of State by the Health and Safety Executive under section 11(3) of the 1974 Act.

Before submitting those proposals to the Secretary of State, the Health and Safety Executive consulted the bodies that appeared to it to be appropriate, as required by section 50(3) of the 1974 Act.

PART 1

INTERPRETATION AND GENERAL

[8.3812]

1 Citation and commencement

These Regulations may be cited as the Genetically Modified Organisms (Contained Use) Regulations 2014 and come into force on 1st October 2014.

[8.3813]

2 Interpretation

(1) In these Regulations—

"the 1974 Act" means the Health and Safety at Work etc Act 1974;

"the 2000 Regulations" means the Genetically Modified Organisms (Contained Use) Regulations 2000;

"accident" means an incident involving a significant and unintended release of genetically modified organisms in the course of a contained use which presents an immediate or delayed hazard to human health or to the environment;

"class" in relation to a contained use involving micro-organisms, means one of the four classes set out in Schedule 1;

"competent authority" means, in relation to premises situated in, or contained use taking place in—

(a) England and Wales, the Secretary of State and the Executive, acting jointly; or

(b) Scotland, the Scottish Ministers and the Executive, acting jointly,

and the expressions "competent authority as regards England and Wales" and "competent authority as regards Scotland" are to be construed accordingly;

"contained use" means an activity in which organisms are genetically modified or in which genetically modified organisms are cultured, stored, transported, destroyed, disposed of or used in any other way and for which physical, chemical or biological barriers, or any combination of such barriers, are used to limit their contact with, and to provide a high level of protection for, humans and the environment;

"emergency plan" means a plan required by regulation 21;

"emergency services" means the police, fire and ambulance services;

"genetic modification" in relation to an organism means the altering of the genetic material in that organism in a way that does not occur naturally by mating or natural recombination (or both) and within the terms of this definition—

(a) genetic modification occurs at least through the use of the techniques listed in Part 1 of Schedule 2; and

(b) the techniques set out in Part 2 of Schedule 2 are not considered to result in genetic modification,

and "genetically modified" is to be construed accordingly;

"joint competent authority" means the competent authority as regards England and Wales and the competent authority as regards Scotland acting jointly;

"larger GMO" means an organism which is genetically modified or is the subject of genetic modification which is not a micro-organism;

"micro-organism" means a microbiological entity, cellular or non-cellular, capable of replication or of transferring genetic material, and includes a virus, a viroid, and an animal or plant cell in culture;

"notifier" means, except in regulation 14, the person who submits or has submitted a notification to the competent authority under regulation 9(2), 10(2), 11(2), 12(2) or 33(3);

"organism" means a biological entity capable of replication or of transferring genetic material and includes a micro-organism, but does not include a human, human embryo or human admixed embryo and for the purposes of this definition—

(a) "human admixed embryo" has the same meaning as in the Human Fertilisation and Embryology Act 1990 by virtue of section 4A(6) and (11) of that Act; and

(b) "human embryo" has the same meaning as "embryo" in the Human Fertilisation and Embryology Act 1990 (apart from section 4A) by virtue of section 1(1) and (6) of that Act;

"person responsible for contained use" or "person responsible for the contained use" means—

(a) a person who has the authority to determine whether a particular contained use takes place; or

(b) a person who has control of the planning or conduct (or both) of that contained use,

and there may be more than one person responsible for the same contained use;

"premises" means both single buildings and a site made up of more than one building;

"risk assessment" means, in the context of contained use involving—

(a) genetically modified micro-organisms, an assessment carried out as required by regulation 5(1); or

(b) larger GMOs, an assessment carried out as required by regulation 6(1);

"transboundary movement" has the meaning assigned to it by Article 3 of Regulation (EC) No 1946/2003 of the European Parliament and the Council on transboundary movements of genetically modified organisms;

"user" means a person who undertakes or proposes to undertake a contained use;

"working day" means any day other than a Saturday, a Sunday, Christmas Day or Good Friday, or a bank holiday specified in Schedule 1 to the Banking and Financial Dealings Act 1971.

(2) A reference in these Regulations to the competent authority is to be construed as a reference to the joint competent authority in relation to premises or contained use where the relevant notification is required to be submitted to the joint competent authority in accordance with regulation 9(5) or 13(1).

(3) In these Regulations—

(a) a reference to an appropriate containment level is a reference to the containment level assigned to a contained use involving micro-organisms in accordance with paragraphs 3(i) and 4 of Part 2 of Schedule 3;

(b) any reference to a contained use in a numbered class is a reference to a contained use involving micro-organisms which has been classified as belonging to the class of that number in accordance with paragraph 3(j) and (k) of Part 2 of Schedule 3.

(4) The measures in—

(a) Part 2 of Schedule 8 are to be applied in accordance with Part 1 of that Schedule; and

(b) Tables 1a, 1b and 1c in Part 2 of Schedule 8 are to be applied in accordance with the notes set out at the end of the table in question.

[8.3814]

3 Application

(1) These Regulations (except regulation 18) do not apply to the genetic modification of organisms solely by any of the techniques referred to in Part 3 of Schedule 2 nor to any organisms so modified.

(2) These Regulations do not apply to any activity in which—

(a) genetically modified organisms are cultured, stored, transported, destroyed, disposed of or used, where such organisms are, or are contained in, a product marketed in accordance with—

(i) the consent of any of the following granted under section 111(1) of the Environmental Protection Act 1990—

(aa) the Secretary of State;

(bb) the Scottish Ministers, as regards Scotland;

(cc) the Welsh Ministers, as regards Wales;

(ii) a consent granted by the Northern Ireland Department of the Environment under article 8(1) of the Genetically Modified Organisms (Northern Ireland) Order 1991;

(iii) a written consent given by the competent authority of an EEA state in accordance with Article 15(3), 17(6), or 18(2) of Directive (EC) No 2001/18 of the European Parliament and the Council on the deliberate release into the environment of genetically modified organisms,

and, in each case, that activity is conducted in accordance with any conditions or limitations attached to that consent;

(b) genetically modified organisms are cultured, stored, transported, destroyed, disposed of or used, where such organisms are, or are contained in—

(i) a medicinal product for human or veterinary use marketed in accordance with Regulation (EC) No 726/2004 of the European Parliament and the Council laying down Community procedures for the authorisation and supervision of medicinal products for human and veterinary use and establishing a European Medicines Agency;

(ii) food or feed authorised in accordance with the provisions of Regulation (EC) No 1829/2003 of the European Parliament and the Council on genetically modified food and feed; or

(iii) food products notified to the Commission in accordance with the provisions of Article 8.1, or feed products notified to the Commission in accordance with the provisions of Article 20.1, of Regulation (EC) No 1829/2003 of the European Parliament and the Council;

(c) genetically modified organisms are released or marketed in cases or circumstances in which—

(i) the consent of any of the following is required under section 111(1) of the Environmental Protection Act 1990—

(aa) the Secretary of State;

(bb) the Scottish Ministers, as regards Scotland;

(cc) the Welsh Ministers, as regards Wales; or

(ii) the consent of the Northern Ireland Department of the Environment is required under article 8(1) of the Genetically Modified Organisms (Northern Ireland) Order 1991.

(3) Regulations 7, 9 to 17, 18(2) and (4), 19, 20, and 23 to 25 do not apply to the transport of genetically modified organisms by road, rail, inland waterway, sea or air.

(4) Regulation 5 applies to the transport of genetically modified organisms by road, rail, inland waterway, sea or air, except that, in making the assessment required by regulation 5(1), the person undertaking that assessment is not required to include the steps set out in paragraph 3(i) to (k) of Part 2 of Schedule 3.

(5) These Regulations do not extend to Northern Ireland.

(6) In this regulation, "product" means a product consisting of, or containing, a genetically modified organism or a combination of genetically modified organisms.

[8.3815]

4 Meaning of "work" and "at work" and modification of the 1974 Act

(1) For the purpose of these Regulations and Part 1 of the 1974 Act, the meaning of "work" is extended to include any contained use and the meaning of "at work" is extended accordingly.

(2) Sections 2(1), (2) and (3) and 7 of the 1974 Act are modified in relation to contained use as follows—

(a) those sections have effect as if a reference to—

(i) an employer includes a reference to an educational establishment providing a course of study; and

(ii) an employee includes a reference to a student undertaking contained use in that educational establishment to the extent that the contained use is under the control of that educational establishment.

(3) Section 3(2) of the 1974 Act is modified in relation to contained use so as to have effect as if the reference in that section—

(a) to a self-employed person were a reference to any person (except a student) undertaking contained use who is not an employer or an employee; and

(b) to that person's undertaking includes a reference to that contained use.

(4) In this regulation—

"educational establishment" means a university, college, school or similar educational or technical institute; and

"student" means any person studying at an educational establishment.

PART 2

RISK ASSESSMENT AND NOTIFICATION OF CONTAINED USE

[8.3816]

5 Risk assessment of contained use involving micro-organisms

(1) Before any contained use involving micro-organisms is commenced, a person responsible for the contained use must ensure that a suitable and sufficient assessment of the risks to human health and the environment created by the contained use is carried out.

(2) The assessment required by paragraph (1) must take into account the matters set out in Part 1, and include the steps set out in Part 2, of Schedule 3.

[8.3817]

6 Risk assessment of contained use involving larger GMOs

(1) Before any contained use involving larger GMOs is commenced, a person responsible for the contained use must ensure that a suitable and sufficient assessment of the risks to human health created by the contained use is carried out.

(2) The assessment required by paragraph (1) must take into account the matters set out in Part 1, and include the steps set out in Part 2, of Schedule 4.

[8.3818]

7 Review and recording of risk assessments

(1) A person responsible for contained use must ensure that the risk assessment is reviewed immediately where—

(a) there is reason to suspect that the risk assessment is no longer valid; or

(b) there has been a significant change in the contained use to which the risk assessment relates.

(2) A person responsible for contained use must—

(a) keep a record of the risk assessment and any review of the risk assessment, for at least 10 years from the date the contained use stops; and

(b) make the record available to the competent authority when requested to do so.

[8.3819]

8 Advice from a genetic modification safety committee

(1) Subject to paragraph (2), a person responsible for contained use must obtain advice on a risk assessment from either—

(a) a person; or

(b) a genetic modification safety committee,

with expertise in risk assessment relating to contained use.

(2) Where the risk assessment indicates that the contained use is classified as class 2 or above the advice must be obtained from a genetic modification safety committee.

[8.3820]

9 Notification of premises to be used for contained use

(1) A user must not use premises for contained use unless the premises have been notified to the competent authority in accordance with this regulation.

(2) Before premises are used for contained use for the first time, a person responsible for the contained use must—

- (a) submit a notification to the competent authority containing the information specified in Schedule 5; and
- (b) have received an acknowledgement of receipt of the notification from the Executive.

(3) The Executive must send an acknowledgement of receipt to the notifier within 10 working days of the competent authority receiving the notification.

(4) A single notification may include more than one premises.

(5) Where a notification includes more than one premises and at least one of those premises is situated in England or Wales and at least one of those premises is situated in Scotland the notification must be submitted to the joint competent authority.

(6) The notifier must nominate one address which is to be the principal address for the purposes of a notification under paragraph (4) or (5).

[8.3821]

10 Notification of class 2 contained use

(1) A user must not undertake a contained use involving micro-organisms classified as class 2 unless the provisions of this regulation have been complied with.

(2) A person responsible for the contained use must submit a notification to the competent authority containing the information specified in Schedule 6.

(3) The Executive must send an acknowledgement of receipt to the notifier within 10 working days of the competent authority receiving the notification.

(4) Where the premises in the notification have not previously been notified for class 2 or a higher class of contained use, a user may undertake the class 2 contained use if—

- (a) 45 days have elapsed since the acknowledgement of receipt was received, provided that the competent authority has not informed the notifier that the class 2 contained use may not be undertaken; or
- (b) the competent authority has agreed in writing that the class 2 contained use may commence sooner.

(5) Where the premises in the notification have—

- (a) previously been notified for class 2 contained use; or
- (b) already been granted consent for class 3 or class 4 contained use,

a user may undertake the class 2 contained use if the notifier has received the acknowledgement of receipt.

(6) Where a notifier submits a notification for a class 2 contained use which is to be undertaken for the second or subsequent time at the premises in the notification, the notifier may request that the competent authority provide a written agreement that the contained use may be undertaken.

(7) The competent authority must make a decision and, if they agree, provide the written agreement requested under paragraph (6), within 45 days of the date on which the acknowledgement of receipt was sent to the notifier.

[8.3822]

11 Notification of class 3 or class 4 contained use

(1) A user must not undertake a contained use involving micro-organisms classified as class 3 or class 4 unless written consent for that contained use has been granted by the competent authority.

(2) A person responsible for the contained use must submit a notification to the competent authority containing the information specified in Schedule 6.

(3) The Executive must send an acknowledgement of receipt to the notifier within 10 working days of the competent authority receiving the notification.

(4) Where the premises in the notification have not previously been notified for class 3 or class 4 contained use, the competent authority must inform the notifier, in writing, of its decision to grant or refuse consent for the class 3 or class 4 contained use, within 90 days of the date on which the acknowledgement of receipt was sent to the notifier.

(5) Where the premises in the notification have previously been notified for class 3 or class 4 contained use and all relevant conditions of existing consents have been complied with, the competent authority must inform the notifier, in writing, of its decision to grant or refuse consent for the class 3 or class 4 contained use, within 45 days of the date on which the acknowledgement of receipt was sent to the notifier.

(6) Before granting consent, the competent authority must ensure that an emergency plan has been prepared where the risk assessment shows an emergency plan is required.

(7) Before deciding whether to grant or refuse consent, the competent authority must take into account any representations made to it by any person within 30 days of the date on which the acknowledgement of receipt was sent to the notifier.

(8) A consent granted under this regulation may be granted subject to conditions.

[8.3823]

12 Notification of contained use involving larger GMOs

(1) A user must not undertake a contained use involving larger GMOs unless the provisions of this regulation have been complied with.

(2) A person responsible for the contained use must submit a notification to the competent authority containing the information specified in Schedule 6.

(3) The Executive must send an acknowledgement of receipt to the notifier within 10 working days of the competent authority receiving the notification.

(4) A user may undertake the contained use if—

(a) 45 days have elapsed since the acknowledgement of receipt was received, provided that the competent authority has not informed the notifier that the contained use may not be undertaken; or

(b) the competent authority has agreed in writing that the contained use may commence sooner.

(5) This regulation does not apply to a contained use which results in a larger GMO that poses no greater risk to humans than its unmodified parental organism.

[8.3824]

13 Single notifications to the joint competent authority and for connected programmes of work

(1) Where a notification is required under regulation 10(2), 11(2) or 12(2) in respect of a contained use which is to take place in premises that fall within regulation 9(5) the notifier must submit the notification for that contained use to the joint competent authority.

(2) A competent authority, or where paragraph (1) applies the joint competent authority, may accept a single notification submitted under regulation 10(2), 11(2) or 12(2) in respect of a connected programme of work undertaken at—

(a) one premises; or

(b) more than one premises.

(3) A competent authority, or where paragraph (1) applies the joint competent authority, may accept a single notification submitted under regulation 10(2), 11(2) or 12(2) in respect of a single contained use undertaken at more than one premises.

(4) In this regulation—

"connected programme of work" means a series of activities involving contained use which form a coherent and integrated programme.

[8.3825]

14 Changes of circumstances relating to notifications

(1) Full details in writing must be sent immediately to the competent authority of—

(a) any change in the information specified in paragraph (a), (d) or (e) of Schedule 5 in relation to premises previously notified in accordance with regulation 9(2);

(b) any new building—

(i) added to premises previously notified in accordance with regulation 9(2); and

(ii) under the notifier's control;

(c) premises notified under regulation 9(2) that will no longer be used for contained use;

(d) any cessation, for the time being, of all contained use at premises notified under regulation 9(2);

(e) any cessation of a contained use notified in accordance with regulation 10(2), 11(2) or 12(2);

(f) any recommencement of contained use at premises in respect of which the notifier had previously given details of a cessation under sub-paragraph (d);

(g) any use of additional premises in connection with a single contained use where a single notification for that contained use was accepted by the competent authority under regulation 13(3);

(h) any change in the information specified in paragraph (b) or (c) of Schedule 5 as provided by the original notifier in accordance with regulation 9(2);

(i) any change in the information specified in paragraph (c) or (d) of Schedule 6 as provided by the original notifier in accordance with regulation 10(2), 11(2) or 12(2).

(2) Where—

(a) a notifier has informed the competent authority of additional premises under paragraph (1)(g); and

(b) that information, taken together with the notification for that single contained use accepted under regulation 13(3), provides all the information required for notification of those premises under regulation 9(2),

the provision of that information will be treated as notification of those premises for the purposes of regulation 9(2).

(3) The details required by paragraph (1) must be provided by—

(a) the original notifier;

(b) a person responsible for the premises notified under regulation 9(2); or

(c) a person responsible for the contained use notified under regulation 10(2), 11(2) or 12(2).

(4) In this regulation—

"notifier" means the person who sends the details required by paragraph (1) to the competent authority; and

"original notifier" means the person who submitted the notification of the premises under regulation 9(2) or the contained use under regulation 10(2), 11(2) or 12(2).

[8.3826]

15 Duty to notify significant changes affecting risks

(1) Where, after submitting a notification, a notifier—

(a) makes a change in the premises or the contained use to which the notification relates which may have significant consequences for the risks arising from the contained use; or

(b) becomes aware of any new information which may have significant consequences for the risks arising from the contained use,

the notifier must immediately send to the competent authority full details in writing of the change or the new information.

(2) As long as the change or new information does not affect the class of the contained use, the notifier need not submit a further notification under regulation 10(2), 11(2) or 12(2), and the change or new information will be treated as a modification of the original notification.

[8.3827]

16 Action of notifier and user on receipt of request for additional information

(1) If additional information relating to a notification is requested by the Executive under regulation 24(1), a user must not commence the contained use that is the subject of the notification until the competent authority has given its approval in writing.

(2) Subject to paragraphs (3) and (4), if the contained use has commenced before the Executive requests additional information, a user may not continue the contained use until the competent authority has given its approval in writing.

(3) The Executive may give the notifier instructions concerning the cessation of the contained use and the notifier and any user undertaking the contained use must comply with the instructions.

(4) Subject to any instructions, the notifier or user may continue the contained use only to the extent necessary to store or destroy all genetically modified organisms resulting from the contained use.

[8.3828]

17 Withdrawal of notification

A notifier may withdraw a notification by giving written notice to the competent authority, provided that the contained use to which the notification related has not commenced.

PART 3

CONDUCT OF CONTAINED USE

[8.3829]

18 Principles of occupational and environmental safety

(1) A user who undertakes a contained use involving micro-organisms must ensure that the risks to human health and the environment arising from the contained use are reduced to the lowest level that is reasonably practicable.

(2) The measures to be taken in order to comply with the duty under paragraph (1) must include the general principles of good microbiological practice and of good occupational safety and hygiene set out in Schedule 7.

(3) A user who undertakes a contained use involving larger GMOs must ensure that the risks to human health arising from the contained use are reduced to the lowest level that is reasonably practicable.

(4) For contained use involving larger GMOs, the general principles set out in Schedule 7 must be applied to the extent that they are appropriate.

[8.3830]

19 Containment and control measures for contained use involving micro-organisms

(1) A user who undertakes a contained use involving micro-organisms must apply the containment measures set out in the applicable table in Part 2 of Schedule 8, where and to the extent required in the column of the appropriate containment level.

(2) A user need not apply a containment measure required for the appropriate containment level where—

- (a) the risk assessment, or any review of the risk assessment, shows that the containment measure is not necessary or practicable for a specific activity;
- (b) the notifier of the contained use has provided justification in writing to the competent authority; and
- (c) the notifier has received the written agreement of the competent authority that the containment measure need not be applied.

(3) A person responsible for the contained use must review the containment measures applied—

- (a) at suitably regular intervals; and
- (b) immediately, if that person suspects that—
  - (i) the containment measures are no longer adequate;
  - (ii) the class assigned to the contained use in the risk assessment is no longer appropriate; or
  - (iii) in the light of new scientific or technical knowledge, the risk assessment is no longer valid.

[8.3831]

20 Containment and control measures for contained use involving larger GMOs

(1) A user who undertakes a contained use involving larger GMOs must apply the containment measures selected in the risk assessment for the contained use.

(2) A person responsible for the contained use must review the containment measures applied—

- (a) at suitably regular intervals; and
- (b) immediately, if that person suspects that—
  - (i) the containment measures are no longer adequate; or
  - (ii) in the light of new scientific or technical knowledge, the risk assessment is no longer valid.

[8.3832]

21 Emergency plans

(1) Where an assessment carried out under regulation 5(1) shows that, as a result of any reasonably foreseeable accident—

(a) the health or safety of persons outside the premises in which the contained use is undertaken is liable to be seriously affected; or

(b) there is a risk of serious damage to the environment from the contained use,

a person responsible for the contained use must ensure that, before the contained use commences, a suitable emergency plan is prepared with a view to securing the health and safety of those persons or the protection of the environment or both.

(2) Where an assessment carried out under regulation 6(1) shows that, as a result of any reasonably foreseeable accident, the health or safety of persons outside the premises in which the contained use is undertaken is liable to be seriously affected, a person responsible for the contained use must ensure that, before the contained use commences, a suitable emergency plan is prepared with a view to securing the health and safety of those persons.

(3) An emergency plan must—

(a) include the measures to be taken in the event of an accident to which the plan relates; and

(b) be reviewed and, where necessary, revised at suitably regular intervals.

(4) A person responsible for the contained use which is the subject of an emergency plan must—

(a) inform the emergency services, and any body or authority liable to be affected by an accident to which the plan relates, of the contents of the plan and of any relevant revisions; and

(b) make information about the plan and any such revisions publicly available.

[8.3833]

22 Information relating to accidents

If an accident occurs, a person responsible for the contained use must immediately inform the competent authority of the accident and must provide the following information—

(a) the circumstances of the accident;

(b) the identity and quantity of the genetically modified organisms concerned;

(c) any information necessary to assess the effects of the accident on the health of the general population and, in the case of a genetically modified micro-organism, on the environment; and

(d) any measures taken in response to the accident.

PART 4

DUTIES AND POWERS OF THE COMPETENT AUTHORITY

[8.3834]

23 Duties of competent authority on receiving a notification

The competent authority must examine a notification and accompanying documentation submitted under regulation 9(2), 10(2), 11(2) or 12(2) for—

(a) conformity with the requirements of these Regulations;

(b) the accuracy and completeness of the information provided;

(c) the adequacy and correctness of the risk assessment or summary of the risk assessment;

(d) the adequacy of the waste management and emergency response measures;

(e) in the case of a notification submitted under regulation 10(2) or 11(2), the correctness of the class assigned to the contained use; and

(f) the inclusion of an emergency plan where the risk assessment indicates that such a plan is necessary.

[8.3835]

24 Requests for additional information

(1) For the purpose of carrying out an examination of a notification in accordance with regulation 23 the Executive may, on behalf of the competent authority, request the notifier to provide such additional information relating to the notification as it may specify.

(2) If requested to do so by the Secretary of State or the Scottish Ministers, the Executive must request additional information under paragraph (1).

(3) A request for additional information must be made in writing.

(4) The Executive must send an acknowledgement of receipt to the notifier within 10 working days of receipt of all of the additional information.

(5) The period of time beginning with the date on which the Executive requests additional information and ending with the date on which the Executive receives all of that additional information will not be taken into account in calculating the period of days referred to in regulation 10(4), 10(7), 11(4), 11(5) or 12(4).

(6) The competent authority may return a notification to the notifier where—

(a) the Executive has requested additional information;

(b) the notifier has not provided all the additional information requested within six months of the date on which the Executive sent the request; and

(i) contained use has not commenced at the premises to which a notification made under regulation 9(2) relates; or

(ii) the contained use referred to in the notification has not commenced.

[8.3836]

25 Powers of competent authority in relation to contained use

The competent authority may at any time by notice in writing to a notifier—

(a) set a time limit for, or impose conditions with regard to, a particular contained use;

(b) require the notifier and any user to suspend, terminate or not to commence a particular contained use;

(c) revoke or vary a consent granted to the notifier under regulation 11,

and the notifier and any user undertaking the contained use must comply with that notice.

[8.3837]

26 Exemption certificates

(1) A competent authority may, by a certificate in writing, exempt—

(a) any person or class of persons; or

(b) any genetically modified organism or class of genetically modified organisms,

from all or any of the requirements of, or prohibitions imposed by, these Regulations.

(2) An exemption may be granted subject to conditions and to a time limit and may be revoked by a certificate in writing at any time.

(3) A competent authority must not grant an exemption unless, having regard to the circumstances of the case and in particular to—

- (a) the conditions, if any, that it proposes to attach to the exemption; and
- (b) any relevant requirements imposed by or under any enactments,

it is satisfied about the matters referred to in paragraph (4).

(4) The matters are—

- (a) that the health or safety of persons who are likely to be affected by the exemption will not be prejudiced in consequence of it; and
- (b) where the exemption relates to a contained use involving a micro-organism, that the environment will not be prejudiced in consequence of the exemption.

[8.3838]

27 Duties of competent authority on receipt of information about accidents

Where the competent authority is informed of an accident in accordance with regulation 22, it must—

- (a) ensure that any necessary measures are taken;
- (b) immediately inform those EEA states which could be affected by the accident;
- (c) collect, where possible, the information necessary for a full analysis of the accident and, where appropriate, make recommendations to avoid similar accidents in the future and to limit their effects; and
- (d) send to the European Commission—
  - (i) the information provided under regulation 22(a), (b) and (d);
  - (ii) information on the effectiveness of the measures taken in response to the accident; and
  - (iii) an analysis of the accident, including recommendations to limit its effects and to avoid similar accidents in the future.

[8.3839]

28 Register of notifications

(1) This regulation is subject to regulation 29.

(2) The competent authority must maintain a register of every notification submitted under regulations 9 to 12.

(3) Subject to paragraph (4) the register must contain—

- (a) in relation to each notification submitted under regulation 9(2), 10(2), 11(2) or 12(2)—
  - (i) the name, address and telephone number and any fax number and any e-mail address of the notifier;
  - (ii) the date on which the Executive acknowledged receipt of the notification; and
  - (iii) where the competent authority receives details of a matter referred to in sub-paragraphs (a) to (g) of regulation 14(1), or in regulation 15(1), confirmation that such details have been received;
- (b) in relation to each notification submitted under regulation 10(2), 11(2) or 12(2), the date of any cessation of the contained use to which the notification related;
- (c) in relation to each notification submitted under regulation 9(2)—

(i) the information specified in paragraphs (d) to (g) and (h)(ii) and (iii) of Schedule 5;

(ii) if applicable, the fact that the competent authority has been informed of an accident at those premises under regulation 22;

(d) in relation to each notification submitted under regulation 10(2), the information specified in paragraphs (e) to (k) and (m)(i) and (ii) of Schedule 6;

(e) in relation to each notification submitted under regulation 11(2)—

(i) the information specified in paragraphs (e) to (j), (1), (m)(i),(iii) and (iv) and (r) of Schedule 6;

(ii) if applicable, confirmation that consent for the contained use has been granted under regulation 11(4) or 11(5);

(f) in relation to each notification submitted under regulation 12(2), the information specified in paragraphs (e) to (j) and (m)(i) of Schedule 6.

(4) The competent authority must omit information from the register where—

(a) the information falls within one of the exceptions to disclosure in—

(i) regulation 12(5) or 13(1) of the Environmental Information Regulations 2004; or

(ii) regulation 10(5) or 11(1) of the Environmental Information (Scotland) Regulations 2004;

(b) the notifier has requested that the competent authority treat the information as confidential; and

(c) the competent authority has decided that the information is to be kept confidential.

(5) The competent authority may not keep the following information confidential if it was submitted in accordance with the requirements of regulation 9(2), 10(2) or 11(2)—

(a) the general characteristics of any genetically modified micro-organisms, the name and address of the notifier, and the location of use;

(b) the class of contained use and the containment measures;

(c) the evaluation of foreseeable effects, in particular any harmful effects on human health and the environment.

(6) Information must be entered in the register within 14 days of its receipt by the competent authority.

(7) The competent authority may remove from the register details of—

(a) premises which are no longer used for contained use, ten years after being informed of this under regulation 14(1)(c);

(b) premises where all contained use has ceased for the time being, ten years after being informed of this under regulation 14(1)(d), provided that no notice of recommencement under regulation 14(1)(f) has been received;

(c) a contained use that has ceased, ten years after being informed of this under regulation 14(1)(e).

(8) A copy of the register must be made available for inspection to members of the public by the Executive, by such means as it considers appropriate, which may include publication on its website.

[8.3840]

29 Information not to be included in the register

(1) No information may be included in the register if and so long as, in the opinion of the Secretary of State, the inclusion in the register of that information, or information of that description, would be contrary to the interests of national security.

(2) For the purpose of securing the exclusion from the register of information to which paragraph (1) applies, the Secretary of State may give the competent authority directions—

- (a) specifying information, or descriptions of information, to be excluded from the register; or
- (b) specifying descriptions of information to be referred to the Secretary of State for his or her determination.

(3) No information referred to the Secretary of State under paragraph (2)(b) may be included in the register unless the Secretary of State determines that it should be included.

(4) The competent authority must notify the Secretary of State of any information it excludes from the register in accordance with directions given to it under paragraph (2).

(5) A person may give a written notice to the Secretary of State specifying information which appears to that person to be information to which paragraph (1) may apply and stating why it should not be included in the register.

(6) If a person gives a written notice under paragraph (5), at the same time that person must give written notice to the competent authority that they have done so.

(7) No information notified under paragraph (5) may be included in the register unless the Secretary of State determines that it should be included.

PART 5

MISCELLANEOUS AND GENERAL

[8.3841]

30 Enforcement

(1) This regulation applies to the extent that any part of these Regulations are not health and safety regulations within the meaning of section 15 of the 1974 Act.

(2) The following provisions apply to the whole of these Regulations as if they were health and safety regulations for the purposes of that Act—

- (a) sections 16 to 26 (approved codes of practice and enforcement) and sections 33 to 42 (provisions as to offences) of the 1974 Act; and
- (b) the Health and Safety (Training for Employment) Regulations 1990.

(3) Every function of the Executive under any provision of the 1974 Act, or under health and safety regulations, is exercisable in relation to these Regulations as if the whole of these Regulations were health and safety regulations for the purposes of that Act.

(4) Despite section 33(1)(c) of the 1974 Act a failure to discharge a duty placed on the competent authority or the Executive by these Regulations is not an offence.

(5) Despite regulation 3 of the Health and Safety (Enforcing Authority) Regulations 1998, the enforcing authority for these Regulations is the Executive.

[8.3842]

31 Appeals

(1) A person responsible for contained use who is aggrieved by any of the following may appeal to the appropriate person—

- (a) a decision by the competent authority—
  - (i) to refuse to provide a written agreement requested under regulation 10(6);
  - (ii) to refuse consent for a class 3 or class 4 contained use notified under regulation 11(2);

(iii) to refuse to provide written agreement under regulation 19(2)(c) that a particular containment measure need not be applied for a specific activity;

(iv) to refuse to grant an exemption certificate under regulation 26(1) or to revoke such a certificate;

(v) to impose a condition or a time limit on an exemption certificate issued under regulation 26(1);

(b) an instruction concerning the cessation of a contained use under regulation 16(3);

(c) a request for additional information by the Executive under regulation 24(1);

(d) a notice from the competent authority under regulation 25.

(2) The appropriate person may direct that an appeal be determined on their behalf by one or more persons appointed for that purpose.

(3) The appropriate person may pay such remuneration and allowances to an appointed person as the appropriate person may determine.

(4) An appointed person may decide the procedure to be followed on the appeal and may give such directions as are appropriate to give effect to the determination of the appeal.

(5) Where an appeal is brought under this regulation—

(a) the following remain valid pending the final determination of the appeal—

(i) a decision of the competent authority referred to in paragraph (1)(a);

(ii) a request for additional information made under regulation 24(1);

(b) the following are not suspended pending the final determination of the appeal—

(i) the operation of regulation 16 and any instructions given under regulation 16(3);

(ii) a notice issued under regulation 25.

(6) The period of time beginning with the date on which an appeal is lodged and ending with the date on which that appeal is determined will not be taken into account in calculating the period of days referred to in regulation 10(4), 10(7), 11(4), 11(5) or 12(4).

(7) In this regulation,

"appointed person" means the person appointed by the appropriate person to determine an appeal;

"appropriate person" means—

(a) the Secretary of State, in the case of—

(i) an appeal under paragraph (1)(a) or (d) against a decision of, or a notice issued by, the competent authority as regards England and Wales; or

(ii) an appeal under paragraph (1)(b) or (c) against a request or instruction relating to—

(aa) the undertaking or proposed undertaking of a contained use; or

(bb) premises which are the subject of a notification under regulation 9(2),

in England or Wales;

(b) the Secretary of State and the Scottish Ministers, acting jointly, in the case of—

(i) an appeal under paragraph (1)(a) or (d) against a decision of, or a notice issued by, the competent authority as regards Scotland or the joint competent authority; or

(ii) an appeal under paragraph (1)(b) or (c) against a request or instruction relating to—
(aa) the undertaking or proposed undertaking of a contained use; or
(bb) premises which are the subject of a notification under regulation 9(2) or 9(5),

in Scotland.

[8.3843]

32 Competent authority address

Anything required to be submitted or sent to a competent authority under these Regulations must be sent to the Executive at the address published for this purpose on its website which may be, or include, an address for submission by electronic means.

[8.3844]

33 Saving and transitional provisions

(1) Subject to paragraph (3) the following continue to have effect and are deemed to have been made, granted or imposed under these Regulations—

(a) a notification made under any of regulations 9 to 13 of the 2000 Regulations, provided that the notification complied with the provisions of those Regulations, as if the notification had been made by a notifier under the corresponding regulation of these Regulations;

(b) a consent granted by the competent authority under regulation 11 of the 2000 Regulations as if it were granted under regulation 11 of these Regulations;

(c) an agreement by the competent authority under regulation 18(2) of the 2000 Regulations that a specific containment measure need not be applied to a contained use, as if it were made under regulation 19(2) of these Regulations;

(d) a request for additional information made under regulation 14(2) of the 2000 Regulations, as if it were made under regulation 24(1) of these Regulations;

(e) a condition, limit of time or other requirement imposed by the competent authority under regulation 15(1) of the 2000 Regulations, as if it were imposed under regulation 25 of these Regulations.

(2) Every record required to be kept under regulation 8(2) of the 2000 Regulations must be kept in the same manner and for the same period as specified in that regulation as if the requirement were imposed under regulation 7(2) of these Regulations.

(3) A person responsible for contained use involving micro-organisms must submit a notification to the competent authority in the following circumstances—

(a) the contained use was being undertaken in accordance with the 2000 Regulations before the date on which these Regulations come into force;

(b) the appropriate containment level for the contained use is different under these Regulations to the appropriate containment level under the 2000 Regulations; and

(c) as a result the contained use is classified under these Regulations at a higher class than under the 2000 Regulations.

(4) The notification must be submitted to the competent authority within the specified period.

(5) Subject to paragraphs (6) to (8) the notification must be treated as a notification required under regulation 10(2) or 11(2) of these Regulations.

(6) The notification must contain the information in Schedule 6 that is specified for the new class of contained use, unless the competent authority exempts the notifier from some or all of the requirements of Schedule 6.

(7) Where a notification is submitted for a contained use that requires consent as class 3 or class 4 contained use, the competent authority must inform the notifier of its decision whether or not to grant consent within 90 days of receipt of the notification.

(8) The contained use referred to in paragraph (3) may continue provided that—

(a) the notification is submitted within the specified period;

(b) the risk assessment shows no increase in the risks to human health or the environment created by the contained use;

(c) the competent authority does not require the notifier to suspend or terminate the contained use under regulation 25 of these Regulations; and

(d) the competent authority has not refused consent for the contained use.

(9) In this regulation—

"specified period" means the 90 days beginning with the date on which these Regulations come into force.

[8.3845]

34 Consequential amendments

(1) The Health and Safety (Fees) Regulations 2012 are amended as follows.

(2) In regulation 13—

(a) in the heading, for "2000" substitute "2014";

(b) in paragraph (1), for "2000" substitute "2014";

(c) for paragraph (2) substitute—

"(2) No fee is to be returned to a notifier where the notifier withdraws a notification under regulation 17 of the 2014 Regulations or the competent authority returns a notification under regulation 24(6) of the 2014 Regulations.";

(d) in paragraph (3) in both instances, for "2000" substitute "2014".

(3) In regulation 24(16)(b) for "2000" substitute "2014".

(4) In Schedule 10—

(a) in the heading, for "2000" substitute "2014";

(b) in column 1 of the table—

(i) for paragraph (a) substitute "Notification of premises to be used for contained use for the first time under regulation 9(2)";

(ii) for paragraph (b) substitute "Notification of class 2 contained use under regulation 10(2)";

(iii) for paragraph (c) substitute "Notification of premises to be used for contained use for the first time under regulation 9(2) at the same time as notification of class 2 contained use under regulation 10(2)";

(iv) for paragraph (d) substitute "Notification of class 3 contained use under regulation 11(2)";

(v) for paragraph (e) substitute "Notification of premises to be used for contained use for the first time under regulation 9(2) at the same time as notification of class 3 contained use under regulation 11(2)";

(vi) for paragraph (f) substitute "Notification of class 4 contained use under regulation 11(2)";

(vii) for paragraph (g) substitute "Notification of premises to be used for contained use for the first time under regulation 9(2) at the same time as notification of class 4 contained use under regulation 11(2)";

(viii) for paragraph (h) substitute "Notification of contained use under regulation 12(2)";

(ix) for paragraph (i) substitute "Notification of premises to be used for contained use for the first time under regulation 9(2) at the same time as notification of contained use under regulation 12(2)";

(x) for paragraph (j) substitute "Notification of a change or new information affecting risks under regulation 15(1)";

(xi) in paragraph (k) for "18(2)" substitute "19(2)" and for "9(1), 10(1), 11(1) or 12(1)" substitute "9(2), 10(2), 11(2) or 12(2)".

[8.3846]

35 Revocations

The following are revoked—

(a) the 2000 Regulations;

(b) the Genetically Modified Organisms (Contained Use) (Amendment) Regulations 2002;

(c) the Genetically Modified Organisms (Contained Use) (Amendment) Regulations 2005;

(d) the Genetically Modified Organisms (Contained Use) (Amendment) Regulations 2010.

SCHEDULE 1

CLASSES OF CONTAINED USE

Regulation 2(1)

[8.3847]

Class	*Description*
1	Contained use of no or negligible risk, for which containment level 1 is appropriate to protect human health and the environment.
2	Contained use of low risk, for which containment level 2 is appropriate to protect human health and the environment.
3	Contained use of moderate risk, for which containment level 3 is appropriate to protect human health and the environment.
4	Contained use of high risk, for which containment level 4 is appropriate to protect human health and the environment.

SCHEDULE 2

Regulations 2(1) and 3(1)

PART 1

TECHNIQUES CONSTITUTING GENETIC MODIFICATION

[8.3848]

1

The techniques which constitute genetic modification referred to in sub-paragraph (a) of the definition of "genetic modification" in regulation 2(1) are—

(a) recombinant nucleic acid techniques involving the formation of new combinations of genetic material by the insertion of nucleic acid molecules, produced by whatever means outside an organism, into any virus, bacterial plasmid or other vector system and their incorporation into a host organism in which they do not naturally occur but in which they are capable of continued propagation;

(b) techniques involving the direct introduction into an organism of heritable genetic material prepared outside the organism, including micro-injection, macro-injection and micro-encapsulation;

(c) cell fusion or hybridization techniques where live cells with new combinations of heritable genetic material are formed through the fusion of two or more cells by means of methods that do not occur naturally.

PART 2

TECHNIQUES WHICH ARE NOT CONSIDERED TO RESULT IN GENETIC MODIFICATION

[8.3849]

2

The following techniques are not considered to result in genetic modification provided that they do not involve the use of recombinant nucleic acid molecules or of genetically modified organisms made by techniques other than those listed in Part 3—

(a) in vitro fertilisation;

(b) natural processes including conjugation, transduction or transformation;

(c) polyploidy induction.

PART 3

TECHNIQUES TO WHICH THESE REGULATIONS DO NOT APPLY

[8.3850]

3

These Regulations (except regulation 18) do not apply to the following techniques of genetic modification, provided that they do not involve the use of recombinant nucleic acid molecules or of genetically modified organisms other than those made by one or more of the following techniques—

(a) mutagenesis;

(b) cell fusion (including protoplast fusion) of prokaryotic species which can exchange genetic material through homologous recombination;

(c) cell fusion (including protoplast fusion) of cells of any eukaryotic species, including production of hybridomas and plant cell fusions;

(d) self-cloning, where the resulting organism is unlikely to cause disease or harm to humans, animals or plants.

4

In paragraph 3—

(a) "self-cloning" means the removal of nucleic acid sequences from a cell of an organism which may or may not be followed by reinsertion of all or part of that nucleic acid (or a synthetic equivalent), whether or not altered by enzymic or mechanical processes, into cells of the same species or into cells of phylogenetically closely related species which can exchange genetic material by homologous recombination; and

(b) self-cloning may include the use of recombinant vectors, with an extended history of safe use in the particular organism, to manipulate and reinsert the nucleic acid sequences, but the vectors must not consist of any genetic elements other than those designed for vector structure, vector replication, vector maintenance or marker genes.

SCHEDULE 3

Regulations 2(3), 3(4) and 5

PART 1

MATTERS TO BE TAKEN INTO ACCOUNT IN CARRYING OUT AN ASSESSMENT FOR THE PURPOSES OF REGULATION 5

[8.3851]

1

The following matters must be taken into account in carrying out an assessment for the purposes of regulation 5—

(a) any potentially harmful effects, in particular those associated with—

(i) the recipient micro-organism;

(ii) the inserted genetic material (originating from the donor organism);

(iii) the vector;

(iv) the donor micro-organism (where that donor micro-organism is used during the contained use);

(v) the resulting genetically modified micro-organism;

(b) the characteristics of the contained use;

(c) the severity of the potentially harmful effects;

(d) the likelihood of the potentially harmful effects being realised;

(e) the disposal of waste and effluents.

2

In paragraph 1, "potentially harmful effects" includes—

(a) disease to humans including allergenic or toxic effects;

(b) disease to animals or plants;

(c) adverse effects resulting from the inability to treat disease or offer an effective prophylaxis;

(d) adverse effects resulting from establishment or dissemination of the genetically modified micro-organisms in the environment;

(e) adverse effects resulting from the natural transfer of genetic material to or from other organisms;

(f) adverse effects resulting from the likely interaction of the genetically modified micro-organism with other organisms at the premises where the contained use is to be conducted.

PART 2

STEPS TO BE INCLUDED WHEN CARRYING OUT AN ASSESSMENT FOR THE PURPOSES OF REGULATION5

[8.3852]

3

An assessment carried out for the purposes of regulation 5 must include—

(a) identification of any harmful properties of the recipient and, where appropriate, the donor micro-organism;

(b) identification of any harmful properties associated with the vector or inserted material, including any alteration in the existing properties of the recipient;

(c) recognition that, in general, only contained use which shows the following characteristics is appropriate for inclusion in class 1 as described in Schedule 1—

   (i) the recipient or parental micro-organism is unlikely to cause disease to humans, animals or plants;

   (ii) the nature of the vector and the insert is such that they do not endow the genetically modified micro-organism with a phenotype likely to cause disease to humans, animals or plants, or likely to cause deleterious effects on the environment; and

   (iii) the genetically modified micro-organism is unlikely to cause disease to humans, animals or plants and is unlikely to have deleterious effects on the environment;

(d) consideration of relevant EU legislation, including Directive (EC) No 2000/54 of the European Parliament and the Council on the protection of workers from risks related to exposure to biological agents at work, other classification schemes referring to plant and animal pathogens, and other international and national classification schemes for genetically modified micro-organisms;

(e) identification of the provisional level of risk associated with the genetically modified micro-organism;

(f) consideration of—

   (i) the characteristics of the environment likely to be exposed;

   (ii) the characteristics of the contained use involving micro-organisms;

   (iii) any contained use of micro-organisms which cannot be controlled adequately by standard laboratory procedures, and which presents risks which require controls for each individual case;

(g) adjustment of the provisional level of risk in the light of the matters referred to in sub-paragraph (f);

(h) selection of the appropriate containment measures from those specified in the applicable table in Schedule 8 on the basis of the provisional level of risk as adjusted in accordance with sub-paragraph (g);

(i) assignment of the contained use to the appropriate containment level, in accordance with paragraph 4;

(j) classification of the contained use in the class of the same number as that of the appropriate containment level;

(k) review and reconsideration of that classification in the light of the completed risk assessment.

4

To assign a contained use to the appropriate containment level for the purposes of paragraph 3(i), the person carrying out the risk assessment must—

(a) first identify for each selected containment measure the column in the applicable table in Schedule 8 having the lowest number in which that selected containment measure is shown as being required, regardless of whether or not such requirement is subject to any qualification;

(b) then select the highest number of all the columns identified in accordance with sub-paragraph (a); and

(c) then assign the contained use to the containment level of that highest number.

5

In paragraph 4, "selected containment measure" means an appropriate containment measure selected in accordance with paragraph 3(h).

SCHEDULE 4

Regulation 6

PART 1

MATTERS TO BE TAKEN INTO ACCOUNT IN CARRYING OUT AN ASSESSMENT FOR THE PURPOSES OF REGULATION 6

[8.3853]

1

The following matters must be taken into account in carrying out an assessment for the purposes of regulation 6—

(a) any potentially harmful effects, in particular those associated with—

(i) the recipient organism;

(ii) the inserted genetic material (originating from the donor organism);

(iii) the vector;

(iv) the donor organism;

(v) the resulting genetically modified organism;

(b) the characteristics of the contained use;

(c) the severity of the potentially harmful effects;

(d) the likelihood of the potentially harmful effects being realised.

2

In paragraph 1, "potentially harmful effects" includes—

(a) disease to humans including allergenic or toxic effects;

(b) acting as a human disease vector or reservoir;

(c) adverse effects to humans arising from change in behaviour or in physical nature;

(d) adverse effects arising from the inability to treat human disease or offer effective prophylaxis.

PART 2

STEPS TO BE INCLUDED WHEN CARRYING OUT AN ASSESSMENT FOR THE PURPOSES OF REGULATION 6

[8.3854]

3

An assessment carried out for the purposes of regulation 6 must include—

(a) identification of any harmful properties of the recipient and, where appropriate, the donor organism;

(b) identification of any harmful properties associated with the vector or inserted material, including any alteration in the existing properties of the recipient;

(c) identification of the provisional level of risk associated with the genetically modified organisms;

(d) selection of containment and other protective measures on the basis of—

(i) the provisional level of risk; and

(ii) the characteristics of the contained use;

(e) adjustment of the level of risk in the light of the matters referred to in sub-paragraph (d);

(f) review and reconsideration of the containment and other protective measures in the light of the steps required by sub-paragraphs (a) to (e).

SCHEDULE 5

INFORMATION REQUIRED FOR A NOTIFICATION UNDER REGULATION 9(2)

Regulations 9(2), 14(1) and 28(3)

[8.3855]

A notification required for the purposes of regulation 9(2) must contain the following information—

(a) the name, address and telephone number and any fax number and any e-mail address of the notifier;

(b) the name of the person with specific responsibility for the supervision and safety of contained use;

(c) information on the training and qualifications of that person;

(d) details of the arrangements for obtaining advice on risk assessments in accordance with regulation 8, including details of any genetic modification safety committee if established;

(e) the address of the premises where the contained use is to be carried out and a general description of the premises, together with, if required by regulation 9(6), the principal address of the premises;

(f) the nature of the work to be undertaken;

(g) the class of any contained use involving micro-organisms;

(h) where the first contained use to be carried out in those premises is a class 1 contained use—

(i) a summary of the risk assessment of that contained use;

(ii) any advice received in relation to the risk assessment from a person or genetic modification safety committee in accordance with regulation 8;

(iii) information on waste management;

(iv) confirmation that the emergency services, and any body or authority liable to be affected by an accident to which any emergency plan relates, will be informed of the contents of the emergency plan and of any relevant revisions;

(i) where the first contained use to be carried out in those premises is a contained use involving larger GMOs and that contained use is not notifiable under regulation 12(2)—

(i) a copy of the risk assessment; and

(ii) confirmation that the emergency services, and any body or authority liable to be affected by an accident to which any emergency plan relates, will be informed of the contents of the emergency plan and of any relevant revisions.

SCHEDULE 6

INFORMATION REQUIRED FOR A NOTIFICATION UNDER REGULATION 10(2), 11(2) OR 12(2)

Regulations 10(2), 11(2), 12(2),14(1), 28(3) and 33(6)

[8.3856]

A notification required for the purposes of regulation 10(2), 11(2) or 12(2) must contain the following information except where it is required only for a specified regulation—

(a) the name, address and telephone number and any fax number and any e-mail address of the notifier;

(b) any centre number allocated by the competent authority in respect of the premises at which the contained use is to be undertaken and the date of the notification required by regulation 9(2) relating to those premises;

(c) the name of the person with specific responsibility for supervision and safety of contained use;

(d) information on the training and qualifications of that person;

(e) the recipient or parental micro-organism to be used;

(f) the donor micro-organism to be used;

(g) where applicable, the host-vector system to be used;

(h) the source and intended function of the genetic material involved in the modification;

(i) the identity and characteristics of the genetically modified organism;

(j) the purpose of the contained use, including its expected results;

(k) for regulation 10(2) the approximate culture volumes to be used;

(l) for regulation 11(2) the culture volumes to be used;

(m) a description of the containment and other protective measures to be applied, including—

(i) information on waste management, including the type and form of wastes to be generated, their treatment, ultimate form and destination; and

(ii) for regulation 10(2) justification for not applying any containment measure at containment level 2;

(iii) for regulation 11(2), for class 3 contained use, justification for not applying any containment measure at containment level 3;

(iv) for regulation 11(2), for class 4 contained use, justification for not applying any containment measure at containment level 4;

(n) for regulations 10(2) and 11(2) a copy of the risk assessment;

(o) for regulations 10(2) and 11(2) the advice received in relation to that assessment from a genetic modification safety committee;
(p) for regulation 12(2) a copy of the risk assessment;
(q) information in relation to any accident prevention and emergency plans including—
  (i) the information necessary for the competent authority to evaluate any emergency plan;
  (ii) confirmation that the emergency services, and any body or authority liable to be affected by an accident to which any emergency plan relates, will be informed of the contents of the emergency plan and of any relevant revisions;
  (iii) for regulation 11(2), in addition—
    (aa) any specific hazards arising from the location of the installation;
    (bb) the preventive measures applied, including safety equipment, alarm systems and containment methods;
    (cc) procedures and plans for verifying the continuing effectiveness of the containment measures;
    (dd) a description of the information provided to workers;
(r) for regulation 11(2) a description of the parts of the installation;
(s) for regulation 11(2) whether the genetically modified organism is likely to be subject to transboundary movement.

SCHEDULE 7

GENERAL PRINCIPLES OF GOOD MICROBIOLOGICAL PRACTICE AND OF GOOD OCCUPATIONAL SAFETY AND HYGIENE

Regulation 18

[8.3857]

The general principles of good microbiological practice and of good occupational safety and hygiene are as follows—
(a) keeping workplace and environmental exposure to any genetically modified micro-organism to the lowest reasonably practicable level;
(b) exercising engineering control measures at source and supplementing these with appropriate personal protective clothing and equipment where necessary;
(c) testing adequately and maintaining control measures and equipment;
(d) testing, where necessary, for the presence of viable process organisms outside the primary physical containment;
(e) providing appropriate training of personnel;
(f) establishing a genetic modification safety committee, if required;
(g) formulating and implementing local codes of practice for the safety of personnel, as required;
(h) displaying biohazard signs where appropriate;
(i) providing washing and decontamination facilities for personnel;
(j) keeping adequate records;
(k) prohibiting in the work area eating, drinking, smoking, applying cosmetics or the storing of food for human consumption;
(l) prohibiting mouth pipetting;
(m) providing written standard operating procedures where appropriate to ensure safety;

(n) having effective disinfectants and specified disinfection procedures available in case of spillage of genetically modified micro-organisms;

(o) providing safe storage for contaminated laboratory equipment and materials where appropriate.

SCHEDULE 8

Regulations 2(2) and 19(1)

PART 1

GENERAL

[8.3858]

1

In this Schedule—

"GMMs" means genetically modified micro-organisms;

"HEPA" means High Efficiency Particulate Air;

"inactivation" means the complete or partial destruction of GMMs so as to ensure that any contact between the GMMs and humans or the environment is limited to an extent commensurate with the risks identified in the risk assessment and to provide a high level of protection for humans and the environment;

"plant growth facilities" means a structure, whether permanent or impermanent, designed and used principally for growing plants in a controlled and protected environment.

2

For the purposes of this Schedule where in the final column of Table 1b or 1c, a measure is specified as—

(a) a modification, it is to be read in substitution for the relevant measure in Table 1a;

(b) additional, it is to be read as an addition to the measures in Table 1a, and any measure which has been substituted for a measure in Table 1a, in accordance with paragraph 2(a).

3

For the purposes of this Schedule—

(a) Table 1a describes containment measures applicable to contained use involving micro-organisms in laboratories;

(b) Table 1a, read with Table 1b, describes containment measures applicable to contained use involving micro-organisms in plant growth facilities;

(c) Table 1a, read with Table 1c, describes containment measures applicable to contained use involving micro-organisms in animal units;

(d) Table 2 describes containment measures applicable to contained use involving micro-organisms in premises other than those referred to in Tables 1a, 1b and 1c.

PART 2

CONTAINMENT MEASURES

[8.3859]

Table 1a

Containment measures applicable to contained use involving micro-organisms in laboratories

	Containment Measures	*Containment Levels*			
		1	2	3	4
Facilities					
1	Laboratory suite: isolation[1]	not required	not required	required	required
2	Laboratory: sealable for fumigation	not required	not required	required	required
Equipment					
3	Surfaces impervious to water, resistant to acids, alkalis, solvents, disinfectants and decontamination agents and easy to clean	required for any bench	required for any bench	required for any bench and floor	required for any bench, floor, ceiling and walls
4	Entry to laboratory via airlock[2]	not required	not required	required where and to extent the risk assessment shows it is required	required
5	Negative pressure relative to the pressure of the immediate surroundings	not required	not required	required except for activities where transmission does not occur by the airborne route	required
6	Extract and input air from the laboratory must be HEPA filtered	not required	not required	HEPA filters required for extract air except for activities where transmission does not occur by the airborne route	HEPA filters required for input and extract air[3]

7	Microbiological safety cabinet/enclosure	not required	required where and to extent the risk assessment shows it is required	required, and all procedures with infective materials required to be contained within a cabinet/enclosure	required, and all procedures with infective materials required to be contained within a cabinet/enclosure
8	Autoclave	required on site	required in the building	required in the laboratory suite[4]	double ended autoclave required in laboratory
System of work					
9	Access restricted to authorised personnel only	not required	required	required	required (via airlock key procedure)
10	Biohazard sign on door	not required	required	required	required
11	Specific measures to control aerosol dissemination	not required	required so as to minimise	required so as to prevent	required so as to prevent

12	Shower	not required	not required	required where and to extent the risk assessment shows it is required	required
13	Protective clothing	suitable protective clothing required	suitable protective clothing required	suitable protective clothing required; footwear required where and to extent the risk assessment shows it is required	complete change of clothing and footwear required before entry and exit
14	Gloves	not required	required where and to extent the risk assessment shows they are required	required	required
15	Efficient control of disease vectors (eg rodents and insects) which could disseminate GMMs	required where and to extent the risk assessment shows it is required	required	required	required
Waste					
16	Inactivation of GMMs in effluent from hand-washing sinks and showers and similar effluents	not required	not required	required where and to extent the risk assessment shows it is required	required

17	Inactivation of GMMs in contaminated material and waste	required by validated means where and to extent the risk assessment shows it is required	required by validated means	required by validated means, with waste inactivated within the laboratory suite	required by validated means, with waste inactivated within the laboratory
Other measures					
18	Laboratory to contain its own equipment	not required	not required	required, so far as is reasonably practicable	required
19	An observation window or alternative is to be present so that occupants can be seen	required where and to extent the risk assessment shows it is required	required where and to extent the risk assessment shows it is required	required where and to extent the risk assessment shows it is required	required
20	Safe storage of GMMs	required where and to extent the risk assessment shows it is required	required	required	secure storage required
21	Written records of staff training	not required	required where and to extent the risk assessment shows it is required	required	required

(1)"isolation" means, in relation to a laboratory, separation of the laboratory from other areas in the same building, or being in a separate building.

(2)Entry must be through an airlock which is a chamber isolated from the laboratory. The clean side of the airlock must be separated from the restricted side by changing or showering facilities and preferably by interlocking doors.

(3)Where viruses are not retained by the HEPA filters, extra requirements will be necessary for extract air.

(4)Where the autoclave is outside the laboratory in which the contained use is being undertaken, but within the laboratory suite, there must be validated procedures for the safe transfer of material into that autoclave, which provide a level of protection equivalent to that which would be achieved by having an autoclave in that laboratory.

Table 1b

Containment measures applicable to contained use involving micro-organisms in plant growth facilities (to be read with Table 1a as indicated in paragraph 3(b) of Part 1)

	Containment Measures	*Containment Levels*				*Additional/ modification*
		1	2	3	4	
Facilities						
1	Permanent structure[5]	required where and to extent the risk assessment shows it is required	required	required	required	modification
Equipment						
2	Entry via a separate room with two interlocking doors	not required	required where and to extent the risk assessment shows it is required	required where and to extent the risk assessment shows it is required	required (via airlock key procedure)	additional
3	Control of contaminated run-off water	required where and to extent the risk assessment shows it is required	required so as to minimise run-off	required so as to prevent run-off	required so as to prevent run-off	additional
System of work						
4	Effective control of disease vectors such as insects, rodents and arthropods which could disseminate GMMs	required	required	required	required	additional

5	Effective control of pollen, seeds and other plant material which could disseminate GMMs	required where and to extent the risk assessment shows it is required	required so as to minimise dissemination	required so as to prevent dissemination	required so as to prevent dissemination	additional
6	Procedures for transfer of living material between the plant growth facilities, protective structure and laboratory must control dissemination of GMMs	required so as to minimise dissemination	required so as to minimise dissemination	required so as to prevent dissemination	required so as to prevent dissemination	additional

(1)A permanent structure refers to a fixed structure with walls, a roof and a floor. Where the permanent structure is a greenhouse, that structure must also have a continuous waterproof covering and self-closing lockable outer doors, and be located on a site designed to prevent the entry of surface run-off water.

Table 1c

Containment measures applicable to contained use involving micro-organisms in animal units (to be read with Table 1a as indicated in paragraph 3(c))

	Containment Measures	*Containment Levels*				*Additional/ modification*
1	2	3				4
Facilities						
1	Isolation of animal unit[6]	required where and to extent the risk assessment shows it is required	required	required	required	modification
2	Animal facilities[7] separated by lockable doors	required where and to extent the risk assessment shows it is required	required	required	required	additional
3	Animal facilities (cages, etc) designed to facilitate decontamination (waterproof and easily washable material)	required where and to extent the risk assessment shows it is required	required where and to extent the risk assessment shows it is required	required	required	additional
4	Floor, walls and ceiling easily washable	required where and to extent the risk assessment shows it is required	required for floor	required for floor and walls	required for floor, walls and ceiling	modification
5	Appropriate filters on isolators or isolated rooms[8]	not required	required where and to extent the risk assessment shows it is required	required	required	additional

6	Appropriate barriers at the room exit, and at drains or ventilation duct work	required	re-quired	re-quired	re-quired	additional
7	Animals kept in appropriate containment facilities, such as cages, pens or tanks but not isolators	required where and to extent the risk assess-ment shows it is required	re-quired where and to extent the risk assess-ment shows it is re-quired	re-quired where and to extent the risk assess-ment shows it is re-quired	re-quired where and to extent the risk assess-ment shows it is re-quired	additional
8	Animals kept in isolators	not re-quired	re-quired where and to extent the risk assess-ment shows it is re-quired	re-quired	re-quired	modification

(1)"animal unit" means a building, or separate area within a building, containing an animal facility and other areas including changing rooms, showers, autoclaves and food storage areas.

(2)"animal facility" means a facility normally used to house stock, breeding or experimental animals or one which is used for the performance of minor surgical procedures on animals.

(3)"isolators" means transparent boxes where small animals are contained within or outside a cage; for large animals, isolated rooms may be more appropriate.

Table 2

Containment measures applicable to contained use involving micro-organisms in premises other than those referred to in Tables 1a, 1b and 1c

	Contain-ment Measure	Containment Levels	
1	2	3	4
General			

1	Viable micro-organisms must be contained in a system which separates the process from the workplace and wider environment (closed system)	required where and to extent the risk assessment shows it is required	required	required	required
2	Closed systems located within a controlled area	not required	required where and to extent the risk assessment shows it is required	required	required
3	Control of exhaust gases from the closed system	not required	required so as to minimise release	required so as to prevent release	required so as to prevent release
4	Control of aerosols during sample collection, addition of material to a closed system or transfer of material to another closed system	required where and to extent the risk assessment shows it is required	required so as to minimise release	required so as to prevent release	required so as to prevent release
5	Inactivation of bulk culture fluids before removal from the closed system	required where and to extent the risk assessment shows it is required	required by validated means	required by validated means	required by validated means

6	Seals must be designed so as to minimise or prevent release	not required	required so as to minimise release	required so as to prevent release	required so as to prevent release
7	The controlled area designed to contain spillage of the entire contents of the closed system	required where and to extent the risk assessment shows it is required	required where and to extent the risk assessment shows it is required	required	required
8	The controlled area sealable to permit fumigation	not required	required where and to extent the risk assessment shows it is required	required where and to extent the risk assessment shows it is required	required
9	Biohazard signs posted	not required	required	required	required
Equipment					
10	Entry via airlock	not required	not required	required where and to extent the risk assessment shows it is required	required
11	Surfaces resistant to water, acids, alkalis, solvents, disinfectants and decontamination agents and easy to clean	required for any bench	required for any bench	required for any bench and floor	required for any bench, floor, ceilings and walls

12	Specific measures to ventilate adequately the controlled areas in order to minimise air contamination	required where and to extent the risk assessment shows they are required	required where and to extent the risk assessment shows they are required	required where and to extent the risk assessment shows they are required	required
13	The controlled area maintained at an air pressure negative to the immediate surroundings	not required	not required	required where and to extent the risk assessment shows it is required	required
14	Extract and input air from the controlled area must be HEPA filtered	not required	not required	required for extract air; required where and to extent the risk assessment shows it is required for input air	required for input and extract air
System of work					
15	Access restricted to authorised personnel only	not required	required	required	required
16	Personnel must shower before leaving the controlled area	not required	not required	required where and to extent the risk assessment shows it is required	required

17	Personnel must wear protective clothing	work clothing required	work clothing required	required	complete change required before exit and entry
18	Written procedures and records of staff training	not required	required where and to extent the risk assessment shows they are required	required	required
Waste					
19	Inactivation of GMMs in effluent from hand-washing sinks and showers or similar effluents	not required	not required	required where and to extent the risk assessment shows it is required	required
20	Inactivation of GMMs in contaminated material and waste including those in process effluent before final discharge	required by validated means where and to extent the risk assessment shows it to be required	required by validated means	required by validated means	required by validated means

PART 9
CONSTRUCTION

Part 9

Construction (Design and Management) Regulations 2007

SI 2007/320

as amended by SI 2013/1471 and SI 2013/1666

PART 3

ADDITIONAL DUTIES WHERE PROJECT IS NOTIFIABLE

[9.269]

19 Additional duties of contractors

(1) Where a project is notifiable, no contractor shall carry out construction work in relation to the project unless—

(a) he has been provided with the names of the CDM co-ordinator and principal contractor;

(b) he has been given access to such part of the construction phase plan as is relevant to the work to be performed by him, containing sufficient detail in relation to such work; and

(c) notice of the project has been given to the Executive, or as the case may be the Office of Rail Regulation, under regulation 21.

(2) Every contractor shall—

(a) promptly provide the principal contractor with any information (including any relevant part of any risk assessment in his possession or control) which—

(i) might affect the health or safety of any person carrying out the construction work or of any person who may be affected by it,

(ii) might justify a review of the construction phase plan, or

(iii) has been identified for inclusion in the health and safety file in pursuance of regulation 22(1)(j);

(b) promptly identify any contractor whom he appoints or engages in his turn in connection with the project to the principal contractor;

(c) comply with—

(i) any directions of the principal contractor given to him under regulation 22(1)(e), and

(ii) any site rules;

(d) promptly provide the principal contractor with the information in relation to any death, injury, condition or dangerous occurrence which the contractor is required to notify or report under [the Reporting of Injuries, Diseases and Dangerous Occurrences Regulations 2013].

(3) Every contractor shall—

(a) in complying with his duty under regulation 13(2) take all reasonable steps to ensure that the construction work is carried out in accordance with the construction phase plan;

(b) take appropriate action to ensure health and safety where it is not possible to comply with the construction phase plan in any particular case; and

(c) notify the principal contractor of any significant finding which requires the construction phase plan to be altered or added to.

[9.270]

Statutory annotations

Amendment

Para (2): in sub-para (d) words 'the Reporting of Injuries, Diseases and Dangerous Occurrences Regulations 2013' in square brackets substituted by SI 2013/1471, reg 18(2), Sch 4.

PART 5

GENERAL

[9.321]

45 . . .

[*Revoked by SI 2013/1666, art 4.*]

PART 10
TRANSPORT

Railways Act 2005

2005 c 14

as amended

SCHEDULE 3

TRANSFER OF SAFETY FUNCTIONS

Section 2

Railway safety purposes

[10.273]

1

(1) Subject to sub-paragraph (4), in this Schedule 'railway safety purposes' means so much of the general purposes of Part 1 of the 1974 Act (within the meaning of section 1 of that Act) as—

(a) relates to risks that are exclusively relevant to one or more of the purposes specified in sub-paragraph (2); or

(b) relates to so much of any risks that are primarily so relevant as arises in connection with anything mentioned in that sub-paragraph.

(2) The purposes mentioned in sub-paragraph (1) are—

(a) securing the proper construction and safe operation of transport systems falling within sub-paragraph (3);

(b) securing the proper construction and safe operation of locomotives, rolling stock or other vehicles used, or to be used, on such systems;

(c) protecting the public (whether or not they are passengers) from personal injury and other risks arising from the construction and operation of such systems;

(d) protecting persons at work from personal injury and other risks so arising.

(3) A transport system falls within this sub-paragraph if it is—

(a) a railway;

(b) a tramway; [or]

(c) . . .

(d) a transport system using any other mode of guided transport [other than a guided bus system],

[but not to the extent that the transport system consists of or is operated for the purposes of fairground equipment].

(4) The Secretary of State may, by regulations, modify the definition of 'railway safety purposes'.

(5) Before making any regulations under sub-paragraph (4), the Secretary of State must consult—

(a) the Office of Rail Regulation;

(b) the [Health and Safety Executive]; and

[(ba) the Office for Nuclear Regulation;] and

(c) such other persons as he considers appropriate.

(6) Regulations under sub-paragraph (4) are subject to the negative resolution procedure.

(7) In this paragraph—

['bus' means a motor vehicle which is designed or adapted to travel along roads and to carry more than eight passengers but which is not a tramcar;

'fairground equipment' has the same meaning as in section 53 of the 1974 Act;

'guided bus system' means a system of transport, used wholly or mainly for the carriage of passengers, that employs buses which for some or all of the time when they are in operation—

(a) travel along roads; and

(b) are guided (whether while on the road or at other times) by means of—

(i) apparatus, a structure or other device which is fixed and not part of the bus; or

(ii) a guidance system which is automatic;]

'guided transport', 'railway', 'tramway', ' . . . ' and 'vehicle' have the same meanings as in the Transport and Works Act 1992 (c 42) except that, for this purpose, paragraph (b) of the definition of 'railway' in section 67(1) of that Act (which includes a condition as to the minimum gauge of track) is to be disregarded; and

'person at work' has the same meaning as it for the time being has in Part 1 of the 1974 Act (see section 52 of that Act);

['road'—

(a) in England and Wales, means any length of highway or of any other road to which the public has access, and includes bridges over which a road passes; and

(b) in Scotland, has the same meaning as in the Roads (Scotland) Act 1984].

ORR's principal railway safety functions

2

(1) It shall be the general duty of the Office of Rail Regulation—

(a) to do such things and make such arrangements as it considers appropriate for the railway safety purposes; and

(b) to assist and encourage persons concerned with matters relevant to any of those purposes to further those purposes.

(2) It shall be the duty of the Office of Rail Regulation—

(a) to make such arrangements as it considers appropriate for the carrying out of research in connection with the railway safety purposes and for the publication of the results of such research; and

(b) to encourage research by others in that connection.

(3) It shall be the duty of the Office of Rail Regulation—

(a) to make such arrangements as it considers appropriate for the provision of training and information in connection with the railway safety purposes;

(b) to encourage the provision by others of training and information in that connection; and

(c) to make such arrangements as it considers appropriate for securing that the persons mentioned in sub-paragraph (4)—

(i) are provided with an information and advisory service with respect to matters relevant to the railway safety purposes;

(ii) are kept informed about such matters; and

(iii) are adequately advised about them.

(4) Those persons are—

(a) government departments;
(b) employers;
(c) employees;
(d) organisations representing employers and employees respectively; and
(e) other persons concerned with matters relevant to any of the railway safety purposes.

(5) It shall be the duty of the Office of Rail Regulation from time to time to submit such proposals as it considers appropriate for the making of regulations for the railway safety purposes to the authorities having power to make regulations for those purposes under any of the relevant statutory provisions.

(6) Before submitting proposals under sub-paragraph (5), the Office of Rail Regulation must[—

(a) if the proposals relate to regulations that are relevant to the ONR's purposes (within the meaning of Part 3 of the Energy Act 2013), consult the Office for Nuclear Regulation;
(b) in any case, consult—
    (i) such government departments, and
    (ii) such other persons,

as it considers appropriate].

(7) Before—

(a) making regulations for the railway safety purposes independently of any proposals submitted under sub-paragraph (5), or
(b) making regulations which give effect to such proposals but with modifications,

the authority having the power to make the regulations must consult the Office of Rail Regulation [and, if the regulations are relevant to the ONR's purposes (within the meaning of Part 3 of the Energy Act 2013), the Office for Nuclear Regulation].

(8) In this paragraph 'employer', 'employee' and 'the relevant statutory provisions' have the same meanings as in Part 1 of the 1974 Act (see section 53(1) of that Act).

Removal of corresponding functions from HSC

3

(1) In section 11 of the 1974 Act (which confers functions on the HSC for general purposes that are equivalent to those conferred on the ORR for railway safety purposes by paragraph 2)—

(a) in subsection (1), for 'subject to subsection (3)' substitute 'subject to subsections (2A) and (3)'; and
(b) after subsection (2) insert the subsection set out in sub-paragraph (2).

(2) The subsection to be inserted is—

'(2A) In subsections (1) and (2) above—

(a) references to the general purposes of this Part do not include references to the railway safety purposes; and
(b) the reference to a power to make regulations under the relevant statutory provisions does not include a reference to any power so far as it is exercisable for the railway safety purposes.'

Reports and investigations

4

(1) The Office of Rail Regulation may authorise a person to investigate and make a special report on any accident, occurrence, situation or other matter of any sort which that Office thinks it necessary or expedient to investigate—

Part 10

(a) for any of the railway safety purposes; or

(b) with a view to the making of regulations for any of those purposes.

(2) The Office of Rail Regulation may cause—

(a) the contents of a special report made under this paragraph, or

(b) so much of them as it considers appropriate,

to be made public at such time, and in such manner, as it thinks fit.

(3) Where a person who is not a member, officer or employee of the Office of Rail Regulation carries out an investigation and makes a special report under this paragraph, that Office may pay him such remuneration and expenses as the Secretary of State determines.

(4) The Office of Rail Regulation may, to such extent as the Secretary of State may determine, defray the other costs (if any) of—

(a) an investigation under this paragraph; or

(b) the making of a special report following such an investigation.

[(4A) The Office of Rail Regulation must consult the Office for Nuclear Regulation before taking any step under sub-paragraph (1) in relation to a matter which appears to the Office of Rail Regulation to be, or likely to be, relevant to the ONR's purposes (within the meaning of Part 3 of the Energy Act 2013).]

(5) In section 14(1) of the 1974 Act (matters about which investigations etc may be required or authorised by the HSC), for 'it is' substitute—

'(a) those general purposes shall be treated as not including the railway safety purposes; but

(b) it is otherwise'.

5

(1) This paragraph applies where, before the commencement of paragraph 4, the Health and Safety Commission directed or authorised a person to investigate and make a special report under section 14(2)(a) of the 1974 Act on any accident, occurrence, situation or other matter which the Commission thought it necessary or expedient to investigate—

(a) for any of the railway safety purposes; or

(b) with a view to the making of regulations for any of those purposes.

(2) For the purposes of the carrying out or continuation of an investigation and the making of a report after the commencement of paragraph 4, the direction or authorisation shall be treated as an authorisation by the Office of Rail Regulation under paragraph 4.

(3) If—

(a) the person to whom the Health and Safety Commission gave the direction or authorisation has made a special report to the Commission, but

(b) the Commission has not yet caused the report, or a part of it, to be made public under section 14(5) of the 1974 Act,

the report shall be treated as if it had been made to the Office of Rail Regulation under paragraph 4.

(4) If the person to whom the Health and Safety Commission gave the direction or authorisation has not yet made a special report to the Commission, the Office of Rail Regulation may direct that person—

(a) to abandon the investigation without making a special report; or

(b) to continue the investigation in such manner as that Office may direct.

(5) Where the Health and Safety Commission agreed before the commencement of paragraph 4 to exercise its power by virtue of section 14(6)(a) or (c) of the 1974 Act to pay remuneration or expenses to any person in respect of the investigation and special report or to defray costs of the investigation and report—

(a) the Commission must pay amounts accruing in respect of the agreed matters before the commencement of paragraph 4; and

(b) the Office of Rail Regulation must pay amounts so accruing after that time;

and paragraph (b) does not affect the power of that Office under that paragraph, in relation to times after the commencement of that paragraph, to make other payments of remuneration or expenses or to defray costs.

6

(1) This paragraph applies where, before the commencement of paragraph 4, the Health and Safety Commission directed an inquiry to be held under section 14(2)(b) of the 1974 Act into any accident, occurrence, situation or other matter which the Commission thought it necessary or expedient to investigate—

(a) for any of the railway safety purposes; or

(b) with a view to the making of regulations for any of those purposes.

(2) Subject to sub-paragraph (3), section 14 of the 1974 Act shall continue to apply in relation to that inquiry after the commencement of paragraph 4 but as if all things done or required to be done by or in relation to the Health and Safety Commission under that section were done or required to be done by or in relation to the Office of Rail Regulation.

(3) Where the Health and Safety Commission agreed before the commencement of paragraph 4 to exercise its power by virtue of section 14(6)(b) or (c) of the 1974 Act to pay remuneration or expenses to any person in respect of the inquiry or to defray costs of the inquiry—

(a) the Commission must pay amounts accruing in respect of the agreed matters before the commencement of paragraph 4; and

(b) the Office of Rail Regulation must pay amounts so accruing after that time;

and paragraph (b) does not affect the power of that Office, in relation to times after the commencement of paragraph 4, to make other payments of remuneration or expenses under section 14(6)(b), or to defray costs under section 14(6)(c).

ORR acting as agent of government departments and other public authorities

7

(1) Where the Secretary of State considers that any of his statutory or other functions can appropriately be carried out by the Office of Rail Regulation in connection with the carrying out of its safety functions, he and that Office may enter into an agreement for that Office to carry out those functions on his behalf.

(2) Where the Secretary of State considers that any of the statutory or other functions of—

(a) a government department, or

(b) any other public authority,

can appropriately be carried out by the Office of Rail Regulation in connection with the carrying out of its safety functions, he may authorise the Office and that department or authority to enter into an agreement for that Office to carry out those functions on behalf of that department or authority.

(3) An agreement under this paragraph—

(a) may include provision for payments to be made to the Office of Rail Regulation for the carrying out of the functions that it is authorised or required to carry out under the agreement; but

(b) may not authorise or require the Office of Rail Regulation to carry out any function consisting in a power to make regulations or other instruments of a legislative character.

[(4) Sub-paragraph (3)(b) does not prevent the Secretary of State and the Office of Rail Regulation from entering into an agreement for that Office to carry out on his behalf the function of making orders under section 1 of the Level Crossings Act 1983.]

Government departments and other public authorities acting as agent of ORR

8

(1) Where the Office of Rail Regulation considers that any of its safety functions can appropriately be carried out—

(a) by a government department, or

(b) by any other public authority,

that Office and the department or authority may enter into an agreement for the department or authority to carry out those functions on that Office's behalf.

(2) An agreement under this paragraph may include provision for payments to be made by the Office of Rail Regulation to a government department or public authority for the carrying out of the functions that it is authorised or required to carry out under the agreement.

Restrictions on codes of practice by HSC

9

(1) In subsection (1) of section 16 of the 1974 Act (codes of practice with respect to the requirements of certain provisions), for 'sections 2 to 7 or of health and safety regulations or of any of the existing statutory provisions' substitute 'any of the enactments or instruments mentioned in subsection (1A) below'.

(2) After that subsection insert—

'(1A) Those enactments and instruments are—

(a) sections 2 to 7 above;

(b) health and safety regulations, except so far as they make provision exclusively in relation to transport systems falling within paragraph 1(3) of Schedule 3 to the Railways Act 2005; and

(c) the existing statutory provisions that are not such provisions by virtue of section 117(4) of the Railways Act 1993.'

Co-operation with the HSC

10

(1) It shall be the duty of the Office of Rail Regulation and the Health and Safety Commission—

(a) as soon as practicable after the commencement of this paragraph, to enter into arrangements with each other for securing co-operation and the exchange of information in connection with the carrying out of safety functions;

(b) to maintain and from time to time to review those arrangements; and

(c) to revise them whenever they consider it appropriate to do so.

(2) In sub-paragraph (1) 'safety functions' means all the functions of . . . the Health and Safety Executive and the safety functions of the Office of Rail Regulation.

(3) In section 18(5)(b) of the 1974 Act (duty of enforcing authority to act in accordance with guidance of HSC), at the beginning insert 'except where that authority is the Office of Rail Regulation,'.

Information powers corresponding to section 27 of the 1974 Act

11

(1) The Office of Rail Regulation may serve a notice under this paragraph on any person for the purpose of obtaining information which that Office needs for the carrying out of its safety functions.

(2) A notice under this paragraph is one requiring the person on whom it is served—

- (a) to provide the Office of Rail Regulation with information about such matters as may be specified in the notice; and
- (b) to do so in the form and manner so specified.

(3) The consent of the Secretary of State is required for the service of a notice under this paragraph.

(4) The consents that may be given for the purposes of sub-paragraph (3) include a general consent relating to notices of a specified description.

(5) Nothing in the Statistics of Trade Act 1947 (c 39) is to be construed as preventing or penalising the disclosure by a Minister of the Crown or government department to—

- (a) the Office of Rail Regulation,
- (b) a member, officer or employee of that Office, or
- (c) a committee established by that Office,

of information falling within sub-paragraph (6) about an undertaking (within the meaning of that Act).

(6) Information about an undertaking falls within this sub-paragraph if it consists of—

- (a) the names and addresses of the persons carrying on the undertaking;
- (b) the nature of the undertaking's activities;
- (c) the number of persons of different descriptions who work in the undertaking;
- (d) the addresses or places where activities of the undertaking are or were carried on;
- (e) the nature of the activities carried on there; or
- (f) the number of persons of different descriptions who work or worked in the undertaking there.

(7) A person to whom a disclosure that is authorised by sub-paragraph (5) is made must not use the information except for the purposes of the safety functions of the Office of Rail Regulation.

(8) It is an offence for a person—

- (a) to contravene a requirement imposed by a notice under this paragraph; or
- (b) to use information in contravention of sub-paragraph (7).

(9) A person guilty of an offence under sub-paragraph (8) shall be liable—

- (a) on summary conviction, to a fine not exceeding the statutory maximum;
- (b) on conviction on indictment—
  - (i) in the case of an offence of contravening a notice, to a fine; and
  - (ii) in the case of an offence of using information in contravention of sub-paragraph (7), to imprisonment for a term not exceeding two years or to a fine, or to both.

(10) Section 52 of the 1974 Act (meaning of 'work' and related expressions) applies for the purposes of this paragraph as it applies for the purposes of Part 1 of that Act.

Railway safety levy

12

(1) Section 43A of the 1974 Act (railway safety levy) is amended as follows.

(2) In subsection (2) (expenses in respect of which levy raised), for paragraphs (a) and (b) substitute—

'(a) in respect of activities undertaken by the Office of Rail Regulation under or by virtue of this Act or Schedule 3 to the Railways Act 2005; or

(b) in respect of activities in relation to a transport system falling within paragraph 1(3) of that Schedule that are undertaken by that Office under or by virtue of any other enactment.'

(3) In subsections (5) and (6), for 'Commission or the Executive' substitute 'Office of Rail Regulation'.

(4) In subsection (9), for 'to which section 117 of the Railways Act 1993 applies' substitute 'falling within paragraph 1(3) of Schedule 3 to the Railways Act 2005'.

Removal of requirement of consultation with HSC for railway safety regulations

13

In section 50 of the 1974 Act (consultation with HSC), after subsection (1) insert—

'(1A) Subsection (1) does not apply to the exercise of a power to make regulations so far as it is exercised—

(a) for giving effect (with or without modifications) to proposals submitted by the Office of Rail Regulation under paragraph 2(5) of Schedule 3 to the Railways Act 2005; or

(b) otherwise for or in connection with the railway safety purposes.'

. . .

14

. . .

Interpretation

15

(1) In this Schedule—

'the 1974 Act' means the Health and Safety at Work etc Act 1974 (c 37); and

'railway safety purposes' has the meaning given by paragraph 1.

(2) In this Schedule a reference to the safety functions of the Office of Rail Regulation is a reference to—

(a) its functions under this Schedule;

(b) its functions under the 1974 Act; and

(c) its other functions so far as carried out for the railway safety purposes.

(3) In section 53(1) of the 1974 Act (interpretation of Part 1), after the definition of 'prohibition notice' insert—

''railway safety purposes' has the same meaning as in Schedule 3 to the Railways Act 2005;'.

[10.274]–[10.550]

Statutory annotations

Amendment

Para 1: sub-para (5)(ba) inserted by the Energy Act 2013, s 116(1), Sch 12, Pt 5, paras 82, 83.

Para 2: in sub-para (6) words from '— (a) if the' to 'it considers appropriate' in square brackets substituted by the Energy Act 2013, s 116(1), Sch 12, Pt 5, paras 82, 84(1), (2); in sub-para (7) words from 'and, if the ' to 'Office for Nuclear Regulation' in square brackets inserted by the Energy Act 2013, s 116(1), Sch 12, Pt 5, paras 82, 84(1), (3).

Para 4: sub-para (4A) inserted by the Energy Act 2013, s 116(1), Sch 12, Pt 5, paras 82, 85.

Railway Safety (Miscellaneous Provisions) Regulations 1997

SI 1997/553

as amended by SIs 1998/494, 1999/2024, 2006/557, 2006/599, 2007/320, 2011/1043 and 2013/1666

Made 28th February 1997

[10.575]

3 Unauthorised access

(1) So far as is reasonably practicable, a person in control of any infrastructure of a transport system to which this regulation applies shall ensure, where and to the extent necessary for safety, that unauthorised access to that infrastructure is prevented.

(2) In paragraph (1) 'access' means access by any person not at work on the transport system or by any animal.

(3) This regulation applies to any transport system except that it does not apply to any part of such a system which—

(a) is within a harbour, harbour area, maintenance or goods depot; or

(b) is part of a factory, mine or quarry,

where access to the harbour, harbour area, maintenance or goods depot, factory, mine or quarry is adequately controlled.

(4) . . .

[10.575A]

Statutory annotations

AMENDMENTS

Para (4): revoked by SI 2013/1666, art 2.

Health and Safety (Enforcing Authority for Railways and Other Guided Transport Systems) Regulations 2006

SI 2006/557

as amended by SI 2006/2739, 2007/320, 2007/1573, 2008/2323 and 2013/950

[10.928]

2 Interpretation

In these Regulations—

'the 1974 Act' means the Health and Safety at Work etc Act 1974;

'bus' means a motor vehicle which is designed or adapted to travel along roads and to carry more than eight passengers but which is not a tramcar;

['bus substitution service' means a service for the carriage of passengers by road that is provided as an alternative to the whole or a part of a railway or tramway passenger service that has been discontinued, reduced or modified (whether temporarily or permanently);]

'cableway installation' means an installation made up of several components that—

- (a) is used or intended to be used for the purpose of providing an operational system for carrying persons in vehicles, on chairs or by towing devices;
- (b) uses cables positioned along the line of travel to provide suspension or traction or both; and
- (c) is one of the following—
  - (i) a cable car (including a gondola and chair lift) where the cabins or chairs are lifted or displaced by one or more carrier cables;
  - (ii) a drag lift, where users with appropriate equipment are dragged by means of a cable; or
  - (iii) a funicular railway or other installation with vehicles mounted on wheels or on other suspension devices where traction is provided by one or more cables;

but does not include rack railways or cable operated tramways;

'carriageway'—

- (a) in England and Wales, has the same meaning as in section 329(1) of the Highways Act 1980; and
- (b) in Scotland, has the same meaning as in section 151 of the Roads (Scotland) Act 1984;

'construction work' has the same meaning as in regulation 2(1) of the Construction (Design and Management) Regulations [2007] except that it does not include the installation, commissioning, maintenance, repair or removal of gas services which are normally fixed within or to a structure;

['contractor' has the same meaning as in regulation 2(1) of the Construction (Design and Management) Regulations 2007;]

. . .

'Enforcing Authority Regulations' means the Health and Safety (Enforcing Authority) Regulations 1998;

Part 10

['entity in charge of maintenance' has the same meaning as in the Railways and Other Guided Transport Systems (Safety) Regulations 2006;]

'the Executive' means the Health and Safety Executive;

'factory' means a factory within the meaning of section 175 of the Factories Act 1961 or a power generating station but does not include—

(a) light maintenance depots;

(b) any premises forming part of a heritage railway which are used exclusively or primarily for the reconstruction, renovation, refurbishment or repair of vehicles for use on the heritage railway;

(c) any premises forming part of a heritage tramway which are used exclusively or primarily for the reconstruction, renovation, refurbishment or repair of vehicles for use on the heritage tramway;

(d) premises owned or operated by, or on behalf of, the Secretary of State for Defence; or

(e) power generating stations which produce power exclusively for use in a system of transport specified in regulation 3(2);

'guided bus system' means a system of transport, used wholly or mainly for the carriage of passengers, that employs buses which for some or all of the time when they are in operation—

(a) travel along roads; and

(b) are guided (whether while on the road or at other times) by means of—

(i) apparatus, a structure or other device which is fixed and not part of the bus; or

(ii) a guidance system which is automatic;

'guided transport' means a system of transport, used wholly or mainly for the carriage of passengers, employing vehicles;

['harbour' means a harbour or harbour area as defined in regulation 2(1) of the Dangerous Substances in Harbour Areas Regulations 1987;]

'heritage railway' means a railway which is operated to—

(a) preserve, re-create or simulate railways of the past; or

(b) demonstrate or operate historical or special types of motive power or rolling stock;

and is exclusively or primarily used for tourist, educational or recreational purposes;

'heritage tramway' means a tramway which is operated to—

(a) preserve, re-create or simulate tramways of the past; or

(b) demonstrate or operate historical or special types of motive power or rolling stock;

and is exclusively or primarily used for tourist, educational or recreational purposes;

'infrastructure' means the following fixed assets used for the operation of any system of transport specified in regulation 3(2)—

(a) the permanent way;

(b) electrical and mechanical installations used for signalling;

(c) electrical installations used for supplying power to vehicles; and

(d) any other plant, equipment or electrical or mechanical installation;

but, where and to the extent that the system is operated on a road, does not include road lighting, road traffic signals and signage or any apparatus located on, under or above the road unless used for the purposes of that system and maintained by the owner, operator or manager of the system or persons acting on behalf of one or more of them;

'intermodal depot' means a freight handling facility which is operated exclusively or primarily for the purpose of transferring containers from railway vehicles onto a different mode of transport or vice versa;

'lift' means an appliance, other than a cableway installation, serving specific levels and having a car moving—

(a) along guides which are rigid; or

(b) along a fixed course even where it does not move along guides which are rigid;

and inclined at an angle of more than 15 degrees to the horizontal and intended wholly or mainly for the transport of persons;

'light maintenance depot' means any premises used exclusively or primarily for, or in connection with, the provision of light maintenance services;

'light maintenance services' means services of any of the following descriptions—

(a) the refuelling of vehicles;

(b) the cleaning of vehicles; and

(c) the carrying out to vehicles of maintenance work of a kind which is normally carried out at regular intervals of twelve months or less to prepare the vehicles for service;

and, for the purposes of paragraph (c), 'maintenance work' includes the detection and rectification of any faults;

['mainline railway' has the same meaning as in the Railways and Other Guided Transport Systems (Safety) Regulations 2006;]

'mine' has the same meaning as in section 180 of the Mines and Quarries Act 1954 but, notwithstanding subsection (5) of that section, does not include any railway serving the mine unless and to the extent that the railway is located within the curtilage of the mine;

['miniature railway' means a system of transport employing parallel rails which—

(a) provide support and guidance for vehicles carried on flanged wheels; and

(b) form a track of a gauge of less than 350 millimetres;]

'nuclear licensed site' means a site in respect of which a nuclear site licence, within the meaning of section 1(1) of the Nuclear Installations Act 1965, has been granted and is in force;

'operation of a railway' includes, in particular,—

(a) use, parking, sheltering, maintenance and repair of a vehicle including its inspection, cleaning, fuelling and preparation for use;

(b) subject to paragraph (k), loading or unloading of goods on or from vehicles at operational premises;

(c) ensuring the fitness of trains to enter into service, in particular, in the case of freight trains, by checking the condition and distribution of goods carried;

(d) use of the infrastructure;

(e) use of those fixed assets of the railway (other than infrastructure) which are both used in its operation and comprise, or are located within, operational premises;
(f) use of signalling control centres, or centres used for the operation of communications systems or systems used for monitoring the operation of the railway, whether or not within operational premises, together with any railway offices within the same premises;
(g) training of railway staff within operational premises or other premises included within paragraph (f);
(h) carrying out construction work to the extent stated in regulation 5; and
(i) reconstruction, renovation, refurbishment and repair of vehicles for use on a heritage railway carried out at premises used exclusively or primarily for these purposes and which form part of the heritage railway;

but does not include—
(j) subject to paragraph (f), office activities relating to the railway which are not carried out within operational premises; or
(k) loading or unloading of goods on or from vehicles at [harbours], intermodal depots or premises owned or operated by, or on behalf of, the Secretary of State for Defence;

'operation of a tramway' and 'operation of any other system of guided transport' includes, in particular,—
(a) use, parking, sheltering, maintenance and repair of vehicles including their inspection, cleaning, fuelling and preparation for use;
(b) ensuring the fitness of vehicles to enter into service;
(c) use of the infrastructure;
(d) use of those fixed assets of the system of transport (other than infrastructure) which are both used in its operation and comprise, or are located within, operational premises;
(e) use of signalling control centres, or centres used for the operation of communications systems or systems used for monitoring the operation of the system of transport, whether or not within operational premises, together with any offices of the system of transport within the same premises;
(f) training of staff within operational premises or other premises included within paragraph (e);
(g) carrying out construction work to the extent stated in regulation 5; and
(h) in the case of tramways, the reconstruction, renovation, refurbishment and repair of vehicles for use on a heritage tramway carried out at premises used exclusively or primarily for these purposes and which form part of the heritage tramway;

but, subject to paragraph (e), does not include office activities relating to the system of transport which are not carried out within operational premises;

'operational premises' means, in relation to any system of transport specified in regulation 3(2),—
(a) the permanent way;
(b) stations;
(c) light maintenance depots; and

(d) any land adjacent to or adjoining the permanent way which is used, occupied or held for the purposes of the system;

but does not include—

(e) parts separately occupied for the carrying on of any trade, business or undertaking other than for the purposes of the system, including shops, offices or premises used for the provision of catering services or temporary residential accommodation; or

(f) where and to the extent that the system is operated on a road, all other parts of the road including bridges, tunnels, culverts, retaining walls or other structures used or to be used for the support of, or otherwise in connection with, the road;

'the permanent way', in relation to any system of transport specified in regulation 3(2), means any land or other property comprising the permanent way of the system, taken together with the ballast, sleepers and metals or other means of guidance laid on, under or over it, whether or not the land or other property is also used for other purposes and includes—

(a) any level crossings, bridges, viaducts, tunnels, culverts, retaining walls or other structures used or to be used for the support of, or otherwise in connection with, the permanent way; and

(b) any walls, fences or other structures bounding the system or bounding any adjacent or adjoining property;

but does not include—

(c) any bridge, structure or other property which spans the permanent way and is maintained by persons other than the owner, operator or manager of the system or persons acting on behalf of one or more of them; or

(d) where and to the extent that the system is operated on a road, all other parts of the road including bridges, tunnels, culverts, retaining walls or other structures used or to be used for the support of, or otherwise in connection with, the road;

'power generating station' means any premises used exclusively or primarily for the purposes of generating electrical energy;

'quarry' has the same meaning as in regulation 3 of the Quarries Regulations 1999;

'railway' means a system of transport employing parallel rails which—

(a) provide support and guidance for vehicles carried on flanged wheels; and

(b) form a track which either is of a gauge of at least 350 millimetres or crosses a carriageway (whether or not on the same level);

but does not include a tramway;

'road'—

(a) in England and Wales, means any length of highway or of any other road to which the public has access, and includes bridges over which a road passes; and

(b) in Scotland, has the same meaning as in the Roads (Scotland) Act 1984;

'station' means any premises used as, or for the purposes of, or otherwise in connection with, a passenger station or terminal of a system of transport specified in regulation 3(2), (including station offices, ticket offices, left luggage and lost property facilities, approaches, forecourt, car parks and cycle stores), whether or not the premises are also used for other purposes;

'tramway' means a system of transport used wholly or mainly for the carriage of passengers—

- (a) which employs parallel rails which—
  - (i) provide support and guidance for vehicles carried on flanged wheels;
  - (ii) are laid wholly or partly along a road or in any other place to which the public has access (including a place to which the public has access only on making a payment); and
- (b) on any part of which the permitted maximum speed is such as to enable the driver to stop a vehicle in the distance he can see to be clear ahead;

'vehicle' means any vehicle which for some or all of the time when it is in operation is guided by means of—

- (a) rails, beams, slots, guides or other apparatus, structures or devices which are fixed and not part of the vehicle; or
- (b) a guidance system which is automatic;

and includes a mobile traction unit;

'warehouse premises' means any premises used exclusively or primarily for the storage of goods for retail or wholesale distribution.

[10.929]

Statutory annotations

Amendment

Definition 'entity in charge of maintenance' inserted by SI 2013/950, reg 2(1)(a); definition 'mainline railway' inserted by SI 2013/950, reg 2(1)(b); definition 'miniature railway' inserted by SI 2008/2323, reg 3(d).

[10.932]

4 Exceptions

(1) The activities specified in regulation 3(2) do not include the operation of—

- (a) a cableway installation; or
- (b) any fairground equipment; [or
- (c) a miniature railway which does not cross a carriageway (whether or not on the same level); or
- (d) a bus substitution service when outside operational premises].

(2) The systems of guided transport specified in regulation 3(2) (c) do not include—

- (a) a guided bus system;
- (b) any other system of guided transport that employs vehicles which for some or all of the time when they are in operation travel along roads; or
- (c) a lift, unless it is used in the operation of any system of transport specified in regulation 3(2).

(3) The operation of a system of transport specified in regulation 3(2) does not include such an operation carried out within—

- [(a) a harbour within premises referred to in any of sub paragraphs (c) to (h);
- (b) any other harbour unless—
  - (i) the system of transport operated is part of a system which is operated outside the harbour and the operation is for the purpose of or related to the carriage of passengers or goods to or from the harbour; or

(ii) the system of transport is not operated outside the harbour but is a railway, tramway or other system of guided transport for the carriage of the public not connected to any other railway;]

(c) an establishment to which the Control of Major Accident Hazards Regulations 1999 applies;

(d) a factory;

(e) a mine;

(f) a nuclear licensed site;

(g) a quarry; or

(h) warehouse premises.

(4) Notwithstanding paragraph (3), the Office of Rail Regulation shall be responsible for the enforcement of orders made under section 1 of the Level Crossings Act 1983 whether or not a level crossing to which such an order relates is within any premises referred to in paragraph (3).

[(4A) Notwithstanding paragraph (3), where an entity in charge of maintenance performs maintenance on a vehicle that is to be placed in service or is used on the mainline railway, the operation of a railway specified in regulation 3(2)(a) shall include such maintenance performed on a vehicle within any premises referred to in paragraph (3).]

(5) The activities specified in regulation 3(2) do not include the undertaking of any work which is prohibited by [the Control of Asbestos Regulations 2012] from being carried out by any employer or self employed person without a licence granted under [regulation 8] of those Regulations relating to that work.

[10.933]

Statutory annotations

Amendment

Para (4A): inserted by SI 2013/950, reg 2(2).

[10.938]

[8 Review]

[(1) Before the end of each review period, the Secretary of State must—

(a) carry out a review of these Regulations;

(b) set out the conclusions of the review in a report; and

(c) publish the report.

(2) The report must in particular—

(a) set out the objectives intended to be achieved by the regulatory system established by these Regulations;

(b) assess the extent to which those objectives are achieved; and

(c) assess whether those objectives remain appropriate and, if so, the extent to which they could be achieved with a system that imposes less regulation.

(3) 'Review period' means—

(a) the period of five years beginning with the day on which the Railways and Other Guided Transport Systems (Miscellaneous Amendments) Regulations 2013 come into force; and

(b) subject to paragraph (4), each successive period of five years.

(4) If a report under this regulation is published before the last day of the review period to which it relates, the following review period is to begin with the day on which that report is published.]

[10.939]

Statutory annotations

Amendment

Inserted by SI 2013/950, reg 2(3).

Railways and Other Guided Transport Systems (Safety) Regulations 2006

SI 2006/599

as amended by SI 2006/1057, SI 2007/3531, SI 2010/439, SI 2011/1860, SI 2011/3066 and 2013/950

PART 1

INTRODUCTION

[10.944]

2 Interpretation and application

(1) In these Regulations—

'building operation' means the—

(a) construction, structural alteration, repair or maintenance of a building and 'maintenance' shall include repointing, redecoration and external cleaning of the structure;

(b) demolition of a building; or

(c) preparation for and laying the foundation of an intended building,

but does not include any operation which is a work of engineering construction;

'bus' means a motor vehicle which is designed or adapted to travel along roads and to carry more than eight passengers but which is not a tramcar;

'cableway installation' means an installation made up of several components that—

(a) is used or intended to be used for the purpose of providing an operational system for carrying persons in vehicles, on chairs or by towing devices;

(b) uses cables positioned along the line of travel to provide suspension or traction or both; and

(c) is one of the following—

(i) cable car (including a gondola and chair lift) where the cabins or chairs are lifted or displaced by one or more carrier cables;

(ii) drag lift, where users with appropriate equipment are dragged by means of a cable; or

(iii) funicular railway or other installation with vehicles mounted on wheels or on other suspension devices where traction is provided by one or more cables;

but does not include cable operated tramways, rack railways or lifts;

'carriageway' has the same meaning as in the Highways Act 1980, or in Scotland the Roads (Scotland) Act 1984;

['certification body' has the same meaning as in the ECM Regulation;]

'common safety methods' ('CSMs') means the methods, developed pursuant to article 6 of the Directive, to describe how—

(a) safety levels;

(b) achievement of safety targets; and

(c) compliance with other safety requirements,

are assessed, as revised and reissued from time to time;

'common safety targets' ('CSTs') means the safety levels, developed pursuant to article 7 of the Directive, that must be reached by—

(a) different parts of the mainline railway system; and

(b) that system as a whole,

expressed in risk acceptance criteria, as revised and reissued from time to time;

'competent person' means, except for the purposes of Part 4, a person who—

(a) has sufficient skills, knowledge, experience and resources to undertake the safety verification in relation to which he is appointed;

(b) has not borne such responsibility in relation to any of the matters he has to consider in undertaking that safety verification that might compromise his objectivity; and

(c) is sufficiently independent of a management system, or a part thereof, which has borne responsibility for any of the matters he has to consider in undertaking the safety verification, to ensure that he will be objective in carrying out the safety verification for which he is appointed;

'conventional Directive' means Council Directive 2001/16 of the European Parliament and of the Council on the interoperability of the conventional rail system;

. . .

. . .

['the Directive' means Directive 2004/49/EC of the European Parliament and of the Council on safety on the Community's railways as amended by the Interoperability Directive, Directive 2008/110/EC of the European Parliament and of the Council on safety on the Community's railways and Commission Directive 2009/149/EC on Common Safety Indicators and common methods to calculate accident costs;]

['ECM certificate' means a certificate issued in accordance with the ECM Regulation to an entity in charge of maintenance for the purposes of Article 14a(4) of the Directive or a certificate or self-declaration recognised as being equivalent for those purposes in accordance with Article 12(3) to (7) of the ECM Regulation;]

['ECM Regulation' means Commission Regulation (EU) No 445/2011 on a system of certification of entities in charge of maintenance for freight wagons and amending Regulation (EC) No 653/2007;]

'engineering possession' means a section of track which is closed to normal traffic and where the closure is for the purpose of carrying out maintenance which shall include any repair alteration, reconditioning, examination or testing of infrastructure;

['entity in charge of maintenance' means an entity in charge of maintenance of a vehicle, and includes a transport undertaking, an infrastructure manager or a keeper;]

'European Railway Agency' means the Community agency for railway safety and interoperability established by Regulation (EC) No 881/2004 of the European Parliament and of the Council establishing a European Railway Agency;

'factory' means a factory within the meaning of section 175 of the Factories Act 1961 and premises to which section 123(1) or (2) or 125(1) of that Act applies;

['freight wagon' means a non-self-propelled-vehicle designed for the purpose of transporting freight or other materials to be used for activities such as construction or infrastructure maintenance;]

'guided bus system' means a system of transport, used wholly or mainly for the carriage of passengers, that employs buses which for some or all of the time when they are in operation—

(a) travel along roads; and

(b) are guided (whether while on the road or at other times) by means of—

 (i) apparatus, a structure or other device which is fixed and not part of the bus; or

 (ii) a guidance system which is automatic;

'guided transport' means a system of transport, used wholly or mainly for the carriage of passengers, employing vehicles which for some or all of the time when they are in operation are guided by means of—

(a) rails, beams, slots, guides or other apparatus, structures or devices which are fixed and not part of the vehicle; or

(b) a guidance system which is automatic;

'harbour' and 'harbour area' have the meanings assigned to them by regulation 2(1) of the Dangerous Substances in Harbour Areas Regulations 1987;

. . .

'high-speed Directive' means Council Directive 96/48/EC on the interoperability of the trans-European high-speed rail system;

'infrastructure' means fixed assets used for the operation of a transport system which shall include, without prejudice to the generality of the foregoing—

(a) its permanent way or other means of guiding or supporting vehicles;

(b) any station; and

(c) plant used for signalling or exclusively for supplying electricity for operational purposes to the transport system;

'infrastructure manager' means the person who—

(a) in relation to infrastructure other than a station, is responsible for developing and maintaining that infrastructure or, in relation to a station, the person who is responsible for managing and operating that station, except that it shall not include any person solely on the basis that he carries out the construction of that infrastructure or station or its maintenance, repair or alteration; and

(b) manages and uses that infrastructure or station, or permits it to be used, for the operation of a vehicle;

['Interoperability Directive' means Directive 2008/57/EC of the European Parliament and of the Council of 17th June 2008 on the interoperability of the rail system within the Community (Recast);]

'Interoperability Regulations' means the Railways (Interoperability) Regulations [2011];

['keeper' means the person who, being the owner of a vehicle or having the right to use it, exploits the vehicle as a means of transport and is registered as being the keeper in the National Vehicle Register;]

'mainline application' means an application for—

(a) a safety certificate or an amended safety certificate; or

(b) a safety authorisation or an amended safety authorisation, made in relation to an operation on the mainline railway;

['mainline railway' means any railway except for any railway or part of a railway—

(a) that the Office of Rail Regulation determines in accordance with regulation 2A (determination of exclusion from the mainline railway) falls within one of the categories listed in paragraph (1) of that regulation; or

(b) which is privately owned infrastructure that exists solely for use by the infrastructure owner for its own freight operations;]

'mainline railway system' means the mainline railway and the management and operation of the mainline railway as a whole;

['maintenance file' means the written file that contains all the technical and management information that is necessary to carry out the maintenance of a vehicle;

'maintenance rules' means any rules, applicable [in] Great Britain, which set out requirements relating to the maintenance of vehicles;]

'material' includes plant;

'military establishment' means an establishment intended for use for naval, military or air force purposes or for the purposes of the Department of the Secretary of State responsible for defence;

'mine' has the meaning assigned to it by section 180 of the Mines and Quarries Act 1954;

'national safety rules' means any legislation and other requirements—

(a) applicable [in] Great Britain; and

[(b) which contain requirements (including common operating rules) relating to railway safety which are imposed on more than one transport undertaking operating on the mainline railway,]

except that where the requirements in sub-paragraph (b) consist of common operating rules of the mainline railway it shall not include such rules which regulate matters which are covered by a TSI;

['National Vehicle Register' means the register of vehicles authorised in Great Britain, required by regulation [36] of the Interoperability Regulations;]

'new' in relation to regulations 5 and 6 means new to the transport system in question;

'non-mainline application' means an application for—

(a) a safety certificate or an amended safety certificate; or

(b) a safety authorisation or an amended safety authorisation,

made in relation to an operation on a transport system other than the mainline railway;

'operator of last resort' means a transport operator appointed by the Secretary of State to provide transport services in accordance with section 30 of the Railways Act 1993;

['owner', in relation to a vehicle, means any person who has an estate or interest in, or a right over, that vehicle, and whose permission is needed before another may use it;]

'Part A of a safety certificate' means that part of a safety certificate certifying the matters set out in regulation 7(4)(b)(i) and related expressions shall be construed accordingly;

'Part B of a safety certificate' means that part of a safety certificate certifying the matters set out in regulation 7(4)(b)(ii) and related expressions shall be construed accordingly;

[‘placed in service’ means when a vehicle or infrastructure, having been constructed, upgraded or renewed, is first operated in the provision of a transport service, and in ascertaining when this takes place no regard shall be had to any trials or testing that take place to the vehicle or infrastructure, and cognate expressions shall be construed accordingly;]

‘quarry’ has the meaning assigned to it by regulation 3 of the Quarries Regulations 1999;

‘railway’ means a system of transport employing parallel rails which—

(a) provide support and guidance for vehicles carried on flanged wheels; and

(b) form a track which either is of a gauge of at least 350 millimetres or crosses a carriageway (whether or not on the same level),

but does not include a tramway;

‘relevant infrastructure manager’ means the infrastructure manager for any infrastructure used in relation to the operation in question;

‘relevant infrastructure or vehicle’ means any new or altered—

(a) infrastructure; or

(b) vehicle,

falling within regulation 5(4) or 6(4) and related expressions shall be construed accordingly;

‘responsible person’ means in relation to any relevant infrastructure or vehicle, any person who—

(a) has contracted with another person for the manufacture or construction by that other person of that infrastructure or vehicle; or

(b) manufactures or constructs that infrastructure or vehicle for his own use, or for sale to, or use by, another person but not where he is contracted to do so by a person falling under sub-paragraph (a),

and includes an authorised representative established in Great Britain of such a person.

‘risk’ means in Parts 1 and 2 a risk to the safety of a person;

‘road’ means in the definition of ‘guided bus system’ and ‘tramway’–

(a) in England and Wales, any length of highway or of any other road to which the public has access, and includes bridges over which a road passes; and

(b) in Scotland, has the same meaning as in the Roads (Scotland) Act 1984;

‘rolling stock’ has the meaning in section 83(1) of the Railways Act 1993;

‘ROTS’ means the Railways and Other Transport Systems (Approval of Works, Plant and Equipment) Regulations 1994;

‘safety authorisation’ means a safety authorisation issued by the Office of Rail Regulation in accordance with regulation 10 or 12;

‘safety authority’ means—

(a) as regards a member State other than the United Kingdom, the authority established in that State in accordance with article 16.1 of the Directive;

(b) as regards Great Britain, the Office of Rail Regulation; or

(c) as regards Northern Ireland, the Department for Regional Development established by article 3(1) of the Departments (Northern Ireland) Order 1999;

'safety certificate' means a safety certificate issued by the Office of Rail Regulation in accordance with regulation 7 or 9;

'safety management system' means the organisation and arrangements established by a transport operator to ensure the safe management of its operation;

'significant safety risk' means, in relation to new or altered infrastructure or a new or altered vehicle the design or construction of which incorporates significant changes compared to any infrastructure or vehicle already in use on the transport system, the capability of significantly increasing an existing safety risk or creating a significant safety risk to—

(a) passengers on the transport system in question; or
(b) members of the public on roads and any other location where the transport system in question operates and to which the public have access (including a place to which the public has access only on making a payment), except a location which is a crossing subject to an Order made under section 1 of the Level Crossings Act 1983;

'station' means a passenger stop, station or terminal on a transport system but does not include any permanent way or other means of guiding or supporting vehicles or plant used for signalling or exclusively for supplying electricity for operational purposes to a transport system;

['subsystem' has the same meaning as in the Interoperability Directive;]

'technical specifications for interoperability' ('TSIs') means technical specifications for interoperability which are published in the Official Journal of the European Communities pursuant to—

(a) Article 6.1 of the high-speed Directive; . . .
(b) Article 6.1 of the conventional Directive[; or
(c) Article 6.1 of the Interoperability Directive],

and in force;

'train' includes any rolling stock;

'tramway' means a system of transport used wholly or mainly for the carriage of passengers—

(a) which employs parallel rails which—
  (i) provide support and guidance for vehicles carried on flanged wheels;
  (ii) are laid wholly or partly along a road or in any other place to which the public has access (including a place to which the public has access only on making a payment); and
(b) on any part of which the permitted maximum speed is such as to enable the driver to stop a vehicle in the distance he can see to be clear ahead;

'transport operator' means any transport undertaking or infrastructure manager;

'transport system' means a railway, a tramway, or any other system using guided transport where that other system is used wholly or mainly for the carriage of passengers but a transport system does not include—

(a) a guided bus system;
(b) a trolley vehicle system;
(c) any part of a transport system—

(i) within a harbour or harbour area or which is part of a factory, mine or quarry;
(ii) used solely for the purpose of carrying out a building operation or work of engineering construction;
(iii) within a maintenance or goods depot;
(iv) within a siding except where Part 4 applies; or
(v) which is within a military establishment;

(d) any fairground equipment;
(e) any cableway installation; or
(f) any transport system where the track forms a gauge of less than 350mm except where such a track crosses a carriageway (whether or not on the same level),

except where the transport system in question forms part of the mainline railway;

'transport undertaking' means any person who operates a vehicle in relation to any infrastructure but shall not include a person who operates a vehicle solely within an engineering possession;

'trolley vehicle system' means a system of transport by vehicles constructed or adapted for use on roads without rails under electric power transmitted to them by overhead wires (whether or not there is in addition a source of power on board the vehicles);

['vehicle'—

(a) includes a mobile traction unit; and
(b) in respect of the mainline railway, means a vehicle that runs on its own wheels on railway lines of a gauge of at least 350 millimetres, with or without traction, and is composed of one or more structural and functional subsystems or parts of such subsystems;]

'work of engineering construction' means the—

(a) construction of any line or siding otherwise than on an existing transport system; and
(b) construction, structural alteration, repair (including repointing and repainting) or demolition of any tunnel, bridge or viaduct except where carried on upon a transport system; and

'writing' apart from its usual meaning includes any text transmitted using electronic communications that is received, or accessible by the person to whom it is sent, in legible form.

(2) Any reference in these Regulations to a person who operates a train or a vehicle is a reference to the person operating the train or vehicle for the time being in the course of a business or other undertaking carried on by him, whether for profit or not, but it does not include a self-employed person by reason only that he drives or otherwise controls the movement of a train or vehicle.

(3) Parts 2 and 3 of these Regulations shall not apply to or in relation to the operation of a train or the management or use of infrastructure in the tunnel system within the meaning of section 1(7) of the Channel Tunnel Act 1987.

[10.944A]

Statutory annotations

Amendment

Para (1): definition 'certification body' inserted by SI 2013/950, reg 3(1), (2)(a); definition 'deemed safety authorisation' (omitted) revoked by SI 2013/950, reg 3(1), (2)(b); definition 'deemed safety certificate' (omitted) revoked by SI 2013/950, reg 3(1), (2)(b); definition 'ECM certificate' inserted by SI 2013/950, reg 3(1), (2)(c); definition 'ECM Regulation' inserted by SI 2013/950, reg 3(1), (2)(c); definition 'freight wagon' inserted by SI 2013/950, reg 3(1), (2)(d); definition 'heritage railway' (omitted) revoked by SI 2013/950, reg 3(1), (2)(e); definition 'mainline railway' substituted by SI 2013/950, reg 3(1), (2)(f); in definition 'maintenance rules' word 'in' in square brackets substituted by SI 2013/950, reg 3(1), (2)(g); in definition 'national safety rules' in para (a) word 'in' in square brackets substituted by SI 2013/950, reg 3(1), (2)(h)(i); in definition 'national safety rules' para (b) substituted by SI 2013/950, reg 3(1), (2)(h)(ii); definition 'placed in service' substituted by SI 2013/950, reg 3(1), (2)(i).

[10.944B]

[2A Determination of exclusion from the mainline railway]

[(1) A railway or part of a railway does not form part of the mainline railway if the Office of Rail Regulation determines that it falls within one or more of these categories—

- (a) metros and other light rail systems;
- (b) networks that are functionally separate from the rest of the mainline railway and intended only for the operation of local, urban or suburban passenger services, as well as transport undertakings operating solely on these networks;
- (c) heritage, museum or tourist railways that operate on their own networks.

(2) The Office of Rail Regulation may determine that a heritage vehicle which operates on the mainline railway and complies with national safety rules is deemed not to operate on the mainline railway for the purposes of these Regulations.

(3) The Office of Rail Regulation may revoke or vary a determination under paragraph (1) or (2).

(4) A person may make an application to the Office of Rail Regulation for a determination under paragraph (1) or (2) or for such a determination to be revoked or varied under paragraph (3) and the Office of Rail Regulation shall consider any such application.

(5) The Office of Rail Regulation shall maintain and publish a list of—

- (a) railways or parts of railways that do not form part of the mainline railway; and
- (b) heritage vehicles which operate on both the mainline railway and heritage railway and are deemed not to operate on the mainline railway, by virtue of determinations made under paragraph (1) or (2).]

[10.944C]

Statutory annotations

Amendment

Inserted by SI 2013/950, reg 3(1), (3).

PART 2

SAFETY MANAGEMENT, CERTIFICATION AND AUTHORISATION

[10.945]

3 Use of infrastructure on the mainline railway

(1) After 30th September 2006 no person shall operate a train in relation to any infrastructure on the mainline railway unless—

- (a) he has established and is maintaining a safety management system which meets the requirements set out in regulation [5(1) to (3)]; and
- (b) he holds a current safety certificate in relation to the operation in question,

except to the extent that he is doing so within an engineering possession.

(2) After 30th September 2006 no person who is responsible for developing and maintaining infrastructure other than a station or who is responsible for managing and operating a station on the mainline railway shall manage and use it, or permit it to be used, for the operation of trains unless—

(a) he has established and is maintaining a safety management system which meets the requirements referred to in regulation 5(7);

(b) he holds a current safety authorisation in relation to the infrastructure in question; and

(c) where he is using it or permitting such use, the person who is to use the infrastructure has complied with paragraph (1)(b).

[10.945A]

Statutory annotations

Amendment

Inserted by SI 2013/950, reg 3(1), (3).

[10.947]

5 Safety management system for the mainline railway

(1) The requirements for a safety management system referred to in regulation 3(1)(a) are that—

(a) subject to paragraph (2), it is established to ensure that the mainline railway system—

(i) can achieve the CSTs; and

(ii) is in conformity with relevant national safety rules and relevant safety requirements laid down in TSIs;

(b) it applies the relevant parts of CSMs;

(c) it meets the requirements and contains the elements set out in Schedule 1, adapted to the character, extent and other characteristics of the operation in question;

(d) subject to paragraph (2), it ensures the control of all categories of risk including new or existing risks associated with the operation in question which, without prejudice to the generality of the foregoing, shall include such risks relating to the—

(i) supply of maintenance and material; [and]

(ii) use of contractors; . . .

(iii) . . .

(e) it takes into account, where appropriate and reasonable, the risks arising as a result of activities carried on by other persons; and

(f) all parts of it are documented.

(2) The requirements in paragraphs (1)(a) and (d) shall be met where the safety management system of a transport operator or of an applicant for a safety certificate or a safety authorisation ('the first operator') taken with that of any relevant transport operator is capable of meeting the requirements of the paragraph in question.

(3) In paragraph (2), 'relevant transport operator' means another transport operator whose operation is capable of materially affecting the safety of the operation carried on by the first operator.

(4) . . .

(5) . . .

(6) . . .

[(6) Paragraph (5) does not apply in respect of an authorisation deemed to be given under the Interoperability Regulations by operation of regulation 44 of those Regulations.]

Part 10

(7) The requirements for a safety management system referred to in regulation 3(2)(a) are the requirements in paragraphs [(1) to (3)] . . . and that—

- (a) . . .
- (b) it takes into account the effects of operations of transport undertakings; and
- (c) it contains provisions to ensure that the way in which the infrastructure manager carries out his operation makes it possible for any transport undertaking to operate in accordance with—
  - (i) relevant TSIs and national safety rules; and
  - (ii) the means adopted by the transport undertaking to meet the requirements referred to in regulation 7(4), of which the Office of Rail Regulation accepted that there was sufficient evidence upon issue or amendment of its safety certificate pursuant to these Regulations; and
- (d) it aims to co-ordinate the emergency procedures of the infrastructure manager or of the applicant for a safety authorisation with those of transport undertakings,

and in each case the requirements in [sub-paragraphs (b) to (d)] shall only apply in relation to transport undertakings that operate or will operate a train in relation to the infrastructure of the infrastructure manager or of the applicant for a safety authorisation in question.

[10.947A]

Statutory annotations

Amendment

Para (1): in sub-para (d)(i) word 'and' in square brackets inserted by SI 2013/950, reg 3(1), (5)(a)(i); in sub-para (d)(ii) word omitted revoked by SI 2013/950, reg 3(1), (5)(a)(ii); sub-para (d)(iii) revoked by SI 2013/950, reg 3(1), (5)(a)(iii).

Paras (4), (5): revoked by SI 2013/950, reg 3(1), (5)(b).

Para (7): words '(1) to (3)' in square brackets substituted by SI 2013/950, reg 3(1), (5)(c)(i); words omitted revoked by SI 2013/950, reg 3(1), (5)(c)(ii); sub-para (a) revoked by SI 2013/950, reg 3(1), (5)(c)(iii); words 'sub-paragraphs (b) to (d)' in square brackets substituted by SI 2013/950, reg 3(1), (5)(c)(iv).

[10.959]

17 General provisions relating to safety certificates and safety authorisations

(1) Where an application is made under these Regulations for a safety certificate or safety authorisation or for an amended safety certificate or safety authorisation which relates to an operation on the mainline railway and on a transport system other than the mainline railway then—

- (a) one application may be made for that operation but it shall be split into separate parts for the mainline railway and the other transport system; and
- (b) these Regulations shall apply to those parts as if they were a mainline application and a non-mainline application,

except that where the same information is required it need not be stated twice.

(2) A transport operator may make one application for an operation in relation to which he requires both a safety certificate and a safety authorisation or an amended safety certificate and an amended safety authorisation but—

- (a) such application shall be split into separate parts relating to the safety authorisation and the safety certificate; and
- (b) these Regulations shall apply to those parts as if they were an application for a safety authorisation and a safety certificate or an amended safety authorisation and an amended safety certificate,

except that where the same information is required it need not be stated twice.

(3) Where—

(a) an applicant sends—

(i) an application for a safety certificate or safety authorisation, an amended safety certificate or safety authorisation; or

(ii) further information to the Office of Rail Regulation pursuant to paragraph (5); or

(b) the holder of a safety certificate or a safety authorisation sends a notice pursuant to regulation 13 . . . ,

then he shall at the same time either copy it to any affected party or notify any affected party without delay that the application or further information has been sent and of the address of the website where those documents may be accessed and how they may be accessed and, in either case, shall notify such a party, where the document in question is an application, of the time for making representations to the Office of Rail Regulation pursuant to paragraph (6).

(4) Where the Office of Rail Regulation issues a—

(a) safety certificate or safety authorisation, other than to an operator of last resort;

(b) notice amending a safety certificate or safety authorisation;

(c) notice refusing an application for a safety certificate or a safety authorisation or an amended safety certificate or safety authorisation;

(d) direction to apply for an amended safety certificate or safety authorisation;

(e) notice that it is considering revoking a safety certificate or a safety authorisation; or

(f) notice revoking a safety certificate or a safety authorisation,

then the Office of Rail Regulation shall at the same time either copy it and the reasons given for the Office of Rail Regulation's decision to any affected party or notify any affected party that the relevant document has been issued and of the address of the website where those documents and the reasons given for the decision may be accessed and how they may be accessed and, in either case, shall notify such a party, where the document in question is a notice that it is considering revocation as mentioned in sub-paragraph (e), of the time for making representations to the Office of Rail Regulation pursuant to paragraph (6).

(5) The Office of Rail Regulation may upon receipt of—

(a) an application for a safety certificate or safety authorisation;

(b) an application for an amended safety certificate or safety authorisation;

(c) any further information requested under this paragraph,

request as soon as reasonably possible such further information as it may reasonably require and the applicant shall provide such information as soon as reasonably possible . . . .

(6) Where an affected party receives a copy of an application or a notice relating to revocation pursuant to paragraph (3)(a) or (4)(e) then—

(a) he may make any representations in writing to the Office of Rail Regulation, which are relevant to the application or notice, within 28 days of the date of issue of the application or notice in question; and

(b) the Office of Rail Regulation shall consider any such representations in making its decision.

[(7) If a request for information is made under paragraph (5), the period of 4 months for the Office of Rail Regulation to make a decision referred to in regulations 7(3), 8(4), 10(2) and 11(3) shall not start to run until the date of receipt of the last information requested.]

(8) . . .

(9) An employer who makes an application for a safety certificate or safety authorisation, an amended safety certificate or safety authorisation or sends a notice to the Office of Rail Regulation under regulation 13 . . . shall, in relation to its preparation, consult—

(a) safety representatives within the meaning of regulation 2(1) of the Safety Representatives and Safety Committees Regulations 1977; and

(b) such other employees as he is required to consult by virtue of regulation 3 of the Health and Safety (Consultation with Employees) Regulations 1996.

(10) In this regulation, 'affected party' means for a document sent or issued in relation to—

(a) a safety certificate or an application for a safety certificate—

(i) any relevant infrastructure manager;

(ii) a trade union which is a recognised trade union within the meaning of regulation 2(1) of the Safety Representatives and Safety Committees Regulations 1977 in relation to employees of the operator or applicant employed in relation to the operation in question; and

(iii) the [Passengers' Council] and the London Transport Users' Committee where, in each case, it represents passengers' interests in relation to the operation in question; and

(b) a safety authorisation or an application for a safety authorisation—

(i) any transport undertaking who is or will be operating on the infrastructure of the applicant or infrastructure manager in question;

(ii) any infrastructure manager who manages infrastructure which interfaces or will interface with the infrastructure of the infrastructure manager in question; and

(iii) any person falling within paragraph (10)(a)(ii) or (iii).

[10.959A]

Statutory annotations

Amendment

Para (3): in sub-para (b) words omitted revoked by SI 2013/950, reg 3(1), (6)(a).

Para (5): words omitted revoked by SI 2013/950, reg 3(1), (6)(b).

Para (7): substituted by SI 2013/950, reg 3(1), (6)(c).

Para (8): revoked by SI 2013/950, reg 3(1), (6)(d).

Para (9): words omitted revoked by SI 2013/950, reg 3(1), (6)(e).

Para (10): in sub-para (a)(iii) words in square brackets substituted by SI 2010/439, art 2, Schedule, Pt 2, para 14.

[10.960A]

[18A Maintenance of vehicles on the mainline railway]

[(1) No person may place in service or use a vehicle on the mainline railway unless that vehicle has an entity in charge of maintenance assigned to it, and that entity in charge of maintenance—

(a) is registered in relation to that vehicle in the National Vehicle Register; and

(b) holds an ECM certificate if the vehicle is a freight wagon.]

(2) Each entity in charge of maintenance must ensure, by means of a system of maintenance, that a vehicle for which it is in charge of maintenance is in a safe state of running.

(3) The requirement for a system of maintenance referred to in paragraph (2) is that a vehicle must be maintained in accordance with—

(a) the maintenance file for the vehicle;
(b) applicable maintenance rules; and
(c) applicable TSIs.]

[10.960B]

Statutory annotations

Amendment

Para (1): substituted by SI 2013/950, reg 3(1), (7).

PART 3

GENERAL DUTIES

[10.962]

20 Annual safety reports

[(1) Subject to paragraph (2), any transport operator who carries out operations on the mainline railway shall send to the Office of Rail Regulation an annual safety report in respect of its operations on the mainline railway relating to the previous calendar year, which shall contain—

(a) information on how the transport operator's safety targets, referred to in paragraph 2(b) of Schedule 1, are met;
(b) the results achieved through putting the transport operator's safety plans, referred to in paragraph 2(b) of Schedule 1, into effect;
(c) statistics for the common safety indicators listed in Schedule 3 insofar as they are relevant to the operations in question;
(d) the findings of safety auditing carried out pursuant to the procedures referred to in paragraph 2(k) of Schedule 1; and
(e) comments on any deficiencies or malfunctions relating to the running of vehicles or the management of infrastructure relating to the operation in question that may be relevant to the safety of that transport system,

and where an operation is carried out in part on the mainline railway and in part on another transport system the report shall include only information in respect of the part carried out on the mainline railway.]

(2) The first annual report required under paragraph (1) shall be sent by 30th June 2007 and subsequent reports by 30th June in each subsequent calendar year.

(3) Subject to paragraph (4), the Office of Rail Regulation shall publish and send to the European Railway Agency an annual report relating to the previous calendar year which shall contain information on the following in relation to the mainline railway—

(a) the development of railway safety including an aggregation of all the statistics reported to the Office of Rail Regulation for the relevant calendar year pursuant to paragraph (1)(c) which relate to an operation or part of an operation which is carried out on the mainline railway;
(b) any important changes in relation to the regulation of railway safety;
(c) the development of the system for safety certification and authorisation; and
(d) the results of and experience relating to the supervision of transport operators,

in Great Britain.

(4) The first annual report required under paragraph (3) shall be sent to the European Railway Agency by 30th September 2007 and subsequent reports by 30th September in each subsequent calendar year.

(5) Where the Office of Rail Regulation discovers, after sending an annual report, that there were errors or omissions in it then it shall send a corrected report for that

year to the European Railway Agency at the first convenient opportunity and in any event by no later than the time the next annual report is due to be sent.

[10.962A]

Statutory annotations

Amendment

Para (1): substituted by SI 2013/950, reg 3(1), (8).

[10.963]

21 Sending, issuing, and keeping of documents and making them available for public inspection

(1) Any application, notice, report or any other information sent to the Office of Rail Regulation or records made pursuant to these Regulations shall be in writing and in English.

(2) Any certificate, authorisation, notice, direction, request for information, statement or report issued by the Office of Rail Regulation pursuant to these Regulations shall be in writing and in English.

(3) An applicant who makes an application in respect of a safety certificate or a safety authorisation pursuant to regulations 7, 9, 10 or 12 shall when sending the application, notify the Office of Rail Regulation of an address in Great Britain for the purposes of this regulation ('notified address').

(4) Subject to paragraphs (5) and (6), a transport operator shall keep at the notified address in relation to the operation in question—

- (a) the safety certificate or safety authorisation issued in response to his application for such certificate or authorisation and the documentation referenced in that safety certificate or safety authorisation;
- (b) any notice of amendment issued pursuant to Part 2 of these Regulations . . . in relation to his safety certificate or safety authorisation;
- (c) any records he is required to make pursuant to regulation 19(4) and (5);
- (d) any annual safety report sent to the Office of Rail Regulation under regulation 20(1);
- (e) any notification of changes or of a revision notified to the Office of Rail Regulation under regulation 13 . . . ; and
- (f) a record of any findings of internal safety auditing carried out pursuant to the procedures referred to in paragraph 2(k) of Schedule 1 and of any action taken in consequence of such auditing,

or a hard or electronic copy of such documents.

(5) The documents referred to in paragraph (4) shall be kept as long as they—

- (a) are or are a copy of the current safety certificate or safety authorisation or a notice of amendment thereof;
- (b) are or are a copy of a notification of a major change or a revision which is relevant to the current operation of the transport undertaking;
- (c) relate to the information on the basis of which the Office of Rail Regulation's acceptance referred to in regulation 7(4) or 10(3), as the case may be, was made in relation to a current safety certificate or safety authorisation; or
- (d) relate to a risk assessment, as reviewed from time to time, carried out pursuant to regulation 19.

(6) The documents kept pursuant to paragraph (4)(d) or (4)(f) shall be kept for 5 years and the documents kept pursuant to sub-paragraphs (a), (b), (d) and (e) of paragraph (4) shall, subject to paragraph (7), be made available for public inspection at the notified address at reasonable times and on reasonable notice.

(7) Nothing in paragraph (6) shall require the disclosure of any information—

(a) relating to a named individual;

(b) which is commercially confidential; or

(c) which is detrimental to national security or to the security of the transport system in question.

(8) A person who has a notified address may subsequently notify the Office of Rail Regulation of a different address in Great Britain and in this case references in this regulation to the notified address shall be construed as a reference to the last address notified under this paragraph.

[10.963A]

Statutory annotations

Amendment

Para (4): in sub-para (b) words omitted revoked by SI 2013/950, reg 3(1), (9)(a); in sub-para (e) words omitted revoked by SI 2013/950, reg 3(1), (9)(b).

PART 4

SAFETY CRITICAL WORK

[10.966]

24 Competence and fitness

(1) Every controller of safety critical work shall, so far as is reasonably practicable, ensure that a person under his management, supervision or control, with the exception of where that person is receiving practical training in a safety critical task, only carries out safety critical work where—

(a) that person has been assessed as being competent and fit to carry out that work following an assessment by an assessor;

(b) there is an accurate and up to date record in writing of that person's competence and fitness which references any criteria for determining competence and fitness against which that assessment of competence was made;

(c) the record, or an accurate summary of the record referred to in sub-paragraph (b) is available for inspection, on reasonable request, by any other controller of safety critical work or any operator who may be affected by any safety critical work carried out or to be carried out by that person, for the purposes of establishing that person's competence and fitness to carry out safety critical work; and

(d) there are in place [suitable and sufficient] arrangements for monitoring the competence and fitness of that person.

(2) Every controller of safety critical work shall without unreasonable delay review any person's competence or fitness assessment where—

(a) they have reason to doubt the competence or fitness of a person to carry out that safety critical work; or

(b) there has been a significant change in the matters to which the assessment relates,

and where, as a result of any such review a reassessment of competence or fitness is required, that reassessment of competence or fitness shall be carried out to ensure that the requirements of paragraph (1) are met.

(3) Where a reassessment of competence or fitness under paragraph (2) is required, the controller of safety critical work shall, so far as is reasonably practicable ensure that, as a result, the health and safety of persons on a transport system is not prejudiced.

[10.966A]

Statutory annotations

Amendment

Para (1): in sub-para (d) words 'suitable and sufficient' in square brackets inserted by SI 2013/950, reg 3(1), (10).

PART 5

MISCELLANEOUS

[10.969]

27 Appeals

(1) A person who is aggrieved by a—

(a) decision of the Office of Rail Regulation to refuse his application for—

[(i) a safety certificate, a safety authorisation or an ECM certificate;]

(ii) an amended safety certificate or safety authorisation;

(b) direction of the Office of Rail Regulation to make an application to amend his safety certificate or safety authorisation; . . .

(c) decision of the Office of Rail Regulation to revoke his—

(i) safety certificate or part of it; or

(ii) safety authorisation,

[(d) decision of the Office of Rail Regulation taken pursuant to Articles 7(3), (4) or (7) of the ECM Regulation concerning his ECM certificate; or

(e) determination of the Office of Rail Regulation under regulation 2A or a decision of the Office of Rail Regulation under that regulation to revoke or vary such a determination,]

may appeal to the Secretary of State.

[(1A) A person who is aggrieved by a decision of a certification body accredited or recognised in Great Britain for the purposes of the ECM Regulation either to refuse an application for an ECM certificate or taken by that body pursuant to Articles 7(3), (4) or (7) of the ECM Regulation, may—

(a) appeal to the Secretary of State; and

(b) for the purposes of such appeal, references to the Office of Rail Regulation in this regulation shall be construed as references to that certification body, except for paragraph (7) which does not apply to such appeal.]

(2) For the purposes of [paragraphs (1) and (1A)] the Secretary of State may, in such cases as he considers it appropriate to do so, having regard to the nature of the questions which appear to him to arise, direct that an appeal under that paragraph shall be determined on his behalf by a person appointed by him for that purpose.

(3) Before the determination of an appeal the Secretary of State shall ask the appellant and the Office of Rail Regulation whether they wish to appear and be heard on the appeal and—

(a) The appeal may be determined without a hearing of the parties if both of them express a wish not to appear and be heard as aforesaid;

(b) The Secretary of State shall, if either of the parties expresses a wish to appear and be heard, afford to both of them an opportunity of doing so.

(4) The Tribunals and Inquiries Act 1992 shall apply to a hearing held by a person appointed in pursuance of paragraph (2) to determine an appeal as it applies to a statutory inquiry held by the Secretary of State, but as if in section 10(1) of that Act (statement of reasons for decisions) the reference to any decision taken by the Secretary of State included a reference to a decision taken on his behalf by that person.

[(4A) . . .]

(5) A person who determines an appeal under this regulation on behalf of the Secretary of State and the Secretary of State, if he determines such an appeal, may give such directions as he considers appropriate to give effect to his determination.

(6) The Secretary of State may pay to any person appointed to hear or determine an appeal under paragraph (2) on his behalf such remuneration and allowances as the Secretary of State may with the approval of the [Treasury] determine.

(7) For the purposes of paragraph (1)(a), a failure by the Office of Rail Regulation to make a decision on whether or not to issue or amend a safety certificate or safety authorisation within the four month period for making a decision calculated in accordance with regulation 17(7) shall be treated as a refusal of the application.

(8) The Health and Safety Licensing Appeals (Hearing Procedure) Rules 1974, as respects England and Wales, and the Health and Safety Licensing Appeals (Hearing Procedure) (Scotland) Rules 1974, as respects Scotland, shall apply to an appeal under paragraph (1) as they apply to an appeal under [section 44(1) of the Health and Safety at Work etc Act 1974], but with the modification that references to a licensing authority are to be read as references to the Office of Rail Regulation.

(9) Where an appeal is made under paragraphs (1)(a) or (1)(c), the decision in question shall be suspended pending the final determination of the appeal.

[10.969A]

Statutory annotations

Amendment

Para (1): sub-para (a)(i) substituted by SI 2013/950, reg 3(1), (11)(a); in sub-para (b) word omitted revoked by SI 2013/950, reg 3(1), (11)(b); sub-paras (d), (e) inserted by SI 2013/950, reg 3(1), (11)(c).

Para (1A): inserted by SI 2013/950, reg 3(1), (11)(d).

Para (2): words 'paragraphs (1) and (1A)' in square brackets substituted by SI 2013/950, reg 3(1), (11)(c).

Para (4A): revoked by SI 2013/2042, art 2(2), Schedule, para 63.

[10.971]

29 Transitional provisions and savings

(1) Any competence and fitness assessments made pursuant to regulation 3 of the Railways (Safety Critical Work) Regulations 1994 shall have effect as if they were made under Part 4, provided that the assessment would, at the time it was made, have met the requirements for impartiality and objectivity in that Part.

(2) Notwithstanding the revocation of ROTS pursuant to regulation 34, and subject to paragraph (6) ROTS shall, up to and including 1st October 2008, continue in force as they had effect on 30th September 2006 for the purposes of—

(a) determining applications for approval made;

(b) issuing a written consent for the purposes set out in regulation 4(4)(b) of ROTS in relation to new or altered works, plant or equipment for which an application for approval has been made;

(c) making notices dispensing with or requiring compliance with certain provisions of ROTS under regulation 10(1)(a) or 11(1)(a) of ROTS in relation to new or altered works, plant or equipment, for which an application for approval has been made,

to the Office of Rail Regulation in relation to a relevant transport system before 1st October 2006.

(3) Where an approval is issued by the Office of Rail Regulation in response to an application for approval made in relation to a transport system—

(a) before 1st October 2006 but where the new or altered works, plant or equipment are placed in service within the meaning of regulation 5(6) and 6(5) on or after that date; or

(b) on or before 1st October 2008 pursuant to paragraph (2),

in relation to new or altered works, plant or equipment that is relevant infrastructure or a vehicle then such works, plant or equipment shall be deemed to satisfy the requirements of regulations 5(4) and 6(4).

(4) Where a written consent is issued by the Office of Rail Regulation in relation to new or altered works, plant or equipment which is relevant infrastructure or a vehicle in relation to a transport system—

(a) under regulation 4(4)(b)(i) of ROTS on or before 1st October 2008 pursuant to paragraph (2) then such relevant infrastructure or vehicle shall be deemed to satisfy the requirements of regulations 5(4) and 6(4); or

(b) under regulation 4(4)(b)(ii) of ROTS on or before 1st October 2008 pursuant to paragraph (2) then no regard shall be taken of the use of such relevant infrastructure or vehicle for the purposes for which the written consent relates when determining whether the relevant infrastructure or vehicle has been placed in service in accordance with regulations 5(6) and 6(5).

(5) In this regulation—

(a) 'application for approval' means an application for approval made under regulation 5, 6 or 7 of ROTS;

(b) 'altered works, plant or equipment' shall have the meaning in regulation 2(a) of ROTS; and

(c) 'relevant transport system' shall have the meaning in regulation 2(a) of ROTS.

(6) For the purposes of heritage railways and tramways, all references in this regulation to 1st October 2006 shall be read as if those references were in each case to 1st October 2008, and all references to 1st October 2008 shall be read as if those references were in each case to 1st October 2010.

(7) . . .

[10.971A]

Statutory annotations

Amendment

Para (7): revoked by SI 2013/950, reg 3(1), (12).

SCHEDULE 2

APPLICATION FOR A SAFETY CERTIFICATE

Regulation 7(1)(b)

PART 1

INFORMATION TO BE INCLUDED FOR A MAINLINE APPLICATION

[10.978]

1

The following information shall be included in relation to Part A of a safety certificate—

(a) particulars of the type and extent of the operation in respect of which the application is made; and

(b) either—

[(i) a copy of a current certificate issued to the applicant by—
(aa) the Office of Rail Regulation . . . ;
(bb) a safety authority in another member State;
(cc) a safety authority in Northern Ireland; or
(dd) the safety authority for the tunnel system within the meaning of section 1(7) of the Channel Tunnel Act 1987,

under provisions giving effect to article 10(2)(a) of the Directive which relates to an equivalent railway operation; or]

(ii) particulars of how the safety management system of the applicant meets the requirements set out in regulation 5(1) to (4).

2

The following information shall be included in relation to Part B of a safety certificate—

(a) information on the TSIs, national safety rules and other safety requirements relevant to the applicant's operation including those relevant to persons carrying out work in relation to the operation and the applicant's vehicles and an explanation of how compliance with these requirements is ensured by the safety management system;

(b) information on the different types of work being carried out by persons directly in relation to the operation including evidence of how the applicant ensures that when such persons are carrying out such work that they are doing so in accordance with the requirements of any relevant TSIs and national safety rules; and

(c) information on the different types of rolling stock used for the operation in question including evidence that they meet any relevant TSIs and national safety rules,

and where information is submitted concerning an interoperability constituent or a subsystem which is subject to and complies with the requirements of the [Interoperability Regulations] then only brief details need be supplied concerning compliance of such constituents or subsystems with relevant TSIs and other requirements of those Regulations and in this paragraph 'interoperability constituent' and 'subsystem' shall have the same meaning as in the [Interoperability Regulations].

[10.979]

Statutory annotations

Amendment

Para (1): sub-para (b)(i) substituted by SI 2007/3531, art 10; in sub-para (b)(i)(aa) words omitted revoked by SI 2013/950, reg 3(1), (13).

[SCHEDULE 3

COMMON SAFETY INDICATORS]

Regulation 20(1)(c)

[PART 1

COMMON SAFETY INDICATORS]

(This Schedule substantially reproduces the provisions of Annex I to the Directive)

[10.981]

[1 Interpretation

In this Part, the definitions for the common safety indicators and the methods used to calculate the economic impact of accidents in Part 2 apply.

2 Indicators relating to accidents

(1) Total and relative, to train-kilometres, number of—

(a) significant accidents and a break-down of the following types of accidents—

(i) collisions of vehicles, including collisions with obstacles within the clearance gauge;

(ii) derailments of trains;

(iii) level crossing accidents which includes accidents involving persons at level crossings;

(iv) accidents to persons caused by rolling stock in motion, except for suicides;

(v) fires in rolling stock; and

(vi) any other types of accidents,

and each significant accident shall be reported under the heading of the primary accident even where the consequences of any secondary accident are more severe, such as where a fire follows a derailment;

(b) persons seriously injured or [persons] killed by type of accident divided into the following categories—

(i) passengers;

(ii) persons carrying out work or voluntary work directly in relation to the operation [at the time of the accident];

(iii) level crossing users;

(iv) unauthorised persons on premises of the transport system; and

(v) any other types of person,

and the number of passengers seriously injured or killed shall also be indicated in relation to the total number of passenger-kilometres.

(2) The provisions of Regulation 91/2003 of the European Parliament and the Council on rail transport statistics shall be applied to any information provided under this paragraph.

3 Indicators relating to dangerous goods

Total and relative, to train-kilometres, number of accidents involving the transport of dangerous goods—

(a) involving at least one vehicle transporting dangerous goods; and

(b) number of such accidents in which dangerous goods are released.

4 Indicators relating to suicides

Total and relative, to train-kilometres, number of suicides.

5 Indicators relating to precursors of accidents

Total and relative, to train-kilometres, number of—

(a) broken rails;
(b) buckled rails;
(c) wrong-side signalling failures;
(d) signals passed at danger; and
(e) broken wheels and axles on vehicles in service,

and all such precursors are to be reported, whether or not they result in accidents and where they result in a significant accident, they shall be reported under paragraph 2 of this Part of Schedule 3.

6 Indicators to calculate the economic impact of accidents

(1) Total in Euros and relative, to train-kilometres—

[(a) number of persons killed and persons seriously injured multiplied by the Value of Preventing a Casualty (VPC);]
(b) cost of damage to the environment;
(c) cost of material damage to rolling stock or infrastructure;
(d) cost of delays as a consequence of accidents.

(2) Each annual safety report submitted by the Office of Rail Regulation in accordance with regulation 20(3) shall include the economic impact of significant accidents only.

(3) The VPC is the value society attributes to the prevention of a casualty but is not intended to be used as a reference for the assessment of compensation between parties involved in accidents.

7 Indicators relating to technical safety of infrastructure and its implementation

The—

(a) percentage of tracks with a train protection system, within the meaning of regulation 2(1) of the Railway Safety Regulations 1999, in operation;
(b) percentage of train-kilometres with a train protection system falling within paragraph (a) in operation;
(c) number of (total per line-kilometre and per track-kilometre)—
(d) percentage of level crossings with automatic or manual protection.
  (i) active level crossings with—
    (aa) automatic user-side warning;
    (bb) automatic user-side protection;
    (cc) automatic user-side protection and warning;
    (dd) automatic user-side protection and warning, and rail side protection;
    (ee) manual user-side warning;
    (ff) manual user-side protection;
    (gg) manual user-side protection and warning; and
  (ii) passive level crossings.

8 Indicators relating to the management of safety

Internal audits carried out by transport operators pursuant to the procedures referred to in paragraph 2(k) of Schedule 1 and the number of such audits which have been carried out and that number expressed as a percentage of the audits which were planned for that year.]

[10.981A]

Statutory annotations

Amendment

Para 2: in sub-para (1)(b) word 'persons' in square brackets inserted by SI 2013/950, reg 3(1), (14)(a)(i); in sub-para (1)(b)(ii) words 'at the time of the accident' in square brackets inserted by SI 2013/950, reg 3(1), (14)(a)(ii).

Para 6: sub-para (1)(a) substituted by SI 2013/950, reg 3(1), (14)(a)(iii).

[PART 2

COMMON DEFINITIONS AND METHODS TO CALCULATE THE ECONOMIC IMPACT OF ACCIDENTS]

[10.981B]

[1 Indicators relating to accidents

(1) 'Significant accident' means any accident involving at least one rail vehicle in motion, resulting in at least one killed or seriously injured person, or in significant damage to stock, track, other installations or environment, or extensive disruptions to traffic. Accidents in workshops, warehouses and depots are excluded.

(2) 'Significant damage to stock, track, other installations or environment' means damage that is equivalent to €150,000 or more.

(3) 'Extensive disruptions to traffic' means that train services on a main railway line are suspended for six hours or more.

(4) 'Train' means one or more railway vehicles hauled by one or more locomotives or railcars, or one railcar travelling alone, running under a given number or specific designation from an initial fixed point to a terminal fixed point. A light engine, ie a locomotive travelling on its own, is considered to be a train.

(5) 'Collision of trains, including collisions with obstacles within the clearance gauge' means a front to front, front to end or a side collision between a part of a train and a part of another train, or with—

- (a) shunting rolling stock; or
- (b) objects fixed or temporarily present on or near the track (except at level crossings if lost by a crossing vehicle or user).

(6) 'Train derailment' means any case in which at least one wheel of a train leaves the rails.

(7) 'Level crossing accidents' means accidents at level crossings involving at least one railway vehicle and one or more crossing vehicles, other crossing users such as pedestrians or other objects temporarily present on or near the track if lost by a crossing vehicle/user.

(8) 'Accidents to persons caused by rolling stock in motion' means accidents to one or more persons who are either hit by a railway vehicle or by an object attached to, or that has become detached from, the vehicle. Persons who fall from railway vehicles are included, as well as persons who fall or are hit by loose objects when travelling on board vehicles.

(9) 'Fires in rolling stock' means fires and explosions that occur in railway vehicles (including their load) when they are running between the departure station and the destination, including when stopped at the departure station, the destination or intermediate stops, as well as during re-marshalling operations.

(10) 'Other types of accidents' means all accidents other than those already mentioned (train collisions, train derailments, at level crossing, to persons caused by rolling stock in motion and fires in rolling stock).

(11) 'Passenger' means any person, excluding members of the train crew, who makes a trip by rail. For accident statistics, passengers trying to embark/disembark onto/from a moving train are included.

(12) . . .

(13) 'Level crossing users' means all persons using a level crossing to cross the railway line by any means of transport or by foot.

(14) 'Unauthorised persons on railway premises' means any person present on railway premises where such presence is forbidden, with the exception of level crossing users.

(15) . . .

(16) [Persons killed] means any person killed immediately or dying within 30 days as a result of an accident, excluding suicides.

(17) [Persons seriously injured] means any person injured who was hospitalised for more than 24 hours as a result of an accident, excluding attempted suicides.

2 Indicators relating to dangerous goods

(1) 'Accident involving the transport of dangerous goods' means any accident or incident that must be reported in accordance with RID section 1.8.5, as revised or reissued from time to time.

(2) 'Dangerous goods' means those substances and articles the carriage of which is prohibited by RID, or authorised only under the conditions prescribed therein.

(3) In this paragraph 'RID' means the Regulations concerning the International Carriage of Dangerous Goods by Rail as adopted under Directive 2008/68/EC of the European Parliament and of the Council of 24th September 2008 on the inland transport of dangerous goods.

3 Indicators relating to suicides

Suicide' means an act to deliberately injure oneself resulting in death, as recorded and classified by the Railway Safety and Standards Board.

4 Indicators relating to precursors of accidents

(1) 'Broken rails' means any rail which is separated in two or more pieces, or any rail from which a piece of metal becomes detached, causing a gap of more than 50mm in length and more than 10mm in depth on the running surface.

(2) 'Track buckles' means faults related to the continuum and the geometry of track, requiring track obstruction or immediate reduction of permitted speed to maintain safety.

(3) 'Wrong side signalling failure' means any failure of a signalling system (either to infrastructure or to rolling stock), resulting in signalling information less restrictive than that demanded.

(4) 'Signal passed at danger (SPAD)' means any occasion when any part of a train proceeds beyond its authorised movement to an unauthorised movement;

'unauthorised movement' means to pass—

- (a) a trackside colour light signal or semaphore at danger, order to STOP, where an Automatic Train Control System (ATCS) or train protection system (as described in paragraph 7(a) of Part 1 of this Schedule) is not operational;
- (b) the end of a safety related movement authority provided in an ATCS or train protection system;
- (c) a point communicated by verbal or written authorisation laid down in regulations; or
- (d) stop boards (buffer stops are not included) or hand signals,

but excludes cases in which—

- (e) vehicles without any traction unit attached or a train that is unattended run away past a signal at danger; or
- (f) for any reason, the signal is not turned to danger in time to allow the driver to stop the train before the signal.

(The Office of Rail Regulation may report separately on items (a) to (d) and shall report at least an aggregate indicator containing data on all four items).

(5) 'Broken wheels and broken axles' means a break affecting the essential parts of the wheel or the axle and creating a risk of accident (derailment or collision).

5 Common methodologies to calculate the economic impact of accidents

(1) The Value of Preventing a Casualty (VPC) is composed of—

Part 10

(a) value of safety per se: Willingness to Pay (WTP) values based on stated preference studies carried out in Great Britain;

(b) direct and indirect economic costs, appraised in Great Britain, composed of—

(i) medical and rehabilitation costs;

(ii) legal and court costs, police, private crash investigations and emergency service costs and administrative costs of insurance;

(iii) production losses: value to society of goods and services that could have been produced by the person if the accident had not occurred.

(2) Common principles to appraise the value of safety per se and direct/indirect economic costs—

(a) for the value of safety per se, the assessment of whether available estimates are appropriate or not shall be based on the following considerations—

(i) estimates shall relate to a system for valuation of mortality risk reduction in the transport sector and follow a WTP approach according to stated preference methods;

(ii) the respondent sample used for the values shall be representative of the population concerned. In particular, the sample has to reflect the age/income distribution along with other relevant socio-economic/demographic characteristics of the population;

(iii) method for eliciting WTP values: survey design shall be such that questions are clear/meaningful to respondents;

(b) direct and indirect economic costs shall be appraised on the basis of the real costs borne by society.

(3) 'Cost of damage to environment' means costs that are to be met by transport undertakings or infrastructure managers, appraised on the basis of their experience, in order to restore the damaged area to its state before the railway accident.

(4) 'Cost of material damage to rolling stock or infrastructure' means the cost of providing new rolling stock or infrastructure, with the same functionalities and technical parameters as that damaged beyond repair, and the cost of restoring repairable rolling stock or infrastructure to its state before the accident. Both are to be estimated by transport undertakings or infrastructure managers on the basis of their experience. Also includes costs related to leasing rolling stock, as a consequence of non availability due to damaged vehicles.

(5) 'Cost of delays as a consequence of accidents' means the monetary value of delays incurred by users of rail transport (passengers and freight customers) as a consequence of accidents, and is calculated by the following model—

VT = monetary value of travel time savings

Value of time for a passenger of a train (an hour)

VTp = [VT of work passengers]*[Average percentage of work passengers per year] + [VT of non-work passengers]*[Average percentage of non-work passengers per year]

VT measured in EUR per passenger per hour

Value of time for a freight train (an hour)

VTF = [VT of freight trains]*[(Tonne-Km)/(Train-Km)]

VT is measured in EUR per freight tonne per hour

Average tonnes of goods transported per train in one year = (Tonne-Km)/(train-Km)

Cm = Cost of 1 minute of delay of a train

Passenger train

CMP = K1 *(VTp/60)*[(Passenger-Km)/(Train Km)]

Average number of passengers per train in one year = (Passenger-Km)/(Train Km)

Freight train

CMF = K2* (VTF/60)

Factors K1 and K2 are between the value of time and the value of delay, as estimated by stated preference studies, to take into account that the time lost as a result of delays is perceived significantly more negatively than normal travel time.

Cost of delays of an accident = CMP*(Minutes of delay of passenger trains) + CMF*(Minutes of delay of freight trains)

Scope of the model

Cost of delays is to be calculated for all accidents, both significant and non-significant.

Delays are to be calculated as follows:

—real delays on the railway lines where accidents occurred,

—real delays or, if not possible, estimated delays on the other affected lines.

6 Indicators relating to technical safety of infrastructure and its implementation

(1) 'Automatic Train Protection' means a system that enforces obedience to signals and speed restrictions by speed supervision, including automatic stop at signals.

(2) 'Level crossing' means any level intersection between the railway and a passage, as recognised by the infrastructure manager and open to public or private users. Passages between platforms within stations are excluded, as well as passages over tracks for the sole use of employees.

(3) 'Passage' means any public or private road, street or highway, including footpaths and bicycle paths, or other route provided for the passage of people, animals, vehicles or machinery.

(4) 'Active level crossing' means a level crossing where the crossing users are protected from or warned of the approaching train by the activation of devices when it is unsafe for the user to traverse the crossing, as follows—

(a) protection by the use of physical devices, including half or full barriers or gates;

(b) warning by the use of fixed equipment at level crossings, including lights, audible devices such as bells, horns or klaxons and physical devices such as vibration due to road bumps;

active level crossings are classified as—

(c) 'level crossing with crossing-user-side automatic protection and/or warning' which means a level crossing where the crossing protection and/or warning are activated by the approaching train. These level crossings are classified as—

(i) automatic user-side warning;

(ii) automatic user-side protection;

(iii) automatic user-side protection and warning;

(iv) automatic user-side protection and warning, and rail-side protection;

'rail-side protection' means a signal or other train protection system that only permits a train to proceed if the level crossing is user-side protected and free from incursion; such freedom from incursion to be achieved by means of surveillance and/or obstacle detection;

(d) 'level crossing with crossing-user-side manual protection and/or warning' which means a level crossing where protection and/or warning is manually activated and there is not an interlocked railway signal showing, to the train, a running aspect only when protection and/or warning of level crossing are activated. These level crossings are classified as—

(i) manual user-side warning;
(ii) manual user-side protection;
(iii) manual user-side protection and warning.

(5) 'Passive level crossing' means a level crossing without any form of warning system and/or protection activated when it is unsafe for the user to traverse the crossing.

7 Indicators relating to the management of safety

'Audit' means a systematic, independent and documented process for obtaining audit evidence and evaluating it objectively to determine the extent to which audit criteria are fulfilled.

8 Definitions of the scaling bases

(1) 'Train-kilometre' (train-km) means the unit of measure representing the movement of a train over one kilometre. The distance used is the distance actually run, if available, otherwise the standard network distance between the origin and destination shall be used.

(2) 'Passenger-kilometre' (passenger-km) means the unit of measure representing the transport of one passenger by rail over a distance of one kilometre.

(3) 'Line-kilometre' (line-km) means the length measured in kilometres of the railway to which these Regulations apply. For multiple-track railway lines, only the distance between origin and destination is to be counted.

(4) 'Track-kilometre' (track-km) means the length measured in kilometres of the railway, to which these Regulations apply. Each track of a multiple-track railway line is to be counted.]

[10.981C]

Statutory annotations

Amendment

Para 1: sub-para (12) revoked by SI 2013/950, reg 3(1), (14)(b)(i); sub-para (15) revoked by SI 2013/950, reg 3(1), (14)(b)(i); in sub-para (16) words 'Persons killed' in square brackets substituted by SI 2013/950, reg 3(1), (14)(b)(ii); in sub-para (17) words 'Persons seriously injured' in square brackets substituted by SI 2013/950, reg 3(1), (14)(b)(iii).

SCHEDULE 4

WRITTEN SAFETY VERIFICATION SCHEME REQUIREMENTS

Regulation . . . 6(4)(a)

INFORMATION TO BE INCLUDED IN A SAFETY VERIFICATION SCHEME

[10.982]

1

(1) The arrangements for the selection, appointment and retention of the competent person, which arrangements should at least provide for:

(a) the appointment of the competent person at an early stage in the design selection process;

(b) the involvement of the competent person in the establishing of the criteria to be applied in the verification process and the design selection process; and

(c) the communication to the competent person of information necessary for the proper implementation, or revision, of the verification scheme and which information is necessary in order for the competent person to undertake the verification.

(2) The arrangements for the examination and testing of new or altered vehicles or infrastructure, which arrangements should at least provide for:

(a) the means of controlling risks that arise during the carrying out of any testing or trials prior to placing in service; and

(b) the standards and criteria to be applied in the verification process.

(3) The arrangements for the review and revision of the verification scheme.

(4) The arrangements for the making and preservation of records showing—

(a) the examination and testing carried out to the new or altered vehicles or infrastructure prior to its being placed in service;

(b) the findings of that examination and testing;

(c) any remedial action recommended as a result of that examination and testing; and

(d) any remedial action performed.

(5) The arrangements for communicating the matters contained in sub-paragraphs (1) to (4) of this Schedule to an appropriate level in the management system of the transport operator or responsible person as the case may be.

[10.982A]

Statutory annotations

Amendment

Heading: words omitted revoked by SI 2013/950, reg 3(1).

SCHEDULE 5

. . .

. . .

[10.983]

[*Revoked by SI 2013/950, reg 3(1), (16).*]

Part 10

Carriage of Dangerous Goods and Use of Transportable Pressure Equipment Regulations 2009

SI 2009/1348

as amended by SIs 2011/1860 and 2013/235

SCHEDULE 2

RADIOLOGICAL EMERGENCIES

Regulation 24(2)

[10.1047]

1 Interpretation

In this Schedule, the expressions mentioned in column 1 of the Table have the meanings given in column 2—

Table

Column 1	***Column 2***
'assist in the intervention'	The taking of such steps, as it is reasonable and practicable in the prevailing circumstances to take, in order to prevent or decrease exposure. The circumstances to be taken into account include— (a) the weather conditions; (b) the time of the occurrence of the emergency; (c) the distribution of the local population; (d) the nature and content of the package involved; (e) the stability of the class 7 goods involved; (f) the nature of the local geography and ecology; (g) any other prevailing hazards; and (h) the relative importance of the emergency in relation to other calls that are being made upon the emergency services.
'intervention'	A human activity that prevents or decreases the exposure of persons to radiation from a radiation emergency or from an event which could lead to a radiation emergency, by acting on the sources of radiation, the paths by which such radiation may be transmitted to persons and on persons themselves.

2 Information to the public about health protection measures

(1) Every consignor, carrier and consignee carrying out the transport of a consignment must—

(a) ensure that any members of the public who are in an area in which, in the opinion of the GB competent authority, they are likely to be affected by a radiological emergency arising from the undertaking of that carrier, consignor or consignee are supplied, in the appropriate manner approved by the GB competent authority and without their having to request it, with at least the information set out in sub-paragraph (2); and

(b) make that information publicly available, which includes endeavouring to enter into an agreement or arrangement with the local authority in the area referred to in Paragraph (a) for the dissemination by that authority of the information required to be supplied to members of the public in accordance with that Paragraph.

(2) The following is the information that is to be supplied and made available—

(a) the basic facts about the radioactivity and its effects on persons and on the environment;

(b) the various types of radiological emergency possible and their consequences for the general public and the environment;

(c) the emergency measures envisaged to alert, protect and assist the general public in the event of the occurrence of a radiological emergency;

(d) appropriate information on action to be taken by the general public in the event of the occurrence of a radiological emergency; and

(e) the appropriate local authority responsible for implementing the emergency measures and action referred to in Paragraphs (c) and (d).

(3) In preparing the information to be supplied and made available, the carrier, consignor or consignee must—

(a) consult the GB competent authority, but remain responsible for the accuracy, completeness and form of the information supplied; and

(b) endeavour to enter into an agreement or arrangement with the local authority in whose area the carrier, consignor or consignee is situated with respect to the dissemination by that authority of the information to members of the public.

(4) The information supplied and made available must be updated—

(a) at regular intervals;

(b) whenever significant changes to any of the matters mentioned in sub-paragraph (2) take place; and

(c) in any event, at least every 3 years.

(5) When information is updated in accordance with sub-paragraph (4) it must again be supplied and made publicly available in accordance with sub-paragraph (2).

3 Duties with respect to the monitoring of particular persons

(1) Any—

(a) employee or agent of a consignor, carrier or consignee; or

(b) person of whose services a consignor, carrier or consignee makes use in the carriage of dangerous goods,

who assists in an intervention and is liable to be subjected to emergency exposure must be treated as being a person classified pursuant to regulation 20 of the Ionising Radiations Regulations 1999 ('the 1999 Regulations') and, accordingly, the consignor, carrier or consignee (as the case may be) has the same duties with regard to the monitoring of such persons as are imposed upon an 'employer' by regulations 21 to 26 of the 1999 Regulations.

(2) To the extent that it is necessary in order to save human lives, an emergency exposure is permitted as a result of which the dose limit specified in paragraph 1, 2, 6, 7 or 8 of Schedule 4 (Dose Limits) to the 1999 Regulations will be exceeded, provided that the person who is proposed to be subjected to a dose in excess of the limit provided for in the relevant paragraph is a volunteer and has been informed of the risks involved in the intervention.

(3) In this paragraph, 'emergency exposure' means an exposure of a person engaged in an activity of, or associated with, the response to a radiation emergency or potential emergency in order to bring help to endangered persons, prevent exposure of a large number of persons or save a valuable installation or goods where one of the individual dose limits referred to in paragraphs 1 or 2 of Part 1 of Schedule 4 to the 1999 Regulations could be exceeded.

4 Duties of consignor and carrier with regard to the preparation of emergency arrangements

(1) Before the carriage of a package begins, the consignor [and the carrier] of that package must ensure that there is a plan in writing setting out such emergency arrangements as are appropriate for the carriage of that package.

(2) The plan must be prepared having regard to—

- (a) the principle that intervention is to be undertaken only if the damage due to the radiation resulting from the radiation emergency is sufficient to justify the potential harm and the potential cost (including the social cost) of that intervention;
- (b) the principle that the form, scale and duration of the intervention should ensure that the benefit to health will be greater than any harm that might be associated with the intervention itself;
- (c) the dose limits provided for in Schedule 4 to the Ionising Radiations Regulations 1999; and
- (d) the levels of radiation dose applicable in an emergency that are specified by [the Secretary of State pursuant to the Secretary of State's functions under section 58 of the Health and Social Care Act 2012].

(3) To the extent that the plan is used in relation to carriage on more than one occasion, the consignor [and the carrier] must review and, whenever necessary, revise the emergency arrangements and must ensure that at suitable intervals they are tested.

5 Duties of drivers, carriers and consignors in the event of the occurrence of a radiological emergency

(1) The driver of a vehicle or train carrying class 7 goods who discovers, or has reason to believe, that a notifiable event has occurred in relation to the transport unit or train must—

- (a) immediately notify—
  - (i) the police;
  - (ii) (where appropriate), as respects England and Wales, the fire and rescue authority and, as respects Scotland, the chief officer of the relevant authority; and
  - (iii) the consignor,

  of that event;
- (b) initiate the emergency arrangements in respect of any radiological emergency; and
- (c) assist in the intervention that is made in connection with that radiological emergency.

(2) The carrier of class 7 goods who becomes aware of the occurrence of a notifiable event in relation to those goods must—

- (a) immediately notify—
  - (i) the police (unless the driver of the transport unit or train has already done so); and
  - (ii) the GB competent authority,

  of that event;

Part 10

(b) assist in the intervention that is made in connection with any radiological emergency; and

(c) as soon as is reasonably practicable, arrange for the examination of the load so as to determine whether contamination has arisen and, if it has, to arrange for the safe disposal of any part of the load that has been contaminated and for the decontamination of the transport unit or train.

(3) If a consignor of class 7 goods becomes aware of the occurrence of a notifiable event in relation to those goods, the consignor must—

(a) immediately notify—

(i) the police; and

(ii) the GB competent authority,

of that event (unless either the driver or the carrier has already done so);

(b) assist in the intervention that is made in connection with any radiological emergency; and

(c) provide the GB competent authority with details of the incident that gave rise to that emergency.

(4) If a consignor of class 7 goods becomes aware that emergency arrangements have been initiated in relation to those goods, the consignor must notify the GB competent authority of the initiation of those arrangements even if, in the event, no intervention was made pursuant to those arrangements.

(5) If a notifiable event occurs the carrier must ensure that a report is made forthwith to the GB competent authority.

(6) The report required under sub-paragraph (5) must be in a form which has been approved by the GB competent authority and must contain all relevant information which the GB competent authority has communicated to the carrier that it considers necessary.

(7) In this paragraph—

(a) 'notifiable event' means—

(i) a radiological emergency,

(ii) the theft or loss of the class 7 goods being carried; or

(iii) an occurrence subject to report as construed in accordance with Sub-section 1.8.5.3; and

(b) 'initiate the emergency arrangements' means the taking of such steps as it is reasonable and practicable to take in order to put into effect the actions that have been planned for in the emergency arrangements.

6 Packages involved in a radiological emergency

A package that has been involved in a radiological emergency must not be carried or caused to be carried unless the consignor or the consignor's agent has examined it and the consignor is satisfied that it complies with the requirements of these Regulations and has issued a certificate to that effect.

7 Power of the competent authority to require documents and require testing, rehearsal and revision of the emergency arrangements

(1) The consignor and carrier must provide to the GB competent authority, within such reasonable period as the GB competent authority may specify, such documents relating to the emergency arrangements as may have been requested by the GB competent authority.

(2) To the extent required by a notice in writing served on the consignor or carrier by the GB competent authority, the consignor or carrier must test, rehearse and revise the emergency arrangements.

[10.1047A]

Statutory annotations

Amendment

Para 4: in sub-para (2)(d) words from 'the Secretary of State' to the end substituted by SI 2013/235, art 11, Sch 2, Pt 1, para 127.

PART 12
OFFSHORE, PIPELINES AND DIVING

Part 12A
Offshore, pipelines and diving: Offshore

Offshore Installations (Safety Representatives and Safety Committees) Regulations 1989

SI 1989/971

as amended by SIs 1992/2885, 1995/738, 1995/743, 1995/3163, 1999/3242, 2005/3117 and 2013/1471

Election of safety representatives etc

[12.443]

17 Powers of safety representatives

(1) To enable him to fulfil his functions under regulation 16 above, a safety representative may exercise the powers set out in paragraphs (2), (3), (4)(b) and (5) of this regulation and he may seek advice and guidance whether from persons on the offshore installation or elsewhere on any matters arising out of regulation 16 above and regulation 22 below, as appropriate.

(2) A safety representative may inspect any part of the offshore installation or its equipment if—

(a) he has given the installation manager and, if his employer is not the [duty holder], his employer, reasonable notice in writing of his intention to do so, and

(b) he has not inspected that part of the installation or its equipment in the previous three months;

and he may carry out more frequent inspections by agreement with the installation manager and, if his employer is not the [duty holder], his employer.

(3) If—

(a) there has been a notifiable incident, [or an over three day injury] and

(b) it is safe for an inspection to be carried out, and

(c) the interests of the members of his constituency might be involved,

a safety representative may inspect the part of the installation or the equipment concerned and, so far as is necessary for the purpose of determining the cause, he may inspect any other part of the installation or its equipment; provided that the safety representative shall first notify of his intention to carry out the inspection—

(i) the installation manager; and

(ii) where his employer is not the [duty holder] and it is reasonably practicable to notify him, his employer.

(4) Where two or more safety representatives consider there is an imminent risk of serious personal injury arising from an activity carried out on the installation—

(a) they shall make representations to the installation manager who shall prepare and send a report in writing on the matter to an inspector appointed under section 6(4) of the 1971 Act as soon as is reasonably practicable; and

(b) a safety representative may make a report in writing by the fastest practicable means to an inspector appointed under section 6(4) of the 1971 Act.

(5) A safety representative may receive information given under section 28(8) of the 1974 Act by an inspector appointed under section 19 of that Act.

[(6) In this regulation—

'notifiable incident' means any death, injury, disease or dangerous occurrence which is required to be reported under the Reporting of Injuries, Diseases or Dangerous Occurrences Regulations 1995; and 'over three day injury' means an injury required to be recorded in accordance with regulation 7(1)(aa) of the Reporting of Injuries, Diseases and Dangerous Occurrences Regulations 1995.]

[12.444]

Statutory annotations

Amendment

Para (3): in sub-para (a) words 'or an over three day injury' in square brackets inserted by SI 2013/1471, reg 18(2), Sch 4.

Para (6): substituted by SI 2013/1471, reg 18(2), Sch 4.

Offshore Installations and Wells (Design and Construction, etc) Regulations 1996

SI 1996/913

as amended by SIs 2005/1093, 2005/1643, 2005/3117 and 2013/1471

PART II

INTEGRITY OF INSTALLATIONS

[12.602]

9 Reporting of danger to an installation

(1) The duty holder shall ensure that, within 10 days after the appearance of evidence of a significant threat to the integrity of an installation, a report is made to the Executive in writing identifying such threat and specifying any action taken or to be taken to avert it.

(2) Paragraph (1) shall not apply to anything of which [the Reporting of Injuries, Diseases and Dangerous Occurrences Regulations 2013] require a report to be made.

[12.602A]

Statutory annotations

Amendment

Para (2): words 'the Reporting of Injuries, Diseases and Dangerous Occurrences Regulations 2013' in square brackets substituted by SI 2013/1471, reg 18(2), Sch 4.

Directive

2013/30/EU

of 12 June 2013

on safety of offshore oil and gas operations and amending Directive 2004/35/EC

[12.901]

THE EUROPEAN PARLIAMENT AND THE COUNCIL OF THE EUROPEAN UNION,

Having regard to the Treaty on the Functioning of the European Union, and in particular Article 192(1) thereof,

Having regard to the proposal from the European Commission,

After transmission of the draft legislative act to the national parliaments,

Having regard to the opinion of the European Economic and Social Committee,

After consulting the Committee of the Regions,

Acting in accordance with the ordinary legislative procedure,

Whereas:

(1) Article 191 of the Treaty on the Functioning of the European Union establishes the objectives of preserving, protecting and improving the quality of the environment and the prudent and rational utilisation of natural resources. It creates an obligation for all Union action to be supported by a high level of protection based on the precautionary principle, and on the principles that preventive action needs to be taken, that environmental damage needs as a matter of priority to be rectified at source and that the polluter must pay.

(2) The objective of this Directive is to reduce as far as possible the occurrence of major accidents relating to offshore oil and gas operations and to limit their consequences, thus increasing the protection of the marine environment and coastal economies against pollution, establishing minimum conditions for safe offshore exploration and exploitation of oil and gas and limiting possible disruptions to Union indigenous energy production, and to improve the response mechanisms in case of an accident.

(3) This Directive should apply not only to future offshore oil and gas installations and operations but, subject to transitional arrangements, also to existing installations.

(4) Major accidents relating to offshore oil and gas operations are likely to have devastating and irreversible consequences on the marine and coastal environment as well as significant negative impacts on coastal economies.

(5) Accidents relating to offshore oil and gas operations, in particular the accident in the Gulf of Mexico in 2010, have raised public awareness of the risks involved in offshore oil and gas operations and have prompted a review of policies aimed at ensuring the safety of such operations. The Commission launched a review of offshore oil and gas operations and expressed its initial views on the safety thereof in its Communication 'Facing the challenge of the safety of offshore oil and gas activities' on 13 October 2010. The European Parliament adopted resolutions on the topic on 7 October 2010 and 13 September 2011. Energy Ministers of the Member States expressed their views in the Conclusions of the Council of 3 December 2010.

(6) The risks relating to major offshore oil or gas accidents are significant. By reducing the risk of pollution of offshore waters, this Directive should therefore contribute to ensuring the protection of the marine environment and in particular to achieving or maintaining good environmental status by 2020 at the latest, an objective set out in Directive 2008/56/EC of the European Parliament and the Council of 17 June 2008 establishing a framework for community action in the field of marine environmental policy (Marine Strategy Framework Directive).

(7) Directive 2008/56/EC aims to address, as one of its central purposes, the cumulative impacts from all activities on the marine environment, and is the

environmental pillar of the Integrated Maritime Policy. That policy is relevant to offshore oil and gas operations as it requires the linking of particular concerns from each economic sector with the general aim of ensuring a comprehensive understanding of the oceans, seas and coastal areas, with the objective of developing a coherent approach to the seas taking into account all economic, environmental and social aspects through the use of maritime spatial planning and marine knowledge.

(8) Offshore oil and gas industries are established in a number of regions of the Union, and there are prospects for new regional developments in offshore waters of Member States, with technological developments allowing for drilling in more challenging environments. Production of offshore oil and gas is a significant element in security of the Union's energy supply.

(9) The existing divergent and fragmented regulatory framework applying to safety of offshore oil and gas operations in the Union and current industry safety practices do not provide a fully adequate assurance that the risk of offshore accidents is minimised throughout the Union, and that in the event of an accident occurring in offshore waters of Member States, the most effective response would be deployed in a timely manner. Under existing liability regimes, the party responsible may not always be clearly identifiable and may not be able, or liable, to pay all the costs to remedy the damage it has caused. The party responsible should always be clearly identifiable before offshore oil and gas operations are commenced.

(10) Pursuant to Directive 94/22/EC of the European Parliament and of the Council of 30 May 1994 on the conditions for granting and using authorizations for the prospection, exploration and production of hydrocarbons offshore oil and gas operations in the Union may be carried out subject to obtaining an authorisation. In this context the licensing authority is required to consider the technical and financial risks, and where appropriate, the previous record of responsibility, of applicants seeking exclusive exploration and production licences. There is the need to ensure that when examining the technical and financial capability of the licensee the licensing authority thoroughly examine also its capability for ensuring continued safe and effective operations under all foreseeable conditions. When assessing the financial capability of entities applying for authorisation pursuant to Directive 94/22/EC, Member States should verify that such entities have provided appropriate evidence that adequate provisions have been or are to be made to cover liabilities deriving from major accidents.

(11) There is a need to clarify that holders of authorisations for offshore oil and gas operations pursuant to Directive 94/22/EC are also the liable 'operators' within the meaning of Directive 2004/35/EC of the European Parliament and the Council of 21 April 2004 on environmental liability with regard to the prevention and remedying of environmental damage, and should not delegate their responsibilities in this regard to third parties contracted by them.

(12) While general authorisations pursuant to Directive 94/22/EC guarantee to the licensees exclusive rights for exploring for or producing oil or gas within a given licensed area, offshore oil and gas operations within that area should be subject to continuous expert regulatory oversight by Member States in order to ensure there are effective controls in place for preventing major accidents, and limiting their impacts to persons, the environment, and security of energy supply.

(13) Offshore oil and gas operations should be conducted only by operators appointed by licensees or licensing authorities. The operator can be a third party or the licensee or one of the licensees depending on commercial arrangements or national administrative requirements. The operator should always be the entity with the primary responsibility for safety of operations and should be at all times competent to act in that regard. That role differs depending on the particular stage of activities covered by the licence. The operator's role is therefore to operate a well at the exploration stage and to operate a production installation at the production stage. It should be possible for the operator of a well at the exploration stage and the operator of a production installation to be the same entity for a given licensed area.

(14) Operators should reduce the risk of a major accident as low as reasonably practicable, to the point where the cost of further risk reduction would be grossly disproportionate to the benefits of such reduction. The reasonable practicability of risk reduction measures should be kept under review in the light of new knowledge and

technology developments. In assessing whether the time, cost and effort would be grossly disproportionate to the benefits of further reducing the risk, regard should be had to best practice risk levels compatible with the operations being conducted.

(15) It is important to ensure that the public is given early and effective opportunity to participate in the decision-making relating to operations that can potentially have significant effects on the environment in the Union. This policy is in line with the Union's international commitments, such as the UN/ECE Convention on Access to Information, Public Participation in Decision-Making and Access to Justice in Environmental Matters (the Aarhus Convention). Article 6 of the Aarhus Convention provides for public participation in decisions on the specific activities listed in Annex I thereto and on activities not listed there which may have a significant effect on the environment. Article 7 of the Aarhus Convention requires public participation concerning plans and programmes relating to the environment.

(16) Relevant requirements exist in Union legal acts in relation to the development of plans and projects, in particular in Directive 2001/42/EC of the European Parliament and of the Council of 27 June 2001 on the assessment of the effects of certain plans and programmes on the environment, Directive 2003/35/EC of the European Parliament and of the Council of 26 May 2003 providing for public participation in respect of the drawing up of certain plans and programmes relating to the environment, Directive 2011/92/EU of the European Parliament and of the Council of 13 December 2011 on the assessment of the effects of certain public and private projects on the environment and Directive 2012/18/EU of the European Parliament and of the Council of 4 July 2012 on the control of major-accident hazards involving dangerous substances. However, not all exploratory offshore oil and gas operations are covered by existing Union requirements on public participation. This applies in particular to the decision-making that aims or could lead to exploration operations being commenced from a non-production installation. However, such exploration operations may in some circumstances potentially have significant effects on the environment and the decision-making should therefore be the subject of public participation as required under the Aarhus Convention.

(17) Within the Union, there are already examples of good standards in national regulatory practices relating to offshore oil and gas operations. However, these are inconsistently applied throughout the Union and no Member State has yet incorporated all of the best regulatory practices in its legislation for preventing major accidents or limiting the consequences for human life and health, and for the environment. Best regulatory practices are necessary to deliver effective regulation which secures the highest safety standards and protects the environment, and can be achieved, inter alia, by integrating related functions into a competent authority that may draw resources from one or more national bodies.

(18) In accordance with Council Directive 92/91/EEC of 3 November 1992 concerning the minimum requirements for improving the safety and health protection of workers in the mineral-extracting industries through drilling (eleventh individual Directive within the meaning of Article 16(1) of Directive 89/391/EEC), workers and/or their representatives should be consulted on matters relating to safety and health at work and be allowed to take part in discussions on all questions relating to safety and health at work. In addition, best practice in the Union is for consultation mechanisms to be formally established by Member States on a tripartite basis comprising the competent authority, operators and owners, and worker representatives. An example of such formal consultation is the International Labour Organisation Tripartite Consultation (International Labour Standards) Convention, 1976 (No 144).

(19) Member States should ensure that the competent authority is legally empowered and adequately resourced to be capable of taking effective, proportionate and transparent enforcement action, including where appropriate cessation of operations, in cases of unsatisfactory safety performance and environmental protection by operators and owners.

(20) The independence and objectivity of the competent authority should be ensured. In this regard, experience gained from major accidents shows clearly that the organisation of administrative competences within a Member State can prevent conflicts of interest by a clear separation between regulatory functions and associated decisions relating to offshore safety and the environment, and to the regulatory functions relating to the economic development of offshore natural resources includ-

ing licensing and revenues management. Such conflicts of interest are best prevented by a complete separation of the competent authority from the functions relating to the economic development of offshore natural resources.

(21) However, complete separation of the competent authority from economic development of offshore natural resources may be disproportionate where there is a low level of offshore oil and gas operations in a Member State. In such a case, the Member State concerned would be expected to make the best alternative arrangements to secure the independence and objectivity of the competent authority.

(22) Specific legislation is needed to address the major hazards relating to the offshore oil and gas industry, specifically in process safety, safe containment of hydro-carbons, structural integrity, prevention of fire and explosion, evacuation, escape and rescue, and limiting environmental impact following a major accident.

(23) This Directive should apply without prejudice to any requirements under any other Union legal acts, especially in the field of safety and health of workers at work, in particular Council Directive 89/391/EEC of 12 June 1989 on the introduction of measures to encourage improvements in the safety and health of workers at work and Directive 92/91/EEC.

(24) An offshore regime needs to apply both to operations carried out on fixed installations and to those on mobile installations, and to the lifecycle of exploration and production activities from design to decommissioning and permanent abandonment.

(25) The best practices currently available for major accident prevention in offshore oil and gas operations are based on a goal-setting approach and on achieving desirable outcomes through thorough risk assessment and reliable management systems.

(26) According to the best practices in the Union, operators and owners are encouraged to establish effective corporate safety and environmental policies and to give effect to them in a comprehensive safety and environmental management system and emergency response plan. In order to make suitable arrangements for major accident prevention, operators and owners should comprehensively and systematically identify all major accident scenarios relating to all hazardous activities that may be carried out on that installation, including impacts on the environment arising from a major accident. Those best practices also require an assessment of the likelihood and consequences and therefore the risk of major accidents, and also the measures necessary to prevent them and the measures necessary for emergency response, should a major accident nonetheless occur. The risk assessments and arrangements for major accident prevention should be clearly described and compiled in the report on major hazards. The report on major hazards should be complementary to the safety and health document referred to in Directive 92/91/EEC. The workers should be consulted at the relevant stages of the preparation of the report on major hazards. The report on major hazards should have to be thoroughly assessed and accepted by the competent authority.

(27) In order to maintain the effectiveness of major hazard controls in offshore waters of Member States, the report on major hazards should be prepared and, as necessary, amended in respect of any significant aspect of the lifecycle of a production installation, including design, operation, operations when combined with other installations, relocation of such installation within the offshore waters of the Member State in question, major modifications, and final abandonment. Similarly, the report on major hazards should also be prepared in respect of non-production installations and amended as necessary to take into account significant changes to the installation. No installation should be operated in offshore waters of Member States unless the competent authority has accepted the report on major hazards submitted by the operator or owner. Acceptance by the competent authority of the report on major hazards should not imply any transfer of responsibility for control of major hazards from the operator or the owner to the competent authority.

(28) Well operations should be undertaken only by an installation which is technically capable of controlling all the foreseeable hazards at the well location, and in respect of which a report on major hazards has been accepted.

(29) In addition to using a suitable installation, the operator should prepare a detailed design plan and an operating plan pertinent to the particular circumstances and hazards of each well operation. In accordance with best practices in the Union, the

operator should provide for independent expert examination of the well design. The operator should send a notification of well plans to the competent authority in sufficient time for the competent authority to take any necessary action in respect of the planned well operation. In this respect, Member States may introduce more stringent national requirements prior to the commencement of a well operation.

(30) To ensure safety in design and continuous safe operations, the industry is required to follow the best practices defined in authoritative standards and guidance. Such standards and guidance should be updated based on new knowledge and invention to ensure continuous improvement. Operators, owners and competent authorities should collaborate to establish priorities for the creation of new or improved standards and guidance in the light of the Deepwater Horizon accident experience and other major accidents. Having due regard to the established priorities the preparation of new or improved standards and guidance should be commissioned without delay.

(31) In view of the complexity of offshore oil and gas operations, the implementation of the best practices by the operators and owners requires a scheme of independent verification of safety and environmental critical elements throughout the lifecycle of the installation, including, in the case of production installations, the design stage.

(32) In so far as mobile offshore drilling units are in transit and are to be considered as ships, they are subject to international maritime conventions, in particular, SOLAS, MARPOL or the equivalent standards of the applicable version of the Code for the construction and equipment of mobile offshore drilling units (MODU Code). Such mobile offshore drilling units when in transit in offshore waters are also subject to Union law concerning port State control and compliance with flag State requirements. This Directive addresses such units when they are stationed in offshore waters for drilling, production or other activities associated with offshore oil and gas operations.

(33) The report on major hazards should, inter alia, take into account risks to the environment, including the impact of climatic conditions and climate change on the long term resilience of the installations. Given that offshore oil and gas operations in one Member State can have significant adverse environmental effects in another Member State, it is necessary to establish and apply specific provisions in accordance with the UN/ECE Convention on Environmental Impact Assessment in a Transboundary Context done at Espoo (Finland), on 25 February 1991. Member States with offshore waters that are inactive in offshore oil and gas operations should appoint contact points in order to facilitate effective cooperation in this regard.

(34) Operators should notify Member States without delay if a major accident occurs, or may be about to occur, so that the Member State can initiate a response as appropriate. Therefore, operators should include in the notification suitable and sufficient particulars concerning the location, magnitude and nature of the actual or imminent major accident, their own response, and the worst case escalation scenario including transboundary potential.

(35) In order to ensure effective response to emergencies, operators should prepare internal emergency response plans that are site specific and based on risks and hazard scenarios identified in the report on major hazards, submit them to their competent authority, and maintain such resources as are necessary for prompt execution of those plans when needed. In the case of mobile offshore drilling units, operators need to ensure that the owners' internal emergency response plans for the installation are amended as necessary to be applicable to the specific location and well operation hazards. Such amendments should be included in the notification of well operations. The adequate availability of emergency response resources should be assessed against the capacity to deploy them at the site of an accident. The readiness and effectiveness of emergency response resources should be assured and regularly tested by the operators. Where duly justified, response arrangements are allowed to be reliant on speedily transporting the response equipment such as capping devices, and other resources, from distant locations.

(36) Best global practice requires licensees, operators and owners to take primary responsibility for controlling the risks they create by their operations, including operations conducted by contractors on their behalf and therefore to establish within a corporate major accident prevention policy the mechanisms and highest level of corporate ownership to implement that policy consistently throughout the organisation in the Union and outside of the Union.

(37) Responsible operators and owners should be expected to conduct their operations worldwide in accordance with best practices and standards. Consistent application of such best practices and standards should become mandatory within the Union, and it would be desirable for operators and owners registered in the territory of a Member State to apply the corporate major accident prevention policy when operating outside offshore waters of Member States as far as possible within the applicable national legal framework.

(38) While recognising that it may not be possible to enforce application of the corporate major accident prevention policy outside of the Union, Member States should ensure that operators and owners include their offshore oil and gas operations outside of the Union in their corporate major accident prevention policy documents.

(39) Information on major accidents in offshore oil and gas operations outside the Union can help in further understanding their potential causes, in promoting learning of key lessons and in further developing the regulatory framework. Therefore, all Member States, including the landlocked Member States and the Member States with offshore waters which do not have offshore oil and gas operations or licensing activities, should require reports on major accidents occurring outside the Union which involve companies registered in their territory, and should share this information at Union level. The reporting requirement should not interfere with emergency response or the legal proceedings relating to an accident. Instead they should focus on the relevance of the accident for further developing the safety of offshore oil and gas operations in the Union.

(40) Member States should expect operators and owners, in following best practices, to establish effective cooperative relationships with the competent authority, supporting best regulatory practice by the competent authority and to proactively ensure the highest levels of safety, including, where necessary, suspending operations without the competent authority needing to intervene.

(41) To ensure that no relevant safety concerns are over-looked or ignored, it is important to establish and encourage adequate means for the confidential reporting of those concerns and the protection of whistleblowers. While Member States are not able to enforce rules outside the Union, those means should enable the reporting of concerns of persons involved in offshore oil and gas operations outside the Union.

(42) The sharing of comparable data between Member States is rendered difficult and unreliable due to the lack of a common data reporting format across all Member States. A common format for the reporting of data by operators and owners to the Member State would provide transparency of the safety and environmental performance of operators and owners and would provide public access to relevant and Union-wide comparable information on safety of offshore oil and gas operations and would facilitate dissemination lessons learned from major accidents and near misses.

(43) In order to ensure uniform conditions for sharing information and encouraging transparency of performance of the offshore oil and gas sector, implementing powers should be conferred on the Commission regarding the format and details of information to be shared and to be made publicly available. Those powers should be exercised in accordance with Regulation (EU) No 182/2011 of the European Parliament and of the Council of 16 February 2011 laying down the rules and general principles concerning mechanisms for control by the Member States of the Commission's exercise of implementing powers.

(44) The advisory procedure should be used for the adoption of relevant implementing acts given that those acts are mainly of a mere practical nature. Therefore, the application of the examination procedure would not be justified.

(45) To facilitate public confidence in the authority and integrity of offshore oil and gas operations in the Union, Member States should provide periodic reports of activity and incidents to the Commission. The Commission should publish reports periodically on levels of Union activity and trends in the safety and environmental performance of the offshore oil and gas sector. Member States should, without delay, inform the Commission, and any other Member State whose territory or offshore waters are affected, as well as the public concerned, of a major accident.

(46) Experience shows that ensuring the confidentiality of sensitive data is necessary in order to foster an open dialogue between the competent authority and the operator and owner. To that effect the dialogue between operators and owners and all

Member States should be based on relevant existing international legal instruments and Union law on access to environmentally relevant information subject to any overriding requirement for safety and environment protection.

(47) The value of collaboration between offshore authorities has been clearly established by the activities of the North Sea Offshore Authorities Forum and the International Regulators Forum. Similar collaboration has been established across the Union in an expert group, the European Union Offshore Oil and Gas Authorities Group (EUOAG), whose task is to promote efficient collaboration between national representatives and the Commission, including disseminating best practices and operational intelligence, establishing priorities for raising standards, and for advising the Commission on regulatory reform.

(48) Emergency response and contingency planning for major accidents should be made more effective by systematic and planned cooperation between Member States and between Member States and the oil and gas industry, as well as by sharing compatible emergency response assets including expertise. Where appropriate, those responses and planning should also make use of the existing resources and assistance available from within the Union, in particular through the European Maritime Safety Agency ('the Agency'), established by Regulation (EC) No 1406/2002, and the Union Civil Protection Mechanism, established by the Council Decision 2007/779/EC, Euratom. Member States should also be allowed to request additional assistance from the Agency through the Union Civil Protection Mechanism.

(49) Pursuant to Regulation (EC) No 1406/2002, the Agency was established for the purpose of ensuring a high, uniform and effective level of maritime safety and prevention of pollution by ships within the Union as well as ensuring a response to marine pollution caused by oil and gas installations.

(50) In implementing the obligations under this Directive, account should be taken of the fact that marine waters covered by the sovereignty or sovereign rights and jurisdiction of Member States form an integral part of the four marine regions identified in Article 4(1) of Directive 2008/56/EC, namely the Baltic Sea, the North-east Atlantic Ocean, the Mediterranean Sea and the Black Sea. For this reason, the Union should, as a matter of priority, strengthen coordination with third countries that have sovereignty or sovereign rights and jurisdiction over marine waters in such marine regions. Appropriate cooperation frameworks include regional sea conventions, as defined in point 10 of Article 3 of Directive 2008/56/EC.

(51) In relation to the Mediterranean Sea, in conjunction with this Directive, the necessary actions were undertaken for the Union to accede to the Protocol for the Protection of the Mediterranean Sea against Pollution Resulting from Exploration and Exploitation of the Continental Shelf and the Seabed and its Subsoil ('the Offshore Protocol') to the Convention for the Protection of the Marine Environment and the Coastal Region of the Mediterranean ('the Barcelona Convention'), which was concluded by Council Decision 77/585/EEC.

(52) The Arctic waters are a neighbouring marine environment of particular importance for the Union, and play an important role in mitigating climate change. The serious environmental concerns relating to the Arctic waters require special attention to ensure the environmental protection of the Arctic in relation to any offshore oil and gas operation, including exploration, taking into account the risk of major accidents and the need for effective response. Member States who are members of the Arctic Council are encouraged to actively promote the highest standards with regard to environmental safety in this vulnerable and unique ecosystem, such as through the creation of international instruments on prevention, preparedness and response to Arctic marine oil pollution, and through building, inter alia, on the work of the Task Force established by the Arctic Council and the existing Arctic Council Offshore Oil and Gas Guidelines.

(53) National external emergency plans should be based on risk assessment, taking into account the reports on major hazards for the installations stationed in the offshore waters concerned. Member States should take into account the most up-to-date Risk Assessment and Mapping Guidelines for Disaster Management as prepared by the Commission.

(54) Effective response to emergencies requires immediate action by the operator and owner and close cooperation with Member States' emergency response organi-

sations which coordinate the introduction of additional emergency response resources as the situation develops. Such response should also include a thorough investigation of the emergency which should commence without delay so as to ensure minimum loss of relevant information and evidence. Following an emergency, Member States should draw up appropriate conclusions and take any necessary measures.

(55) It is crucial that all relevant information, including the technical data and parameters, are available for the later investigation. Member States should ensure that relevant data are collected during the offshore oil and gas operations and that in the event of a major accident, relevant data are secured and data collection is intensified appropriately. In this context, Member States should encourage the use of suitable technical means in order to promote the reliability and recording of relevant data and to prevent possible manipulation thereof.

(56) In order to ensure effective implementation of the requirements of this Directive, effective, proportionate and dissuasive penalties for infringements should be put in place.

(57) In order to adapt certain Annexes to include additional information which may become necessary in light of technical progress, the power to adopt acts in accordance with Article 290 of the Treaty on the Functioning of the European Union should be delegated to the Commission in respect of amending the requirements in certain Annexes to this Directive. It is of particular importance that the Commission carry out appropriate consultations during its preparatory work, including at expert level. The Commission, when preparing and drawing up delegated acts, should ensure a simultaneous, timely and appropriate transmission of relevant documents to the European Parliament and to the Council.

(58) The definition of water damage in Directive 2004/35/EC should be amended to ensure that the liability of licensees under that Directive applies to marine waters of Member States as defined in Directive 2008/56/EC.

(59) Many provisions of this Directive are not relevant for the landlocked Member States, namely Austria, the Czech Republic, Hungary, Luxembourg and Slovakia. It is nonetheless desirable that those Member States promote the principles and high standards existing in Union law for the safety of offshore oil and gas operations in their bilateral contacts with third countries and with relevant international organisations.

(60) Not all Member States with offshore waters allow for offshore oil and gas operations under their jurisdiction. Those Member States are not engaged in the licensing and prevention of major accidents of such operations. It would therefore be a disproportionate and unnecessary obligation if those Member States had to transpose and implement all provisions of this Directive. However, accidents during offshore oil and gas operations may affect their shores. Therefore, those Member States should, inter alia, be prepared to respond to and investigate major accidents and should cooperate through contact points with other Member States concerned and with relevant third countries.

(61) Given their geographical location, landlocked Member States are neither engaged in the licensing of, and prevention of major accidents in, offshore oil and gas operations nor are they potentially affected by such accidents in offshore waters of other Member States. Therefore, they should not have to transpose the majority of provisions of this Directive. However, where a company that is active, itself or through subsidiaries, in offshore oil and gas operations outside the Union is registered in a landlocked Member State, that Member State should request that company to provide a report on accidents occurring in such operations, which can be shared at Union level, in order for all the interested parties in the Union to benefit from the experience gained from such accidents.

(62) Apart from the measures introduced by this Directive, the Commission should explore other appropriate means of improving the prevention of major accidents and limiting their consequences.

(63) Operators should ensure they have access to sufficient physical, human and financial resources to prevent major accidents and limit the consequences of such accidents. However, as no existing financial security instruments, including risk pooling arrangements, can accommodate all possible consequences of major accidents, the Commission should undertake further analysis and studies of the appropriate measures to ensure an adequately robust liability regime for damages relating

to offshore oil and gas operations, requirements on financial capacity including availability of appropriated financial security instruments or other arrangements. This may include an examination of the feasibility of a mutual compensation scheme. The Commission should submit a report to the European Parliament and to the Council on its findings, accompanied if appropriate, by proposals.

(64) At Union level, it is important that technical standards are complemented by a corresponding legal framework of product safety legislation and that such standards apply to all offshore installations in offshore waters of Member States, and not just non-mobile production installations. The Commission should therefore undertake further analysis of the product safety standards applicable to offshore oil and gas operations.

(65) Since the objective of this Directive, namely establishing minimum requirements for preventing major accidents in offshore oil and gas operations and limiting the consequences of such accidents, cannot be sufficiently achieved by the Member States and can therefore, by reason of the scale and effects of the proposed action, be better achieved at Union level, the Union may adopt measures, in accordance with the principle of subsidiarity as set out in Article 5 of the Treaty on European Union. In accordance with the principle of proportionality, as set out in that Article, this Directive does not go beyond what is necessary in order to achieve that objective,

HAS ADOPTED THIS DIRECTIVE:

CHAPTER I

INTRODUCTORY PROVISIONS

[12.902]

Article 1 Subject and scope

1. This Directive establishes minimum requirements for preventing major accidents in offshore oil and gas operations and limiting the consequences of such accidents.

2. This Directive shall be without prejudice to Union law concerning safety and health of workers at work, in particular Directives 89/391/EEC and 92/91/EEC.

3. This Directive shall be without prejudice to Directives 94/22/EC, 2001/42/EC, 2003/4/EC, 2003/35/EC, 2010/75/EU and 2011/92/EU.

[12.903]

Article 2 Definitions

For the purpose of this Directive:

(1) 'major accident' means, in relation to an installation or connected infrastructure:

- (a) an incident involving an explosion, fire, loss of well control, or release of oil, gas or dangerous substances involving, or with a significant potential to cause, fatalities or serious personal injury;
- (b) an incident leading to serious damage to the installation or connected infrastructure involving, or with a significant potential to cause, fatalities or serious personal injury;
- (c) any other incident leading to fatalities or serious injury to five or more persons who are on the offshore installation where the source of danger occurs or who are engaged in an offshore oil and gas operation in connection with the installation or connected infrastructure; or
- (d) any major environmental incident resulting from incidents referred to in points (a), (b) and (c).

For the purposes of determining whether an incident constitutes a major accident under points (a), (b) or (d), an installation that is normally unattended shall be treated as if it were attended;

(2) 'offshore' means situated in the territorial sea, the Exclusive Economic Zone or the continental shelf of a Member State within the meaning of the United Nations Convention on the Law of the Sea;

(3) 'offshore oil and gas operations' means all activities associated with an installation or connected infrastructure, including design, planning, construction, operation and decommissioning thereof, relating to exploration and production of oil or gas, but excluding conveyance of oil and gas from one coast to another;

(4) 'risk' means the combination of the probability of an event and the consequences of that event;

(5) 'operator' means the entity appointed by the licensee or licensing authority to conduct offshore oil and gas operations, including planning and executing a well operation or managing and controlling the functions of a production installation;

(6) 'suitable' means right or fully appropriate, including consideration of proportionate effort and cost, for a given requirement or situation, based on objective evidence and demonstrated by an analysis, comparison with appropriate standards or other solutions used in comparable situations by other authorities or industry;

(7) 'entity' means any natural or legal person or any group of such persons;

(8) 'acceptable', in relation to a risk, means a level of risk for which the time, cost or effort of further reducing it would be grossly disproportionate to the benefits of such reduction. In assessing whether the time, cost or effort would be grossly disproportionate to the benefits of further reducing the risk, regard shall be had to best practice risk levels compatible with the undertaking;

(9) 'licence' means an authorisation for offshore oil and gas operations pursuant to Directive 94/22/EC;

(10) 'licensed area' means the geographical area covered by the licence;

(11) 'licensee' means the holder or joint holders of a licence;

(12) 'contractor' means any entity contracted by the operator or owner to perform specific tasks on behalf of the operator or owner;

(13) 'licensing authority' means the public authority which is responsible for granting authorisations or for monitoring the use of authorisations as provided for in Directive 94/22/EC;

(14) 'competent authority' means the public authority, appointed pursuant to this Directive and responsible for the duties assigned to it in this Directive. The competent authority may be comprised of one or more public bodies;

(15) 'exploration' means drilling into a prospect and all related offshore oil and gas operations necessary prior to production-related operations;

(16) 'production' means offshore extraction of oil and gas from the underground strata of the licensed area including offshore processing of oil and gas and its conveyance through connected infrastructure;

(17) 'non-production installation' means an installation other than an installation used for production of oil and gas;

(18) 'the public' means one or more entities and, in accordance with national legislation or practice, their associations, organisations or groups;

(19) 'installation' means a stationary, fixed or mobile facility, or a combination of facilities permanently inter-connected by bridges or other structures, used for offshore oil and gas operations or in connection with such operations. Installations include mobile offshore drilling units only when they are stationed in offshore waters for drilling, production or other activities associated with offshore oil and gas operations;

(20) 'production installation' means an installation used for production;

(21) 'connected infrastructure' means, within the safety zone or within a nearby zone of a greater distance from the installation at the discretion of the Member State:

(a) any well and associated structures, supplementary units and devices connected to the installation;

(b) any apparatus or works on or fixed to the main structure of the installation;

(c) any attached pipeline apparatus or works;

(22) 'acceptance', in relation to the report on major hazards, means the communication in writing by the competent authority to the operator or the owner that the report, if implemented as set out therein, meets the requirements of this Directive. Acceptance does not imply any transfer of responsibility for control of major hazards to the competent authority;

(23) 'major hazard' means a situation with the potential to result in a major accident;

(24) 'well operation' means any operation concerning a well that could result in the accidental release of materials that has the potential to lead to a major accident, including the drilling of a well, the repair or modification of a well, the suspension of well operations and the permanent abandonment of a well;

(25) 'combined operation' means an operation carried out from an installation with another installation or installations for purposes related to the other installation(s) which thereby materially affects the risks to the safety of persons or the protection of the environment on any or all of the installations;

(26) 'safety zone' means the area within a distance of 500 metres from any part of the installation, established by the Member State;

(27) 'owner' means an entity legally entitled to control the operation of a non-production installation;

(28) 'internal emergency response plan' means a plan prepared by the operator or owner pursuant to the requirements of this Directive concerning the measures to prevent escalation or limit the consequences of a major accident relating to offshore oil and gas operations;

(29) 'independent verification' means an assessment and confirmation of the validity of particular written statements by an entity or an organisational part of the operator or the owner that is not under the control of or influenced by, the entity or the organisational part using those statements;

(30) 'material change' means:

(a) in the case of a report on major hazards, a change to the basis on which the original report was accepted including, inter alia, physical modifications, availability of new knowledge or technology and operational management changes;

(b) in the case of a notification of well operations or combined operations, a change to the basis on which the original notification was submitted including, inter alia, physical modifications, replacement of one installation with another, availability of new knowledge or technology and operational management changes;

(31) 'commencement of operations' means the point in time when the installation or connected infrastructure is involved for the first time in the operations for which it is designed;

(32) 'oil spill response effectiveness' means the effectiveness of spill response systems in responding to an oil spill, on the basis of an analysis of the frequency, duration, and timing of environmental conditions that would preclude a response. The assessment of oil spill response effectiveness is to be expressed as a percentage of time that such conditions are not present and is to include a description of the operating limitations placed on the installations concerned as a result of that assessment;

(33) 'safety and environmental critical elements' means parts of an installation, including computer programmes, the purpose of which is to prevent or limit the consequences of a major accident, or the failure of which could cause or contribute substantially to a major accident;

(34) 'tripartite consultation' means a formal arrangement to enable dialogue and cooperation between the competent authority, operators and owners, and workers' representatives;

(35) 'industry' means entities that are directly involved in offshore oil and gas operations covered by this Directive or whose activities are closely related to those operations;

(36) 'external emergency response plan' means a local, national or regional strategy to prevent escalation or limit the consequences of a major accident relating to offshore oil and gas operations using all resources available to the operator as described in the relevant internal emergency response plan, and any supplementary resources made available by the Member States;

(37) 'major environmental incident' means an incident which results, or is likely to result, in significant adverse effects on the environment in accordance with Directive 2004/35/EC.

CHAPTER II

PREVENTION OF MAJOR ACCIDENTS RELATING TO OFFSHORE OIL AND GAS OPERATIONS

[12.904]

Article 3 General principles of risk management in offshore oil and gas operations

1. Member States shall require operators to ensure that all suitable measures are taken to prevent major accidents in offshore oil and gas operations.

2. Member States shall ensure that operators are not relieved of their duties under this Directive by the fact that actions or omissions leading or contributing to major accidents were carried out by contractors.

3. In the case of a major accident, Member States shall ensure that operators take all suitable measures to limit its consequences for human health and for the environment.

4. Member States shall require operators to ensure that offshore oil and gas operations are carried out on the basis of systematic risk management so that the residual risks of major accidents to persons, the environment and offshore installations are acceptable.

[12.905]

Article 4 Safety and environmental considerations relating to licences

1. Member States shall ensure that decisions on granting or transferring licences to carry out offshore oil and gas operations take into account the capability of an applicant for such a licence to meet the requirements for operations within the framework of the licence as required by the relevant provisions of Union law, in particular this Directive.

2. In particular, when assessing the technical and financial capability of the applicant for a licence, due account shall be taken of the following:

(a) the risk, the hazards and any other relevant information relating to the licensed area concerned, including, where appropriate, the cost of degradation of the marine environment referred to in point (c) of Article 8(1) of Directive 2008/56/EC;

(b) the particular stage of offshore oil and gas operations;

Part 12

(c) the applicant's financial capabilities, including any financial security, to cover liabilities potentially deriving from the offshore oil and gas operations in question including liability for potential economic damages where such liability is provided for by national law;

(d) the available information relating to the safety and environmental performance of the applicant, including in relation to major accidents, as may be appropriate to the operations for which the licence was requested.

Before granting or transferring a licence for offshore oil and gas operations, the licensing authority shall consult, where appropriate, the competent authority.

3. Member States shall ensure that the licensing authority does not grant a licence unless it is satisfied with evidence from the applicant that the applicant has made or will make adequate provision, on the basis of arrangements to be decided by Member States, to cover liabilities potentially deriving from the applicant's offshore oil and gas operations. Such provision shall be valid and effective from the start of offshore oil and gas operations. Member State shall require applicants to provide, in an appropriate manner, evidence of technical and financial capacity and any other relevant information relating to the area covered by the licence and the particular stage of offshore oil and gas operations. Member States shall assess the adequacy of provisions referred to in the first subparagraph in order to establish whether the applicant has sufficient financial resources for the immediate launch and uninterrupted continuation of all measures necessary for effective emergency response and subsequent remediation.

Member States shall facilitate the deployment of sustainable financial instruments and other arrangements to assist applicants for licences in demonstrating their financial capacity pursuant to the first subparagraph.

Member States shall, as a minimum, establish procedures for ensuring prompt and adequate handling of compensation claims including in respect of compensation payments for trans-boundary incidents.

The Member States shall require the licensee to maintain sufficient capacity to meet their financial obligations resulting from liabilities for offshore oil and gas operations.

4. The licensing authority or the licensee shall appoint the operator. Where the operator is to be appointed by the licensee, the licensing authority shall be notified of the appointment in advance. In such cases, the licensing authority, if necessary in consultation with the competent authority, may object to the appointment of the operator. Where such an objection is raised, the Member States shall require the licensee to appoint a suitable alternative operator or assume the responsibilities of the operator under this Directive.

5. The licensing procedures for offshore oil and gas operations relating to a given licensed area shall be organised in such a way that information collected as a result of exploration can be considered by the Member State prior to production commencing.

6. When assessing the technical and financial capabilities of an applicant for a licence, special attention shall be paid to any environmentally sensitive marine and coastal environments, in particular ecosystems which play an important role in mitigation and adaptation to climate change, such as salt marshes and sea grass beds, and marine protected areas, such as special areas of conservation pursuant to the Council Directive 92/43/EEC of 21 May 1992 on the conservation of natural habitats and of wild fauna and flora, special protection areas pursuant to the Directive 2009/147/EC of the European Parliament and of the Council of 30 November 2009 on the conservation of wild birds, and marine protected areas as agreed by the Union or Member States concerned within the framework of any international or regional agreements to which they are a party.

[12.906]

Article 5 Public participation relating to the effects of planned offshore oil and gas exploration operations on the environment

1. The drilling of an exploration well from a non-production installation shall not be commenced unless the relevant authorities of the Member State have previously ensured that early and effective public participation on the possible effects of planned offshore oil and gas operations on the environment pursuant to other Union legal acts, in particular Directive 2001/42/EC or 2011/92/EU as appropriate, has been undertaken.

2. Where public participation has not been undertaken pursuant to paragraph 1, Member States shall ensure that the following arrangements are made:

(a) the public is informed, whether by public notices or other appropriate means such as electronic media, where it is planned to allow exploration operations;

(b) the public concerned is identified, including the public affected or likely to be affected by, or having an interest in, the decision to allow exploration operations, including relevant non-governmental organisations such as those promoting environmental protection, and other relevant organisations;

(c) relevant information about such planned operations is made available to the public including, inter alia, information about the right to participate in decision-making, and to whom comments or questions may be submitted;

(d) the public is entitled to express comments and opinions at a time when all options are open before decisions to allow exploration are taken;

(e) when decisions under point (d) are taken, due account is taken of the results of the public participation; and

(f) the Member State in question promptly informs the public, after examining the comments and opinions expressed by them, about the decisions taken and the reasons therefor and considerations upon which those decisions are based, including information about the public participation process.

Reasonable time-frames shall be provided allowing sufficient time for each of the different stages of public participation.

3. This Article does not apply in respect of areas licensed before 18 July 2013.

[12.907]

Article 6 Offshore oil and gas operations within licensed areas

1. Member States shall ensure that production installations and connected infrastructure are operated only in licensed areas and only by operators appointed for that purpose pursuant to Article 4(4).

2. Member States shall require the licensee to ensure that the operator has the capacity to meet the requirements for specific operations within the framework of the licence.

3. Throughout all offshore oil and gas operations, Member States shall require the licensee to take all reasonable steps to ensure that the operator meets the requirements, carries out its functions and discharges its duties under this Directive.

4. Where the competent authority determines that the operator no longer has the capacity to meet the relevant requirements under this Directive, the licensing authority shall be informed. The licensing authority shall then notify the licensee thereof and the licensee shall assume responsibility for the discharge of the duties concerned and shall, without delay, propose a replacement operator to the licensing authority.

5. Member States shall ensure that operations relating to production and non-production installations are not commenced or continued until the report on major hazards has been accepted by the competent authority in accordance with this Directive.

6. Member States shall ensure that well operations or combined operations are not commenced or continued until the report on major hazards for the installations involved has been accepted in accordance with this Directive. Furthermore, such operations shall not be commenced or continued where a notification of well operations or a notification of combined operations has not been submitted pursuant to point (h) or (i) of Article 11(1) respectively to the competent authority or where the competent authority expresses objections to the content of a notification.

7. Member States shall ensure that a safety zone is established around an installation and that vessels are prohibited from entering or remaining in that safety zone. However, that prohibition shall not apply to a vessel entering or remaining in the safety zone:

(a) in connection with the laying, inspection, testing, repair, maintenance, alteration, renewal or removal of any submarine cable or pipeline in or near that safety zone;
(b) to provide services for or to transport persons or goods to or from any installation in that safety zone;
(c) to inspect any installation or connected infrastructure in that safety zone under the authority of the Member State;
(d) in connection with saving or attempting to save life or property;
(e) owing to stress of weather;
(f) when in distress; or
(g) if there is consent from the operator, owner or the Member State in which the safety zone is located.

8. Member States shall establish a mechanism for effective participation in tripartite consultation between the competent authority, operators and owners, and worker representatives in the formulation of standards and policies dealing with major accident prevention.

[12.908]

Article 7 Liability for environmental damage

Without prejudice to the existing scope of liability relating to the prevention and remediation of environmental damage pursuant to Directive 2004/35/EC, Member States shall ensure that the licensee is financially liable for the prevention and remediation of environmental damage as defined in that Directive, caused by offshore oil and gas operations carried out by, or on behalf of, the licensee or the operator.

[12.909]

Article 8 Appointment of the competent authority

1. Member States shall appoint a competent authority responsible for the following regulatory functions:

(a) assessing and accepting reports on major hazards, assessing design notifications, and assessing notifications of well operations or combined operations, and other similar documents that are submitted to it;
(b) overseeing compliance by operators and owners with this Directive, including inspections, investigations and enforcement actions;
(c) advising other authorities or bodies, including the licensing authority;
(d) making annual plans pursuant to Article 21;

(e) producing reports;

(f) cooperating with the competent authorities or contact points pursuant to Article 27.

2. Member States shall at all times ensure the independence and objectivity of the competent authority in carrying out its regulatory functions and particularly in respect of points (a), (b) and (c) of paragraph 1. Accordingly, conflicts of interest shall be prevented between, on the one hand, the regulatory functions of the competent authority and, on the other hand, the regulatory functions relating to the economic development of the offshore natural resources and licensing of offshore oil and gas operations within the Member State and the collection and management of revenues from those operations.

3. In order to achieve the objectives set out in paragraph 2, Member States shall require the regulatory functions of the competent authority to be carried out within an authority that is independent of any of the functions of the Member State relating to the economic development of the offshore natural resources and licensing of offshore oil and gas operations within the Member State and the collection and management of revenues from those operations. However, where the total number of normally attended installations is below six, the Member State concerned may decide not to apply the first subparagraph. Such a decision shall be without prejudice to its obligations under paragraph 2.

4. Member States shall make available to the public a description of how the competent authority is organised, including why they have established the competent authority in such a way, and how they have ensured that the regulatory functions set out in paragraph 1 are carried out and that the obligations set out in paragraph 2 are complied with.

5. Member States shall ensure that the competent authority has adequate human and financial resources to carry out its duties under this Directive. Those resources shall be commensurate with the extent of offshore oil and gas operations of the Member States.

6. Member States may enter into formal agreements with appropriate Union agencies or other suitable bodies where available for the provision of specialist expertise to support the competent authority in carrying out its regulatory functions. For the purposes of this paragraph a body shall not be deemed suitable where its objectivity may be compromised by conflicts of interest.

7. Member States may establish mechanisms according to which the financial costs to the competent authority in carrying out its duties under this Directive may be recovered from licensees, operators or owners.

8. Where the competent authority is comprised of more than one body, Member States shall make every effort to avoid duplication of regulatory functions between the bodies. Member States may designate one of the constituent bodies as the lead body with responsibility for the coordination of the regulatory functions under this Directive and for reporting to the Commission.

9. Member States shall review the activities of the competent authority and shall take any necessary measures to improve its effectiveness in carrying out the regulatory functions set out in paragraph 1.

[12.910]

Article 9 Functioning of the competent authority

Member States shall ensure that the competent authority:

(a) acts independently of policies, regulatory decisions or other considerations unrelated to its duties under this Directive;

(b) makes clear the extent of its responsibilities and the responsibilities of the operator and the owner for the control of major accident risks under this Directive;

(c) establishes a policy, process and procedures for thorough assessment of reports on major hazards and notifications submitted pursuant to Article 11 as well as for overseeing compliance with this Directive within the jurisdiction of the Member State, including inspection, investigation and enforcement actions;

(d) makes the policy, process and procedures pursuant to point (c) available to operators and owners and makes summaries thereof available to the public;

(e) where necessary, prepares and implements coordinated or joint procedures with other authorities in the Member State to undertake the duties under this Directive; and

(f) bases its policy, organisation and operational procedures on the principles set out in Annex III.

[12.911]

Article 10 Tasks of the European Maritime Safety Agency

1. The European Maritime Safety Agency (EMSA, hereinafter 'Agency') shall provide the Member States and Commission with technical and scientific assistance in accordance with its mandate under Regulation (EC) No 1406/2002.

2. Within the framework of its mandate, the Agency shall:

(a) assist the Commission and the affected Member State, on its request, in detecting and monitoring the extent of an oil or gas spill;

(b) assist Member States, at their request, with the preparation and execution of external emergency response plans, especially when there are transboundary impacts within and beyond offshore waters of Member States;

(c) on the basis of the Member States' external and internal emergency response plans, develop with Member States and operators a catalogue of emergency equipment and services available.

3. The Agency may, if requested:

(a) assist the Commission in assessing the external emergency response plans of Member States to check whether the plans are in conformity with this Directive;

(b) review exercises that focus on testing transboundary and Union emergency mechanisms.

CHAPTER III

PREPARING AND CARRYING OUT OFFSHORE OIL AND GAS OPERATIONS

[12.912]

Article 11 Documents to be submitted for carrying out offshore oil and gas operations

1. Member States shall ensure that the operator or the owner submit to the competent authority the following documents:

(a) the corporate major accident prevention policy or an adequate description thereof, in accordance with Article 19(1) and (5);

(b) the safety and environmental management system applicable to the installation, or an adequate description thereof, in accordance with Article 19(3) and (5);

(c) in the case of a planned production installation, a design notification in accordance with the requirements of Annex I, Part 1;

(d) a description of the scheme of independent verification in accordance with Article 17;

(e) a report on major hazards, in accordance with Articles 12 and 13;

(f) in the event of a material change or dismantling of an installation, an amended report on major hazards in accordance with Articles 12 and 13;

(g) the internal emergency response plan or an adequate description thereof, in accordance with Articles 14 and 28;

(h) in the case of a well operation, a notification of that well operation and information on that well operation in accordance with Article 15;

(i) in the case of a combined operation, a notification of combined operations in accordance with Article 16;

(j) in the case of an existing production installation which is to be moved to a new production location where it is to be operated, a relocation notification in accordance with Annex I, Part 1;

(k) any other relevant document requested by the competent authority.

2. The documents to be submitted under points (a), (b), (d) and (g) of paragraph 1 shall be included with the report on major hazards required under point (e) of paragraph 1. The corporate major accident prevention policy of an operator of a well shall, where not previously submitted, be included with the notification of well operations to be submitted under point (h) of paragraph 1.

3. The design notification required pursuant to point (c) of paragraph 1 shall be submitted to the competent authority by a deadline set by the competent authority before the intended submission of the report on major hazards for the planned operation. The competent authority shall respond to the design notification with comments to be taken into account in the report on major hazards.

4. Where an existing production installation is to enter or leave the offshore waters of a Member State, the operator shall notify the competent authority in writing prior to the date on which the production installation is due to enter or leave the offshore waters of the Member State.

5. The relocation notification required pursuant to point (j) of paragraph 1 shall be submitted to the competent authority at a stage that is sufficiently early in the proposed development to enable the operator to take into account any matters raised by the competent authority during the preparation of the report on major hazards.

6. Where there is a material change affecting the design notification or the relocation notification prior to the submission of the report on major hazards, the competent authority shall be notified of that change as soon as possible.

7. The report on major hazards required pursuant to point (e) of paragraph 1 shall be submitted to the competent authority by a deadline set by the competent authority that is before the planned commencement of the operations.

[12.913]

Article 12 Report on major hazards for a production installation

1. Member States shall ensure that the operator prepares a report on major hazards for a production installation, to be submitted pursuant to point (e) of Article 11(1). That report shall contain the information specified in Annex I, Parts 2 and 5 and shall be updated whenever appropriate or when so required by the competent authority.

2. Member States shall ensure that workers' representatives are consulted at the relevant stages in the preparation of the report on major hazards for a production installation, and that evidence is provided to this effect in accordance with Annex I, Part 2, point 3.

3. The report on major hazards for a production installation may be prepared in relation to a group of installations, subject to the agreement of the competent authority.

4. Where further information is necessary before a report on major hazards can be accepted, Member States shall ensure that the operator provides, at the request of the competent authority, such information and makes any necessary changes to the submitted report on major hazards.

5. Where modifications are to be made to the production installation that entail a material change, or it is intended to dismantle a fixed production installation, the operator shall prepare an amended report on major hazards, to be submitted pursuant to point (f) of Article 11(1) by a deadline specified by the competent authority, in accordance with Annex I, Part 6.

6. Member States shall ensure that the planned modifications are not brought into use nor any dismantlement commenced until the competent authority has accepted the amended report on major hazards for the production installation.

7. The report on major hazards for a production installation shall be subject to a thorough periodic review by the operator at least every five years or earlier when so required by the competent authority. The results of the review shall be notified to the competent authority.

[12.914]

Article 13 Report on major hazards for a non-production installation

1. Member States shall ensure that the owner prepares a report on major hazards for a non-production installation, to be submitted pursuant to point (e) of Article 11(1). That report shall contain the information specified in Annex I, Parts 3 and 5 and shall be updated whenever appropriate or when so required by the competent authority.

2. Member States shall ensure that workers' representatives are consulted at the relevant stages in the preparation of the report on major hazards for a non-production installation, and that evidence is provided to this effect in accordance with Annex I, Part 3, point 2.

3. Where further information is necessary before a report on major hazards for a non-production installation can be accepted, Member States shall require the owner to provide, at the request of the competent authority, such information and to make any necessary changes to the submitted report on major hazards.

4. Where modifications are to be made to the non-production installation that entail a material change, or it is intended to dismantle a fixed non-production installation, the owner shall prepare an amended report on major hazards, to be submitted pursuant to point (f) of Article 11(1) by a deadline specified by the competent authority, in accordance with Annex I, Part 6, points 1, 2 and 3.

5. For a fixed non-production installation, Member States shall ensure that the planned modifications are not brought into use nor any dismantlement commenced until the competent authority has accepted the amended report on major hazards for the fixed non-production installation.

6. For a mobile non-production installation, Member States shall ensure that the planned modifications are not brought into use until the competent authority has accepted the amended report on major hazards for the mobile non-production installation.

7. The report on major hazards for a non-production installation shall be subject to a thorough periodic review by the owner at least every five years or earlier when so required by the competent authority. The results of the review shall be notified to the competent authority.

[12.915]

Article 14 Internal emergency response plans

1. Member States shall ensure that operators or owners, as appropriate, prepare internal emergency response plans to be submitted pursuant to point (g) of Article

11(1). The plans shall be prepared in accordance with Article 28 taking into account the major accident risk assessment undertaken during preparation of the most recent report on major hazards. The plan shall include an analysis of the oil spill response effectiveness.

2. In the event that a mobile non-production installation is to be used for carrying out well operations, the internal emergency response plan for the installation shall take into account the risk assessment undertaken during the preparation of the notification of well operations to be submitted pursuant to point (h) of Article 11(1). Where the internal emergency response plan has to be amended due to the particular nature or location of the well, Member States shall ensure that the operator of the well submits the amended internal emergency response plan, or an adequate description thereof, to the competent authority to complement the relevant notification of well operations.

3. In the event that a non-production installation is to be used for carrying out combined operations, the internal emergency response plan shall be amended to cover the combined operations and shall be submitted to the competent authority to complement the relevant notification of the combined operations.

[12.916]

Article 15 Notification of and information on well operations

1. Member States shall ensure that the operator of a well prepares the notification to be submitted pursuant to point (h) of Article 11(1) to the competent authority. It shall be submitted by a deadline set by the competent authority that is before the commencement of the well operation. That notification of well operations shall contain details of the design of the well and the proposed well operations in accordance with Annex I, Part 4. This shall include an analysis of the oil spill response effectiveness.

2. The competent authority shall consider the notification and, if deemed necessary, take appropriate action before the well operations are commenced, which may include prohibiting the operation from being commenced.

3. Member States shall ensure that the operator of the well involves the independent verifier in planning and preparation of a material change to the submitted notification of well operations pursuant to point (b) of Article 17(4) and that it immediately informs the competent authority of any material change to the submitted notification of well operations. The competent authority shall consider those changes and, if deemed necessary, take appropriate action.

4. Member States shall ensure that the operator of the well submits reports of well operations to the competent authority in accordance with the requirements of Annex II. The reports shall be submitted at weekly intervals, starting on the day of commencement of the well operations, or at intervals specified by the competent authority.

[12.917]

Article 16 Notification of combined operations

1. Member States shall ensure that operators and owners involved in a combined operation jointly prepare the notification to be submitted pursuant to point (i) of Article 11(1). The notification shall contain the information specified in Annex I, Part 7. Member States shall ensure that one of the operators concerned submits the notification of combined operations to the competent authority. The notification shall be submitted by a deadline set by the competent authority before combined operations are commenced.

2. The competent authority shall consider the notification and, if deemed necessary, take appropriate action before the combined operations are commenced, which may include prohibiting the operation from being commenced.

3. Member States shall ensure that the operator who submitted the notification informs, without delay, the competent authority of any material change to the submitted notification. The competent authority shall consider those changes and, if deemed necessary, take appropriate action.

[12.918]

Article 17 Independent verification

1. Member States shall ensure that operators and owners establish schemes for independent verification and that they prepare a description of such schemes, to be submitted pursuant to point (d) of Article 11(1) and included within the safety and environmental management system submitted pursuant to point (b) of Article 11(1). The description shall contain the information specified in Annex I, Part 5.

2. The results of the independent verification shall be without prejudice to the responsibility of the operator or the owner for the correct and safe functioning of the equipment and systems under verification.

3. The selection of the independent verifier and the design of schemes for independent verification shall meet the criteria of Annex V.

4. The schemes for independent verification shall be established:

(a) in respect of installations, to give independent assurance that the safety and environmental critical elements identified in the risk assessment for the installation, as described in the report on major hazards, are suitable and that the schedule of examination and testing of the safety and environmental critical elements is suitable, up-to-date and operating as intended;

(b) in respect of notifications of well operations, to give independent assurance that the well design and well control measures are suitable for the anticipated well conditions at all times.

5. Member States shall ensure that operators and owners respond to and take appropriate action based on the advice of the independent verifier.

6. Member States shall require operators and owners to ensure that advice received from the independent verifier pursuant to point (a) of paragraph 4 and records of action taken on the basis of such advice are made available to the competent authority and retained by the operator or the owner for a period of six months after completion of the offshore oil and gas operations to which they relate.

7. Member States shall require operators of wells to ensure that the findings and comments of the independent verifier pursuant to point (b) of paragraph 4 of this Article and their actions in response to those findings and comments are presented in the notification of well operations prepared in accordance with Article 15.

8. For a production installation, the verification scheme shall be in place prior to the completion of the design. For a non-production installation, the scheme shall be in place prior to the commencement of operations in the offshore waters of Member States.

[12.919]

Article 18 Power of the competent authority in relation to operations on installations

Member States shall ensure that the competent authority:

(a) prohibits the operation or commencement of operations on any installation or any connected infrastructure where the measures proposed in the report on major hazards for the prevention or limiting the consequences of major accidents or notifications of well operations or combined operations submitted pursuant to points (h) or (i) of Article 11(1) respectively are considered insufficient to fulfil the requirements set out in this Directive;

(b) in exceptional situations and where it considers that safety and environmental protection are not compromised, shortens the time interval required between the submission of the report on major hazards or other documents to be submitted pursuant to Article 11 and the commencement of operations;

(c) requires the operator to take such proportionate measures as the competent authority considers necessary to ensure compliance with Article 3(1);

(d) where Article 6(4) applies, takes adequate measures to ensure the continuing safety of operations;

(e) is empowered to require improvements and, if necessary, prohibit the continued operation of any installation or any part thereof or any connected infrastructure where it is shown by the outcome of an inspection, a determination pursuant to Article 6(4), a periodic review of the report on major hazards submitted pursuant to point (e) of Article 11(1) or by changes to notifications submitted pursuant to Article 11, that the requirements of this Directive are not being fulfilled or there are reasonable concerns about the safety of offshore oil and gas operations or installations.

CHAPTER IV

PREVENTION POLICY

[12.920]

Article 19 Major accident prevention by operators and owners

1. Member States shall require operators and owners to prepare a document setting out their corporate major accident prevention policy which is to be submitted pursuant to point (a) of Article 11(1), and to ensure that it is implemented throughout their offshore oil and gas operations, including by setting up appropriate monitoring arrangements to assure effectiveness of the policy. The document shall contain the information specified in Annex I, Part 8.

2. The corporate major accident prevention policy shall take account of the operators' primary responsibility for, inter alia, the control of risks of a major accident that are a result of its operations and for continuously improving control of those risks so as to ensure a high level of protection at all times.

3. Member States shall ensure that operators and owners prepare a document setting out their safety and environmental management system which is to be submitted pursuant to point (b) of Article 11(1). That document shall include a description of the:

(a) organisational arrangements for control of major hazards;

(b) arrangements for preparing and submitting reports on major hazards, and other documents as appropriate, pursuant to this Directive; and

(c) schemes for independent verification established pursuant to Article 17.

4. Member States shall create opportunities for operators and owners to contribute to mechanisms for effective tripartite consultation established pursuant to Article 6(8). When appropriate, an operator's and owner's commitment to such mechanisms may be outlined in the corporate major accident prevention policy.

5. The corporate major accident prevention policy and the safety and environmental management systems shall be prepared in accordance with Annex I, Parts 8 and 9 and Annex IV. The following conditions shall apply:

(a) the corporate major accident prevention policy shall be in writing and shall establish the overall aims and arrangements for controlling the risk of a major accident, and how those aims are to be achieved and arrangements put into effect at corporate level;

(b) the safety and environmental management system shall be integrated within the overall management system of the operator or owner and shall include organisational structure, responsibilities, practices, procedures, processes and resources for determining and implementing the corporate major accident prevention policy.

6. Member States shall ensure that operators and owners prepare and maintain a complete inventory of emergency response equipment pertinent to their offshore oil and gas operation.

7. Member States shall ensure that operators and, owners in consultation with the competent authority and making use of the exchanges of knowledge, information and experience provided for in Article 27(1), prepare and revise standards and guidance on best practice in relation to the control of major hazards throughout the design and operational lifecycle of offshore oil and gas operations, and that as a minimum they follow the outline in Annex VI.

8. Member States shall require operators and owners to ensure that their corporate major accident prevention policy document referred to in paragraph 1 also covers their production and non-production installations outside of the Union.

9. Where an activity carried out by an operator or an owner poses an immediate danger to human health or significantly increases the risk of a major accident, Member States shall ensure that the operator or the owner takes suitable measures which may include, if deemed necessary, suspending the relevant activity until the danger or risk is adequately controlled. Member States shall ensure that where such measures are taken, the operator or the owner notifies the competent authority accordingly without delay and no later than 24 hours after taking those measures.

10. Member States shall ensure that, where appropriate, operators and owners take suitable measures to use suitable technical means or procedures in order to promote the reliability of the collection and recording of relevant data and to prevent possible manipulation thereof.

[12.921]

Article 20 Offshore oil and gas operations conducted outside the Union

1. Member States shall require companies registered in their territory and conducting, themselves or through subsidiaries, offshore oil and gas operations outside the Union as licence holders or operators to report to them, on request, the circumstances of any major accident in which they have been involved.

2. In the request for a report pursuant to paragraph 1 of this Article, the relevant Member State shall specify the details of the information required. Such reports shall be exchanged in accordance with Article 27(1). Member States which have neither a competent authority nor a contact point shall submit the reports received to the Commission.

[12.922]

Article 21 Securing compliance with the regulatory framework for major accident prevention

1. Member States shall ensure that operators and owners comply with the measures established in the report on major hazards and in the plans referred to in the notification of well operations and notification of combined operations, submitted pursuant to points (e), (h) and (i) of Article 11(1) respectively.

2. Member States shall ensure that operators and owners provide the competent authority, or any other persons acting under the direction of the competent authority, with transport to or from an installation or vessel associated with oil and gas operations, including the conveyance of their equipment, at any reasonable time, and

with accommodation, meals and other subsistence in connection with the visits to the installations, for the purpose of facilitating competent authority oversight, including inspections, investigations and enforcement of compliance with this Directive.

3. Member States shall ensure that the competent authority develops annual plans for effective oversight, including inspections, of major hazards based on risk management and with particular regard to compliance with the report on major hazards and other documents submitted pursuant to Article 11. The effectiveness of the plans shall be regularly reviewed and the competent authority shall take any necessary measures to improve them.

[12.923]

Article 22 Confidential reporting of safety concerns

1. Member States shall ensure that the competent authority establishes mechanisms:

(a) for confidential reporting of safety and environmental concerns relating to offshore oil and gas operations from any source; and

(b) for investigation of such reports while maintaining the anonymity of the individuals concerned.

2. Member States shall require operators and owners to communicate details of the national arrangements for the mechanisms referred to in paragraph 1 to their employees and contractors connected with the operation and their employees, and to ensure that reference to confidential reporting is included in relevant training and notices.

CHAPTER V

TRANSPARENCY AND SHARING OF INFORMATION

[12.924]

Article 23 Sharing of information

1. Member States shall ensure that operators and owners provide the competent authority, as a minimum, with the information described in Annex IX.

2. The Commission shall by means of an implementing act determine a common data reporting format and the details of information to be shared. That implementing act shall be adopted in accordance with the advisory procedure referred to in Article 37(2).

[12.925]

Article 24 Transparency

1. Member States shall make the information referred to in Annex IX publicly available.

2. The Commission shall by means of an implementing act determine a common publication format that enables easy cross-border comparison of data. That implementing act shall be adopted in accordance with the advisory procedure referred to in Article 37(2). The common publication format shall allow for a reliable comparison of national practices under this Article and Article 25.

[12.926]

Article 25 Reporting on safety and environmental impact

1. Member States shall submit an annual report to the Commission containing the information specified in Annex IX, point 3.

2. Member States shall designate an authority to be responsible for exchanging information pursuant to Article 23 and for publication of information pursuant to Article 24.

3. The Commission shall publish an annual report based on the information reported to it by Member States pursuant to paragraph 1.

[12.927]

Article 26 Investigation following a major accident

1. Member States shall initiate thorough investigations of major accidents occurring in their jurisdiction.

2. A summary of the findings pursuant to paragraph 1 shall be made available to the Commission either at the conclusion of the investigation or at the conclusion of legal proceedings as appropriate. The Member States shall make a non-confidential version of the findings publicly available.

3. Member States shall ensure that, following the investigations pursuant to paragraph 1, the competent authority implements any recommendations of the investigation that are within its powers to act.

CHAPTER VI

COOPERATION

[12.928]

Article 27 Cooperation between Member States

1. Each Member State shall ensure that its competent authority regularly exchanges knowledge, information and experience with other competent authorities, inter alia, through the European Union Offshore Oil and Gas Authorities Group (EUOAG), and that it engages in consultations on the application of relevant national and Union law with the industry, other stakeholders and the Commission. For Member States without offshore oil and gas operations under their jurisdiction, the information referred to in the first subparagraph shall be received by the contact points appointed pursuant to Article 32(1).

2. Knowledge, information and experience exchanged pursuant to paragraph 1 shall concern, in particular, the functioning of the measures for risk management, major accident prevention, verification of compliance and emergency response relating to offshore oil and gas operations within the Union, as well as outside of the Union where appropriate.

3. Each Member State shall ensure that its competent authority participates in establishing clear joint priorities for the preparation and updating of standards and guidance in order to identify and facilitate the implementation and consistent application of best practices in offshore oil and gas operations.

4. By 19 July 2014, the Commission shall present to the Member States a report on the adequacy of national expert resources for complying with the regulatory functions pursuant to this Directive which, if necessary, shall include proposals for ensuring all Member States have access to adequate expert resources.

5. By 19 July 2016, the Member States shall notify the Commission of the national measures they have in place regarding access to knowledge, assets and expert resources, including formal agreements pursuant to Article 8(6).

CHAPTER VII

EMERGENCY PREPAREDNESS AND RESPONSE

[12.929]

Article 28 Requirements for internal emergency response plans

1. Member States shall ensure that the internal emergency response plans to be prepared by the operator or the owner in accordance with Article 14 and submitted pursuant to point (g) of Article 11(1) are:

(a) put into action without delay to respond to any major accident or a situation where there is an immediate risk of a major accident; and

(b) consistent with the external emergency response plan referred to in Article 29.

2. Member States shall ensure that the operator and the owner maintain equipment and expertise relevant to the internal emergency response plan in order for that equipment and expertise to be available at all times and to be made available as necessary to the authorities responsible for the execution of the external emergency response plan of the Member State where the internal emergency response plan applies.

3. The internal emergency response plan shall be prepared in accordance with Annex I, Part 10, and updated as a consequence of any material change to the report on major hazards or notifications submitted pursuant to Article 11 Any such updates shall be submitted to the competent authority pursuant to point (g) of Article 11(1) and notified to the relevant authority or authorities responsible for preparing the external emergency response plans for the area concerned.

4. The internal emergency response plan shall be integrated with other measures relating to protection and rescue of personnel from the stricken installation so as to secure a good prospect of personal safety and survival.

[12.930]

Article 29 External emergency response plans and emergency preparedness

1. Member States shall prepare external emergency response plans covering all offshore oil and gas installations or connected infrastructure and potentially affected areas within their jurisdiction. Member States shall specify the role and financial obligation of licensees and operators in the external emergency response plans.

2. External emergency response plans shall be prepared by the Member State in cooperation with relevant operators and owners and, as appropriate, licensees and the competent authority, and shall take into account the most up to date version of the internal emergency response plans of the existing or planned installations or connected infrastructure in the area covered by the external emergency response plan.

3. External emergency response plans shall be prepared in accordance with Annex VII, and shall be made available to the Commission, other potentially affected Member States and the public. When making available their external emergency response plans, the Member States shall ensure that disclosed information does not pose risks to the safety and security of offshore oil and gas installations and their operation and does not harm the economic interests of the Member States or the personal safety and well-being of officials of Member States.

4. Member States shall take suitable measures to achieve a high level of compatibility and interoperability of response equipment and expertise between all Member States in a geographical region, and further afield where appropriate. Member States shall encourage industry to develop response equipment and contracted services that are compatible and interoperable throughout the geographical region.

5. Member States shall keep records of emergency response equipment and services in accordance with Annex VIII, point 1. Those records shall be available to the other potentially affected Member States and the Commission and, on a reciprocal basis, to neighbouring third countries.

6. Member States shall ensure that operators and owners regularly test their preparedness to respond effectively to major accidents in close cooperation with the relevant authorities of the Member States.

7. Member States shall ensure that competent authorities or, where appropriate, contact points develop cooperation scenarios for emergencies. Such scenarios shall be regularly assessed and updated as necessary.

[12.931]

Article 30 Emergency response

1. Member States shall ensure that the operator or, if appropriate, the owner notifies without delay the relevant authorities of a major accident or of a situation where there is an immediate risk of a major accident. That notification shall describe the circumstances, including, where possible, the origin, the potential impacts on the environment and the potential major consequences.

2. Member States shall ensure that in the event of a major accident, the operator or the owner takes all suitable measures to prevent its escalation and to limit its consequences. The relevant authorities of the Member States may assist the operator or owner, including with the supply of additional resources.

3. In the course of the emergency response, the Member State shall collect the information necessary for thorough investigation pursuant to Article 26(1).

CHAPTER VIII

TRANSBOUNDARY EFFECTS

[12.932]

Article 31 Transboundary emergency preparedness and response of Member States with offshore oil and gas operations under their jurisdiction

1. Where a Member State considers that a major hazard relating to offshore oil and gas operations that are to take place under its jurisdiction is likely to have significant effects on the environment in another Member State, it shall, prior to the commencement of operations, forward the relevant information to the potentially affected Member State and shall endeavour, jointly with that Member State, to adopt measures to prevent damage.

Member States that consider themselves to be potentially affected may request the Member State in whose jurisdiction the offshore oil and gas operation is to take place, to forward all relevant information to them.

Those Member States may jointly assess the effectiveness of the measures, without prejudice to the regulatory functions of the competent authority with jurisdiction for the operation concerned under points (a), (b) and (c) of Article 8(1).

2. The major hazards identified pursuant to paragraph 1 shall be taken into account in internal and external emergency response plans to facilitate joint effective response to a major accident.

3. Where there is a risk of the foreseeable transboundary effects of major accidents affecting third countries, Member States shall, on a reciprocal basis, make information available to the third countries.

4. Member States shall coordinate between themselves measures relating to areas outside of the Union in order to prevent potential negative effects of offshore oil and gas operations.

5. Member States shall regularly test their preparedness to respond effectively to major accidents in cooperation with potentially affected Member States, relevant Union agencies and, on a reciprocal basis, potentially affected third countries. The Commission may contribute to exercises focused on testing transboundary emergency mechanisms.

6. In the event of a major accident, or of an imminent threat thereof, which has or is capable of having transboundary effects, the Member State under whose jurisdiction the situation occurs shall, without delay, notify the Commission and those Member States or third countries which may be affected by the situation and shall continuously provide information relevant for an effective emergency response.

[12.933]

Article 32 Transboundary emergency preparedness and response of Member States without offshore oil and gas operations under their jurisdiction

1. Member States without offshore oil and gas operations under their jurisdiction shall appoint a contact point in order to exchange information with relevant adjacent Member States.

2. Member States without offshore oil and gas operations under their jurisdiction shall apply Article 29(4) and (7) so as to ensure that adequate response capacity is in place in the event that they are affected by a major accident.

3. Member States without offshore oil and gas operations under their jurisdiction shall coordinate their national contingency planning in the marine environment with other relevant Member States to the extent necessary to ensure the most effective response to a major accident.

4. Where a Member State without offshore oil and gas operations under its jurisdiction is affected by a major accident, it shall:

(a) take all suitable measures, in line with the national contingency planning referred to in paragraph 3;

(b) ensure that any information which is under its control and available within its jurisdiction and which may be relevant for a full investigation of the major accident is provided or made accessible on request to the Member State conducting the investigation pursuant to Article 26.

[12.934]

Article 33 Coordinated approach towards the safety of offshore oil and gas operations at international level

1. The Commission shall, in close cooperation with the Member State and without prejudice to relevant international agreements, promote cooperation with third countries that undertake offshore oil and gas operations in the same marine regions as Member States.

2. The Commission shall facilitate the exchange of information between Member States with offshore oil and gas operations and adjacent third countries with similar operations in order to promote preventive measures and regional emergency response plans.

3. The Commission shall promote high safety standards for offshore oil and gas operations at international level in relevant global and regional fora, including those relating to Arctic waters.

CHAPTER IX

FINAL PROVISIONS

[12.935]

Article 34 Penalties

Member States shall lay down the rules on penalties applicable to infringements of the national provisions adopted pursuant to this Directive and shall take all measures necessary to ensure that they are implemented. The penalties provided for shall be effective, proportionate and dissuasive. Member States shall notify those provisions to the Commission by 19 July 2015 and shall notify it without delay of any subsequent amendment affecting them.

[12.936]

Article 35 Delegated powers of the Commission

The Commission shall be empowered to adopt delegated acts in accordance with Article 36 in order to adapt Annexes I, II, VI and VII to include additional information which may become necessary in light of technical progress. Such adaptations shall not result in substantial changes in the obligations laid down in this Directive.

[12.937]

Article 36 Exercise of the delegation

1. The power to adopt delegated acts is conferred on the Commission subject to the conditions laid down in this Article.

2. The power to adopt delegated acts referred to in Article 35 shall be conferred on the Commission for a period of five years from 18 July 2013. The Commission shall draw up a report in respect of the delegation of power no later than nine months before the end of the five-year period. The delegation of power shall be tacitly extended for periods of an identical duration, unless the European Parliament or the Council opposes such extension not later than four months before the end of each period.

3. The delegation of power referred to in Article 35 may be revoked at any time by the European Parliament or by the Council. A decision to revoke shall put an end to the delegation of the power specified in that decision. It shall take effect the day following the publication of the decision in the *Official Journal of the European Union* or at a later date specified therein. It shall not affect the validity of any delegated acts already in force.

4. As soon as it adopts a delegated act, the Commission shall notify it simultaneously to the European Parliament and to the Council.

5. A delegated act adopted pursuant to Article 35 shall enter into force only if no objection has been expressed either by the European Parliament or the Council within a period of two months of notification of that act to the European Parliament and the Council or if, before the expiry of that period, the European Parliament and the Council have both informed the Commission that they will not object. That period shall be extended by two months at the initiative of the European Parliament or of the Council.

[12.938]

Article 37 Committee procedure

1. The Commission shall be assisted by a committee. That committee shall be a committee within the meaning of Regulation (EU) No 182/2011.

2. Where reference is made to this paragraph, Article 4 of Regulation (EU) No 182/2011 shall apply.

[12.939]

Article 38 Amendment to Directive 2004/35/EC

1. In Article 2(1) of Directive 2004/35/EC, point (b) shall be replaced by the following:

'(b) "water damage", which is any damage that significantly adversely affects:

(i) the ecological, chemical or quantitative status or the ecological potential, as defined in Directive 2000/60/EC, of the waters concerned, with the exception of adverse effects where Article 4(7) of that Directive applies; or

(ii) the environmental status of the marine waters concerned, as defined in Directive 2008/56/EC, in so far as particular aspects of the environmental status of the marine environment are not already addressed through Directive 2000/60/EC;'.

2. Member States shall bring into force the laws, regulations and administrative provisions necessary to comply with paragraph 1 by 19 July 2015. They shall forthwith inform the Commission thereof.

[12.940]

Article 39 Reports to the European Parliament and to the Council

1. The Commission shall, by 31 December 2014, submit to the European Parliament and to the Council a report on the availability of financial security instruments, and on the handling of compensation claims, where appropriate, accompanied by proposals.

2. The Commission shall, by 19 July 2015, submit to the European Parliament and to the Council a report on its assessment of the effectiveness of the liability regimes in the Union in respect of the damage caused by offshore oil and gas operations. That report shall include an assessment of the appropriateness of broadening liability provisions. The report shall be accompanied, where appropriate, by proposals.

3. The Commission shall examine the appropriateness of bringing certain conduct leading to a major accident within the scope of Directive 2008/99/EC of the European Parliament and of the Council of 19 November 2008 on the protection of the environment through criminal law. The Commission shall, by 31 December 2014, submit a report on its findings to the European Parliament and the Council, accompanied, where appropriate, by legislative proposals, subject to appropriate information being made available by Member States.

[12.941]

Article 40 Report and review

1. No later than 19 July 2019, the Commission shall, taking due account of the efforts and experiences of competent authorities, assess the experience of implementing this Directive.

2. The Commission shall submit a report to the European Parliament and to the Council with the result of that assessment. That report shall include any appropriate proposals for amending this Directive.

[12.942]

Article 41 Transposition

1. Member States shall bring into force the laws, regulations, and administrative provisions necessary to comply with this Directive by 19 July 2015. They shall immediately inform the Commission thereof. When Member States adopt those measures, they shall contain a reference to this Directive or shall be accompanied by

such a reference on the occasion of their official publication. The methods of making such reference shall be laid down by Member States.

2. Member States shall communicate to the Commission the texts of the main measures of national law which they adopt in the field covered by this Directive.

3. By way of derogation from the first subparagraph of paragraph 1 and subject to paragraph 5, Member States with offshore waters that do not have offshore oil and gas operations under their jurisdiction, and which do not plan to license such operations, shall inform the Commission thereof and shall be obliged to bring into force, by 19 July 2015, only those measures which are necessary to ensure compliance with Articles 20, 32 and 34. Such Member States may not license such operations until they have transposed and implemented the remaining provisions of this Directive and have informed the Commission thereof.

4. By way of derogation from the first subparagraph of paragraph 1 and subject to paragraph 5, landlocked Member States shall be obliged to bring into force, by 19 July 2015, only those measures which are necessary to ensure compliance with Article 20.

5. Where, on 18 July 2013, no company conducting operations covered by Article 20 is registered in a Member State falling within paragraph 3 or 4, that Member State shall be obliged to bring into force those measures which are necessary to ensure compliance with Article 20 only as from 12 months following any later registration of such a company in that Member State or by 19 July 2015, whichever is the later.

[12.943]

Article 42 Transitional provisions

1. In relation to owners, operators of planned production installations and operators planning or executing well operations, Member States shall apply the laws, regulations and administrative provisions adopted pursuant to Article 41 by 19 July 2016.

2. In relation to existing installations, Member States shall apply the laws, regulations and administrative provisions adopted pursuant to Article 41 from the date of scheduled regulatory review of risk assessment documentation and no later than by 19 July 2018.

[12.944]

Article 43 Entry into force

This Directive shall enter into force on the twentieth day following that of its publication in the *Official Journal of the European Union*.

[12.945]

Article 44 Addressees

This Directive is addressed to the Member States.

Done at Strasbourg, 12 June 2013.

ANNEX I

INFORMATION TO BE INCLUDED IN DOCUMENTS SUBMITTED TO THE COMPETENT AUTHORITY PURSUANT TO ARTICLE 11

[12.946]

1. Information to be submitted in a design or relocation notification for a production installation

The design notification and the relocation notification for a production installation to be submitted pursuant to points (c) and (j) of Article 11(1) respectively shall contain at least the following information:

(1) the name and address of the operator of the installation;

(2) a description of the design process for the production operations and systems, from an initial concept to the submitted design or selection of an existing installation, the relevant standards used, and the design concepts included in the process;

(3) a description of the selected design concept in relation to the major hazard scenarios for the particular installation and its location, and the primary risk control features;

(4) a demonstration that the concept contributes to reducing major hazard risks to an acceptable level;

(5) a description of the installation and the conditions at its intended location;

(6) a description of any environmental, meteorological and seabed limitations on safe operations, and the arrangements for identifying risks from seabed and marine hazards such as pipelines and the moorings of adjacent installations;

(7) a description of the types of major hazard operations to be carried out;

(8) a general description of the safety and environmental management system by which the intended major accident risk control measures are to be maintained in good effect;

(9) a description of the independent verification schemes and an initial list of safety and environmental critical elements and their required performance;

(10) where an existing production installation is to be moved to a new location to serve a different production operation, a demonstration that the installation is suitable for the proposed production operation;

(11) where a non-production installation is to be converted for use as a production installation, a justification demonstrating that the installation is suitable for such conversion.

2. Information to be submitted in a report on major hazards for operation of a production installation

Reports on major hazards for a production installation to be prepared in accordance with Article 12 and submitted pursuant to point (e) of Article 11(1) shall contain at least the following information:

(1) a description of the account taken of the competent authority's response to the design notification;

(2) the name and address of the operator of the installation;

(3) a summary of any worker involvement in the preparation of the report on major hazards;

(4) a description of the installation and any association with other installations or connected infrastructure, including wells;

Part 12

(5) demonstration that all the major hazards have been identified, their likelihood and consequences assessed, including any environmental, meteorological and seabed limitations on safe operations, and that their control measures including associated safety and environmental critical elements are suitable so as to reduce the risk of a major accident to an acceptable level; this demonstration shall include an assessment of oil spill response effectiveness;

(6) a description of the types of operations with major hazard potential to be carried out, and the maximum number of persons that can be on the installation at any time;

(7) a description of equipment and arrangements to ensure well control, process safety, containment of hazardous substances, prevention of fire and explosion, protection of the workers from hazardous substances, and protection of the environment from an incipient major accident;

(8) a description of the arrangements to protect persons on the installation from major hazards, and to ensure their safe escape, evacuation and rescue, and arrangements for the maintenance of control systems to prevent damage to the installation and the environment in the event that all personnel are evacuated;

(9) relevant codes, standards and guidance used in the construction and commissioning of the installation;

(10) information, regarding the operator's safety and environmental management system, that is relevant to the production installation;

(11) an internal emergency response plan or an adequate description thereof;

(12) a description of the independent verification scheme;

(13) any other relevant details, for example where two or more installations operate in combination in a way which affects the major hazard potential of either or all installations;

(14) the information relevant to other requirements under this Directive obtained pursuant to the major accident prevention requirements of Directive 92/91/EEC;

(15) in respect of operations to be conducted from the installation, any information relating to the prevention of major accidents resulting in significant or serious damage to the environment relevant to other requirements under this Directive, obtained pursuant to Directive 2011/92/EU;

(16) an assessment of the identified potential environmental effects resulting from the loss of containment of pollutants arising from a major accident, and a description of the technical and non-technical measures envisaged to prevent, reduce or offset them, including monitoring.

3. Information to be submitted in a report on major hazards for a non-production installation

Reports on major hazards for a non-production installation to be prepared in accordance with Article 13 and submitted pursuant to point (e) of Article 11(1) shall contain at least the following information:

(1) the name and address of the owner;

(2) a summary of any worker involvement in the preparation of the report on major hazards;

(3) a description of the installation and, in the case of a mobile installation, a description of its means of transfer between locations, and its stationing system;

(4) a description of the types of operations with major hazard potential that the installation is capable of performing, and the maximum number of persons that can be on the installation at any time;

(5) demonstration that all the major hazards have been identified, their likelihood and consequences assessed, including any environmental, meteorological and seabed limitations on safe operations and that their control measures including associated safety and environmental critical elements are suitable so as to reduce the risk of a major accident to an acceptable level; this demonstration shall include an assessment of any oil spill response effectiveness;

(6) a description of the plant and arrangements to ensure well control, process safety, containment of hazardous substances, prevention of fire and explosion, protection of the workers from hazardous substances, and protection of the environment from a major accident;

(7) a description of the arrangements to protect persons on the installation from major hazards, and to ensure their safe escape, evacuation and rescue, and arrangements for the maintenance of control systems to prevent damage to the installation and the environment in the event that all personnel are evacuated;

(8) relevant codes, standards and guidance used in the construction and commissioning of the installation;

(9) demonstration that all the major hazards have been identified for all operations the installation is capable of performing, and that the risk of a major accident is reduced to an acceptable level;

(10) a description of any environmental, meteorological and seabed limitations on safe operations, and the arrangements for identifying risks from seabed and marine hazards such as pipelines and the moorings of adjacent installations;

(11) information, regarding the safety and environmental management system, that is relevant to the non-production installation;

(12) an internal emergency response plan or an adequate description thereof;

(13) a description of the independent verification scheme;

(14) any other relevant details, for example where two or more installations operate in combination in a way which affects the major hazard potential of either or all installations;

(15) in respect of operations to be conducted from the installation, any information obtained pursuant to Directive 2011/92/EU relating to the prevention of major accidents resulting in significant or serious damage to the environment relevant to other requirements under this Directive;

(16) an assessment of the identified potential environmental effects resulting from the loss of containment of pollutants arising from a major accident, and a description of the technical and non-technical measures envisaged to prevent, reduce or offset them, including monitoring.

4. Information to be submitted in a notification of well operations

Notifications of well operations to be prepared in accordance with Article 15 and submitted pursuant to point (h) of Article 11(1) shall contain at least the following information:

(1) the name and address of the operator of the well;

(2) the name of the installation to be used and the name and address of the owner or, in the case of a production installation, the contractor undertaking drilling activities;

(3) details that identify the well and any association with installations and connected infrastructure;

(4) information on the well work programme, including the period of its operation, details and verification of barriers against loss of well control (equipment, drilling fluids and cement etc), directional control of the well path, and limitations on safe operations in keeping with the risk management;

(5) in the case of an existing well, information regarding its history and condition;

(6) any details concerning safety equipment to be deployed that are not described in the current report on major hazards for the installation;

(7) a risk assessment incorporating a description of:

(a) the particular hazards associated with the well operation including any environmental, meteorological and seabed limitations on safe operations;

(b) the subsurface hazards;

(c) any surface or subsea operations which introduce simultaneous major hazard potential;

(d) suitable control measures;

(8) a description of the well configuration at the end of operations – ie permanently or temporarily abandoned; and whether production equipment has been placed into the well for future use;

(9) in the case of a modification to a previously submitted notification of well operations, sufficient details to fully update the notification;

(10) where a well is to be constructed, modified or maintained by means of a non-production installation, additional information as follows:

(a) a description of any environmental, meteorological and seabed limitations on safe operations, and arrangements for identifying risks from seabed and marine hazards such as pipelines and the moorings of adjacent installations;

(b) a description of environmental conditions that have been taken into account within the internal emergency response plan for the installation;

(c) a description of emergency response arrangements including arrangements for responding in cases of environmental incidents that are not described in the report on major hazards; and

(d) a description of how the management systems of the operator of the well and the owner are to be coordinated to ensure effective control of major hazards at all times;

(11) a report with findings of the independent well examination, including a statement by the operator of the well that, after considering the report and findings of independent well examination by the independent verifier, the risk management relating to well design and its barriers to loss of control are suitable for all anticipated conditions and circumstances;

(12) the information relevant to this Directive obtained pursuant to the major accident prevention requirements of Directive 92/91/EEC;

(13) in respect of the well operations to be conducted, any information relevant to other requirements under this Directive obtained pursuant to Directive 2011/92/EU relating to the prevention of major accidents resulting in significant or serious damage to the environment.

5. Information to be submitted relation to a vertification scheme

Descriptions to be submitted pursuant to point (d) of Article 11(1) in relation to schemes of independent verification to be established pursuant to Article 17(1) shall include:

(a) a statement by the operator or owner, made after considering the report of the independent verifier, that the record of safety critical elements and their scheme of maintenance as specified in the report on major hazards are or will be suitable;

(b) a description of the verification scheme including the selection of independent verifiers, the means of verification that safety and environmental critical elements and any specified plant in the scheme remain in good repair and condition;

(c) a description of the means of verification referred to in point (b) that shall include details of the principles that will be applied to carry out the functions under the scheme and to keep the scheme under review throughout the lifecycle of the installation including:
   (i) the examination and testing of the safety and environmental critical elements by independent and competent verifiers;
   (ii) verification of the design, standard, certification or other system of conformity of the safety and environmental critical elements;
   (iii) examination of work in progress;
   (iv) the reporting of any instances of non-compliance;
   (v) remedial actions taken by the operator or owner.

6. Information to be provided in respect of a material change to an installation, including removal of a fixed installation

Where material changes are to be made to the installation as referred to in Article 12(5) and Article 13(4), the amended report on major hazards incorporating the material changes to be submitted pursuant to point (f) of Article 11(1) shall contain at least the following information:

(1) the name and address of the operator or the owner;

(2) a summary of any worker involvement in the preparation of the revised report on major hazards;

(3) sufficient details to fully update the earlier report on major hazards and associated internal emergency response plan for the installation and to demonstrate major hazard risks are reduced to an acceptable level;

(4) in the case of taking a fixed production installation out of use:
   (a) means of isolating all hazardous substances and in the case of wells connected to the installation, the permanent sealing of the wells from the installation and the environment;
   (b) a description of major hazard risks associated with the decommissioning of the installation to workers and the environment, the total exposed population, and the risk control measures;
   (c) emergency response arrangements to secure safe evacuation and rescue of personnel and to maintain control systems for preventing a major accident to the environment.

7. Information to be submitted in a notification of combined operations

The notification of combined operations to be prepared pursuant to Article 16 and submitted pursuant to point (i) of Article 11(1) shall contain at least the following information:

(1) the name and address of the operator submitting the notification;

(2) in the event that other operators or owners are involved in the combined operations their names and addresses, including a confirmation that they agree with the contents of the notification;

(3) a description, in the form of a bridging document authorised by all parties to the document, of how the management systems for the

installations involved in the combined operation will be coordinated so as to reduce the risk of a major accident to an acceptable level;

(4) a description of any equipment to be used in connection with the combined operation but which is not described in the current report on major hazards for any of the installations involved in the combined operations;

(5) a summary of the risk assessment carried out by all operators and owners involved in the combined operations, which shall include:

   (a) a description of any operation during the combined operation which may involve hazards with the potential to cause a major accident on or in connection with an installation;

   (b) a description of any risk control measures introduced as a result of the risk assessment;

(6) a description of the combined operation and a programme of work.

8. Information to be submitted in respect of a corporate major accident prevention policy

The corporate major accident prevention policy to be prepared in accordance with Article 19(1) and submitted pursuant to point (a) of Article 11(1) shall include but not be limited to:

(1) the responsibility at corporate board level for ensuring, on a continuous basis, that the corporate major accident prevention policy is suitable, implemented, and operating as intended;

(2) measures for building and maintaining a strong safety culture with a high likelihood of continuous safe operation;

(3) the extent and intensity of process auditing;

(4) measures for rewarding and recognising desired behaviours;

(5) the evaluation of the company's capabilities and goals;

(6) measures for maintenance of safety and environmental protection standards as a corporate core value;

(7) formal command and control systems that include board members and senior management of the company;

(8) the approach to competency at all levels of the company;

(9) the extent to which particulars (1)–(8) are applied in the company's offshore oil and gas operations conducted outside the Union.

9. Information to be provided in respect of a safety and environmental management system

The safety and environmental management system to be prepared pursuant to Article 19(3) and submitted pursuant to point (b) of Article 11(1) shall include but not be limited to:

(1) organisation structure and personnel roles and responsibilities;

(2) identification and evaluation of major hazards as well as their likelihood and potential consequences;

(3) integration of environmental impact into major accident risk assessments in the report on major hazards;

(4) controls of the major hazards during normal operations;

(5) management of change;

(6) emergency planning and response;

(7) limitation of damage to the environment;

(8) monitoring of performance;

(9) audit and review arrangements; and

(10) the measures in place for participating in tripartite consultations and how actions resulting from those consultations are put into effect.

10. Information to be provided in an internal emergency response plan

Internal emergency response plans to be prepared pursuant to Article 14 and submitted pursuant to point (g) of Article 11(1) shall include but not be limited to:

(1) names and positions of persons authorised to initiate emergency response procedures and the person directing the internal emergency response;

(2) name or position of the person with responsibility for liaising with the authority or authorities responsible for the external emergency response plan;

(3) a description of all foreseeable conditions or events which could cause a major accident, as described in the report on major hazards to which the plan is attached;

(4) a description of the actions that will be taken to control conditions or events which could cause a major accident and to limit their consequences;

(5) a description of the equipment and the resources available, including for capping any potential spill;

(6) arrangements for limiting the risks to persons on the installation and the environment, including how warnings are to be given and the actions persons are expected to take on receipt of a warning;

(7) in the case of combined operation, arrangements for coordinating escape, evacuation and rescue between the installations concerned, to secure a good prospect of survival for persons on the installations during a major accident;

(8) an estimate of oil spill response effectiveness. Environmental conditions to be considered in this response analysis shall include:

   (i) weather, including wind, visibility, precipitation and temperature;

   (ii) states, tides, and currents;

   (iii) presence of ice and debris;

   (iv) hours of daylight; and

   (v) other known environmental conditions that might influence the efficiency of the response equipment or the overall effectiveness of a response effort;

(9) arrangements for providing early warning of a major accident to the authority or authorities responsible for initiating the external emergency response plan, the type of information which shall be contained in an initial warning and the arrangements for the provision of more detailed information as it becomes available;

(10) arrangements for training personnel in the duties they will be expected to carry out, and where necessary coordinating this with external emergency responders;

(11) arrangements for coordinating internal emergency response with external emergency response;

(12) evidence of prior assessments of any chemicals used as dispersants that have been carried out to minimise public health implications and any further environmental damage.

ANNEX II

REPORTS OF WELL OPERATIONS TO BE SUBMITTED PURSUANT TO ARTICLE 15(4)

[12.947]

The reports to be submitted to the competent authority pursuant to Article 15(4) shall contain at least the following information:

(1) the name and address of the operator of the well;
(2) the name of the installation and the name and address of the operator or owner;
(3) details that identify the well and any association with installations or connected infrastructure;
(4) a summary of the operations undertaken since the commencement of operations or since the previous report;
(5) the diameter and true vertical and measured depths of:
   (a) any hole drilled; and
   (b) any casing installed;
(6) the drilling fluid density at the time of making the report; and
(7) in the case of operations relating to an existing well, its current operational state.

ANNEX III

PROVISIONS RELATING TO THE APPOINTMENT AND FUNCTIONING OF THE COMPETENT AUTHORITY PURSUANT TO ARTICLES 8 AND 9

[12.948]

1. Provisions relating to member states

(1) For the purposes of appointing a competent authority responsible for the duties set out in Article 8, Member States shall as a minimum undertake the following:

(a) make organisational arrangements which allow for the duties assigned to the competent authority in this Directive to be effectively discharged, including arrangements for regulating safety and environmental protection in an equitable manner;

(b) prepare a policy statement describing the aims of oversight and enforcement, and the obligations on the competent authority to achieve transparency, consistency, proportionality and objectivity in its regulation of offshore oil and gas operations.

(2) Member States shall make the necessary provisions to bring the arrangements in point 1 into effect, including:

(a) funding sufficient specialist expertise available internally or by formal agreements with third parties or both in order that the competent authority may inspect and investigate operations, take enforcement action, and to handle reports on major hazards and notifications;

(b) where there is reliance on external sources of expertise, funding the preparation of sufficient written guidance and oversight to maintain consistency of approach and to ensure the legally appointed competent authority retains full responsibility under this Directive;

(c) funding essential training, communication, access to technology, travel and subsistence of competent authority personnel for the carrying out of their duties, and to facilitate the active cooperation between competent authorities pursuant to Article 27;

(d) where appropriate, requiring operators or owners to reimburse the competent authority for the cost of carrying out its duties pursuant to this Directive;

(e) funding and encouraging research pursuant to the competent authority's duties under this Directive;

(f) providing funding for reports by the competent authority.

2. Provisions relating to the functioning of the competent authority

(1) For the purposes of carrying out its duties pursuant to Article 9 effectively, the competent authority shall prepare:

(a) a written strategy that describes its duties, priorities for action ie in design and operation of installations, integrity management and in emergency preparedness and response, and how it is organised;

(b) operating procedures that describe how it will inspect and enforce the execution of the duties of operators and owners under this Directive, including how it will handle, assess and accept reports on major hazards, handle notifications of well operations and how the intervals between inspection of major hazard risk control measures, including to the environment, for a given installation or activity are to be determined;

(c) procedures for carrying out its duties without prejudice to other responsibilities, for example onshore oil and gas operations, and arrangements pursuant to Directive 92/91/EEC;

(d) where the competent authority is comprised of more than one body, a formal agreement establishing the necessary mechanisms for joint operation of the competent authority, including senior management oversight and monitoring and reviews, joint planning and inspection, division of responsibilities for handling reports on major hazards, joint investigation, internal communications, and reports to be published jointly externally.

(2) The detailed procedures for assessment of reports on major hazards shall require all factual information and other particulars required under this Directive to be provided by the operator or the owner. As a minimum the competent authority shall ensure that the requirements for the following information are clearly specified in guidance to operators and owners:

(a) all foreseeable hazards with the potential to cause a major accident, including to the environment, have been identified, their risks evaluated and measures identified, including emergency responses, to control the risks;

(b) the safety and environmental management system is adequately described to demonstrate compliance with this Directive;

(c) adequate arrangements have been described for independent verification, and for audit by the operator or owner.

(3) In undertaking a thorough assessment of reports on major hazards, the competent authority shall ensure that:

(a) all factual information required is provided;

(b) the operator or the owner has identified all reasonably foreseeable major accident hazards that apply to the installation and its functions, together with potential initiating events, and that the methodology and evaluation criteria adopted for major accident risk management are clearly explained, including factors for uncertainty in the analysis;

(c) the risk management have taken into consideration all relevant stages in the lifecycle of the installation and have anticipated all foreseeable situations including:

(i) how the design decisions described in the design notification have taken account of risk management so as to ensure inherent safety and environmental principles are incorporated;

(ii) how well operations are to be conducted from the installation when operating;

(iii) how well operations are to be undertaken and temporarily suspended before production is commenced from a production installation;

Part 12

(iv) how combined operations are to be undertaken with other installation;

(v) how the decommissioning of the installation will be undertaken;

(d) how risk reduction measures identified as part of the risk management are intended to be implemented if necessary to reduce risks to an acceptable level;

(e) whether, in determining the necessary measures to achieve acceptable levels of risk, the operator or owner has clearly demonstrated how relevant good practice and judgment based on sound engineering, best management practice, and human and organisational factors principles have been taken into account;

(f) whether the measures and arrangements for the detection of, and the rapid and effective response to, an emergency are clearly identified and justified;

(g) how escape, evacuation and rescue arrangements and measures to limit escalation of an emergency and reduce its impact on the environment are integrated in a logical and systematic manner, taking account of the likely emergency conditions in which they will be operated;

(h) how the requirements are incorporated in the internal emergency response plans and whether a copy or an adequate description of the internal emergency response plan has been submitted to the competent authority;

(i) whether the safety and environmental management system described in the report on major hazards is adequate to ensure control of the major hazard risks at each stage of the installation lifecycle, and ensures compliance with all relevant legal provisions, and provides for auditing and implementing audit recommendations;

(j) whether the scheme for independent verification is clearly explained.

ANNEX IV

PROVISIONS BY OPERATORS AND OWNERS FOR PREVENTION OF MAJOR ACCIDENTS PURSUANT TO ARTICLE 19

[12.949]

1. Member States shall ensure that operators and owners:

(a) pay particular attention to evaluation of the reliability and integrity requirements of all safety and environmental critical systems and base their inspection and maintenance systems on achieving the required level of safety and environmental integrity;

(b) take appropriate measures to ensure as far as reasonably practicable that there is no unplanned escape of hazardous substances from pipelines, vessels and systems intended for their safe confinement. In addition, operators and owners shall ensure that no single failure of a containment barrier can lead to a major accident;

(c) prepare an inventory of available equipment, its ownership, location, transport to and mode of deployment at the installation and any entities relevant to the implementation of the internal emergency response plan. The inventory shall identify measures in place to ensure equipment and procedures are maintained in operable condition;

(d) ensure they have a suitable framework for monitoring compliance with all relevant statutory provisions by incorporating their statutory duties in respect of major hazards control and environmental protection into their standard operating procedures; and

(e) pay particular attention to building and maintaining a strong safety culture with a high likelihood of continuous safe operation, including with regard to securing cooperation of the workers through, inter alia:

(i) visible commitment to tripartite consultations and actions arising therefrom;

(ii) encouraging and rewarding reporting of accidents and near-misses;

(iii) working effectively with elected safety representatives;

(iv) protecting whistleblowers.

[12.949]

2. Member States shall ensure that industry cooperates with competent authorities to establish and implement a priority plan for the development of standards, guidance and rules which will give effect to best practice in major accident prevention, and limitation of consequences of major accidents should they nonetheless occur.

ANNEX V

SELECTION OF THE INDEPENDENT VERIFIER AND THE DESIGN OF SCHEMES FOR INDEPENDENT VERIFICATION PURSUANT TO ARTICLE 17(3)

[12.950]

1. Member States shall require the operator or owner to ensure the following conditions are fulfilled with regard to the verifier's independence from the operator and the owner:

(a) the function does not require the independent verifier to consider any aspect of a safety and environmental critical element or any part of an installation or a well or a well design in which the verifier was previously involved prior to the verification activity or where his or her objectivity might be compromised;

(b) the independent verifier is sufficiently independent of a management system which has, or has had, any responsibility for any aspect of a component covered by the scheme for independent verification or well examination so as to ensure objectivity in carrying out his or her functions under the scheme.

2. Member States shall require the operator or the owner to ensure that, in respect of the scheme for independent verification relating to an installation or a well, the following conditions are fulfilled:

(a) the independent verifier has suitable technical competence, including where necessary, suitably qualified and experienced personnel in adequate numbers who fulfil the requirements of point 1 of this Annex;

(b) tasks under the scheme for independent verification are appropriately allocated by the independent verifier to personnel qualified to undertake them;

(c) suitable arrangements are in place for the flow of information between the operator or owner and the independent verifier;

(d) the independent verifier is given suitable authority to be able to carry out the functions effectively.

3. Material changes shall be referred to the independent verifier for further verification in accordance with the scheme for independent verification, and the outcomes of such further verification shall be communicated to the competent authority, if requested.

ANNEX VI

INFORMATION RELATING TO PRIORITIES FOR COOPERATION BETWEEN OPERATORS AND OWNERS AND COMPETENT AUTHORITIES PURSUANT TO ARTICLE 19(7)

[12.951]

The matters to be considered for establishing priorities for the development of standards and guidance shall give practical effect to major accident prevention and limitation of their consequences. The matters shall include:

(a) improving well integrity, well control equipment and barriers and monitoring their effectiveness;
(b) improving primary containment;
(c) improving secondary containment that restricts escalation of an incipient major accident, including well blow-outs;
(d) reliable decision making;
(e) management and supervision of major hazard operations;
(f) competency of key post holders;
(g) effective risk management;
(h) reliability assessment for safety and environmental critical systems;
(i) key performance indicators;
(j) effectively integrating safety and environmental management systems between operators and owners and other entities involved in oil and gas operations.

ANNEX VII

INFORMATION TO BE PROVIDED IN EXTERNAL EMERGENCY RESPONSE PLANS PURSUANT TO ARTICLE 29

[12.952]

External emergency response plans prepared pursuant to Article 29 shall include but not be limited to:

(a) names and positions of persons authorised to initiate emergency procedures, and of persons authorised to direct the external emergency response;
(b) arrangements for receiving early warning of major accidents, and the associated alert and emergency response procedures;
(c) arrangements for coordinating resources necessary to implement the external emergency response plan;
(d) arrangements for providing assistance to the internal emergency response;
(e) a detailed description of the external emergency response arrangements;
(f) arrangements for providing persons and organisations that may be affected by the major accident with suitable information and advice relating to it;
(g) arrangements for the provision of information to the emergency services of other Member States and the Commission in the event of a major accident with possible transboundary consequences;

(h) arrangements for the mitigation of the negative impacts on wildlife both onshore and offshore including the situations where oiled animals reach shore earlier than the actual spill.

ANNEX VIII

PARTICULARS TO BE INCLUDED IN THE PREPARATION OF EXTERNAL EMERGENCY RESPONSE PLANS PURSUANT TO ARTICLE 29

[12.953]

1. The authority or authorities responsible for coordinating emergency response shall make the following available:
   (a) an inventory of available equipment, its ownership, location, means of transport to and mode of deployment at the site of the major accident;
   (b) a description of the measures in place to ensure equipment and procedures are maintained in operable condition;
   (c) an inventory of industry-owned equipment that can be made available in an emergency;
   (d) a description of the general arrangements for responding to major accidents, including competencies and responsibilities of all involved parties and the bodies responsible for maintaining such arrangements;
   (e) measures to ensure that equipment, personnel and procedures are available and up to date and sufficient members of trained personnel are available at all times;
   (f) evidence of prior environment and health assessments of any chemicals foreseen for use as dispersants.
2. External emergency response plans shall clearly explain the role of the authorities, emergency responders, coordinators and other subjects active in emergency response, so that cooperation is ensured in responding to major accidents.
3. Arrangements shall include provisions for responding to a major accident that potentially overwhelms the Member State or exceeds its boundaries by:
   (a) sharing external emergency response plans with adjacent Member States and the Commission;
   (b) compiling at cross-border level the inventories of response assets, both industry and publicly owned and all necessary adaptations to make equipment and procedures compatible between adjacent countries and Member States;
   (c) procedures for invoking the Union Civil Protection Mechanism;
   (d) arranging transboundary exercises of external emergency response.

ANNEX IX

SHARING OF INFORMATION AND TRANSPARENCY

[12.954]

1. The common data reporting format for major hazard indicators shall make it possible to compare information from competent authorities and to compare information from individual operators and owners.
2. The information to be shared by the competent authority and operators and owners shall include information relating to:
   (a) unintended release of oil, gas or other hazardous substances, whether or not ignited;

Part 12

(b) loss of well control requiring actuation of well control equipment, or failure of a well barrier requiring its replacement or repair;
(c) failure of a safety and environmental critical element;
(d) significant loss of structural integrity, or loss of protection against the effects of fire or explosion, or loss of station keeping in relation to a mobile installation;
(e) vessels on collision course and actual vessel collisions with an offshore installation;
(f) helicopter accidents, on or near offshore installations;
(g) any fatal accident;
(h) any serious injuries to 5 or more persons in the same accident;
(i) any evacuation of personnel;
(j) a major environmental incident.

3. The annual reports to be submitted by Member States pursuant to Article 25 shall contain as a minimum the following information:
(a) the number, age and location of installations;
(b) the number and type of inspections and investigations carried out, any enforcement actions or convictions;
(c) incident data pursuant to the common reporting system required in Article 23;
(d) any major change in the offshore regulatory framework;
(e) the performance of offshore oil and gas operations in relation to prevention of major accidents and the limiting of consequences of major accidents that do occur.

4. The information referred to in point 2 shall consist of both factual information and analytical data regarding oil and gas operations, and shall be unambiguous. The information and data provided shall be such that the performance of individual operators and owners can be compared within the Member State and the performance of the industry as a whole can be compared between Member States.

5. The information collected and assembled referred to in point 2 shall enable Member States to provide advanced warning of potential deterioration of safety and environmentally critical barriers, and shall enable them to take preventive action. The information shall also demonstrate the overall effectiveness of measures and controls implemented by individual operators and owners, and industry as a whole, in particular to prevent major accidents and to minimise risks for the environment.

6. In order to meet the requirements of Article 24, a simplified format shall be developed to facilitate publication of relevant data pursuant to point 2 of this Annex and preparation of reports pursuant to Article 25 in a way that is easily accessible to the public and facilitates transboundary comparison of data.

STATEMENT BY THE COMMISSION

[12.955]–[12.1000]

1. The Commission regrets that under paragraphs 3 and 5 of Article 41 some Member States are partially exempted from the obligation to transpose the Directive and considers that such derogations shall not be regarded as a precedent in order not to affect the integrity of EU law.

2. The Commission notes that Member States may use the option not to transpose and apply Article 20 of the Directive because of the current absence of any company registered in their jurisdiction which has offshore activities outside the territory of the Union.

In order to ensure effective enforcement of this Directive, the Commission underlines that it is incumbent on these Member States to ensure that companies already registered with them do not circumvent the aims of the Directive by extending their business objects to include offshore activities without notification of this extension to the competent national authorities so that they can take the necessary steps to ensure full application of Article 20.

The Commission will take all necessary measures against any circumvention which may be brought to its attention.

Part 12B

Offshore, pipelines and diving: Pipelines and diving

Part 12

Pipelines Safety Regulations 1996

SI 1996/825

as amended by 1999 c 29, 2004 c 36, and SIs 2003/2563, 2009/716 and 2013/448

SCHEDULE 6

Regulation 31

PART I

REVOCATION OF INSTRUMENTS

[12.1044]–[12.1399]

(1)	(2)	(3)
Title	***Reference***	***Extent of revocation***
The Gas Safety Regulations 1972	SI 1972/1178	The whole Regulations.
The Gas (Metrication) Regulations 1980	SI 1980/1851	Regulation 3(1).
The Submarine Pipe-lines Safety Regulations 1982	SI 1982/1513	The whole Regulations except regulations 1(1) and 11.
The Submarine Pipe-lines Safety (Amendment) Regulations 1986	SI 1986/1985	The whole Regulations.
The Offshore Installations (Emergency Pipe-line Valve) Regulations 1989	SI 1989/1029	The whole Regulations.
The Submarine Pipe-lines (Inspectors and Safety) (Amendment) Regulations 1991	SI 1991/680	Regulation 3.

PART II

[*Revoked by SI 2013/448, reg 2 and Sch.*]

Diving at Work Regulations 1997

SI 1997/2776

[12.1401]

General note

These Regulations were revoked by SI 2013/1471.

PART 13
WORKING TIME

Working Time Regulations 1998

SI 1998/1833

as amended by SIs 1999/3242, 1999/3372, 2001/3256, 2002/3128, 2003/1684, 2003/3049, 2004/752, 2004/1713, 2004/2516, 2005/2241, 2006/99, 2006/557, 2006/594, 2006/2389, 2007/2079, 2008/960, 2008/1660, 2009/1567, 2009/2766, 2009/3348, 2011/1133, 2013/1956, 2013/2228 and 2014/480

PART II

RIGHTS AND OBLIGATIONS CONCERNING WORKING TIME

[13.28]

13 Entitlement to annual leave

[(1) Subject to paragraph (5), a worker is entitled to four weeks' annual leave in each leave year.]

(2) . . .

(3) A worker's leave year, for the purposes of this regulation, begins—

- (a) on such date during the calendar year as may be provided for in a relevant agreement; or
- (b) where there are no provisions of a relevant agreement which apply—
  - (i) if the worker's employment began on or before 1st October 1998, on that date and each subsequent anniversary of that date; or
  - (ii) if the worker's employment begins after 1st October 1998, on the date on which that employment begins and each subsequent anniversary of that date.

(4) Paragraph (3) does not apply to a worker to whom Schedule 2 applies (workers employed in agriculture [in Wales or Scotland]) except where, in the case of a worker partly employed in agriculture [in Wales or Scotland], a relevant agreement so provides.

(5) Where the date on which a worker's employment begins is later than the date on which (by virtue of a relevant agreement) his first leave year begins, the leave to which he is entitled in that leave year is a proportion of the period applicable under [paragraph (1)] equal to the proportion of that leave year remaining on the date on which his employment begins.

(6) . . .

(7) . . .

(8) . . .

(9) Leave to which a worker is entitled under this regulation may be taken in instalments, but—

- (a) it may only be taken in the leave year in respect of which it is due, and
- (b) it may not be replaced by a payment in lieu except where the worker's employment is terminated.

[13.29]

Statutory annotations

Amendment

Para (4): words 'in Wales or Scotland' in square brackets in both places they occur inserted by SI 2013/2228, reg 2(1), (2).

[13.34]

15 Dates on which leave is taken

(1) A worker may take leave to which he is entitled under [regulation 13] [and regulation 13A] on such days as he may elect by giving notice to his employer in accordance with paragraph (3), subject to any requirement imposed on him by his employer under paragraph (2).

(2) A worker's employer may require the worker—

(a) to take leave to which the worker is entitled under [regulation 13] [or regulation 13A]; or

(b) not to take such leave,

on particular days, by giving notice to the worker in accordance with paragraph (3).

(3) A notice under paragraph (1) or (2)—

(a) may relate to all or part of the leave to which a worker is entitled in a leave year;

(b) shall specify the days on which leave is or (as the case may be) is not to be taken and, where the leave on a particular day is to be in respect of only part of the day, its duration; and

(c) shall be given to the employer or, as the case may be, the worker before the relevant date.

(4) The relevant date, for the purposes of paragraph (3), is the date—

(a) in the case of a notice under paragraph (1) or (2)(a), twice as many days in advance of the earliest day specified in the notice as the number of days or part-days to which the notice relates, and

(b) in the case of a notice under paragraph (2)(b), as many days in advance of the earliest day so specified as the number of days or part-days to which the notice relates.

(5) Any right or obligation under paragraphs (1) to (4) may be varied or excluded by a relevant agreement.

(6) This regulation does not apply to a worker to whom Schedule 2 applies (workers employed in agriculture [in Wales or Scotland]) except where, in the case of a worker partly employed in agriculture [in Wales or Scotland], a relevant agreement so provides.

[13.35]

Statutory annotations

Amendment

Para (6): words 'in Wales or Scotland' in square brackets in both places they occur inserted by SI 2013/2228, reg 2(1), (3).

PART III

EXCEPTIONS

[13.41]

[18 Excluded sectors

(1) These Regulations do not apply—

[(a) to workers to whom the Merchant Shipping (Hours of Work) Regulations 2002 apply;]

[(b) to workers to whom the Fishing Vessels (Working Time: Sea-fishermen) Regulations 2004 apply;] or

[(c) to workers to whom the Merchant Shipping (Working Time: Inland Waterways) Regulations 2003 apply].

(2) Regulations 4(1) and (2), 6(1), (2) and (7), 7(1) and (6), 8, 10(1), 11(1) and (2), 12(1), 13[, 13A] and 16 do not apply—

(a) where characteristics peculiar to certain specific services such as the armed forces or the police, or to certain specific activities in the civil protection services, inevitably conflict with the provisions of these Regulations;

(b) to workers to whom the European Agreement on the organisation of working time of mobile staff in civil aviation concluded on 22nd March 2000 and implemented by Council Directive 2000/79/EC of 27th November 2000 applies; or

(c) to the activities of workers who are doctors in training.

(3) Paragraph (2)(c) has effect only until 31st July 2004.

(4) Regulations 4(1) and (2), 6(1), (2) and (7), 8, 10(1), 11(1) and (2) and 12(1) do not apply to workers to whom Directive 2002/15/EC of the European Parliament and of the Council on the organisation of the working time of persons performing mobile road transport activities, dated 11th March 2002 applies.

[(5) Regulation 24 does not apply to workers to whom the Cross-border Railways Services (Working Time) Regulations 2008 apply.]]

[13.42]

Statutory annotations

Amendment

Para (1): sub-para (a) substituted by SI 2014/308, reg 3, Schedule, para 5.

PART IV

MISCELLANEOUS

[13.67]

[28 Enforcement

(1) In this regulation, regulations 29–29E and Schedule 3—

'the 1974 Act' means the Health and Safety at Work etc Act 1974;

'the Civil Aviation Authority' means the authority referred to in section 2(1) of the Civil Aviation Act 1982;

'code of practice' includes a standard, a specification and any other documentary form of practical guidance;

. . .

['DVSA' means the Driver and Vehicle Standards Agency;]

'enforcement authority' means the Executive, a local authority, the Civil Aviation Authority[, [DVSA] or the Office of Rail Regulation];

'the Executive' means the Health and Safety Executive referred to in [section 10(1)] of the 1974 Act;

'local authority' means—

(a) in relation to England, a county council so far as they are the council for an area for which there are no district councils, a district council, a London borough council, the Common Council of the City of London, the Sub-Treasurer of the Inner Temple or the Under-Treasurer of the Middle Temple;

(b) in relation to Wales, a county council or a county borough council;

(c) in relation to Scotland, a council constituted under section 2 of the Local Government etc (Scotland) Act 1994;

'premises' includes any place and, in particular, includes—

(a) any vehicle, vessel, aircraft or hovercraft;

(b) any installation on land (including the foreshore and other land intermittently covered by water), any offshore installation, and any other installation (whether floating, or resting on the seabed or the subsoil thereof, or resting on other land covered with water or the subsoil thereof) and

(c) any tent or movable structure;

'relevant civil aviation worker' means a mobile worker who works mainly on board civil aircraft, excluding any worker to whom regulation 18(2)(b) applies;

'the relevant requirements' means the following provisions—

(a) regulations 4(2), 5A(4), 6(2) and (7), 6A, 7(1), (2) and (6), 8, 9 and 27A(4)(a);

(b) regulation 24, in so far as it applies where regulation 6(1), (2) or (7) is modified or excluded, and

(c) regulation 24A(2), in so far as it applies where regulations 6(1), (2) or (7) is excluded;

'relevant road transport worker' means a mobile worker to whom one or more of the following applies—

(a) Council Regulation (EEC) 3820/85,

(b) the European Agreement concerning the Work of Crews of Vehicles engaged in International Road Transport (AETR) of 1st July 1970, and

(c) the United Kingdom domestic driver's hours code, which is set out in Part VI of the Transport Act 1968;

'the relevant statutory provisions' means—

(a) the provisions of the 1974 Act and of any regulations made under powers contained in that Act; and

(b) while and to the extent that they remain in force, the provisions of the Acts mentioned in Schedule 1 to the 1974 Act and which are specified in the third column of that Schedule and the regulations, orders or other instruments of a legislative character made or having effect under a provision so specified;

. . .

. . .

(2) It shall be the duty of the Executive to make adequate arrangements for the enforcement of the relevant requirements except to the extent that—

(a) a local authority is made responsible for their enforcement by paragraph (3);

(b) the Civil Aviation Authority is made responsible for their enforcement by paragraph (5); . . .

(c) [DVSA] is made responsible for their enforcement by paragraph (6);

[(d) the Office of Rail Regulation is made responsible for their enforcement by paragraph (3A)].

(3) Where the relevant requirements apply in relation to workers employed in premises in respect of which a local authority is responsible, under the Health and Safety (Enforcing Authority) Regulations 1998, for enforcing any of the relevant statutory provisions, it shall be the duty of that authority to enforce those requirements.

[(3A) Where the relevant requirements apply in relation to workers employed in the carrying out of any of the activities specified in regulation 3(2) of the Health and Safety (Enforcing Authority for Railways and Other Guided Transport Systems) Regulations 2006 it shall be the duty of the Office of Rail Regulation to enforce those requirements.]

(4) The duty imposed on local authorities by paragraph (3) shall be performed in accordance with such guidance as may be given to them by [the Executive].

(5) It shall be the duty of the Civil Aviation Authority to enforce the relevant requirements in relation to relevant civil aviation workers.

(6) It shall be the duty of [DVSA] to enforce the relevant requirements in relation to relevant road transport workers.

(7) The provisions of Schedule 3 shall apply in relation to the enforcement of the relevant requirements.

(8) . . .]

[13.68]

Statutory annotations

Amendment

Para (1): definition 'DVSA' inserted by SI 2014/480, reg 9(1), (2)(c); in definition 'enforcement authority' reference to 'DVSA' in square brackets substituted by SI 2014/480, reg 9(1), (2)(a); in definition 'the relevant statutory provisions' in para (b) word omitted revoked by SI 2014/480, reg 9(1), (2)(b); definition 'VOSA' (omitted) revoked by SI 2014/480, reg 9(1), (2)(b).

Para (2): in sub-para (c) reference to 'DVSA' in square brackets substituted by SI 2014/480, reg 9(1), (3).

Para (6): reference to 'DVSA' in square brackets substituted by SI 2014/480, reg 9(1), (4).

[13.81]

30 Remedies

(1) A worker may present a complaint to an employment tribunal that his employer—

(a) has refused to permit him to exercise any right he has under—

[(i) regulation 10(1) or (2), 11(1), (2) or (3), 12(1) or (4), 13 or 13A;]

(ii) regulation 24, in so far as it applies where regulation 10(1), 11(1) or (2) or 12(1) is modified or excluded; . . .

[(iii) regulation 24A, in so far as it applies where regulation 10(1), 11(1) or (2) or 12(1) is excluded; or

(iv) regulation 25(3), 27A(4)(b) or 27(2); or]

(b) has failed to pay him the whole or any part of any amount due to him under regulation 14(2) or 16(1).

(2) [Subject to [regulations 30A and 30B], an employment tribunal] shall not consider a complaint under this regulation unless it is presented—

(a) before the end of the period of three months (or, in a case to which regulation 38(2) applies, six months) beginning with the date on which it is alleged that the exercise of the right should have been permitted (or in the case of a rest period or leave extending over more than one day, the date on which it should have been permitted to begin) or, as the case may be, the payment should have been made;

(b) within such further period as the tribunal considers reasonable in a case where it is satisfied that it was not reasonably practicable for the complaint to be presented before the end of that period of three or, as the case may be, six months.

[(2A) Where the period within which a complaint must be presented in accordance with paragraph (2) is extended by regulation 15 of the Employment Act 2002 (Dispute Resolution) Regulations 2004, the period within which the complaint must be presented shall be the extended period rather than the period in paragraph (2).]

(3) Where an employment tribunal finds a complaint under paragraph (1)(a) well-founded, the tribunal—

(a) shall make a declaration to that effect, and

(b) may make an award of compensation to be paid by the employer to the worker.

(4) The amount of the compensation shall be such as the tribunal considers just and equitable in all the circumstances having regard to—

(a) the employer's default in refusing to permit the worker to exercise his right, and

(b) any loss sustained by the worker which is attributable to the matters complained of.

(5) Where on a complaint under paragraph (1)(b) an employment tribunal finds that an employer has failed to pay a worker in accordance with regulation 14(2) or 16(1), it shall order the employer to pay to the worker the amount which it finds to be due to him.

[13.82]

Statutory annotations

Amendment

Para (2): words 'regulations 30A and 30B' in square brackets substituted by SI 2014/386, art 2, Schedule, paras 2, 3.

[13.82C]

[30B Extension of time limit to facilitate conciliation before institution of proceedings]

[(1) In this regulation—

(a) Day A is the day on which the worker concerned complies with the requirement in subsection (1) of section 18A of the Employment Tribunals Act 1996 (requirement to contact ACAS before instituting proceedings) in relation to the matter in respect of which the proceedings are brought, and

(b) Day B is the day on which the worker concerned receives or, if earlier, is treated as receiving (by virtue of regulations made under subsection (11) of that section) the certificate issued under subsection (4) of that section.

(2) In working out when the time limit set by regulation 30(2)(a) expires the period beginning with the day after Day A and ending with Day B is not to be counted.

(3) If the time limit set by regulation 30(2)(a) would (if not extended by this paragraph) expire during the period beginning with Day A and ending one month after Day B, the time limit expires instead at the end of that period.

(4) The power conferred on the employment tribunal by regulation 30(2)(b) to extend the time limit set by paragraph (2)(a) of that regulation is exercisable in relation to that time limit as extended by this regulation.]

[13.82D]

Amendment

Inserted by SI 2014/386, art 2, Schedule, paras 2, 4.

[13.87]

35 Restrictions on contracting out

(1) Any provision in an agreement (whether a contract of employment or not) is void in so far as it purports—

(a) to exclude or limit the operation of any provision of these Regulations, save in so far as these Regulations provide for an agreement to have that effect, or

(b) to preclude a person from bringing proceedings under these Regulations before an employment tribunal.

(2) Paragraph (1) does not apply to—

(a) any agreement to refrain from instituting or continuing proceedings where a conciliation officer has taken action under [any of sections 18A to 18C] of the Employment Tribunals Act 1996 (conciliation); or

(b) any agreement to refrain from instituting or continuing proceedings within [section 18(1)(j)] of the Employment Tribunals Act 1996 (proceedings under these Regulations where conciliation is available), if the conditions regulating [settlement] agreements under these Regulations are satisfied in relation to the agreement.

(3) For the purposes of paragraph (2)(b) the conditions regulating [settlement] agreements under these Regulations are that—

(a) the agreement must be in writing,

(b) the agreement must relate to the particular complaint,

(c) the worker must have received advice from a relevant independent adviser as to the terms and effect of the proposed agreement and, in particular, its effect on his ability to pursue his rights before an employment tribunal,

(d) there must be in force, when the adviser gives the advice, a contract of insurance, or an indemnity provided for members of a profession or professional body, covering the risk of a claim by the worker in respect of loss arising in consequence of the advice,

(e) the agreement must identify the adviser, and

(f) the agreement must state that the conditions regulating [settlement] agreements under these Regulations are satisfied.

(4) A person is a relevant independent adviser for the purposes of paragraph (3)(c)—

(a) if he is a qualified lawyer,

(b) if he is an officer, official, employee or member of an independent trade union who has been certified in writing by the trade union as competent to give advice and as authorised to do so on behalf of the trade union, or

(c) if he works at an advice centre (whether as an employee or as a volunteer) and has been certified in writing by the centre as competent to give advice and as authorised to do so on behalf of the centre.

(5) But a person is not a relevant independent adviser for the purposes of paragraph (3)(c) in relation to the worker—

(a) if he, is employed by or is acting in the matter for the employer or an associated employer,

(b) in the case of a person within paragraph (4)(b) or (c), if the trade union or advice centre is the employer or an associated employer, or

(c) in the case of a person within paragraph (4)(c), if the worker makes a payment for the advice received from him.

(6) In paragraph (4)(a), 'qualified lawyer' means—

(a) as respects England and Wales, [a person who, for the purposes of the Legal Services Act 2007), is an authorised person in relation to an activity which constitutes the exercise of a right of audience or the conduct of litigation (within the meaning of that Act)]; and

(b) as respects Scotland, an advocate (whether in practice as such or employed to give legal advice), or a solicitor who holds a practising certificate.

[(6A) A person shall be treated as being a qualified lawyer within paragraph (6)(a) if he is a Fellow of the Institute of Legal Executives [practising in a solicitor's practice (including a body recognised under section 9 of the Administration of Justice Act 1985)].]

(7) For the purposes of paragraph (5) any two employers shall be treated as associated if—

(a) one is a company of which the other (directly or indirectly) has control; or

(b) both are companies of which a third person (directly or indirectly) has control; and 'associated employer' shall be construed accordingly.

[13.88]

Statutory annotations

Amendment

Para (2): in sub-para (a) words 'any of sections 18A to 18C' in square brackets substituted by SI 2014/386, art 2, Schedule, paras 2, 5; in sub-para (b) words 'section 18(1)(j)' in square brackets substituted by SI 2014/431, art 3, Schedule, para 19; in sub-para (b) word 'settlement' in square brackets substituted by SI 2013/1956, art 2, Schedule, para 2.

Para (3): word 'settlement' in square brackets substituted by SI 2013/1956, art 2, Schedule, para 2; in sub-para (f) word 'settlement' in square brackets substituted by SI 2013/1956, art 2, Schedule, para 2.

Para (6A) inserted by SI 2004/2516, reg 2.

PART V

SPECIAL CLASSES OF PERSON

[13.97]

41 Police service

(1) [Subject to paragraph (1A),] for the purposes of these Regulations, the holding, otherwise than under a contract of employment, of the office of constable or an appointment as a police cadet shall be treated as employment, under a worker's contract, by the relevant officer.

[(1A) For the purposes of these Regulations, any constable who has been seconded to [the National Crime Agency] to serve as a member of its staff shall be treated as employed by [the National Crime Agency].]

(2) Any matter relating to the employment of a worker which may be provided for the purposes of these Regulations in a workforce agreement may be provided for the same purposes in relation to the service of a person holding the office of constable or an appointment as a police cadet by an agreement between the relevant officer and a joint branch board.

(3) In this regulation—

'a joint branch board' means a joint branch board constituted in accordance with regulation 7(3) of the Police Federation Regulations 1969 or regulation 7(3) of the Police Federation (Scotland) Regulations 1985, and

'the relevant officer' means—

(a) in relation to a member of a police force or a special constable or police cadet appointed for a police area, the chief officer of police (or, in Scotland, the chief constable);

(b) . . . and

(c) in relation to any other person holding the office of constable or an appointment as a police cadet, the person who has the direction and control of the body of constables or cadets in question.

[(4) For the purposes of these Regulations the relevant officer, as defined by paragraph (3), shall be treated as a corporation sole.

(5) Where, in a case in which the relevant officer, as so defined, is guilty of an offence under these Regulations, it is proved—

(a) that the office-holder personally consented to the commission of the offence;

(b) that he personally connived in its commission; or

(c) that the commission of the offence was attributable to personal neglect on his part,

the office-holder (as well as the corporation sole) shall be guilty of an offence and shall be liable to be proceeded against and punished accordingly.

(6) In paragraph (5) above 'the office-holder', in relation to the relevant officer, means an individual who, at the time of the consent, connivance or neglect—

(a) held the office or other position mentioned in paragraph (3) above as the office or position of that officer; or

(b) was for the time being responsible for exercising and performing the powers and duties of that office or position.

(7) In the application of this regulation to Scotland—

(a) paragraph (4) shall have effect as if for the words 'corporation sole' there were substituted 'distinct juristic person (that is to say, as a juristic person distinct from the individual who for the time being is the office-holder)';

(b) paragraph (5) shall have effect as if for the words 'corporation sole' there were substituted 'juristic person'; and

(c) paragraph (6) shall have effect as if for the words 'paragraph (5)' there were substituted 'paragraphs (4) and (5)'.]

[13.98]

Statutory annotations

Amendment

Para (1A): words 'the National Crime Agency' in square brackets in both places they occur substituted by virtue of the Crime and Courts Act 2013, s 15(3), Sch 8, Pt 4, para 190; for transitional provisions and savings see s 15(3), Sch 8, Pt 1 thereto.

[13.100]

43 [Workers employed in agriculture in Wales or Scotland]

The provisions of Schedule 2 have effect in relation to workers employed in agriculture [in Wales or Scotland].

[13.100A]

Statutory annotations

Amendment

Provision heading: substituted by SI 2013/2228, reg 2(1), (4)(a).

Words 'in Wales or Scotland' in square brackets inserted by SI 2013/2228, reg 2(1), (4)(b).

SCHEDULE 2

WORKERS EMPLOYED IN AGRICULTURE [IN WALES OR SCOTLAND]

Regulations 13(4), 15(6) and 43

[13.102]

1

Except where, in the case of a worker partly employed in agriculture [in Wales or Scotland], different provision is made by a relevant agreement—

(a) for the purposes of regulation 13 [and regulation 13A], the leave year of a worker employed in agriculture [in Wales or Scotland] begins on 6th April each year or such other date as may be specified in an agricultural wages order which applies to him; and

(b) the dates on which leave is taken by a worker employed in agriculture [in Wales or Scotland] shall be determined in accordance with an agricultural wages order which applies to him.

2

Where, in the case referred to in paragraph 1 above, a relevant agreement makes provision different from sub-paragraph (a) or (b) of that paragraph—

(a) neither section 11 of the Agricultural Wages Act 1948 nor section 11 of the Agricultural Wages (Scotland) Act 1949 shall apply to that provision; and

(b) an employer giving effect to that provision shall not thereby be taken to have failed to comply with the requirements of an agricultural wages order.

3

In this Schedule, 'an agricultural wages order' means an order under section 3 of the Agricultural Wages Act 1948 or section 3 of the Agricultural Wages (Scotland) Act 1949.

[13.103]

Statutory annotations

Amendment

Schedule heading: words 'in Wales or Scotland' in square brackets inserted by SI 2013/2228, reg 2(1), (5)(a).

Para 1: words 'in Wales or Scotland' in square brackets in each place they occur inserted by SI 2013/2228, reg 2(1), (5)(b).

[SCHEDULE 3

ENFORCEMENT

Regulation 28(7)]

[13.106]

1 [Appointment of inspectors

(1) Each enforcement authority may appoint as inspectors (under whatever title it may from time to time determine) such persons having suitable qualifications as it thinks necessary for carrying into effect these Regulations within its field of responsibility, and may terminate any appointment made under this paragraph.

(2) Every appointment of a person as an inspector under this paragraph shall be made by an instrument in writing specifying which of the powers conferred on inspectors by these Regulations are to be exercisable by the person appointed; and an inspector shall in right of his appointment under this paragraph—

(a) be entitled to exercise only such of those powers as are so specified; and

(b) be entitled to exercise the powers so specified only within the field of responsibility of the authority which appointed him.

(3) So much of an inspector's instrument of appointment as specifies the powers which he is entitled to exercise may be varied by the enforcement authority which appointed him.

(4) An inspector shall, if so required when exercising or seeking to exercise any power conferred on him by these Regulations, produce his instrument of appointment or a duly authenticated copy thereof.

2 Powers of inspectors

(1) Subject to the provisions of paragraph 1 and this sub-paragraph, an inspector may, for the purpose of carrying into effect these Regulations within the field of responsibility of the enforcement authority which appointed him, exercise the powers set out in sub-paragraph (2) below.

(2) The powers of an inspector referred to in the preceding sub-paragraph are the following, namely—

(a) at any reasonable time (or, in a situation which in his opinion is or may be dangerous, at any time) to enter any premises which he has reason to believe it is necessary for him to enter for the purpose mentioned in sub-paragraph (1) above;

(b) to take with him a constable if he has reasonable cause to apprehend any serious obstruction in the execution of his duty;

(c) without prejudice to the preceding sub-paragraph, on entering any premises by virtue of paragraph (a) above to take with him—

(i) any other person duly authorised by the inspector's enforcement authority; and

(ii) any equipment or materials required for any purpose for which the power of entry is being exercised;

(d) to make such examination and investigation as may in any circumstances be necessary for the purpose mentioned in sub-paragraph (1) above;

(e) to require any person whom he has reasonable cause to believe to be able to give any information relevant to any examination or investigation under paragraph (d) above to answer (in the absence of persons other than a person nominated by him to be present and any persons whom the inspector may allow to be present) such questions as the inspector thinks fit to ask and to sign a declaration of the truth of his answers;

(f) to require the production of, inspect, and take copies of or of any entry in—

(i) any records which by virtue of these Regulations are required to be kept, and

(ii) any other books, records or documents which it is necessary for him to see for the purposes of any examination or investigation under paragraph (d) above;

(g) to require any person to afford him such facilities and assistance with respect to any matters or things within that person's control or in relation to which that person has responsibilities as are necessary to enable the inspector to exercise any of the powers conferred on him by this paragraph;

(h) any other power which is necessary for the purpose mentioned in sub-paragraph (1) above.

(3) No answer given by a person in pursuance of a requirement imposed under sub-paragraph (2)(e) above shall be admissible in evidence against that person or the [spouse or civil partner] of that person in any proceedings..

(4) Nothing in this paragraph shall be taken to compel the production by any person of a document of which he would on grounds of legal professional privilege be entitled to withhold production on an order for discovery in an action in the High Court or, as the case may be, on an order for the production of documents in an action in the Court of Session.

3 Improvement notices

If an inspector is of the opinion that a person—

(a) is contravening one or more of these Regulations; or

(b) has contravened one or more of these Regulations in circumstances that make it likely that the contravention will continue or be repeated,

he may serve on him a notice (in this Schedule referred to as 'an improvement notice') stating that he is of that opinion, specifying the provision or provisions as to which he is of that opinion, giving particulars of the reasons why he is of that opinion, and requiring that person to remedy the contravention or, as the case may be, the matters occasioning it within such period (ending not earlier than the period within which an appeal against the notice can be brought under paragraph 6) as may be specified in the notice.

4 Prohibition notices

(1) This paragraph applies to any activities which are being or are likely to be carried on by or under the control of any person, being activities to or in relation to which any of these Regulations apply or will, if the activities are so carried on, apply.

(2) If as regards any activities to which this paragraph applies an inspector is of the opinion that, as carried on or likely to be carried on by or under the control of the person in question, the activities involve or, as the case may be, will involve a risk of serious personal injury, the inspector may serve on that person a notice (in this Schedule referred to as 'a prohibition notice').

(3) A prohibition notice shall—

(a) state that the inspector is of the said opinion;

(b) specify the matters which in his opinion give or, as the case may be, will give rise to the said risk;

(c) where in his opinion any of those matters involves or, as the case may be, will involve a contravention of any of these Regulations, state that he is of that opinion, specify the regulation or regulations as to which he is of that opinion, and give particulars of the reasons why he is of that opinion; and

(d) direct that the activities to which the notice relates shall not be carried on by or under the control of the person on whom the notice is served unless the matters specified in the notice in pursuance of paragraph (b) above and any associated contraventions of provisions so specified in pursuance of paragraph (c) above have been remedied.

(4) A direction contained in a prohibition notice in pursuance of sub-paragraph (3)(d) above shall take effect—

(a) at the end of the period specified in the notice; or

(b) if the notice so declares, immediately.

5 Provisions supplementary to paragraphs 3 and 4

(1) In this paragraph 'a notice' means an improvement notice or a prohibition notice.

(2) A notice may (but need not) include directions as to the measures to be taken to remedy any contravention or matter to which the notice relates; and any such directions—

(a) may be framed to any extent by reference to any approved code of practice; and

(b) may be framed so as to afford the person on whom the notice is served a choice between different ways of remedying the contravention or matter.

(3) Where an improvement notice or a prohibition notice which is not to take immediate effect has been served—

(a) the notice may be withdrawn by an inspector at any time before the end of the period specified therein in pursuance of paragraph 3 or paragraph 4(4) as the case may be; and

(b) the period so specified may be extended or further extended by an inspector at any time when an appeal against the notice is not pending.

6 Appeal against improvement or prohibition notice

(1) In this paragraph 'a notice' means an improvement or a prohibition notice.

(2) A person on whom a notice is served may within 21 days from the date of its service appeal to an employment tribunal; and on such an appeal the tribunal may either cancel or affirm the notice and, if it affirms it, may do so either in its original form or with such modifications as the tribunal may in the circumstances think fit.

(3) Where an appeal under this paragraph is brought against a notice within the period allowed under the preceding sub-paragraph, then—

(a) in the case of an improvement notice, the bringing of the appeal shall have the effect of suspending the operation of the notice until the appeal is finally disposed of or, if the appeal is withdrawn, until the withdrawal of the appeal;

(b) in the case of a prohibition notice, the bringing of the appeal shall have the like effect if, but only if, on the application of the appellant the tribunal so directs (and then only from the giving of the direction).

(4) One or more assessors may be appointed for the purposes of any proceedings brought before an employment tribunal under this paragraph.

7 Power of enforcement authority to indemnify inspectors

Where an action has been brought against an inspector in respect of an act done in the execution or purported execution of these Regulations and the circumstances are such that he is not legally entitled to require the enforcement authority to indemnify him, that authority may, nevertheless, indemnify him against the whole or part of any damages and costs or expenses which he may have been ordered to pay or may have incurred, if the authority is satisfied that the inspector honestly believed that the act complained of was within his powers and that his duty as an inspector required or entitled him to do it.

8 Restrictions on disclosure of information

(1) In this and the two following sub-paragraphs—

(a) 'relevant information' means information obtained by an inspector in pursuance of a requirement imposed under paragraph 2(2)(e) or (f); and

(b) 'the recipient', in relation to any relevant information, means the person by whom that information was so obtained or to whom that information was so furnished, as the case may be.

(2) Subject to the following sub-paragraph, no relevant information shall be disclosed without the consent of the person by whom it was furnished.

(3) The preceding sub-paragraph shall not apply to—

(a) disclosure of information to the Commission, a government department or any enforcement authority;

(b) without prejudice to paragraph (a) above, disclosure by the recipient of information to any person for the purpose of any function conferred on the recipient by or under any of the relevant statutory provisions or under these Regulations;

(c) without prejudice to paragraph (a) above, disclosure by the recipient of information to—

(i) an officer of a local authority who is authorised by that authority to receive it; or

(ii) a constable authorised by a chief officer of police to receive it; or

(d) disclosure by the recipient of information in a form calculated to prevent it from being identified as relating to a particular person or case.

(4) In the preceding sub-paragraph any reference to the Commission, a government department or an enforcement authority includes respectively a reference to an officer of that body or authority (including in the case of an enforcement authority, any inspector appointed by it), and also, in the case of a reference to the Commission, includes a reference to—

(a) a person performing any functions of the Commission or the Executive on its behalf by virtue of section 13(1)(a) of the 1974 Act;

(b) an officer of a body which is so performing any such functions; and

(c) an adviser appointed in pursuance of section 13(1)(d) of the 1974 Act.

(5) A person to whom information is disclosed in pursuance of sub-paragraph (3) above shall not use the information for a purpose other than—

(a) in a case falling within sub-paragraph (3)(a), a purpose of the Commission, of the government department, or of the enforcement authority in question in connection with these Regulations or with the relevant statutory provisions, as the case may be;

(b) in the case of information given to an officer of a body which is a local authority, the purposes of the body in connection with the relevant statutory provisions or any enactment whatsoever relating to working time, public health, public safety or the protection of the environment;

(c) in the case of information given to a constable, the purposes of the police in connection with these Regulations, the relevant statutory provisions or any enactment whatsoever relating to working time, public health, public safety or the safety of the State.

(6) A person shall not disclose any information obtained by him as a result of the exercise of any power conferred by paragraph 2 of this Schedule (including in particular any information with respect to any trade secret obtained by him in any premises entered by him by virtue of any such power) except—

(a) for the purposes of his functions;

(b) for the purposes of any legal proceedings; or

(c) with the relevant consent.

In this sub-paragraph 'the relevant consent' means the consent of the person who furnished it, and, in any other case, the consent of a person having responsibilities in relation to the premises where the information was obtained.

(7) Notwithstanding anything in the preceding sub-paragraph an inspector shall, in circumstances in which it is necessary to do so for the purpose of assisting in keeping persons (or the representatives of persons) employed at any premises adequately informed about matters affecting their health, safety and welfare or working time, give to such persons or their representatives the following descriptions of information, that is to say—

(a) factual information obtained by him as mentioned in that sub-paragraph which relates to those premises or anything which was or is therein or was or is being done therein; and

(b) information with respect to any action which he has taken or proposes to take in or in connection with those premises in the performance of his functions;

and, where an inspector does as aforesaid, he shall give the like information to the employer of the first-mentioned persons.

(8) Notwithstanding anything in sub-paragraph (6) above, a person who has obtained such information as is referred to in that sub-paragraph may furnish to a person who appears to him to be likely to be a party to any civil proceedings arising out of any accident, occurrence, situation or other matter, a written statement of the relevant facts observed by him in the course of exercising any of the powers referred to in that sub-paragraph.]

[13.107]–[13.500]

Statutory annotations

Amendment

Para 2: in sub-para (3) words 'spouse or civil partner' in square brackets substituted, in relation to England and Wales, by SI 2014/107, art 2, Sch 1, para 25.

PART 14
INDUSTRIAL RELATIONS

Employment Rights Act 1996

1996 chapter 18

as amended

[PART IVA

PROTECTED DISCLOSURES

[14.5]

[43B Disclosures qualifying for protection

(1) In this Part a 'qualifying disclosure' means any disclosure of information which, in the reasonable belief of the worker making the disclosure, [is made in the public interest and] tends to show one or more of the following—

(a) that a criminal offence has been committed, is being committed or is likely to be committed,

(b) that a person has failed, is failing or is likely to fail to comply with any legal obligation to which he is subject,

(c) that a miscarriage of justice has occurred, is occurring or is likely to occur,

(d) that the health or safety of any individual has been, is being or is likely to be endangered,

(e) that the environment has been, is being or is likely to be damaged, or

(f) that information tending to show any matter falling within any one of the preceding paragraphs has been, or is likely to be deliberately concealed.

(2) For the purposes of subsection (1), it is immaterial whether the relevant failure occurred, occurs or would occur in the United Kingdom or elsewhere, and whether the law applying to it is that of the United Kingdom or of any other country or territory.

(3) A disclosure of information is not a qualifying disclosure if the person making the disclosure commits an offence by making it.

(4) A disclosure of information in respect of which a claim to legal professional privilege (or, in Scotland, to confidentiality as between client and professional legal adviser) could be maintained in legal proceedings is not a qualifying disclosure if it is made by a person to whom the information had been disclosed in the course of obtaining legal advice.

(5) In this Part 'the relevant failure', in relation to a qualifying disclosure, means the matter falling within paragraphs (a) to (f) of subsection (1).]

[14.6]

Statutory annotations

Amendment

Sub-s (1): words 'is made in the public interest and' in square brackets inserted by the Enterprise and Regulatory Reform Act 2013, s 17; for transitional provision see s 24(6) thereof.

[14.7]

[43C Disclosure to employer or other responsible person

(1) A qualifying disclosure is made in accordance with this section if the worker makes the disclosure . . . —

(a) to his employer, or

(b) where the worker reasonably believes that the relevant failure relates solely or mainly to—

(i) the conduct of a person other than his employer, or
(ii) any other matter for which a person other than his employer has legal responsibility,

to that other person.

(2) A worker who, in accordance with a procedure whose use by him is authorised by his employer, makes a qualifying disclosure to a person other than his employer, is to be treated for the purposes of this Part as making the qualifying disclosure to his employer.]

[14.8]

Statutory annotations

Amendment

Sub-s (1): words omitted repealed by the Enterprise and Regulatory Reform Act 2013, s 18(1)(a); for transitional provision see s 24(6) thereof.

[14.11]

[43E Disclosure to Minister of the Crown

A qualifying disclosure is made in accordance with this section if—

(a) the worker's employer is—
   (i) an individual appointed under any enactment [(including any enactment comprised in, or in an instrument made under, an Act of the Scottish Parliament)] by a Minister of the Crown, or
   (ii) a body any of whose members are so appointed, and
(b) the disclosure is made . . . to a Minister of the Crown [or a member of the Scottish executive].]

[14.12]

Statutory annotations

Amendment

In para (b) words omitted repealed by the Enterprise and Regulatory Reform Act 2013, s 18(1)(b); for transitional provision see s 24(6) thereof.

[14.13]

[43F Disclosure to prescribed person

(1) A qualifying disclosure is made in accordance with this section if the worker—

(a) makes the disclosure . . . to a person prescribed by an order made by the Secretary of State for the purposes of this section, and
(b) reasonably believes—
   (i) that the relevant failure falls within any description of matters in respect of which that person is so prescribed, and
   (ii) that the information disclosed, and any allegation contained in it, are substantially true.

(2) An order prescribing persons for the purposes of this section may specify persons or descriptions of persons, and shall specify the descriptions of matters in respect of which each person, or persons of each description, is or are prescribed.]

[14.14]

Statutory annotations

Amendment

Sub-s (1): in para (a) words omitted repealed by the Enterprise and Regulatory Reform Act 2013, s 18(1)(c); for transitional provision see s 24(6) thereof.

[14.15]

[43G Disclosure in other cases

(1) A qualifying disclosure is made in accordance with this section if—

(a) . . .

(b) [the worker] reasonably believes that the information disclosed, and any allegation contained in it, are substantially true,

(c) he does not make the disclosure for purposes of personal gain,

(d) any of the conditions in subsection (2) is met, and

(e) in all the circumstances of the case, it is reasonable for him to make the disclosure.

(2) The conditions referred to in subsection (1)(d) are—

(a) that, at the time he makes the disclosure, the worker reasonably believes that he will be subjected to a detriment by his employer if he makes a disclosure to his employer or in accordance with section 43F,

(b) that, in a case where no person is prescribed for the purposes of section 43F in relation to the relevant failure, the worker reasonably believes that it is likely that evidence relating to the relevant failure will be concealed or destroyed if he makes a disclosure to his employer, or

(c) that the worker has previously made a disclosure of substantially the same information—

(i) to his employer, or

(ii) in accordance with section 43F.

(3) In determining for the purposes of subsection (1)(e) whether it is reasonable for the worker to make the disclosure, regard shall be had, in particular, to—

(a) the identity of the person to whom the disclosure is made,

(b) the seriousness of the relevant failure,

(c) whether the relevant failure is continuing or is likely to occur in the future,

(d) whether the disclosure is made in breach of a duty of confidentiality owed by the employer to any other person,

(e) in a case falling within subsection (2)(c)(i) or (ii), any action which the employer or the person to whom the previous disclosure in accordance with section 43F was made has taken or might reasonably be expected to have taken as a result of the previous disclosure, and

(f) in a case falling within subsection (2)(c)(i), whether in making the disclosure to the employer the worker complied with any procedure whose use by him was authorised by the employer.

(3) For the purposes of this section a subsequent disclosure may be regarded as a disclosure of substantially the same information as that disclosed by a previous disclosure as mentioned in subsection (2)(c) even though the subsequent disclosure extends to information about action taken or not taken by any person as a result of the previous disclosure.]

[14.16]

Statutory annotations

Amendment

Sub-s (1): para (a) repealed by the Enterprise and Regulatory Reform Act 2013, s 18(2)(a); for transitional provision see s 24(6) thereof; in para (b) words 'the worker' in square brackets substituted by the Enterprise and Regulatory Reform Act 2013, s 18(2)(b); for transitional provision see s 24(6) thereof.

Part 14

[14.17]

[43H Disclosure of exceptionally serious failure

(1) A qualifying disclosure is made in accordance with this section if—

(a) . . .

(b) [the worker] reasonably believes that the information disclosed, and any allegation contained in it, are substantially true,

(c) he does not make the disclosure for purposes of personal gain,

(d) the relevant failure is of an exceptionally serious nature, and

(e) in all the circumstances of the case, it is reasonable for him to make the disclosure.

(2) In determining for the purposes of subsection (1)(e) whether it is reasonable for the worker to make the disclosure, regard shall be had, in particular, to the identity of the person to whom the disclosure is made.]

[14.18]

Statutory annotations

Amendment

Sub-s (1): para (a) repealed by the Enterprise and Regulatory Reform Act 2013, s 18(3)(a); for transitional provision see s 24(6) thereof; in para (b) words 'the worker' in square brackets substituted by the Enterprise and Regulatory Reform Act 2013, s 18(3)(b); for transitional provision see s 24(6) thereof.

[14.21]

[43K Extension of meaning of 'worker' etc for Part IVA

(1) For the purposes of this Part 'worker' includes an individual who is not a worker as defined by section 230(3) but who—

(a) works or worked for a person in circumstances in which—

(i) he is or was introduced or supplied to do that work by a third person, and

(ii) the terms on which he is or was engaged to do the work are or were in practice substantially determined not by him but by the person for whom he works or worked, by the third person or by both of them,

(b) contracts or contracted with a person, for the purposes of that person's business, for the execution of work to be done in a place not under the control or management of that person and would fall within section 230(3)(b) if for 'personally' in that provision there were substituted '(whether personally or otherwise)',

[(ba) works or worked as a person performing services under a contract entered into by him with [the National Health Service Commissioning Board] [under [section 83(2), 84, 92, 100, 107, 115(4), 117 or 134 of, or Schedule 12 to,] the National Health Service Act 2006 or with a Local Health Board under [section 41(2)(b), 42, 50, 57, 64 or 92 of, or Schedule 7 to,] the National Health Service (Wales) Act 2006] . . . ,]

[(bb) works or worked as a person performing services under a contract entered into by him with a Health Board under section 17J [or 17Q] of the National Health Service (Scotland) Act 1978,]

(c) [works or worked as a person providing services] in accordance with arrangements made—

(i) by [the National Health Service Commissioning Board] [[under section 126 of the National Health Service Act 2006,] or] [Local Health Board] under [section 71 or 80 of the National Health Service (Wales) Act 2006], or

(ii) by a Health Board under section [2C, 17AA, 17C,] . . . 25, 26 or 27 of the National Health Service (Scotland) Act 1978, . . .

(d) is or was provided with work experience provided pursuant to a training course or programme or with training for employment (or with both) otherwise than—

(i) under a contract of employment, or

(ii) by an educational establishment on a course run by that establishment;

and any reference to a worker's contract, to employment or to a worker being 'employed' shall be construed accordingly.

(2) For the purposes of this Part 'employer' includes—

(a) in relation to a worker falling within paragraph (a) of subsection (1), the person who substantially determines or determined the terms on which he is or was engaged,

[(aa) in relation to a worker falling within paragraph (ba) of that subsection, [the National Health Service Commissioning Board, or the] Local Health Board referred to in that paragraph,]

[(ab) in relation to a worker falling within paragraph (bb) of that subsection, the Health Board referred to in that paragraph,]

(b) in relation to a worker falling within paragraph (c) of that subsection, the authority or board referred to in that paragraph, and

(c) in relation to a worker falling within paragraph (d) of that subsection, the person providing the work experience or training.

(3) In this section 'educational establishment' includes any university, college, school or other educational establishment.

[(4) The Secretary of State may by order make amendments to this section as to what individuals count as 'workers' for the purposes of this Part (despite not being within the definition in section 230(3)).

(5) An order under subsection (4) may not make an amendment that has the effect of removing a category of individual unless the Secretary of State is satisfied that there are no longer any individuals in that category.]]

Part 14

[14.22]

Statutory annotations

Amendment

Sub-s (1): in para (ba) words 'the National Health Service Commissioning Board' in square brackets in both places they occur substituted by the Health and Social Care Act 2012, s 55(2), Sch 5, paras 72, 73(a); in para (ba) words from 'section 83(2), 84,' to 'Schedule 12 to,' in square brackets substituted by the Enterprise and Regulatory Reform Act 2013, s 20(1), (2)(a); for transitional provision see s 24(6) thereof; in para (ba) words from 'section 41(2)(b), 42,' to 'Schedule 7 to,' in square brackets substituted by the Enterprise and Regulatory Reform Act 2013, s 20(1), (2)(b); for transitional provision see s 24(6) thereof; in para (ba) words omitted repealed by the Enterprise and Regulatory Reform Act 2013, s 20(1), (2)(c); for transitional provision see s 24(6) thereof; in para (bb) words 'or 17Q' in square brackets inserted by the Enterprise and Regulatory Reform Act 2013, s 20(1), (3); for transitional provision see s 24(6) thereof; in sub-para (c) words 'works or worked as a person providing services' in square brackets substituted by the Enterprise and Regulatory Reform Act 2013, s 20(1), (4)(a); for transitional provision see s 24(6) thereof; in para (c)(i) words 'the National Health Service Commissioning Board' in square brackets substituted by the Health and Social Care Act 2012, s 55(2), Sch 5, paras 72, 73(b); in para (c)(ii) reference to '2C, 17AA, 17C,' in square brackets inserted by the Enterprise and Regulatory Reform Act 2013, s 20(1), (4)(b); for effect see s 20(10) thereof and for transitional provision see s 24(6) thereof; para (ca) and word omitted immediately preceding it repealed by the Enterprise and Regulatory Reform Act 2013, s 20(1), (5); for transitional provision see s 24(6) thereof.

Sub-s (2): in para (aa) words 'the National Health Service Commissioning Board, or the' in square brackets substituted by the Health and Social Care Act 2012, s 55(2), Sch 5, paras 72,

73(c); para (ba) repealed by the Enterprise and Regulatory Reform Act 2013, s 20(1), (6); for transitional provision see s 24(6) thereof.

Sub-ss (4), (5): inserted by the Enterprise and Regulatory Reform Act 2013, s 20(1), (7); for transitional provision see s 24(6) thereof.

[14.23]

[43KA Application of this Part and related provisions to police

(1) For the purposes of—

(a) this Part,

(b) section 47B and sections 48 and 49 so far as relating to that section, and

(c) section 103A and the other provisions of Part 10 so far as relating to the right not to be unfairly dismissed in a case where the dismissal is unfair by virtue of section 103A,

a person who holds, otherwise than under a contract of employment, the office of constable or an appointment as a police cadet shall be treated as an employee employed by the relevant officer under a contract of employment; and any reference to a worker being 'employed' and to his 'employer' shall be construed accordingly.

(2) In this section 'the relevant officer' means—

(a) in relation to a member of a police force or a special constable appointed for a police area, the chief officer of police;

[(b) in relation to a member of a police force seconded to the [National Crime Agency to serve as a National Crime Agency officer], that Agency; and]

(d) in relation to any other person holding the office of constable or an appointment as police cadet, the person who has the direction and control of the body of constables or cadets in question.]

[14.24]

Statutory annotations

Amendment

Sub-s (2): in para (b) words 'National Crime Agency to serve as a National Crime Agency officer' in square brackets substituted by the Crime and Courts Act 2013, s 15(3), Sch 8, Pt 2, paras 49, 50.

PART V

PROTECTION FROM SUFFERING DETRIMENT IN EMPLOYMENT

Rights not to suffer detriment

[14.29]

[45A Working time cases

(1) A worker has the right not to be subjected to any detriment by any act, or any deliberate failure to act, by his employer done on the ground that the worker—

(a) refused (or proposed to refuse) to comply with a requirement which the employer imposed (or proposed to impose) in contravention of the Working Time Regulations 1998,

(b) refused (or proposed to refuse) to forgo a right conferred on him by those Regulations,

(c) failed to sign a workforce agreement for the purposes of those Regulations, or to enter into, or agree to vary or extend, any other agreement with his employer which is provided for in those Regulations,

(d) being—

(i) a representative of members of the workforce for the purposes of Schedule 1 to those Regulations, or

(ii) a candidate in an election in which any person elected will, on being elected, be such a representative,

performed (or proposed to perform) any functions or activities as such a representative or candidate,

(e) brought proceedings against the employer to enforce a right conferred on him by those Regulations, or

(f) alleged that the employer had infringed such a right.

(2) It is immaterial for the purposes of subsection (1)(e) or (f)—

(a) whether or not the worker has the right, or

(b) whether or not the right has been infringed,

but, for those provisions to apply, the claim to the right and that it has been infringed must be made in good faith.

(3) It is sufficient for subsection (1)(f) to apply that the worker, without specifying the right, made it reasonably clear to the employer what the right claimed to have been infringed was.

(4) This section does not apply where a worker is an employee and the detriment in question amounts to dismissal within the meaning of Part X . . .]

[(5) A reference in this section to the Working Time Regulations 1998 includes a reference to [—

(a)] the Merchant Shipping (Working Time: Inland Waterways) Regulations 2003;

[(b) the Fishing Vessels (Working Time: Sea-fishermen) Regulations 2004];

[(c) the Cross-border Railway Services (Working Time) Regulations 2008];

[(d) the Merchant Shipping (Hours of Work) Regulations 2002].]]

[14.30]

Statutory annotations

Amendment

Sub-s (5): para (d) inserted by SI 2014/308, reg 3, Schedule, para 2(1), (2).

[14.31]

[47B Protected disclosures

(1) A worker has the right not to be subjected to any detriment by any act, or any deliberate failure to act, by his employer done on the ground that the worker has made a protected disclosure.

[(1A) A worker ('W') has the right not to be subjected to any detriment by any act, or any deliberate failure to act, done—

(a) by another worker of W's employer in the course of that other worker's employment, or

(b) by an agent of W's employer with the employer's authority,

on the ground that W has made a protected disclosure.

(1B) Where a worker is subjected to detriment by anything done as mentioned in subsection (1A), that thing is treated as also done by the worker's employer.

(1C) For the purposes of subsection (1B), it is immaterial whether the thing is done with the knowledge or approval of the worker's employer.

(1D) In proceedings against W's employer in respect of anything alleged to have been done as mentioned in subsection (1A)(a), it is a defence for the employer to show that the employer took all reasonable steps to prevent the other worker—

(a) from doing that thing, or

(b) from doing anything of that description.

(1E) A worker or agent of W's employer is not liable by reason of subsection (1A) for doing something that subjects W to detriment if—

(a) the worker or agent does that thing in reliance on a statement by the employer that doing it does not contravene this Act, and

(b) it is reasonable for the worker or agent to rely on the statement.

But this does not prevent the employer from being liable by reason of subsection (1B).]

(2) . . . this section does not apply where—

(a) the worker is an employee, and

(b) the detriment in question amounts to dismissal (within the meaning of [Part X]).

(3) For the purposes of this section, and of sections 48 and 49 so far as relating to this section, 'worker', 'worker's contract', 'employment' and 'employer' have the extended meaning given by section 43K.]

[14.32]

Statutory annotations

Amendment

Sub-ss (1A)–(1E): inserted by the Enterprise and Regulatory Reform Act 2013, s 19(1); for transitional provision see s 24(6) thereof.

Enforcement

[14.33]

48 Complaints to [employment tribunals]

(1) An employee may present a complaint to an [employment tribunal] that he has been subjected to a detriment in contravention of section [43M], 44, 45, [46, 47[, 47A[, 47C or [*47E*] [47E[, 47F or 47G]]]]].

[(1ZA) A worker may present a complaint to an employment tribunal that he has been subjected to a detriment in contravention of section 45A.]

[(1A) A worker may present a complaint to an employment tribunal that he has been subjected to a detriment in contravention of section 47B.]

[(1B) A person may present a complaint to an employment tribunal that he has been subjected to a detriment in contravention of section 47D.]

(2) On such a complaint it is for the employer to show the ground on which any act, or deliberate failure to act, was done.

(3) An [employment tribunal] shall not consider a complaint under this section unless it is presented—

(a) before the end of the period of three months beginning with the date of the act or failure to act to which the complaint relates or, where that act or failure is part of a series of similar acts or failures, the last of them, or

(b) within such further period as the tribunal considers reasonable in a case where it is satisfied that it was not reasonably practicable for the complaint to be presented before the end of that period of three months.

[(4A) Section 207A(3) (extension because of mediation in certain European cross-border disputes) [and section 207B (extension of time limits to facilitate conciliation before institution of proceedings) apply] for the purposes of subsection (3)(a).]

(4) For the purposes of subsection (3)—

(a) where an act extends over a period, the 'date of the act' means the last day of that period, and

(b) a deliberate failure to act shall be treated as done when it was decided on;

and, in the absence of evidence establishing the contrary, an employer shall be taken to decide on a failure to act when he does an act inconsistent with doing the failed act or, if he has done no such inconsistent act, when the period expires within which he might reasonably have been expected do the failed act if it was to be done.

[(5) In this section and section 49 any reference to the employer [includes—

(a) where] a person complains that he has been subjected to a detriment in contravention of section 47A, the principal (within the meaning of section 63A(3));

[(b) in the case of proceedings against a worker or agent under section 47B(1A), the worker or agent].]

[14.34]

Statutory annotations

Amendment

Sub-s (1): words ', 47F or 47G' in square brackets substituted by the Growth and Infrastructure Act 2013, s 31(3).

Sub-s (4A): inserted by SI 2011/1133, regs 30, 34; words in square brackets substituted by the Enterprise and Regulatory Reform Act 2013, s 8, Sch 2, paras 15, 19.

Sub-s (5): words 'includes— (a) where' in square brackets substituted by the Enterprise and Regulatory Reform Act 2013, s 19(2)(a); para (b) inserted by the Enterprise and Regulatory Reform Act 2013, s 19(2)(b).

[14.35]

49 Remedies

(1) Where an [employment tribunal] finds a complaint under section 48 well-founded, the tribunal—

(a) shall make a declaration to that effect, and

(b) may make an award of compensation to be paid by the employer to the complainant in respect of the act or failure to act to which the complaint relates.

(2) [Subject to [subsections (5A) and (6)]] The amount of the compensation awarded shall be such as the tribunal considers just and equitable in all the circumstances having regard to—

(a) the infringement to which the complaint relates, and

(b) any loss which is attributable to the act, or failure to act, which infringed the complainant's right.

(3) The loss shall be taken to include—

(a) any expenses reasonably incurred by the complainant in consequence of the act, or failure to act, to which the complaint relates, and

(b) loss of any benefit which he might reasonably be expected to have had but for that act or failure to act.

(4) In ascertaining the loss the tribunal shall apply the same rule concerning the duty of a person to mitigate his loss as applies to damages recoverable under the common law of England and Wales or (as the case may be) Scotland.

(5) Where the tribunal finds that the act, or failure to act, to which the complaint relates was to any extent caused or contributed to by action of the complainant, it shall reduce the amount of the compensation by such proportion as it considers just and equitable having regard to that finding.

[(5A) Where—

(a) the complaint is made under section 48(1ZA),

(b) the detriment to which the worker is subjected is the termination of his worker's contract, and

(c) that contract is not a contract of employment,

any compensation must not exceed the compensation that would be payable under Chapter II of Part X if the worker had been an employee and had been dismissed for the reason specified in section 101A.]

[(6) Where—

(a) the complaint is made under section 48(1A),

(b) The detriment to which the worker is subjected is the termination of his worker's contract, and

(c) That contract is not a contract of employment,

Any compensation must not exceed the compensation that would be payable under Chapter II of Part X if the worker had been an employee and had been dismissed for the reason specified in section 103A.]

[(6A) Where—

(a) the complaint is made under section 48(1A), and

(b) it appears to the tribunal that the protected disclosure was not made in good faith,

the tribunal may, if it considers it just and equitable in all the circumstances to do so, reduce any award it makes to the worker by no more than 25%.]

[(7) Where—

(a) the complaint is made under section 48(1B) by a person who is not an employee, and

(b) the detriment to which he is subjected is the termination of his contract with the person who is his employer for the purposes of section 25 of the Tax Credits Act 2002,

any compensation must not exceed the compensation that would be payable under Chapter 2 of Part 10 if the complainant had been an employee and had been dismissed for the reason specified in section 104B.]

[14.36]

Statutory annotations

Amendment

Sub-s (6A): inserted by the Enterprise and Regulatory Reform Act 2013, s 18(4); for transitional provision see s 24(6) thereof.

[Other dismissals]

[14.41]

[101A Working time cases

[(1)] An employee who is dismissed shall be regarded for the purposes of this Part as unfairly dismissed if the reason (or, if more than one, the principal reason) for the dismissal is that the employee—

(a) refused (or proposed to refuse) to comply with a requirement which the employer imposed (or proposed to impose) in contravention of the Working Time Regulations 1998,

(b) refused (or proposed to refuse) to forgo a right conferred on him by those Regulations,

(c) failed to sign a workforce agreement for the purposes of those Regulations, or to enter into, or agree to vary or extend, any other agreement with his employer which is provided for in those Regulations, or

(d) being—

(i) a representative of members of the workforce for the purposes of Schedule 1 to those Regulations, or

(ii) a candidate in an election in which any person elected will, on being elected, be such a representative,

performed (or proposed to perform) any functions or activities as such a representative or candidate.]

[(2) A reference in this section to the Working Time Regulations 1998 includes a reference to[—

(a)] the Merchant Shipping (Working Time: Inland Waterways) Regulations 2003;]

[(b) the Fishing Vessels (Working Time: Sea-fishermen) Regulations 2004];

[(c) the Cross-border Railway Services (Working Time) Regulations 2008];

[(d) the Merchant Shipping (Hours of Work) Regulations 2002].]]

[14.42]

Statutory annotations

Amendment

Sub-s (2): para (d) inserted by SI 2014/308, reg 3, Schedule, para 2(1), (3).

CHAPTER III

SUPPLEMENTARY

[14.45]

[134A Application to police

(1) For the purposes of section 100, and of the other provisions of this Part so far as relating to the right not to be unfairly dismissed in a case where the dismissal is unfair by virtue of section 100, the holding, otherwise than under a contract of employment, of the office of constable or an appointment as police cadet shall be treated as employment by the relevant officer under a contract of employment.

[(2) In this section 'the relevant officer', in relation to—

(a) a person holding the office of constable, or

(b) a person holding an appointment as a police cadet,

means the person who under section 51A of the Health and Safety at Work etc Act 1974 is to be treated as his employer for the purposes of Part 1 of that Act.]

[(3) Subsection (1) does not apply to the holding of the office of constable by a member of a police force on secondment to the [National Crime Agency].]]

[14.46]–[14.500]

Statutory annotations

Amendment

Sub-s (3): words 'National Crime Agency' in square brackets substituted by the Crime and Courts Act 2013, s 15(3), Sch 8, Pt 2, paras 49, 51.

Safety Representatives and Safety Committees Regulations 1977

SI 1977/500

as amended by 1998 c 8, SIs 1992/2051, 1996/1513, 1999/860, 1999/2024, 1999/3242, 2005/1541, 2006/594, 2008/960, 2013/1471 and SSI 2006/457

[14.515]

6 Inspections following notifiable accidents, occurrences and diseases

(1) Where there has been [an over three day injury,] notifiable accident or dangerous occurrence in [an over three day injury,] workplace or a notifiable disease has been contracted there and—

(a) it is safe for an inspection to be carried out; and

(b) the interests of employees in the group or groups which safety representatives are appointed to represent might be involved.

those safety representatives may carry out an inspection of the part of the workplace concerned and so far as is necessary for the purpose of determining the cause they may inspect any other part of the workplace; where it is reasonably practicable to do so they shall notify the employer or his representative of their intention to carry out the inspection.

(2) The employer shall provide such facilities and assistance as the safety representatives may reasonably require (including facilities for independent investigation by them and private discussion with the employees) for the purpose of carrying out an inspection under this Regulation; but nothing in this paragraph shall preclude the employer or his representative from being present in the workplace during the inspection.

[(3) In this regulation—

'notifiable accident or dangerous occurrence' and 'notifiable disease' mean any accident, dangerous occurrence or disease, as the case may be, notice of which is required to be given by virtue of any of the relevant statutory provisions within the meaning of section 53(1) of the 1974 Act; and

'over three day injury' means an injury required to be recorded in accordance with regulation 12(1)(b) of the Reporting of Injuries, Diseases and Dangerous Occurrences Regulations 2013.]

[14.516]

Statutory annotations

Amendment

Para (1): words 'an over three day injury,' in square brackets substituted by SI 2012/199, reg 6(1), (2)(a); words 'an over three day injury,' in square brackets substituted by SI 2013/1471, reg 18(2), Sch 4.

Para (3): substituted by SI 2013/1471, reg 18(2), Sch 4.

[14.525]

11 Provisions as to [employment tribunals]

(1) A safety representative may, in accordance with the jurisdiction conferred on [employment tribunals] by paragraph 16(2) of Schedule 1 to the Trade Union and Labour Relations Act 1974, present a complaint to an [employment tribunal] that—

(a) the employer has failed to permit him to take time off in accordance with Regulation 4(2) of these Regulations; or

(b) the employer has failed to pay him in accordance with Regulation 4(2) of and the Schedule to these Regulations.

(2) An [employment tribunal] shall not consider a complaint under paragraph (1) above unless it is presented within three months of the date when the failure occurred or within such further period as the tribunal considers reasonable in a case where it is satisfied that it was not reasonably practicable for the complaint to be presented within the period of three months.

[(2A) Regulation 12 (extension of time limits to facilitate conciliation before institution of proceedings) applies for the purposes of paragraph (2).]

(3) Where an [employment tribunal] finds a complaint under paragraph (1)(a) above well-founded the tribunal shall make a declaration to that effect and may make an award of compensation to be paid by the employer to the employee which shall be of such amount as the tribunal considers just and equitable in all the circumstances having regard to the employer's default in failing to permit time off to be taken by the employee and to any loss sustained by the employee which is attributable to the matters complained of.

(4) Where on a complaint under paragraph (1)(b) above an [employment tribunal] finds that the employer has failed to pay the employee the whole or part of the amount required to be paid under paragraph (1)(b), the tribunal shall order the employer to pay the employee the amount which it finds due to him.

(5) . . .

[14.526]

Statutory annotations

Amendment

Para (2A): inserted by SI 2014/431, art 3, Schedule, paras 7, 8.

Health and Safety (Consultation With Employees) Regulations 1996

SI 1996/1513

as amended by 1998 c 8, SIs 1997/1840, 1999/3242, 2005/1541, 2013/1471 and SSI 2006/457

[14.538]

5 Duty of employer to provide information

(1) Where an employer consults employees directly he shall, subject to paragraph (3), make available to those employees such information, within the employer's knowledge, as is necessary to enable them to participate fully and effectively in the consultation.

(2) Where an employer consults representatives of employee safety he shall, subject to paragraph (3), make available to those representatives such information, within the employer's knowledge, as is—

- (a) necessary to enable them to participate fully and effectively in the consultation and in the carrying out of their functions under these Regulations;
- (b) contained in any record which he is required to keep by [regulation 12 of the Reporting of Injuries, Diseases and Dangerous Occurrences Regulations 2013] and which relates to the workplace or the group of employees represented by those representatives.

(3) Nothing in paragraph (1) or (2) shall require an employer to make available any information—

- (a) the disclosure of which would be against the interests of national security;
- (b) which he could not disclose without contravening a prohibition imposed by or under any enactment;
- (c) relating specifically to an individual, unless he has consented to its being disclosed;
- (d) the disclosure of which would, for reasons other than its effect on health or safety, cause substantial injury to the employer's undertaking or, where the information was supplied to him by some other person, to the undertaking of that other person; or
- (e) obtained by the employer for the purpose of bringing, prosecuting or defending any legal proceedings;

or to provide or allow the inspection of any document or part of a document which is not related to health or safety.

[14.539]

Statutory annotations

Amendments

Para (2): in sub-para (b) words 'regulation 12 of the Reporting of Injuries, Diseases and Dangerous Occurrences Regulations 2013' in square brackets substituted by SI 2013/1471, reg 18(2), Sch 4.

SCHEDULE 2

PROVISIONS AS TO [EMPLOYMENT TRIBUNALS]

Regulation 7(3)

[14.557]

1

An [employment tribunal] shall have jurisdiction to determine complaints in accordance with the following provisions of this Schedule.

2

A person (referred to in this Schedule as the 'complainant') may present a complaint to an [employment tribunal] that—

(a) his employer has failed to permit him to take time off in accordance with regulation 7(1)(b) or 7(2); or

(b) his employer has failed to pay him in accordance with regulation 7(1)(b) or 7(2) and Schedule 1.

[3A

(1) In this paragraph—

(a) Day A is the day on which the worker concerned complies with the requirement in subsection (1) of section 18A of the Employment Tribunals Act 1996 (requirement to contact ACAS before instituting proceedings) in relation to the matter in respect of which the proceedings are brought, and

(b) Day B is the day on which the worker concerned receives or, if earlier, is treated as receiving (by virtue of regulations made under subsection (11) of that section) the certificate issued under subsection (4) of that section.

(2) In working out when the three month time limit set by paragraph 3 expires the period beginning with the day after Day A and ending with Day B is not to be counted.

(3) If the three month time limit set by paragraph 3 would (if not extended by this sub-paragraph) expire during the period beginning with Day A and ending one month after Day B, the time limit expires instead at the end of that period.

(4) The power conferred on the employment tribunal by paragraph 3 to extend the three month time limit set by that paragraph is exercisable in relation to that time limit as extended by this paragraph.]

3

[Subject to paragraph 3A an employment tribunal] shall not consider a complaint under paragraph 2 unless it is presented within three months of the date when the failure occurred or within such further period as the tribunal considers reasonable in a case where it is satisfied that it was not reasonably practicable for the complaint to be presented within the period of three months.

4

Where an [employment tribunal] finds a complaint under paragraph 2(a) well-founded the tribunal shall make a declaration to that effect and may make an award of compensation to be paid by the employer to the complainant which shall be of such amount as the tribunal considers just and equitable in all the circumstances having regard to the employer's default in failing to permit time off to be taken by the complainant and to any loss sustained by the complainant which is attributable to the matters complained of.

5

Where on a complaint under paragraph 2(b) an [employment tribunal] finds that the employer has failed to pay the complainant the whole or part of the amount required to be paid in accordance with regulation 7(1)(b) or 7(2) and Schedule 1, the tribunal shall order the employer to pay the complainant the amount which it finds due to him.

[14.558]–[14.578]

Statutory annotations

Amendment

Para 3: words 'Subject to paragraph 3A an employment tribunal' in square brackets substituted by SI 2014/431, art 3, Schedule, paras 16, 17.

Para 3A: inserted by SI 2014/431, art 3, Schedule, paras 16, 18.

Public Interest Disclosure (Prescribed Persons) Order 1999

SI 1999/1549

as amended by 2000 c 36, SIs 2002/1555, 2003/1993 and 2004/664, 2004/3265, 2005/2464, 2005/2035, 2005/3172, 2007/3224, 2008/531, 2008/2831, 2009/462, 2009/2457, 2009/2748, 2010/7, 2010/671, 2011/2581, 2012/462, 2012/725, 2012/1479, 2012/1641 and 2012/2400

[SCHEDULE

Article 2]

[14.581]

[First Column **Persons and descriptions of people**	*Second Column* **Descriptions of matters**
Accounts Commission for Scotland and auditors appointed by the Commission to audit the accounts of local government bodies.	The proper conduct of public business, value for money, fraud and corruption in local government bodies.
Audit Commission for England and Wales and auditors appointed by the Commission to audit the accounts of local government, and health service, bodies [(including local policing bodies, and the chief constables and Commissioner of Police of the Metropolis appointed under the Police Reform and Social Responsibility Act 2011)].	The proper conduct of public business, value for money, fraud and corruption in local government, and health service, bodies [(including local policing bodies, and the chief constables and Commissioner of Police of the Metropolis appointed under the Police Reform and Social Responsibility Act 2011)].
Certification Officer.	Fraud, and other irregularities, relating to the financial affairs of trade unions and employers' associations.
Charity Commissioners for England and Wales.	The proper administration of charities and of funds given or held for charitable purposes.
The Scottish Ministers.	The proper administration of charities and of funds given or held for charitable purposes.
Chief Executive of the Criminal Cases Review Commission.	Actual or potential miscarriages of justice.
Chief Executive of the Scottish Criminal Cases Review Commission.	Actual or potential miscarriages of justice.
Civil Aviation Authority.	Compliance with the requirements of civil aviation legislation, including aviation safety [and aviation security].
[Office of Communications.	Matters relating to— (a) the provision of electronic communications networks and services and the use of the electro-magnetic spectrum; (b) broadcasting and the provision of television and radio services; (c) media ownership and control; and (d) competition in communications markets.]

[First Column Persons and descriptions of people	Second Column Descriptions of matters
. . .	. . .
Commissioners of Customs and Excise.	Value added tax, insurance premium tax, excise duties and landfill tax. The import and export of prohibited or restricted goods.
Commissioners of the Inland Revenue.	Income tax, corporation tax, capital gains tax, petroleum revenue tax, inheritance tax, stamp duties, national insurance contributions, statutory maternity pay, statutory sick pay, tax credits, child benefits, collection of student loans and the enforcement of the national minimum wage.
[Comptroller and Auditor General].	The proper conduct of public business, value for money, fraud and corruption in relation to the provision of centrally-funded public services.
Auditor General for Wales.	The proper conduct of public business, value for money, fraud and corruption in relation to the provision of public services.
Auditor General for Scotland and persons appointed by or on his behalf under the Public Finance and Accountability (Scotland) Act 2000 to act as auditors or examiners for the purposes of sections 21 to 24 of that Act.	The proper conduct of public business, value for money, fraud and corruption in relation to the provision of public services.
Audit Scotland.	The proper conduct of public business, value for money, fraud and corruption in public bodies.
[Gas and Electricity Markets Authority.	The generation, transmission, distribution and supply of electricity, participation in the operation of an electricity interconnector (as defined in section 4(3E) of the Electricity Act 1989) [or providing a smart meter communication service (as defined in section 4(3G) of that Act)] and activities ancillary to these matters. The transportation, shipping and supply of gas through pipes, participation in the operation of a gas interconnector (as defined in section 5(8) of the Gas Act 1986) [or providing a smart meter communication service (as defined in section 5(11) of that Act)] and activities ancillary to these matters.]
. . .	. . .
. . .	. . .
. . .	. . .
[Water Services Regulation Authority].	The supply of water and the provision of sewerage services.
[. . .	. . .
Water Industry Commission for Scotland.	The supply of water and the provision of sewerage services.]

[*First Column* **Persons and descriptions of people**	***Second Column*** **Descriptions of matters**
Water Industry Commissioner for Scotland.	The supply of water and the provision of sewerage services.
Director of the Serious Fraud Office.	Serious or complex fraud.
Lord Advocate, Scotland.	Serious or complex fraud.
Environment Agency.	Acts or omissions which have an actual or potential effect on the environment or the management or regulation of the environment, including those relating to pollution, abstraction of water, flooding, the flow in rivers, inland fisheries and migratory salmon or trout.
Scottish Environment Protection Agency.	Acts or omissions which have an actual or potential effect on the environment or the management or regulation of the environment, including those relating to flood warning systems and pollution.
Food Standards Agency.	Matters which may affect the health of any member of the public in relation to the consumption of food and other matters concerning the protection of the interests of consumers in relation to food.
[Financial Conduct Authority	The listing of securities on a stock exchange; prospectuses on offers of transferable securities to the public; the carrying on of investment business or of insurance business; the operation of banks and building societies, deposit-taking businesses and wholesale money market regimes; the operation of friendly societies, benevolent societies, working men's clubs, specially authorised societies, and industrial and provident societies; the functioning of financial markets and investment exchanges; money laundering, financial crime, and other serious financial misconduct, in connection with activities regulated by the Financial Conduct Authority.
Prudential Regulation Authority	The carrying on of investment business or of insurance business; the operation of banks and building societies; deposit-taking businesses and wholesale money market regimes; the operation of friendly societies, benevolent societies, working men's clubs, specially authorised societies, and industrial and provident societies.
The Bank of England	The functioning of clearing houses.]
[The Financial Reporting Council Limited and its operating bodies the Professional Oversight Board, the Financial Reporting Review Panel and the Accountancy and Actuarial Discipline Board	Matters relating to—

[First Column **Persons and descriptions of people**	***Second Column*** **Descriptions of matters**
	(a) the independent oversight of the regulation of the accountancy, auditing and actuarial professions; (b) the independent supervision of Auditors General (as defined in section 1226 of the Companies Act 2006); (c) the monitoring of major audits (as defined in section 525 of that Act for the purposes of that section or section 522, or paragraph 13(10) of Schedule 10 to that Act for the purposes of that paragraph); (d) the registration of third country auditors (as defined in section 1261 of that Act); (e) compliance with the requirements of legislation relating to accounting and reporting; and (f) the investigation of the conduct of auditors, accountants and actuaries and the holding of disciplinary hearings in public interest cases (as defined in paragraph 24 of Schedule 10 to that Act).]
. . .	. . .
Care Council for Wales.	Matters relating to the registration of social care workers under the Care Standards Act 2000.
Scottish Social Services Council.	Matters relating to the registration of the social services workforce by the Scottish Social Services Council [under the Regulation of Care (Scotland) Act 2001].
[Children's Commissioner	Matters relating to the views and interests of children.
Commissioner for Children and Young People in Scotland	Matters relating to promoting and safeguarding the rights of children and young people.]
Children's Commissioner for Wales.	Matters relating to the rights and welfare of children.
[General Chiropractic Council	Matters relating to— (a) the registration and fitness to practise of a member of a profession regulated by the Council; and (b) any activities not covered by (a) in relation to which the Council has functions.]
[General Dental Council	Matters relating to— (a) the registration and fitness to practise of a member of a profession regulated by the Council; and

[*First Column* **Persons and descriptions of people**	***Second Column*** **Descriptions of matters**
	(b) any activities not covered by (a) in relation to which the Council has functions.]
[General Medical Council	Matters relating to— (a) the registration and fitness to practise of a member of a profession regulated by the Council; and (b) any activities not covered by (a) in relation to which the Council has functions.]
[General Optical Council	Matters relating to— (a) the registration and fitness to practise of a member of a profession regulated by the Council; and (b) any activities not covered by (a) in relation to which the Council has functions.]
[General Osteopathic Council	Matters relating to— (a) the registration and fitness to practise of a member of a profession regulated by the Council; and (b) any activities not covered by (a) in relation to which the Council has functions.]
[General Pharmaceutical Council	Matters relating to— (a) the registration and fitness to practise of a member of a profession regulated by the Council; and (b) any activities not covered by (a) in relation to which the Council has functions.]
[Health and Care Professions Council	Matters relating to— (a) the registration and fitness to practise of a member of a profession regulated by the Council; and (b) any activities not covered by (a) in relation to which the Council has functions.]
[Nursing and Midwifery Council	Matters relating to— (a) the registration and fitness to practise of a member of a profession regulated by the Council; and (b) any activities not covered by (a) in relation to which the Council has functions.]
Health and Safety Executive.	Matters which may affect the health or safety of any individual at work; matters, which may affect the health and safety of any member of the public, arising out of or in connection with the activities of persons at work.

[First Column **Persons and descriptions of people**	***Second Column*** **Descriptions of matters**
[Regulator of Social Housing.]	The registration and operation of [private registered providers of social housing], including their administration of public and private funds and management of their housing stock.
Local authorities which are responsible for the enforcement of health and safety legislation.	Matters which may affect the health or safety of any individual at work; matters, which may affect the health and safety of any member of the public, arising out of or in connection with the activities of persons at work.
[Independent Police Complaints Commission.	Matters relating to the conduct of a person serving with the police (as defined in section 12(7) of the Police Reform Act 2002) or of any other person in relation to whose conduct the Independent Police Complaints Commission exercises functions in or under any legislation.]
Information Commissioner.	Compliance with the requirements of legislation relating to data protection and to freedom of information.
Scottish Information Commissioner.	Compliance with the requirements of legislation relating to freedom of information.
[. . .	
[Care Quality Commission.]	[Matters relating to— (a) the registration and provision of a regulated activity as defined in section 8 of the Health and Social Care Act 2008 and the carrying out of any reviews and investigations under Part 1 of that Act; or (b) any activities not covered by (a) in relation to which the Care Quality Commission exercises its functions.]]
[The [Monitor]	Matters relating to— (a) the regulation and performance of NHS foundation trusts; and (b) any activities not covered by (a) in relation to which the [Monitor] exercises its functions . . .]
[National Assembly for Wales.	Matters relating to the provision of Part II services as defined in section 8 of the Care Standards Act 2000 and the Children Act 1989. Matters relating to the inspection and performance assessment of Welsh local authority social services as defined in section 148 of the Health and Social Care (Community Heath and Standards) Act 2003.

[First Column **Persons and descriptions of people**	*Second Column* **Descriptions of matters**
	Matters relating to the review of, and investigation into, the provision of health care by and for Welsh NHS bodies as defined under the Health and Social Care (Community Health and Standards) Act 2003.
	The registration and operation of registered social landlords, including their administration of public and private funds and management of their housing stock.]
[Social Care and Social Work Improvement Scotland.	Matters relating to the provision of care services, as defined in the Public Services Reform (Scotland) Act 2010.
Healthcare Improvement Scotland.	[Matters relating to— (a) furthering the improvement in the quality of health care (within the meaning of section 10A of the National Health Service (Scotland) Act 1978); and (b) any activities not covered by (a) in relation to which Healthcare Improvement Scotland has functions.]
[Pensions Regulator].	Matters relating to occupational pension schemes and other private pension arrangements [including matters relating to the Regulator's objective of maximising compliance with the duties under Chapter 1 of Part 1 (and the safeguards in sections 50 and 54) of the Pensions Act 2008].
Office of Fair Trading.	Matters concerning the sale of goods or the supply of services, which adversely affect the interests of consumers.
	Competition affecting markets in the United Kingdom.
[Office of Qualifications and Examinations Regulation	Matters in relation to which the Office of Qualifications and Examinations Regulation exercises functions under the Apprenticeships, Skills, Children and Learning Act 2009]
[Office of Rail Regulation].	The provision and supply of railway services.
[Police and Crime Panels	Matters relating to— (a) any complaint about the conduct of a relevant office holder, within the meaning of section 31 of the Police Reform and Social Responsibility Act 2011; and (b) any report made in accordance with section 114 of the Local Government Finance Act 1988 or section 4 or 5 of the Local Government and Housing Act 1989 in relation to the elected local policing body for the Police and Crime Panel's area.]

[First Column Persons and descriptions of people	Second Column Descriptions of matters
[Elected local policing bodies	Matters relating to any report made in accordance with section 114 of the Local Government Finance Act 1988 in relation to— (a) the chief constable for the elected local policing body's area; or (b) where the elected local policing body is the Mayor's Office for Policing and Crime, the Commissioner of Police of the Metropolis.
Standards Board for England.	Breaches by a member or co-opted member of a relevant authority (as defined in section 49(6) of the Local Government Act 2000) of that authority's code of conduct.
Local Commissioner in Wales.	Breaches by a member or co-opted member of a relevant authority (as defined in section 49(6) of the Local Government Act 2000) of that authority's code of conduct.
Standards Commission for Scotland and the Chief Investigating Officer.	Breaches by a councillor or a member of a devolved public body (as defined in section 28 of the Ethical Standards in Public Life etc (Scotland) Act 2000) of the code of conduct applicable to that councillor or member under that Act.
Treasury.	The carrying on of insurance business.
[Secretary of State for Business, Innovation and Skills.]	Fraud, and other misconduct, in relation to companies, investment business, insurance business, or multi-level marketing schemes (and similar trading schemes); insider dealing. Consumer safety.
Secretary of State for Transport.	Compliance with merchant shipping law, including maritime safety.
Local authorities which are responsible for the enforcement of consumer protection legislation.	Compliance with the requirements of consumer protection legislation.
Local authorities which are responsible for the enforcement of food standards.	Compliance with the requirements of food safety legislation.
[A member of the House of Commons.	Any matter specified in this column.]
A person ('person A') carrying out functions, by virtue of legislation, relating to relevant failures falling within one or more matters within a description of matters in respect of which another person ('person B') is prescribed by this Order, where person B was previously responsible for carrying out the same or substantially similar functions and has ceased to be so responsible.	Matters falling within the description of matters in respect of which person B is prescribed by this Order, to the extent that those matters relate to functions currently carried out by person A.]

[14.582]–[14.584]

Statutory annotations

Amendment

In entry beginning with the words 'Audit Commission for England' in columns (1) and (2) words from '(including local policing' to 'Police Reform and Social Responsibility Act 2011)' in square brackets inserted by SI 2013/2213, arts 2, 3.

In entry relating to 'Civil Aviation Authority' in column (2) words 'and aviation security' in square brackets inserted by SI 2013/2213, arts 2, 4.

Entry relating to 'The competent authority under Part IV of the Financial Services and Markets Act 2000' (omitted) revoked by SI 2013/472, art 3, Sch 2, para 25(a).

In entry relating to 'Comptroller and Auditor General' in column (1) words 'Comptroller and Auditor General' in square brackets substituted by SI 2012/725, art 2(4).

Entry relating to the 'Gas and Electricity Markets Authority' inserted by SI 2005/2464, arts 2, 3(3), Schedule; in entry relating to 'Gas and Electricity Markets Authority' in column (2) words 'or providing a smart meter communication service (as defined in section 4(3G) of that Act)' in square brackets inserted by SI 2012/2400, art 34(a); in entry relating to 'Gas and Electricity Markets Authority' in column (2) words 'or providing a smart meter communication service (as defined in section 5(11) of that Act)' in square brackets inserted by SI 2012/2400, art 34(b).

Entries relating to 'Financial Conduct Authority', 'Prudential Regulation Authority' and 'The Bank of England' substituted for entry relating to 'Financial Services Authority', by SI 2013/472, art 2, Sch 2, para 25(b).

Entry relating to the 'General Social Care Council' (omitted) revoked by SI 2012/1479, art 11, Schedule, Pt 2, para 71.

In entry relating to 'Scottish Social Services Council' in column (2) words 'under the Regulation of Care (Scotland) Act 2001' in square brackets inserted by SI 2013/2213, arts 2, 5/

Entry relating to 'General Chiropractic Council' inserted by SI 2013/2213, arts 2, 6, Schedule.

Entry relating to 'General Dental Council' inserted by SI 2013/2213, arts 2, 6, Schedule.

Entry relating to 'General Medical Council' inserted by SI 2013/2213, arts 2, 6, Schedule.

Entry relating to 'General Optical Council' inserted by SI 2013/2213, arts 2, 6, Schedule.

Entry relating to 'General Osteopathic Council' inserted by SI 2013/2213, arts 2, 6, Schedule.

Entry relating to 'General Pharmaceutical Council' inserted by SI 2013/2213, arts 2, 6, Schedule.

Entry relating to 'Health and Care Professions Council' inserted by SI 2013/2213, arts 2, 6, Schedule.

Entry relating to 'Nursing and Midwifery Council' inserted by SI 2013/2213, arts 2, 6, Schedule.

Entry relating to the 'Office of Qualifications and Examinations Regulation' inserted by SI 2012/462, art 2, Schedule.

Entry relating to 'Police and Crime Panels' inserted by SI 2013/2213, arts 2, 7, Schedule.

Entry relating to 'Elected local policing bodies' inserted by SI 2013/2213, arts 2, 7, Schedule.

Entry relating to 'The Monitor' (entry relating to 'The Independent Regulator of NHS Foundation Trusts' as enacted) inserted by SI 2009/2457, arts 2, 3, Schedule; in entry relating to 'The Monitor' (entry relating to 'The Independent Regulator of NHS Foundation Trusts' as enacted) in column 1 word 'Monitor' in square brackets substituted by SI 2012/1641, art 10, Sch 4, para 5(1)(a); in entry relating to 'The Monitor' (entry relating to 'The Independent Regulator of NHS Foundation Trusts' as enacted) in column 2 in para (b) word 'Monitor' in square brackets substituted by SI 2012/1641, art 10, Sch 4, para 5(1)(a); in entry relating to 'The Monitor' (entry relating to 'The Independent Regulator of NHS Foundation Trusts' as enacted) in column 2 in para (b) words omitted revoked by SI 2012/1641, art 10, Sch 4, para 5(1)(b).

Entry relating to 'A member of the House of Commons' inserted by SI 2014/596, art 2.

Part 14

